Environmental Issues and Solutions

Environmental Issues and Solutions

Editors

Professor (Dr.) Arvind Kumar

FLS (London), FASc (Swiss), FISEC, FSESc, FAZ, FZA (Gold Medalist)

Pro-vice Chancellor

S.K.M. University, Dumka – 814 101 (Jharkhand)

&

Dr. Pashupati Kumar Roy

Reader & Head

Department of Political Science,

A.S. College, Deoghar (Jharkhand)

2008

DAYA PUBLISHING HOUSE

Delhi - 110 035

ISBN10 81-7035-491-9
ISBN13 978-81-7035-491-8
EAN: 9788170354918

Published by : **Daya Publishing House**
1123/74, Deva Ram Park
Tri Nagar, Delhi - 110 035
Phone: 27383999
Fax: (011) 23244987
e-mail : dayabooks@vsnl.com
website : www.dayabooks.com

Showroom : 4760-61/23, Ansari Road, Darya Ganj,
New Delhi - 110 002
Phone: 23245578, 23244987

Laser Typesetting : **Classic Computer Services**
Delhi - 110 035

Printed at : **Chawla Offset Printers**
Delhi - 110 052

PRINTED IN INDIA

Preface

The concept of improved quality of life in urban habitat was earlier confined to proper solid waste disposal, wastewater treatment, water supply, roads and transport, access to institutional, commercial and trading facilities. But presently, the need for integrating environmental management by way of plantation, setting up of parks and gardens and maintenance of urban wetlands is getting adequate attention. A green city is not merely for a visual, aesthetic pleasure but to give the citizens a chance to breath clear air. It also reduces the impact of air pollution on human health. Additionally, with urban greening a significant shelter for indicators of biodiversity can be created *viz.* birds and butterflies. Urban greening can also help in controlling microclimate and recharging of ground water and above all it can have immense bio-aesthetic value to augment quality of life. Therefore, the urban/industrial greening programme has to combine the need for meeting the demand of reducing pollution load and increasing aesthetic values. Keeping this in view, the present endeavour has been undertaken to decipher the trenchant environmental issues with their solutions for the sustainable development of mankind.

This book is the unique compilation of esteemed articles of internationally acknowledged experts in the field of Environmental sciences with the intention of providing a sufficient depth of the subject to satisfy the needs at a level which will be comprehensive and interesting. This book will be useful to the students, research scholars, scientists in the field of Environmental management and ecoplanners, politicians and other people with similar interest.

My special thanks and appreciation go to the scientists whose contributions have enriched this volume. I wish to express my sincere gratitude to Dr. P.C. Hembram, Hon'ble Vice Chancellor, S.K.M. University, Dumka, Professor M.C. Dash, Hon'ble former Vice Chancellor of Sambalpur University, Professor N.C. Datta of Calcutta University, Professor K.C. Pandey of Lucknow University, Professor S.K. Konar of Kalyani University, Professor D.K. Belsare of Bhopal University, Professor U.C. Goswami of Gauhati University, Professor P.C. Mishra of Sambalpur University,

Professor P. Natarajan of Kerala University, Professor P. S. Murthy of Bangalore University, Professor Ajit Varma of JNU, Professor A.L. Bhatia of Jaipur University, Professor A.K. Mittal of BHU, Professor S.P. Hosmani of Mysore University, Professor K. Kapoor of Udaipur University, Professor K.B. Reddy of Nagarjuna University, Professor M. Vikram Reddy of Pondicherry University, Professor B.K. Tiwari of NEHU, Professor G. Tripathi of Jodhpur University, Professor R. Ramalingam of Annamalai University, Professor Sharif U. Ahmad of Nagaland University, Professor G.K. Kulkarni of Aurangabad University, Professor S.U. Mehram of Nagpur University, Professor S.K. Battish of PAU, Professor B.M. Sharma of Manipur University, Professor B.D. Joshi of Hardwar University, Professor G.C. Pandey of Faizabad University, Professor K. C. Sharma of Ajmer University, Professor M. Raziuddin of Hazaribag University, Professor U.S. Bagde of Mumbai University, Professor Gurdeep Singh of I.S.M., Dhanbad, Dr. P.K. Goel of Karad, Professor S.P. Roy and Shri Tribhuwan Poddar of Bhagalpur University, Bhagalpur for encouragements.

I also express my deep sense of gratitude to my parents whose blessings have always prompted me to pursue academic activities deeply. I am also thankful to my wife, *Kumari Bimla* and my two lovely sons, *Kumar Pallav Shivshankaran* and *Kumar Prasun Ramakrishnan* whose natural smiles extended to me relief all through this tiresome endeavour.

Last but not the least, I am also thankful to Mr. Anil Mittal, Proprietor, Daya Publishing House, New Delhi for taking keen interest in bringing out of this book. Finally, I will always remain a debtor to all my well-wishers for their blessings, without which this book would not have come into existence.

Professor A. Kumar

Contents

Chapter 1

Noise Pollution: An Overview

Preeti and Arvind Kumar
Environmental Science Research Unit, Post Graduate Department of Zoology, S.K.M. University, Dumka, India

Noise is an Environment Pollutant, but it differs from air and water pollution in that it disappears fast and thus does not remain in environment long. A particular sound may be musical to one but noise to another, pleasant when soft, but noise when loud. So the question is why the sound which at one place or time gives soothing touch to the ear becomes noise and irritating on another occasion.

Prolonged loud sound is generally considered as noise which is mostly caused because of industries, vehicles, airplanes etc. Since such a noise is found to cause some of physical/psychological stress to human beings, it is also considered as a pollutant. Noise is a by-product of human activity and the background noise levels have been increasing year by year for the last few decades. Excessive noise can result in loss of hearing increase in accidents and decrease of productivity and working loss is not only an occupational risk but may becomes environmental risk as well. Hence, it is essential to take suitable precautions to reduce noise pollution particularly in urban and industrial area.

Source of Noise Pollution

Noise is either Natural such as thunder or man made. The main contribution to noise is factories and industries, transportation (air, rail and road), communities and religious activities. There should not be exaggeration to say that we Indians are noisy people and every sentiment and occasion is manifested in a noisy manner–be it a religious occasion, elections or a family celebration.

The main menace of noise, however, comes from transport (Road, Rail and Aircraft) in addition to vehicle horns. The number of road vehicles (two/four wheelers) has increased enormously in the recent years (Delhi, Mumbai and Kolkata 93 dB). Ever since the industrial revolution, there has been doubling every 10 yrs of environmental noise.

Characteristics of Sound

Noise is transmitted as wave motion in an elastic medium like air, water or steel by the vibrations of the molecules in the medium. There are two important measures which describe sound pressure, (a) Frequency (Hz)., (b) amplitude ($W/m^2/N/m^2$).

The sound intensity is measured in decibel (dB). Loudness may be described as listener's auditory expression which is not synonymous with sound pressure. In normal conversation 60 dB has been suggested but when exceeds this level it is being considered as hazardous.

Noise are only interferes with communication but also affects our peace of mind, health and behaviour. The various effects of noise beings may be classified as Auditory (affecting hearing faculty) effects and non auditory effects.

Physiological Effects

Loud sounds can cause increased secretion of many hormones of the pituitary gland e.g. ACTH (adenocorticotropic hormone) in turn stimulates the adrenal gland, which secretes several other hormones. Through a variety of influences these hormones in turn triggers various such as:

(a) Enhancement of the sensitivity, (b) Increase of blood sugar levels, (c) Suppression of immune system and (d) decreasing the efficiency of liver to detoxify blood.

Epidemiologic environmental noise studies on changes in BP and increased risk for ischemic heart disease are limited in adults while in children higher systolic and diastolic BP were observed (> 60 dBA).

Temporary deafness occurs at 4000-6000 Hz and this effect in known as TTS (Temporary Threshold Shift-recovered). Permanent loss of hearing occurs threshold shift is called PTS (Permanent Threshold Shift-not recovered). TTS may be converted in PTS when person exposed continuously by high noise levels.

Psychosocial Effects

Psychosocial effects due to exposure to environmental noise that have been studied in epidemiologic investigations include annoyance, psychosocial well being and psychiatric hospitalization of which annoyance is the main.

Noise annoyance is a feeling of resentment, displeasure, discomfort dissatisfaction, or offense when noise interferes with some one's thoughts, feeling or actual activities. Reaction to stressor can be psychological (feeling of fear, depression, sorrow), behavioural (Social isolation, aggression, excessive use of alcohol, tobacco, food, drugs) and somatic (cardiovascular, and gastro-intestinal, respiratory illness) in nature.

Remedial Steps for Noise Pollution

In the electronic age it is almost impossible to completely get rid of the malady of noise pollution. However, there can be ways and means to reduce noise intensity so that its influence can be minimized. The means of noise control are:

1. Planting more trees along highways, streets and near industrial areas they absorb and dissipate sound especially high frequency and act as buffer zone. In the metropolitan cities green belt vegetation may have a great value in minimizing sound e.g. Ashok, Bougainvillea and Neem.

2. Imposing strict legal restrictions on noise polluting bodies.
3. Promoting technologies to develop quieter type of diesel engines to reduce the noise pollution produced by vehicles. Noise producing vehicles should use silencers, and if their silencers are not working they should not be allowed to play on the roads.
4. Standards for noise control measures should be set up for industry and community.
5. Building codes should enforce sound proofing in the construction of buildings.
6. Nobody should be permitted to create noise in silent zones (educational, hospital, court).
7. Personal protection can be achieved by holding our hands over ears under noisy situations, or by running away from the source, or by simply stuffing in the ears kit of cotton. Use of ear plugs can reduce noise almost, by 40-50 dB (Helmet, ear muffs).

In India, there is law exclusively dealing with the problem of noise. Noise made by vehicle can be reduced by banning honking and stream traffic flow besides ensuring good body and silencer designs. More than anything else, the rules framed in keeping noise levels low should be implemented strictly. All products should be labeled according to noise standards (Ecomark). We believe that two subjects should be priorities in research; the study of cardiovascular effects and the underlying mechanisms and the study of the effects of noise on children. Attention on children in urgent, as such effects in term of years of life with a reduced quality of life or a handicap are greater for children than for adults.

It is high time the authorities lent a sympathetic ear to the ever increasing problem of noise pollution. But then, we all can in our own way do that little bit and try to keep down the level of the irritating and potentially harmful noise that we all produce in one way or the other.

Chapter 2

Water and Wastewater Management: A Membrane Bioreactor Technology: A Review

K.L. Prakash and R.K. Somashekar
Department of Environmental Sciences, Bangalore University, Bangalore – 56

ABSTRACT

MBRs provide a reliable, high-quality, reusable effluent. Its particle-free effluent allows more effective post-disinfection, as required before reuse. MBRs provide excellent pretreatment when reverse osmosis (RO) is needed to generate very high-quality reclaimed water. MBRs may also remove fouling fractions of organic matter more effectively than micro filtration prior to RO.

***Keywords**: Wastewater Treatment, Membrane Bioreactor, Pollutants.*

Introduction

In the evolution of life on earth, the membrane was an essential component in that it allowed the formation of cells, and later followed by the compartmentalization of processes in cells. As humans have progressively learned how to produce more complicated and efficient synthetic membranes, so also have developed the ability to compartmentalise various processes. In this way, for example membranes can be used to filter cells from wastewater. If the filtered cells play a role in breaking down additional waste flowing through such loaded membranes, a membrane bioreactor may be said to have been created. A membrane bioreactor consists of some biological constituent or constituents in association with a membrane. A membrane is a surface that has the ability to selectively let some components through it while blocking some others (Crites and Tchobanoglous, 1998).

Since the early 1960s the membrane technology encountered a broad pattern of applications in the field of water, wastewater treatment and other systems of solid-liquid and solid-gas separation (Cornel and Krause, 2003 and Gunder, 2001). The maximum application was the Reverse-Osmosis (RO) high-pressure process for seawater desalination and the high-flux however, low pressure systems for saline water desalination. The Reverse-Osmosis technology allows removing nearly all dissolved solids contained in the liquid. Membrane treatment of groundwater, surface waters and wastewater is based on the removal of particles, bacteria and cysts, and, consequently turbidity reduction.

Micro Filtration (MF) and Ultra Filtration (UF) can also maintain these tasks. The advantage of MF or UF for organic removal with the selective salt removal by Reverse-Osmosis membranes turn integrated systems into a very promising and challenging development approach.

Recently membrane technology is developed for low quality water and wastewater treatment in both industrial and agricultural applications (Stephenson, *et al.*, 2000). Many groups have conducted studies on performance of pilot and full-scale wastewater; develop and encourage international cooperation towards optimal membrane performance and performance evaluation; support operational research studies and management modelling of membrane systems; identify research needs and develop international technology transfer processes related to polishing low quality water for unlimited uses (Femandez *et al.*, 2000 and Daigger, *et al.*, 2001).

In order to maintain selective solute removal, anew type of membranes was developed finally. The intermittent size Nano Filtration (NF) membranes enable to maintain an improved selective dissolved solids removal. The NF membranes retain hazardous materials and allow other favorable constituents to flow with the permeate. An attractive area of application for NF membranes application is groundwater and surface water treatment that contain pollutants such as soluble organic compounds and heavy metals.

Objectives

The objectives of the Membrane Bioreactor technique are to advance knowledge and experience in the area of development and implementation of membrane technology for low quality water and wastewater treatment. It is associated with the development of membranes and adjustment to the low quality to be treated and subject to the required qualities.

Salient Features of MBR

The use of different membrane processes in combination with biological and biotechnological treatment processes has been developed intensively over the last decade. The use of different membrane clusters can optimize removal of organic constituents such as BOD5, COD, TOC, and TSS and reuse criteria for reclaimed effluent for irrigation. This new concept concerning various membrane combinations in compact vessels generates the efficient Membrane Bioreactor. The MBR is the new advanced trend in wastewater treatment technology. Several treatment systems based on the use of the MBR technology are already in operation. The simple procedure involves:

1. Combine the activated sludge found in high throughput sewerage treatment plants with membrane filtration (see image below), to facilitate removal of biodegradable organics, suspended solids and inorganic nutrients (such as nitrogen and phosphorus).

2. MBR retains particulate and slow-growing organisms (thereby treating more slowly biodegraded organics) and removes a very high percentage of pathogens (thereby reducing chemical disinfection requirements).
3. They also require less space than traditional activated sludge systems because less hydraulic residence time (HRT) is needed to achieve a given solids retention time (SRT).
4. In addition, MBRs are more automated, making them ideal for decentralized treatment because they are simpler to operate.

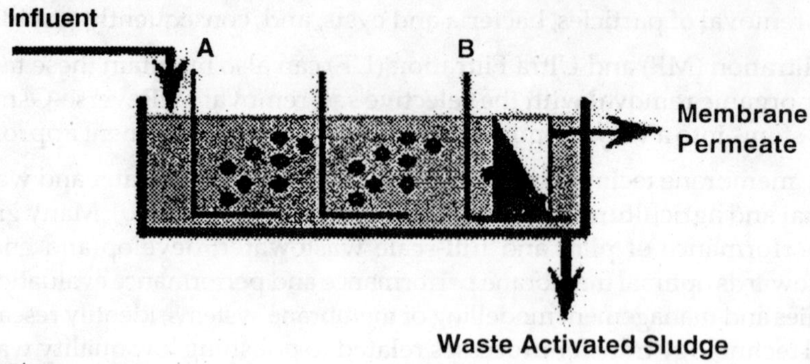

(A) Fine Bubble Aeration
(B) Coarse Bubble Aeration

Description of MBR Technology

An MBR is a combination of the activated sludge process, a wastewater treatment process characterized by a suspended growth of biomass, with a micro- or ultra-filtration membrane system that rejects particles. The membrane system replaces the traditional gravity sedimentation unit (clarifier) in the activated sludge process. The turbidity and suspended solids concentration of the effluent is far lower than in conventional treatment. All biomass is retained and becomes returned activated sludge. Biological growth leaves the system as waste activated sludge.

MBR Technology is Based on the Following Principles:

The engineering principles underlying MBRs are familiar enough to ensure reliability. Because MBRs combines two familiar technologies:

1. Activated sludge and
2. Membrane filtration

Significant engineering expertise can be applied to MBR design and operation. Several studies already have applied activated-sludge-related biology to MBRs, although current investigations suggest potentially important differences in growth, population diversity, cell activity, and competition. One noticeable difference is that MBR membranes have to be cleaned periodically to minimize biological and chemical fouling, and MBR manufacturers have already developed some cleaning methods.

MBRs have been used to treat a wide range of municipal and industrial wastewaters, and currently are installed at more than 1000 sites in Asia, Europe, and North America, according to a database assembled by the Water Environment Research Foundation. Most units currently treat a few hundred m^3/d (the largest treats less than 40,000 m^3/d). But plans are underway to build MBRs that will treat 30,000 to 150,000 m^3/d, and the technology could be used to treat 300,000 to 800,000 m^3/d.

Enough reliable equipment and technological support are commercially available to meet existing and developing demand. Membrane-manufacturing capacity is expanding, consequently unit costs are declining. The long-term trend is a "virtuous cycle" in which declining costs spur more demand, enabling further cost reduction.

Management Aspects

The key tasks in wastewater treatment are associated with generating high quality waters under cost effective methods. Particularly, construction and maintenance of treatment facilities plays a central role in the planning and operating process of wastewater treatment system for disposal and/or reuses (Tchobanoglous, 2003 and Van der Roest *et al.*, 2002). Optimal condition can be obtained by adopting the holistic approach. The integrative approach is based on trying to encompass all relevant aspects of the membrane units under consideration. The various aspects of membrane treatment plants can be viewed at the following levels:

1. The local level of the isolated treatment and the Point of Use (POU) of the reclaimed effluent;
2. The regional level which refers to the entire picture of the water resources, namely treatment options that include also the membrane alternative, and;
3. Other phases which include environmental issue, health risks, regulative and traditional aspects. The membrane element is only one link in a multi-component treatment system.

Intangible benefits and disadvantages of membrane bioreactor are those features that can hardly be quantified by financial terms, nevertheless are relevant in the decision-making procedure. Although they are difficult (or impossible) to quantify, the intangible benefits and/or disadvantages of a technology that can outweigh the results of the tangible cost evaluation, thus affecting the go/no-go decision of potential users. For example, potential users may find it beneficial to implement the new technology (concerning increased plant safety and reliability, reduced future liability, promoting public perception, etc.) even it is more expensive than the baseline technology.

There are two basic approaches concerning water management:

1. Water is a public good–consequently the prices are subject to national considerations.
2. Water prices are subject to the market forces. The impact of the market forces played major role in the UK where the farmers halted the irrigation of cereal crops and preferred to irrigate crops (mainly potatoes) which provide high shadow-price yields. Current use of effluent for low-grade agricultural products is often justified mainly on social grounds, however, when there will be a clear substitute in imported products, the outlook for these investments does not look bright.

The Challenge for Sustainable Sanitation Services in Developing Countries

This is emerging as a major problem so far as developing countries are considered. It is noticed that:

1. Half of the world's people (about 3 billion) live on less than US$1 per day.
2. About 800 million people lack access to health care.
3. About 10 million children under 5 years died in 1999, mostly from preventable diseases.
4. In 2002, an estimated 1.1 billion people lacked access to a safe water supply and 2.4 billion to improved sanitation.

5. Africa has 38 per cent of its population unserved by safe water and 40 per cent by sanitation, the figures for Asia are 19 per cent and 52 per cent, and 15 per cent and 22 per cent for Latin America and Caribbean.
6. Over the next 25 years, the urban populations of Africa and Asia will almost double; the urban populations of Latin American and the Caribbean will increase by nearly 50 per cent.
7. Delegates to the 2002 Johannesburg Summit agreed to cut in half the proportion of people without basic sanitation; this means providing sanitation to 2 billion more people.
8. The provision of full water and wastewater connections and primary wastewater treatment to the urban population would entail an annual cost of US$ 17 billion for water and US$ 32 billion for sanitation. To serve 2 billion more people by 2015 will require connections for more than 350,000 individuals each year.
9. The recent Third World Water Forum highlighted the fact that there are a further 3 billion people who only use pit toilets, flush toilets, or sewers without any treatment before discharge to the environment (World Water Forum, Rich Nations Get Wealth by Polluting Poor Nations, 17th March, Kyoto, 2003).

References

Cornel, P. and Krause, S., 2003. State of the art on MBRs in Europe. In: *Proceedings of Applications and Perspectives of MBR in Wastewater Treatment and Reuse*, 28–29 April 2003, Cremona, Italy.

Crites, R.W. and Tchobanoglous, G., 1998. *Small and Decentralized Wastewater Management Systems.* McGraw-Hill Book Company, New York, NY.

Daigger, G.T., Crawford, G., Fernandez, A., Lozier, J.C. and Fleischer, E., 2001. WERF Project: Feasibility of Membrane Technology for Biological Wastewater Treatment B Identification of Issues and MBR Technology Assessment Tool. In: *Proceedings of the Water Environment Federation 74th Annual Conference and Exposition*, Atlanta, GA, CD-ROM.

Fernandez, A., Lozier, J.C. and Daigger, G.T., 2000. Investigating Membrane Bioreactor Operation for Domestic Wastewater Treatment: A Case Study. In: *Proceedings of the Water Environment Federation 73rd Annual Conference and Exposition on Water Quality and Wastewater Treatment*, Anaheim, CA, CD-ROM.

Gunder, B., 2001. *The Membrane-Coupled Activated Sludge Process in Municipal Wastewater Treatment*, Technomic Publishing Co, Lancaster, PA.

Stephenson, T., Judd, S., Jefferson, B., and Brindle, K., *2000. Membrane Bioreactors for Wastewater Treatment.* IWA Publishing, London.

Tchobanoglous, G., 2003. The strategic importance of decentralized wastewater management in the twenty-first century. In: *Proceedings of IDA Conference on Water Reuse and Desalination*, Singapore (CD-Rom).

Van der Roest, H.F., Lawrence, D.P. and van Bentem, A.G.N., 2002. *Membrane Bioreactors for Municipal Wastewater Treatment*, IWA Publishing, London.

Chapter 3

Effect of Growth Regulators on Chlorophyll Content, Growth and Yield Parameters of Berseem (*Trifolium alexandrinum* L.) Under Punjab Condition

J.S. Kang, Avtar Singh and Jyoti Gulati
Department of Agronomy and Agrometeorology,
Punjab Agriculture University, Ludhiana, Punjab

ABSTRACT

Various growth regulators as Gibberelic acid (GA), Mepiquat chloride (MC), Succinic acid (SA), Garden glow (GG, a mixture of essential plant nutrient) and water spray were tested to see the effect on the seed yield production of berseem at Punjab Agricultural University, Ludhiana during the year 2001–02 and 2002–2003. Results revealed that seed yield enhanced significantly with the spray of GA (20 to 40 ppm), SA (400 to 600 ppm) and GA (5g/l) at flower initiation stage over the control. The increase in seed yield under all concentration of GA and SA was 18.21 to 39.28 per cent than the water spray. The leaf length and breadth, dry matter, photosynthetic pigment fragment, number of heads/sq. ft. Progressively increased with the increase in concentration of GA from 20 to 40 ppm over the control. However, the similar trend was observed in number of heads/sq. ft under the application of Succinic acid. With the application of GA and SA, Plants took less number of days to complete flowering over the control. The data on seed yield, growth and growth contributing characters showed that application of Mepiquat chloride was ineffective to enhance the seed yield of berseem.

Application of wide variety of both naturally occurring and synthetic chemical growth regulators has been extensively used in order to ascertain their beneficial effect on plants. A number of synthetic growth retardant have been discovered and proved to be of considerable importance in Agriculture (Yadava *et al.*, 1984).

Gibberelic acid, Succinic acid, Mepiquat chloride have been reported to be playing a significant role in seed production of Persian clover (*Trifolium resupinatum* L.) (Kang *et al.*, 2002). But the information on growth, flowering behaviour chlorophyll content in lacking which is also a tool for judging the crop productivity. Accordingly, the present investigation was undertaken to explore the possibility of improving seed yield in tetraploid berseem (*Trifolium alexandrinum* L.) crop in relation to leaf growth, dry matter accumulation, flowering behaviour and photosynthetic pigment.

Materials and Methods

A field trial was carried out at Punjab Agricultural University, Ludhiana during 2001–2002 and 2002–2003. The experimental soil was loamy sand, low in organic carbon and available nitrogen and medium in available phosphorus. Berseem variety BL-42 was sown on October 5. The first cutting was taken around 55 days and subsequent cutting were taken at 25/30 days interval. Flowering behaviour was noted from four spots of 2 × 2 feet in each plot. Various growth regulators like Gibberelic acid (GA), Mepiquat chloride (MC) Succinic acid (SA), Garden glow (GG) (mixture of essential plant nutrients) and water spray as control were tested. First spray of growth chemicals was given at 30 days after sowing and second spray was given at the time of initiation of flowering.

Leaf growth in term of length, breadth, dry matter accumulation in leaves and photosynthetic pigments (Chlorophyll) were recorded at 3-sucessive cuts. Pigments were extracted by aqueous acetone (Mackinney, 1941) and concentration was measured in duplicate samples at 660 and 640 mμ wavelength. The fractions of pigment were calculated from the absorption data using the equation of Comar (1942). Days to complete flowering were recorded keeping the initation of flowering as starting first day (day one). Number of heads/sq.feet were recorded one month before harvest from specified spots. Total seed yield was recorded after threshing of the harvested produce and the percentage increase or decrease over control was also worked out.

Results

Average length and breadth of fully expanded terminal leaf lamina in untreated and growth regulator treated plants are presented in Table 3.1. It was observed that leaves treated with Mepiquat chloride were shorter and narrower than those in control where as Gibberelic acid treated plants showed increasing trend in length and breadth of leaves with the increase in concentration up to 40 ppm. Observation from Succinic acid treated plots showed that with increase in concentration the leaf length and breadth rapidly decreased. Garden glow positively also affected the leaf growth.

Percentage of dry matter accumulation in leaves treated with Gibberelic acid (GA), Succinic acid (SA), Garden glow (GG) was recorded higher as compared to control (Table 3.1). However, Mepiquat Chloride (MC) up to 400 ppm improved the production but thereafter it showed declining trend. GA, SA and GG treatments were superior than control. The affect of treatment was not proportional to the concentration used.

Data on the photosynthesis pigments (Chlorophyll and its fractions) of berseem leaves under the various treatments are presented in Table 3.2. Total chlorophyll content increased with the application of GA from 20 to 40 ppm. However, it decreased with increasing concentration of Succinic acid (SA) and Gibberelic acid (GA). Except 600 ppm of MC, all the treatment of chemicals induced more total chlorophyll content than control. Highest chlorophyll-a content (0.823) was observed from plots treated with Garden gow and among the chemicals, GA (40 ppm) treated plots gave the maximum contents (0.780). Overall, chlorophyll-a was more than then chlorophyll-b, marked differences were also noticed in the ratio of chlorophyll-a and chlorophyll-b.

Table 3.1: Effect of Growth Regulators on Leaf Growth and Dry Matter (%) of Berseem Leaves (Means of 4 cuts)*

Treatments	*Leaf Length (cm)*	*Leaf Breadth (cm)*	*Dry Matter (%)*
Gibberelic Acid (ppm)			
20	5.30	2.12	15.68
30	5.38	2.19	16.24
40	5.59	2.32	16.98
Succinic Acid (ppm)			
400	5.30	2.12	16.19
500	5.16	2.00	15.81
600	5.10	1.96	15.67
Mepiquat Chloride (ppm)			
400	5.02	1.91	15.98
500	4.87	1.89	15.07
600	4.21	1.72	14.10
Garden Glow (5 g/l)	5.45	2.19	16.32
Water Spray (control)	5.06	2.00	15.09
CD at 5%	0.23	0.11	0.39

* Based on the 2 years mean data.

Data on the time (days) taken for complete flowering after last cut is presented in Table 3.3. It was observed that the number of days taken for complete flowering varied from 17 to 26 under different treatment. The flowering was advanced under the application of GA and SA and it was delayed under the Mepiquat chloride spray and garden glow. However, under Garden glow the differences were minor when compared with GA and SA.

Seed yield increased (Table 3.3) significantly with the application of GA, SA and garden glow over the control. However seed yield was not influenced with the application of Mepiquat chloride and it was lower than the control under all the treatments of MC. The increase in seed yield was 23.57, 25.71, 31.96, 24.46, 35.89, 39.28 and 18.21 per cent higher under the application of GA, SA and garden glow at 20, 30, 40 ppm, 400, 500, 600 ppm and 5g/1, respectively over the control. This might be due to the more flower heads under all the treatment over the control. Data on the seed yield indicated that application of GA, SA and Garden Glow were found effective to enhance the seed yield of berseem.

Discussion

Mepiquat chloride (MC) induced shorter and narrow leaves than those in control but Gibberelic acid treated plots showed increasing trend in length and breadth of leaf and no phyotoxic symptoms were observed on the foliage. The results in conformity with earlier reports on other plants (Kang *et al.*, 2002, Brar *et a1.*, 1998 and Jain and Yadava, 1980). Pandey *et al.* (2001) observed reduction in plant height, leaf growth with IAA and CCC in rice crop. MC reduced the length and breadth of the leaves but increased the dry matter production in comparison to control where length and breadth from GA and SA treated plots increased with increasing dry matter, therefore it is conceivable that Mepiquat chloride, cause cells to be smaller with a more dense cytoplasm. The view point has been some what substantiated by an increase in chlorophyll content with more dry matter accumulation from Gibberelic acid and Succinic acid treated plants.

Table 3.2: Effect of Growth Regulators on Photosynthetic Pigments of Berseem Leaves (Means of 4 cuts)*

Treatments	*Chlorophyll-a (mg/l) Fresh Wt.*	*Chlorophyll-b (mg/l) Fresh Wt.*	*Total Chlorophyll (mg/l) Fresh Wt.*	*Ratio (a : b)*
Gibberelic Acid (ppm)				
20	0.768	0.469	1.237	1.63
30	0.776	0.495	1.271	1.56
40	0.780	0.521	1.301	1.49
Succinic Acid (ppm)				
400	0.812	0.483	1.295	1.68
500	0.791	0.480	1.271	1.64
600	0.725	0.456	1.181	1.58
Mepiquat Chloride (ppm)				
400	0.708	0.455	1.163	1.55
500	0.695	0.463	1.158	1.50
600	0.597	0.403	1.000	1.48
Garden Glow				
(5g/l)	0.823	0.480	1.303	1.71
Water Spray (control)	0.693	0.349	1.042	1.98

* Based on the 2 years mean data.

Minimum time to reach the complete flowering stage was observed under GA and SA treated plots and the corresponding data on seed yield showed higher yield from GA and SA treated plots over the control. Higher seed yield from GA and SA treated plots over the control might be due to the mobilization of metabolites from the vegetative to the reproductive part which resulted in higher seed yield. The treatments which took more days to reach complete flowering might have been having a slow pace of metabolic reaction due to stress caused by chemical spray as seen from the data on leaf length, breadth and chlorophyll content and therefore less of the metabolites reached the sink site leading to many more days for attaining the complete flowering stage. The delay in complete flowering and with more rise in temperature during April and May month there is drastic reduction in the number of flower heads/sq. ft an ultimately lesser seed yield as seen from the data from Table 3.3.

Table 3.3: Effect of Growth Regulators on Complete Flowering, Numbers of Heads/Sq.ft and Seed Yield (Mean of 2 years)

Treatments	*Days to Complete Flowering After Spray at Initiation Stage*	*Number of Flower Heads (Heads/Sq.ft.)*	*Seed Yield (q/ha)*	*Per cent Increase/ Decrease Over Control*
Gibberelic Acid (ppm)				
20	18	290	6.92	+ 23.57
30	17	295	7.04	+ 25.71
40	18	308	7.39	+ 31.96
Succinic Acid (ppm)				
400	20	290	6.97	+ 24.46
500	17	312	7.61	+ 35.89
600	18	314	7.80	+ 39.28
Mepiquat Chloride (ppm)				
400	26	231	5.36	– 4.28
500	23	220	5.10	– 8.92
600	24	209	5.16	– 7.85
Garden Glow				
(5g/l)	21	286	6.62	+ 18.21
Water Spray (Control)	22	236	5.60	–
CD at 5 per cent	–	35	0.84	–

Mepiquat chloride (400 and 500 ppm) had increased the chlorophyll-'a' content when compared with control but at the highest concentration (600 ppm) the chlorophyll-'a' were lower as compared to control. Chlorophyll-'b', contents were higher at all levels of MC treated plots over control. However, in spite of this, plots treated with MC took more days to reach complete flowering and which might have been having a slow pace of metabolic reaction due to stress caused by chemical spray as seen from data on leaf length, breadth and chlorophyll content and therefore less of the metabolites reached the sink site.

Wasnik and Bagga (1996) were of the view that increased chlorophyll content condition in chick pea could not be harnessed because of lack of any increase in sink size. The delay in flowering and rise in temperature during April and May months caused drastic reduction in the number of flower heads/sq.ft. which ultimately caused lesser seed yield. Similar findings have been reported by Kang *et al.* (2002), Kang and Tiwana (2001), Pandey *et al.* (2001), Bidlack and Roxton (1995), Mohamed *et al.* (1992), and Dhasmana and Aggarwal (1991).

References

Bidlack, J.E. and Roxton, D.R., 1995. Chemical regulators of yield, growth and digestibility of alfalfa. *J. Plant Growth Regulators*, 14: 1–7.

Brar, Z.S., Singh, M. and Singh, G., 1998. Effect of planting dates and growth regulators on production of mung bean. *J. Res. Punjab Agric. Univ.*, Ludhiana, 25(4): 515–520.

Comar, C.L., 1942. Chloroplast substance of spinach leaves. *Bot. Gaz.*, 104: 122–127.

Dhasmana, R. and Aggarwal, A.K., 1991. Effect of foliar spray of GA, IAA and 2,4-D on production of clover (*Trifolium alexandrinum* L.). *Res and Dev. Rep.*, 8: 39–46.

Jain, N. and Yadava, R.B.R., 1980. Effect of Phospon-D on seed yield and quality of cowpea (*Vigna unguiculata* L.) plants. *Seed Res.*, 8(1): 42–45.

Kang, J.S., Tiwana, M.S. and Thind, I.S., 2002. Seed yield response to growth regulators in Persian clover (*Trifolium resupinatum* L.). *J. Res. Punjab Agric. Univ.*, Ludhiana, 39(3): 334–336.

Mohamed, N.A., Ahmed, M.K., Abed, E.L., Halim, A.Z. and Moausa, M.E., 1992. Effect of Alar and micronutrients on yield and morphological characters of berseem. *Egyptian J. Agric. Res.*, 70(4): 177–190.

Pandey, A.K., Tripathi, R.S. and Yadava, R.S., 2001. Effect of certain growth substances on growth, yield and quality of rice (*Oryza sativa* L.). *Indian J. Agric. Res.*, 35(2): 118–120.

Xu, X. and Taylor, H.M., 1982. Increase in drought resistance of cotton seedlings treated with Mepiquat chloride. *Agron. J.*, 84: 569–574.

Wasnik, K.G. and Bagga, A.K., 1996. Effect of Mepiquat chloride on growth and yield of chickpea (*Cicer arietinum* L.). *Indian J. Plant Physiol.*, (4): 239–242.

Chapter 4

Estimation of Parameters in $E_k/G/1$ Queueing System Full Vs Reduced Likelihood Functions

P. Raja Rao[1], K. Chandan[1] and C. Raghavendara Rao[2]
[1]Department of Statistics, Acharya Nagarjuna University, Guntur – 522 510, Andhra Pradesh, India
[2]School of M.C.I.S, University of Hyderabad – 500 075, Andhra Pradesh, India

ABSTRACT

The problem of estimating parameter of arrival and service rates based on reduced and full likelihood methods are considered. Monte Carlo simulations are carried out to study the relative efficiency, for different values of traffic intensity (ρ), varying ρ = 0.5 (0.1) 0.9, sample size n = 15, 25, 50, 100, 200, 400 and 500. In general, it can be seen that the loss of efficiency is small while estimating the arrival rate using the reduced likelihood method when compared with that of full likelihood method, even for moderate samples of size 50 with a minimum efficiency of 92.30 per cent.

Keywords: *$E_k/G/1$ Queueing system, Full and reduced likelihood functions, Poisson approximation.*

Introduction

Most of the queueing models are quite complex and cannot be easily understood. In almost all the queueing situations, the element of uncertainty is there. This element of uncertainty may be of the following types:

1. We may not know the form of theoretical probability distribution which suits in a given situation.
2. We might not know the parameters of the process even though we have a prior knowledge about the form of the distribution.
3. We would simply be knowing only the probability distribution of outcomes and not the distribution of actual outcomes even though (1) and (2) are known and so on.

Some of the above mentioned factors, usually complicates the analysis of queueing systems. Many authors paid their attention in this direction. For example, Clarks (1957) considered the problem of estimation of traffic intensity in M/M/1 systems with the help of method of maximum likelihood. The problem of Statistical inference for birth and death process was studied by Wolf (1965). Billingsley (1961) has applied large sample theory based on M.L.E's to make inferences of arrival and service rates in Markov processes. Similarly, Jenkins (1972), Basava and Prabhu (1981, 1988), Harris (1974), Benes (1957), Goyal and Harris (1972), Crane and, Lemoine (1977) extensively applied estimation theory Queueing systems.

On the lines of Sudha Jain (1992), here an attempt is made to study the relative efficiency in estimating parameters in $E_k/G/1$ queueing systems based on reduced and full likelihood methods using Monte Carlo simulation. The numerical results of the simulation are presented at the end of this study. It can be seen that the minimum relative efficiency is 0.7987841 even for small samples of size equal to 15.

Estimation of the Parameters Using Full and Reduced Likelihood Functions

Denote the interarrival times by $\{t_r, r \geq 1\}$ and the service times by $\{X_s, s \geq 1\}$. Their densities are denoted by $f(t, \lambda)$ and $g(x, \mu)$ respectively.

Initially, customers arrive at time t equal to zero and observe the queue until the first n customers having departed. Let the service times of these customers be $x_1, x_2, \ldots\ldots\ldots x_n$. Let the nth departure occur at the epoch D_n. Observe the interarrival times of all customers during the interval $(0, D_n)$. Let their interarrival times be

$$t_1, t_2, \ldots\ldots\ldots t_{NA},$$

where,

$$N_A = N_A(D_n) = \text{Max}\{r : t_1 + t_2 + \ldots\ldots\ldots + t_r \leq D_n\}.$$

Here $N_A \geq n$.

Full and Reduced likelihood functions respectively are given by,

$$L(\lambda, \mu) = \left[\prod_{i=1}^{N_A} f(t_i; \lambda)\right]\left[\prod_{i=1}^{n} g(x_i; \mu)\right][1 - F(Z_n; \lambda)] \quad \ldots\ldots (1)$$

$$L^R(\lambda, \mu) = \left[\prod_{i=1}^{N_A} f(t_i; \lambda)\right]\left[\prod_{i=1}^{n} g(x_i; \mu)\right] \quad \ldots\ldots (2)$$

where,

$$Z_n = D_n - \sum_{i=1}^{N_A} t_i$$

F is the distributions function of Erlang (k, λ), here in this work, it is approximated by Poisson distribution function. Let $\hat{\lambda}$ and $\hat{\mu}$ be the estimates of λ and μ obtained by applying full likelihood approach. Similarly, let λ^* and μ^* be the estimates based on reduced likelihood approach, and are obtained follows:

$\hat{\lambda}$ is the iterative solution of the equation

$$\frac{N_A}{\lambda} + \frac{Z_n B}{1 + B} = D_n$$

where,

$$B = \sum_{j=1}^{k-1} \frac{(k_\lambda Z_n)^{j-1}}{(j-1)!} \qquad \text{......(3)}$$

$$\lambda^* = \frac{N_A}{\sum_{i=1}^{N_A} t_i} \qquad \text{......(4)}$$

$$\hat{\mu} = \lambda^* = \frac{n}{\sum_{j=1}^{n} x_j} \qquad \text{......(5)}$$

From (5), it is clear that maximum likelihood estimates of μ under the full and reduced likelihood methods are identical and hence, the relative efficiency for the estimation of μ is unity for any sample size. Therefore, to study the performance of $\hat{\lambda}$ and λ^* based on relative efficiency, a simple simulation study is carried out and the methodology is presented below:

Methodology and Results of the Simulation Study

The simulation study is conducted fixing the sample size n (15, 25, 50, 100, 200, 400 and 500) and traffic intensity ρ [0.5(0.1), 0.9]. In each case the empirical sampling distribution of $\hat{\lambda}$ and λ^* and hence the relative efficiency is obtained based on 300 runs. The numerical results obtained are given towards the end of this article.

Conclusion

1. The impact of traffic intensity on the relative efficiency with increasing number of observations is approximately the same for $E_k/M/1$ and $E_k/E_2/1$ for $k \geq 3$.
2. Efficiency for various traffic intensities with increasing number of observations for $E_k/E_m/1$ for k < 3 and $m < 3$ are as shown in Tables 4.1 and 4.2.

   As would be expected, the sample estimate of relative efficiency tens to one as $n \rightarrow \infty$ can be seen from the tables.

The minimal efficiency even for small values of n=15 is 79.88 per cent, suggests that there is not much loss of efficiency while estimating λ using reduced likelihood method when compared with full likelihood.

If the distribution of interarrivals is Erlangin then its parameter λ can be estimated using reduced likelihood method without much loss in efficiency and the method of estimation has no influence in estimating the parameter of the distribution of service times.

Table 4.1: Relative Efficiency for E_k/M/1 Queueing model

n	*k/p*	*0.5*	*0.6*	*0.7*	*0.8*	*0.9*	*Min.eff*
15	1	0.8335581	0.8528731	0.7987841	0.8457098	0.8632942	0.7987841
	2	0.9583402	0.9503756	0.9381929	0.9434395	0.9306441	0.9306441
25	1	0.8433532	0.8536280	0.8805692	0.8654204	0.9145220	0.8433532
	2	0.9753398	0.9746120	0.9647560	0.9644838	0.9594773	0.9594773
50	1	0.9563008	0.9347615	0.9528631	0.9576596	0.9487243	0.9347615
	2	0.9846970	0.9819526	0.9762438	0.9778696	0.9755160	0.9755165
100	1	0.9836449	0.9726178	0.9701495	0.9618783	0.9604666	0.9604666
	2	0.9947774	0.9919042	0.9905181	0.9872656	0.9893526	0.9872656
200	1	0.9895873	0.9842166	0.9801233	0.9835947	0.9885762	0.9801233
	2	0.9988771	0.9973802	0.9963119	0.9960754	0.9962264	0.9960754
400	1	0.9932950	0.9945760	0.9980874	0.9920691	0.9859098	0.9859098
	2	0.9984075	0.9975168	0.9980487	0.9966760	0.9975719	0.9966760
500	1	0.9968913	0.9970092	0.9911672	0.9969994	0.9975424	0.9911672
	2	0.9997157	0.9994.176	0.9989957	0.9990512	0.9995813	0.9989957

Table 4.2: IC Relative Efficiency for E_k/E_2/1 Queueing Model

n	*k/p*	*0.5*	*0.6*	*0.7*	*0.8*	*0.9*	*Min.eff*
15	1	0.8470323	0.8550126	0.9225488	0.8826121	0.8681330	0.8470323
	2	0.9409723	0.9241418	0.9150893	0.8968290	0.9289149	0.89698290
25	1	0.9260871	0.9314642	0.9590673	0.9821219	0.9023365	0.9023365
	2	0.9614693	0.9476824	0.9513674	0.9481175	0.9594715	0.9476824
50	1	0.9486561	0.9515157	0.9743427	0.98629784	0.92309598	0.9230598
	2	0.9797688	0.9768381	0.9738789	0.9787709	0.9689788	0.9689788
100	1	0.9732689	0.9932962	0.9922978	0.9866788	0.9294362	0.9294362
	2	0.9907874	0.9933741	0.9883832	0.9790151	0.9886633	0.9790151
200	1	0.9835034	0.9942181	0.9978029	0.9913411	0.9920730	0.9835034
	2	0.9957043	0.9960262	0.9925704	0.9860026	0.9916946	0.9860026
400	1	0.9890128	0.9944886	0.9951223	0.9927416	0.9931053	0.9890128
	2	0.9995695	0.9964384	0.9956747	0.9992695	0.9959501	0.9929501
500	1	0.9906976	0.9976304	0.9953794	0.9975428	0.9955084	0.9906976
	2	0.9999950	0.9988606	0.9965516	0.9994335	0.9998353	0.9965516

References

Basawa, I.V. and Prabhu, N.U., 1981. Estimation in single server queues. *Naval Res. Log. Quart.*, 28: 475–487.

Basawa, I.V. and Prabhu, N.U., 1988. Large sample inference from single server queues. *Queueing Systems*, 3, 298–304.

Belles, V.E., 1957. A sufficient set of statistics for a simple telephone exchange model. *Bell System Journal*, 36: 939–964.

Billingsley, P., 1961. *Statistical Inference for Markov Processes*. University of Chicago Press.

Clarke, A.B., 1957. Maximum likelihood estimates in a simple queue. *Annuals of Math Statist*, 28: 1036–1040.

Crane, M.A. and A.J. Lemoine, 1977. *An Introduction to the Regenerative Method for Simulation Analysis.* Springer Verlag, New York.

Goyal, T.L. and C.M. Harris, 1972. Maximum likelihood estimation for queues with state dependent service. *Sankhaya*, A34: 65–80.

Harris, C.M., 1974. Some new results in the statistical analysis of queues. In: *Mathematical Methods in Queueing Theory*, (Ed.) A.B. Clarke. Springer Verlag, Berlin.

Jenkins, J.H., 1972. The relative efficiency of direct and maximum likelihood estimators of mean waiting time in the simple queue M/M/1. *J. of Appl. Prob.*, 9: 396–403.

Sudha Jain, 1992. Relative efficiency of a parameter for a M/G/1 Queueing system based on reduced and full likelihood functions. *Commun. Statist. Simula.*, 21(2): 597–606.

Wolff, R.W., 1965. Problems of statistical inference for birth-death queing models. *Oper. Res.*, 13: 343–357.

Chapter 5

Detection of Trace Elements in Dahod Reservoir Water, Raisen District, M.P.

Arun Singh, H.C. Kataria and Subhash Pandey
Assistant Professors, P.G. Department of Chemistry,
Government Geetanjali Girls P.G. College, Bhopal – 462 038, M.P., India

ABSTRACT

The water of 'Dahod' reservoir is polluted by man made activities, there are various reasons of pollution *e.g.*, sewage, industrial wastes and natural contaminants, Dahod reservoir has annul rainfall of 170.18 cm, catchment area 52 sq.km, live storage capacity 604 million calf and constructed on river Bagwani. To evaluate the water quality status, different trace elements Fe, Cu, Zn, Cd and Mn that contaminate in Dahod-reservoir and silt sediments has analyzed by Atomic absorption spectrophotometer model (2380) with air-acetylene flame. Water samples are acidified to pH 2 with nitric acid and then analysed on Perkin Elmer AAS. In the present study Fe, Cu, Cr, Zn, Cd and Mn has ranged from 0.010–0.50, 0.040–6.8, 0.010–0.30, 0.020–8.4, 0.01–0.06, 0.050–0.80 mg/l at different sampling stations. The concentrations in surface waters are lower than that of silt-sediments.

Introduction

Degradation of good water quality due to trace elements has received considerable attention during the recent years. Pollution in surface and groundwater system through human activities and natural process may enter into water reservoir. Solid waste from industrial units is being dumped near factories and sewage water is subjected to reaction with percolating rainwater and reaches the surface water level.

The source of metal pollution in the environment and their health hazards has briefly given by Handa (1994). Higher content of iron is causing a serious problem in water supplies from groundwater.

Pure water is never found in nature, as pure water is that contains only 2 parts of H_2 and 1 part of O_2 by volume water in nature contains a number of impurities in varying amounts. The rainwater, which is originally pure, also absorbs various gases, dust and other impurities. This water when moves on the ground further carry silts organic impurities etc.

The present study aimed to see the suitability of surface-reservoir water as a source of drinking and irrigation purpose. The study will help the authorities for planning in drinking and irrigation supply schemes in coastal catchment area of Raisen district of M.P.

Inappropriate disposal and accidental release into environment of toxic and hazardous substances from chemical and industrial activities have created environment and public health problem all over the world. Detection of traces of toxic chemicals in surface water has boasted the public awareness of the risks posed by hazardous wastes (HWs) to human health and quality of environment.

Dahod village is situated in Tehsil-Gohar Ganj, District Raisen of Madhya Pradesh at longitude 77°20′30″ and latitude 23°2′0″. Dahod dam is the main source of water to fulfil the requirement for drinking and irrigation of Raisen district, the distance of dam is 25 kms far from Bhopal, state of M.P. Dahod reservoir is under environmental stress due to Siltation (by annual rainfall of 170.18 cm), human encroachment, high macrophytic population and sewage input from various resources. Due to decreasing in water quality day-by-day, the present study becomes necessary to assess the trace elements for water quality index of this dam. The soil inflows by rain run-off water from catchment are (52 sq.km) and basin directly affects the water quality of reservoir. Sampling Stations has chosen are:

1. Near Village Side
2. 1 km Down stream
3. 2 km Down stream
4. Tank Bund
5. Waste-wear
6. Spill Channel
7. Dahod Village–Guest House
8. Dhobi Ghat

Materials and Methods

Eight sampling stations were chosen at different points of Dahod reservoir. Water samples are acidified 2 or 3 times and evaporated up to dryness and finally dissolved in HNO_3 and the samples then used to Perkin Elmer, Atomic Absorption Spectrophotometer with air-acetylene flame to detect the trace elements Fe, Cu, Zn, Cr, Cd, and Mn at different wavelength of light and width of specified slit.

The instruments are calibrated for each metal to be detected, with known standard solution and programming aspirate the pretreated samples to the flame. Set zero with double distilled water before each detection. Results are displayed in μ gm/l from which the actual concentration in the water sample can be calculated and total metal content of the water samples can be obtained in (mg/l). The method is similar to APHA (1985), NEERI (1986).

Results and Discussion

Iron

Iron may be present in groundwater in varying quantities from 0.5 to 100 ppm (Ernst, 1991), ISI 1983, and WHO (1984) have set a desirable limit of 0.3 mg/l and maximum permissible limit of

1.0 mg/l for drinking purpose. In the present study iron found in the range of 0.020–0.56 iron in silt sediments, 0.020–0.036 in sewage disposal and 0.018–0.036 ppm in domestic sewage confluence points, lower values of 0.020 ppm at SS_4, SS_8 while higher values are reported at SS_4 and SS_2, SS_7 and SS_3, SS_8, SS_6 has lower values. Higher values have found in silt sediments, sewage disposal and domestic confluence points.

Table 5.1: Detection of Heavy Metals in Dahod Reservoir Water

(A) Concentration of heavy metals in Dahod Silt–Sediments sampling metals in mg/w during 2003–04

Stations	*Fe*	*Cu*	*Cr*	*Zn*	*Cd*	*Mn*
1	0.50	6.80**	BDL	8.4**	3.8**	0.80
2	0.56**	4.2	0.08	8.0	1.0*	0.62*
3	0.48	2.6	0.06*	4.3	1.2	1.2**
4	0.02*	1.4*	BDL	2.4*	nf	BDL
5	0.026	5.6	0.09**	6.2	1.6	1.0
6	0.028	6.2	0.06	6.8	2.4	BDL
7	0.030	4.7	nf	5.4	3.0	0.90
8	0.020*	4.1	nf	5.0	2.8	0.86

(B) Heavy metals near sewages disposal point

Stations	*Fe*	*Cu*	*Cr*	*Zn*	*Cd*	*Mn*
1	0.030	3.0**	0.08**	0.52	BDL	0.14
2	0.036**	1.20	0.07*	0.70	BDL	BDL
3	0.30	0.50	nf	0.72	0.010*	0.80
4	0.026	1.10	nf	0.03*	0.014	0.84**
5	0.022	0.14	BDL	1.20	0.016	0.42
6	0.030	0.05	BDL	2.40**	0.02	0.48
7	0.020*	0.06	BDL	2.00	0.03*	nf
8	0.024	0.04*	BDL	1.10	0.010*	0.08*

(C) Heavy metals near sewages disposal point

Stations	*Fe*	*Cu*	*Cr*	*Zn*	*Cd*	*Mn*
1	0.036**	2.40	BDL	0.60	BDL	0.14
2	0.024	2.30	BDL	0.70	BDL	BDL
3	0.018*	2.50	nf	0.04*	0.010*	0.80**
4	0.026	2.80**	0.05*	0.08	0.03*	0.75
5	0.030	1.60	0.06**	2.40	0.02	0.48
6	0.021	1.80	BDL	2.80**	BDL	nf
7	0.022	0.30	BDL	1.20	BDL	0.06*
8	0.018*	0.08*	BDL	1.16*	nf	

nf: Not found; BDL: Below detection limit.

TLV mg/m³ Cu: 0.2; Cr: 0.01; Zn: 1.0; Mn: 5.0.

Copper

Copper is ubiquitous in the environment and hence frequently present in the water. Since copper is both an essential and potentially toxic element there may be risk to living beings if there is too little or too excess of copper in the environment. Large dose of Cu irritate the Stomach (Ross, 1955). ISI has fixed a desirable limit of 0.05 mg/l for Cu. In the present study copper present in the range of 1.4–6.8, 0.10–3.0 and 0.08–2.80 ppm in Dahod dam silt sediments, near sewage disposal and in domestic confluence points respectively are well within the permissible limit. Copper is essential component of Key–metalloenzyme that maintains the vascular and nervous system. Its deficiency causes livestocks disease. Excessive accumulation of copper in tissues occurs in Wilson's disease, causes neurological dysfunction, progressive cirrhosis.

Chromium

Chromium salts are widely used in industrial processes and the products used by human beings, they enter into water bodies through discharge of waters. The normal level of Cr (V1) in drinking water is 3 to 40 ppm. Natural water contains traces of Cr (Cr^{3+}) under strong oxidizing conditions. It is converted into Cr^{6+} and occurs as CrO_4^{2-}, which is usually results pollution from industrial and domestic effluents and wastes. The maximum contaminant level of Cr is 0.05 mg/l. In the present study Cr has found in BDL or nil 0.06–0.08, BDL–0.08, BDL–0.06 ppm at SS_4, SS_7, SS_8–SS_2 in silt sediments, sewage disposal and domestic confluences respectively.

Zinc

The toxicity has been reported with dizziness, muscular incardination, dizziness, renal failure and anaemia. Most of the cases of Zinc toxicity have resulted from the irrigation of acid food kept in galvanized metal container. Galvanized pipes are contaminated by the flow of waste from industrial sources or toxic waste sites. Bio-chemically, Zinc is present in over 160 enzymes with many functions. Zinc deficiency is associated with impaired growth, parakeratosis, failure of sexual maturation and malnutrition. The. permissible limit for Zn is 5.0 mg/L by ISI: 10500. Zn conc. In the present study, lower values of Zn concentration is recorded at SS_4, SS_3 and higher at SS_1, SS_2, SS_6, SS_5 in silt sediments. Lower values of Zn near sewage and domestic effluents are at SS_4, SS_1 and at SS_3, SS_1 while the higher values are observed at SS_8, SS_5, SS_7, and SS_6 and at SS_5, SS_6 Sampling stations due to sewage.

Cadmium

Cd is a non-essential non-beneficial element known to have high toxic potential. The Cd content in the study area varies from 1.0–3.8, 0.010–0.016 and 0.01–0.03 ppm at A, B, C, points. Desirable limit for Cd is 0.01 mg/l prescribed by ISI for drinking water (BIS, 1991). Beyond the limit, water becomes toxic. Cadmium can cause bronchitis, anaemia and renal stone formation in animals. The water is contaminated with Cd by industrial discharges and leaches from land filled areas. Drinking water is generally contaminated with galvanized iron pipe and plated plumbing fittings of water distribution system.

Manganese

Manganese is a vital micro-nutrient for both plants and animals. The deficiency state of Mn has not been found in human. In other mammals, Mn deficiency is characterized by defective growth, bone abnormalities, reproductive dysfunction. The concentration of Manganese recorded a maximum level of 1.2 in silt sediments at SS_3, and 0.84 and 0.80 ppm in SS_4 and SS_3 in domestic sewage. Mn is an essential element, which does not occur naturally as a metal but is found in various salts and minerals frequently in association with iron-compounds. A concentration of 1.0 mg/l has been recommended as a desirable limit and 0.3 mg/l as the maximum permissible limit for drinking water (BIS, 1991). The presence of Mn beyond the permissible limit imparts alien taste to water. Mn also has adverse effects on domestic uses and water supply structures. Higher concentration may be attributed to the reducing conditions of the water and dissolution of Mn bearing salts from the soil strata (Jain *et al.*, 2004).

Acknowledgement

We are thankful to M.P. Pollution Control Board to help in this study and EPCO, Bhopal to facilitate the Library Consultation.

References

APHA. 1986. *Standard Methods for the Examination of Water and Wastewater*. APHA, AWWA, WPCF, New York.

BIS. 1983. *Indian Standards Specification for Drinking Water*, IS: 10500. Bureau of Indian Standards, New Delhi.

BIS. 1991. *Specifications for Drinking Water*, IS: 10500. Bureau of Indian Standards, New Delhi.

Drinking Water and Health, Vol. 1. National Academy of Sciences, Washington, DC; National Academy Press.

EPA (Environmental Protection Agency). 1980. Ambient water quality criteria for zinc, publication 440/5–80–79.

Ernst, M. 1991. *Metals and their Compounds in the Environment*. VCH, Weinheim, p. 894–1332.

Jain, C.K., Sudhir Kumar and Y.R. Rao. 2004. Trace element contamination in a coastal aquifer of Andhra Pradesh, *Poll. Res.*, 23(1): 3–23.

Kaplary, R.D. *et al.* 2003. Trace elements on groundwater of Tuppa Area New Nanded, Maharashtra. *Poll. Res.*, 22(4): 595–599.

Kataria, H.C. 1994. Heavy metals contamination and pollution in Betwa River. *IJEP*, 15(1): 34–38.

Kataria, H.C. 1995. Analysis of trace elements in lower lake water of Bhopal. *Oriental J. of Chemistry*, 11(3): 288–289.

Kataria, H.C. 1995. Heavy metal water pollution of Pipariya Township of Madhya Pradesh. *J. Freshwater Biol.*, 8(2): 105–107.

More, T.G. *et al.* 2003. Impact of heavy metals on DNA contents *Lamillidenes marginalis. Poll. Res.*, 22(4): 605–611.

NEERI. 1986. *Manual on Water and Wastewater Analysis*. National Environmental Engineering Research Institute, Nagpur, pp. 340.

Ross, A.I. 1995. Lancet, 11,87, e.f. Holaen, W.S. 1970.

Srikanth, R.A., Madhumohan Rao, C.H., Shravan Kumar and Anees Khanum. 1993. Pb, Calcium, Ni and Zn contamination of groundwater around Hussein Sugar lake, Hyderabad. *Bull. Env. Contam. Toxicology*, 50(1): 138–143.

Warrant-Viessman, Jr. and Mark J. Hammer. 1985. *Water Supply and Pollution Control*, 4th ed., Harper and Row, Publ. Inc., p. 220–223.

WHO (World Health Organization). 1984. *Guidelines for Drinking Water Quality*, Vol. 1.

WHO. 1972. *Geochemical Environments, Trace Elements and Cardiovascular Diseases Bull.*, pp. 47.

Chapter 6

Studies on the Effect of Graded Levels of NaCl of *in vitro* IAA Production, Wheat Germination and Salinity Associated Soil Parameters

N.S. Kulkarni[1], M.G. Bodhankar[2], R.B. Somani[3] and J.V. Jaiswal[1]

[1]Department of Microbiology, R.A. College, Washim

[2]Department of Microbiology, Bharti Vidyapeeth's M. B. S. K. College, Kadegaon, Sangli

[3]Department of Plant Pathology, Dr. P.D.K.V., Akola

ABSTRACT

Effect of NaCl concentration on *in vitro* IAA production by *Azotobacter* isolates and their interaction with different salt concentration on seed germination in wheat were investigated, simultaneously, the soil salinity associated parameters were studied. Improvement in *in vitro* IAA production by *Azotobacter* isolates was noticed only at low levels *i.e.* up to 1.5 per cent of NaCl concentration. However, above 2 per cent NaCl, decreased IAA production was recorded. *Azotobacter* species at salt concentration at less than 1.5 per cent, showed significant improvement in seed germination. However, reduction in germination at 1.5 per cent and above salt concentration was noticed. It was also observed that as salt concentration in soil increases the salinity associated properties also fluctuates which may retard the germination in seed.

Keywords: Azotobacter species, NaCl per cent, IAA production, Seed germination, Soil characteristics.

Introduction

Indian farmers since long back had adopted physical and chemical methods for soil management. However, due to several constraints *viz.*, loss of soil fertility, high cost and increasing pollution, chemical methods proved set back for their use (Malik *et al.*, 1998) in soil management.

Recent developments in agricultural research are concerned with enhanced crop yield leaned on the use of microbial products, which not only acts as a source and sink of plant nutrients but synthesizes growth promoting substances.

Materials and Methods

Lates of *Azotobacter* were further tentatively identified as *A. vinclandii, A. chroococcum, A. nigricans, A. beijerinckii* etc. on the basis of conventional morphological, biochemical and cultural studies (Bergey's 1989). IAA production was done by inoculating all the isolates separately, in 0.2 per cent tryptophan enriched Asbhy's media (Castillo, 2000) containing graded concentration of NaCl *viz.* (0.5 per cent, 1.0 per cent to 3.5 per cent). All the isolates were qualitatively analysed for IAA production using TCL (Sadashivam *et al.*, 1982); 2,6-dichlorophenol indophenol was used as a locating agent. The culture supernatant was separated using cooling centrifuge and quantitative analysis was done by adopting spectrophotometric method (Dubey, 2002). In 2 ml of supernatant, 2–3 drops of orthophosphoric acid, and 4 ml of Salper's reagent was added and kept for incubation at 37°C for 25 min. The absorbance was recorded at 560 nm. IAA concentration was estimated by comparing O.D. values with standard caliberation curve.

The pot culture experiments were conducted to assess the expression of Azotobacter species at different salt concentration. The vertisol soil was collected from local wheat farm and prepared by fortifying the calculated dose of NaCl *viz.* (0.5 per cent, 1.0 per cent to 3.5 per cent) in the soil. Pot without NaCl was maintained as control. Seed treatments were done separately on wheat (cv. Lok-1) as test crop using *Azotobacter* isolates. Wheat seeds (100 gm) were first surface sterilized with acidic alcohol followed by bacterization using charcoal based bio inoculant and jaggery solution (10 per cent) as an adhesive for each isolate separately so as to have 10^4 bacterial cells per seed. Sowing was done at the rate of 10 seeds per plot and all plots were irrigated as and when required and germination count as plant stand was noted on 20th DAS.

The soil sample used in pot cultures was analyzed for salinity associated parameters. The water holding capacity (WHC) of soil samples was analyzed adopting KR Box method suggested by Chopra (1991). Electrical conductivity was determined at 25°C (Black, 1965). pH, extractable chloride and soluble Na per cent was estimated as described by Jackson (1973).

Results and Discussion

The soil salinity had induced the major problem in the productivity of agriculture. Hence efforts have been made to investigate the effect of different salt concentration on *in vitro* Indole-3-acetic acid (IAA) production, germination percentage and some salinity associated parameters of soil.

In vitro IAA Production (Table 6.1)

It was observed that, the IAA production increases along with increase in salt concentration upto 1.5 per cent w/w NaCl. However, above 2.0 per cent level of salt concentration the IAA (37.76 µg/ml) was recorded in the treatment having 1.5 per cent NaCl and found statistically significant over control at which the IAA recorded was (35.12 µg/ml) where as the minimum *i.e.* (1.62 µg/ml) IAA was recorded at 3.5 per cent NaCl concentration. The result reveals that salt concentration at low levels *i.e.* upto 1.5 per cent w/w may positively modulates the IAA production. However, at higher levels *i.e.* above 2 per cent w/w, the negative modulation in IAA synthesis may be due to the salt stress on cell physiological activities.

Table 6.1: IAA Production by *Azotobacter* Isolates at Different Salt Concentration

Salt Conc. in Soil w/w (%)	Conc. of IAA (µg/ml) Produced				Mean
	A. vinelandii	*A. chroococcum*	*A. nigricans*	*A. beigernickii*	
Control	45.5	34.0	30.0	31.0	35.12
0.5 per cent	46.0	32.0	34.0	39.0	37.75
1.0 per cent	41.0	46.0	30.0	25.0	35.5
1.5 per cent	13.0	22.0	29.0	22.0	21.5
2.0 per cent	9.5	12.0	11.0	9.0	10.37
2.5 per cent	7.0	8.0	5.0	3.0	5.75
3.0 per cent	2.5	5.0	2.0	1.0	2.62
3.5 per cent	1.0	2.0	2.5	1.0	1.62

For comparing mean IAA (µg/ml) S.E.: 0.40, CD at 5 per cent: 1.15.

Germination Studies (Table 6.2)

The germination study on wheat cv. Lok-1 was carried out in pots. It was observed that, with the increase in salt concentration up to certain level there was increase in germination. However, at salt concentration of 1.5 per cent w/w and above germination per cent decreases. The maximum germination (87.7 per cent) followed by (84 per cent) was recorded in the treatment with 1.0 per cent and 0.5 per cent salt concentration respectively and proved to be statistically significant over control. The treatment with NaCl above 1.5 per cent proved to be statistically insignificant for wheat germination over control. Our results are in accordance with experimental findings of Docuyucu *et al.* (2005). They reported the decrease in callus weight due to increasing residual effects of NaCl concentration. The reduction in germination per cent may be due to reduction in growth hormone secreting microflora or it may be due to alteration in soil characteristics that interfere with the physiological activities in germination seeds. Soil salinity adversely affects the soil microbial activities and microbial transformation of nutrients was reported by Patra *et al.* (1987).

Table 6.2: Effect or *Azotobacter* Inoculation on Germination (per cent) in Wheat at Different Salt Concentration

Salt Conc. in Soil w/w (%)	Germination % in Wheat at 20th DAS				Mean
	A. vinelandii	*A. chroococcum*	*A. nigricans*	*A. beigernickii*	
Control	87.0	94.0	82.0	69.0	83.0
0.5 per cent	87.0	95.0	84.0	70.0	84.0
1.0 per cent	89.0	98.0	86.0	78.0	87.7
1.5 per cent	78.0	83.0	73.0	72.0	76.5
2.0 per cent	64.0	70.0	59.0	35.0	57.0
2.5 per cent	40.0	29.0	18.0	21.0	27.0
3.0 per cent	23.0	27.0	14.0	16.0	20.0
3.5 per cent	20.0	20.0	10.0	16.0	16.5

For comparing mean Germination (per cent) S.E.: 1.22, CD at 5 per cent: 3.51.

Salinity Associated Parameters of Soil (Table 6.3)

It was observed that the parameters *viz.*, water holding capacity (WHC), electrical conductivity (EC), pH, Cl_2, soluble Na per cent increased with the increase in salt concentration as compared to control. The average WHC per cent of salt fortified soil was in the range of 40 to 48.57 per cent, where as 33.85 per cent WHC was detected in control. This might be due to higher accumulation of soluble salt in salt affected soils than normal soil. The pH of salt affected soils was detected in the range of 7.84 to 8.4. However, the treatment with 1.5 per cent NaCl and above showed increase in soil alkalinity over control. The electrical conductivity of salt treated soil varied from 2.17 to 11.55 dsm^{-1} where as it was only 1.26 dsm^{-1} in control treatment, which indicates the fluctuation in soil electrical conductivity due to the salt accumulation. The anions *i.e.* chlorides increases with increase in salt concentration. The average soluble–Na per cent value in salt treated soil was 11.12 to 44.42 per cent, which was more as compared to normal soil (8.70 per cent). This indicates that NaCl induces the changes in soil salinity associated parameters and which may ultimately disequilibrates the microbial activity in soil and hence, retardation in plant growth. Fluctuation in the values of pH, EC, Na per cent and Cl_2 due to saline water-irrigation significantly reduces the growth of the crop at growth, shooting and harvesting stages respectively, was reported by Palaniappan *et al.* (1996). Hence, there is an urgent need to screen the specific microbial consortia for each plant with special interest to their salt tolerance.

Table 6.3: Effect of Graded Doses of NaCl Per cent on Salinity Associated Parameters of Soil at 20th DAS

Salt Conc. in Soil w/w (%)	*WHC%*	*pH*	*E.C. dsm^{-1}*	*Cl_2 m'eq*	*Na%*
Control	33.85	7.86	1.26	7.89	8.70
0.5 per cent	40.0	7.86	2.70	10.16	11.12
1.0 per cent	43.07	7.86	3.14	11.27	12.33
1.5 per cent	43.15	8.03	4.2	32.2	39.81
2.0 per cent	44.81	8.2	6.62	33.93	40.12
2.5 per cent	44.98	8.5	7.75	34.85	40.18
3.0 per cent	46.92	8.3	9.45	35.4	42.59
3.5 per cent	48.57	8.4	11.15	35.76	44.42

References

Agrawal, N. and Singh, H.P., 2002. Antibiotic resistance and inhibitory effect of Azotobacter on soil borne plant pathogens. *Ind. J. Microbiol.*, 42: 245–246.

Black, C.A., Evan, D.D., Ensminger, L.E., White, I.L., Clark, F.C. and Dinarer, R.C. (Eds.), 1965. *Methods of Soil Analysis*, Part-II. Amer. Soc. Agron. Inc., Publisher, Madison, Wisconsin, USA.

Chopra, S.L. and Kanwar, I.S., 1991. *Analytical Agriculture Chemistry*, 4th Edition.

Docuyutu, T., Akkececi, S., Akkaya, A. and Kara, R., 2005. Investigation of the response of bread wheat cultivars to salinity by using callus cultures. *J. Environ. Biol.*, 26(2): 251–255.

Dubey, R.C. and Maheshwary, D.K., 2000. Cultivation and isolation of microorganism soil microbiology. In: *Practical Microbiology*. S. Chand and Company Limited, pp. 90–94, 197–198.

Jackson, M.L., 1973. *Soil Chemical Analysis*. Prentice Hall, New Delhi.

Loper, J. and Schroth, M., 1986. Influence of bacterial sources of Indole-3-acetic acid on root elongation of sugar beet. *Physiol. Biochem.*, 76: 386–398.

Martinez-Toledo, M.V., Saimeron, V. and Gonzalex-Lopez, J., 1988. Root exudate of Zea Mays and production of auxins, gibberelins and cytokinis by *Azotobacter chroococcum. Plant Soil*, 110: 149–152.

Narula, N., Lakshminarayana, K. and Tauro, P., 1981. Ammonia excretion by *Azotobacter chroococcum. Biotech. Bioengg.*, 23: 467–470.

Nileema, 1995. Inhibitory effect of *Azotobacter chroococcum* on a range of rhizosphoric fungi. *Ind. J. Expl. Biol.*, 33: 52–54.

Palaniappan, R. and Yerriswamy, R.M., 1996. Effect of saline water irrigation on growth, yield, quality and nutrition of Robusta Banana. *J. Ind. Soc. Soil. Sci.*, 44(1): 143–146.

Patra, D., Prasad, A., Nirpendra, K. and Singh, D.V., 1996. Nitrogen transformation in salt affected soil. *J. Ind. Soc. Soil Sci.*, 44(1): 151–153.

Sadashivam and Manikam, 1987. *Methods in Biochemistry*, pp. 232.

Torres-Rubio, M.G., Valencia-Plata, S.A., Bernal-Castitlo, J. and Martinex-Neito, P., 2000. Isolation of enterobacteria, *Azotobacter* spp. and *Pseudomona* spp. producers of indole-3-acetic acid and siderphores, from colombian rice rhizosphere. *Ravisto Latinoamerica de Microbiologia*. 42: 171–176.

Verma, A., Kukerja, K., Pathak, D.V., Suneja, S. and Narula, N., 2001. *In vitro* production of plant growth regulators by *Azotobacter chroococcum. Ind. J. Microbiol.*, 41: 305–307.

Yao-Yseng Tachan, 1933. *Azotobacteriacae in Burgey's Manual of Systematic Bacteriology*. William and Wilkins, Baltimore, pp. 218–229.

Chapter 7

Medicinal Plants of Ethnobotanical Importance Curing Diabetes from Namakkal District (Tamil Nadu), India

M. Rajasekara Pandian[1], G. Sharmila Banu[1], G. Kumar[2] and K.H. Smila[1]

***Centre for Biotechnology, Muthayammal College of Arts and Sciences, Kakkaveri, Rasipuram, Namakkal, Tamil Nadu – 637 408, India**

****Department of Biochemistry, Selvamm Arts and Science College, Namakkal, Tamil Nadu – 637 003, India**

ABSTRACT

A survey of medicinal plants has been carried out in different locations of Namakkal District, which are used in the treatment of Diabetes. Some of which are less known, but very effective. This study deals with botanical names, local names and plant parts used in diabetes. During the recent exploration about 100 plants belonging to 53 different families have been recorded.

Keywords: Medicinal plants, Diabetes, Namakkal.

Introduction

Plants have been used as source of medicine throughout the world for more than 5000 years and still continue to occupy an important place in traditional as well as modern systems or medicine. There are about 2000 plant species, which have been found to possess the medicinal value in all these four indigenous systems of medicine *viz.*, Ayurveda, Unani, Siddha and Homeopathy (Farooqi *et al.*,

2003). According to the report of the WHO over 80 per cent of the world population rely on traditional systems of medicines, largely plant based, to meet primary health care needs (Jain, 1967; Tumane *et al.*, 2001). Over 7500 species of plants are estimated to be used by 4635 communities for human and veterinary health care (Beulah Jerlin *et al.*, 2004). Efforts have been taken to assess the medicinal plants throughout the country, but a complete inventory is not available so far (Singh *et al.*, 1997).

Most of the supply of drug is obtained from wild plants (Mundappa and Ommen, 1998). Ethnomedicines are receiving great attention all over the world; therefore a great emphasis has been laid to revive the heritable knowledge on the medicinal plants. But some plants have already been extinct and many are facing a danger of extinction. Hence necessity was felt to conduct the survey of medicinal plants carried out from locality of Namakkal district. Especially the plant survey was made to cure diabetes. Because the disease manifesting in the form of silent epidemic is increasing at a very fast rate with estimated 80 million diabetes in the world today (Singh and Sharma, 2001).

Materials and Methods

Ethnobotanical surveys (2003–2004) were conducted in various parts of Namakkal Districts. During field survey certain plants were observed which are used as diabetes. Voucher specimens were collected photographs were taken; plants were properly identified and compared with standard literature and herbarium. The plant species are arranged in alphabetical order according to botanical names followed by family, local name and part used. The ethnobotanical information has been described as collected from the folklore.

The medicinal uses of plants were compiled with the help of earlier publications (Kirtikar and Basu, 1933; Nadkarni, 1954; Jain, 1967; Maheswari *et al.*, 1993; Krishnan, 1992).

Results and Discussion

100 medicinal plants belonging to 53 families recorded in this study with their botanical and local name and use in the folk medicines.

Scientific studies may elaborate the prospects of growing more and more medicinal plants successively. The distribution of some medicinal plants having good scope for commercial exploitation of pharmaceutically important medicinal plants (Anonymous, 1978), study also revealed that mostly uses of the plants are not recorded in published and compiled literature (Saklani and Jain, 1994; Sharma, 1999; Chandola, 1999; Shriranjini and Acharya, 2001; Chandola and Singh, 2001; Kar *et al.*, 2001; Mohanty and Dwivedi, 2002) which warn for preservation of such valuable knowledge in welfare of mankind.

The present findings indicated that different folk populations inhabiting in the remote villages of Namakkal district are concions for their health. These findings may need phytochemical investigation and proper assessment to establish a crude drug industry for ethnomedicine.

Hence, there is necessity to generate all round awareness in society regarding the conservation of a flora of the plant treasure. For instances, students and local communities must be told that if a particular habitat were destroyed along with the species, what the negative consequences for them would be. Prime importance should be given for multiplication of endangered plant species in botanical garden by *ex situ* conservation or through tissue culture techniques. For this to be successful, habitat restoration will also have to be a top priority. Several medicinal plants domesticated within the last decenniums but much is to be done in this field in the near future. According to Altieri, 1989; Cunningham, 1991 some of the plants in forest as medicinal plants can also be attributed to increased concern with the relationship of indigenous plants knowledge to genetics evolutional and ecological security.

Table 7.1: Anti-diabetic Plants with Botanical Names, Family and Plant Parts Used in Diabetes

Sl.No.	Name of the Plant	Family	Local Name	Parts Used
1.	*Acacia arabica*	Leguminosae	Karuvelum	Gum
2.	*Acorus calamus*	Araceae	Vashambu	Rhizome
3.	*Achyrunthus aspera*	Amaranthaceae	Naaurivi	Leaf
4.	*Adhatoda vasica* Nees	Acanthaceae	Adadodai	Leaf, Root
5.	*Aegle marmeoles*	Rutaceae	Vilvam	Fruit
6.	*Allium cepa*	Liliaceae	Vengayam	Bulb
7.	*Allium sativum*	Liliaceae	Pundu	Bulb
8.	*Aloe vera*	Liliaceae	Katralai	Leaf
9.	*Alpinia galanga*	Zingiberaceae	Peratattai	Rhizome
10.	*Annona squamosa*	Annonaceae	Sitapalam	Leaf
11.	*Asparagus racemosus*	Liliaceae	Shimi shadavari	Root
12.	*Asteracantha longifolia*	Acanthaceae	Nir mulli	Whole plant
13.	*Azadirachta indica*	Meliaceae	Vembu	Leaf, Fruit
14.	*Bauhinia purpurea*	Leguminaceae	Mandarai	Bark
15.	*Bauhinia variegata*	Leguminaceae	Segappu mandaraii	Bark
16.	*Boerhavia diffusa*	Nyctaginaceae	Mukkurattai	Whole plant
17.	*Boswellia serrata*	Burseraceae	Parangisambirani	Gum, Bark, Resin
18.	*Butea monosperma*	Leguminaceae	Parasa	Gum, Leaf, Bark
19.	*Brassica juncea*	Cruciferae	Kadugu	Stem
20.	*Caesalpinia bonducella*	Leguminaceae	Kaliccikkai	Seed
21.	*Cajanus cajan*	Fabaceae	Thuvarai	Seed
22.	*Camellia sinensis*	Theaceae; Fernstroemiaceae	Thayilai	Leaf
23.	*Capsicum annum*	Solanaceae	Milagai	Fruit
24.	*Calophyllum inophyllum*	Guttiferae	Punnagam	Oil of seed
25.	*Cassia auriculata*	Ceasalpiniaceae	Aavarai	Leaves, seed
26.	*Cassia angustifolia*	Ceasalpiniaceae	Nila aavarai	Leaves
27.	*Cassia obovata*	Ceasalpiniaceae	Kattu-nila aavarai	Leaf
28.	*Cassia indica*	Ceasalpiniaceae	Aavarai	Seed
29.	*Cassia fistula*	Leguminaceae	Konnai	Seed
30.	*Casearia esculenta*	Samydaceae	Kottargovai	Root
31.	*Catharanthus roseus*	Apocynaceae	Sudukattu malli	Leaf, Flower
32.	*Cinnamomum tamala* Nees	Lauraceae	Tamalabattirii	Leaf
33.	*Citrus lemon*	Rutaceae	Periya yelumichai	Fruit peel
34.	*Clerodendron phlomidis*	Verbenaceae	Taludalai	Herb
35.	*Clerodendron serratum*	Verbenaceae	Angaravalli	Root
36.	*Coccinia indica*	Cucurbitaceae	Kovaikai	Root, leaf
37.	*Cucumis sativus*	Cucurbitaceae	Vellarikkkay	Fruit

Contd...

Table 7.1–Contd...

Sl.No.	Name of the Plant	Family	Local Name	Parts Used
38.	*Commiphora mukul*	Burseraceae	Gukkulu	Resin
39.	*Curcuma longa*	Zingiberaceae	Viral manjal	Rhizome
40.	*Cyamopsis tetragonoloba*	Fabaceae	Kottavarai	Seed
41.	*Cynadon dactylon*	Poaceae	Arugampillu	Whole plant
42.	*Dolichos lablab*	Fabaceae	Avarai	Seed
43.	*Emblica officinalis*	Euphorbiaceae	Nelli	Fruit bulb
44.	*Enicostemma littorale*	Gentianaceae	Vallari	Whole plant
45.	*Euphorbia hirta*	Euphorbiaceae	Amman pacharchi	Whole plant
46.	*Ficus bengalensis*	Moraceae	Alamaram	Bark
47.	*Ficus religiosa*	Moraceae	Arasu	Bark
48.	*Ficus racemosa*	Moraceae	Atti	Stem bark
49.	*Fumaria parviflora*	Fumaraceae	Tusa	Leaf
50.	*Glycyrrhiza glabra*	Fabaceae	Atimaduram	Root
51.	*Gymnema sylvestre*	Asclepiadaceae	Sirukurunjan	Leaf
52.	*Hemidesmus indicus*	Asclepiadaceae	Nannarai	Root
53.	*Holarrhena antidysentrica*	Apocynaceae	Kudasappalai	Seed
54.	*Indigofera tinctoria*	Fabaceae	Avari	Whole plant
55.	*Mangifera indica*	Anacardiaceae	Mamaram	Tender leaf
56.	*Madhuca latifolia*	Sapotaceae	Kattiluppai	Leaf
57.	*Memecylon umbellatum*	Melastomaceae	Kasai	Leaf
58.	*Mimosa pudica*	Mimosaceae	Thottachinungi	Leaf, Root
59.	*Momordica charantia*	Cucurbitaceae	Pagal	Seed, Fruit, Leaf
60.	*Morus alba*	Moraceae	Musukottai	Root
61.	*Murraya konigii*	Rutaceae	Karuvembu	Leaf
62.	*Musa paradisica*	Musaceae	Kadali valai	Flower
63.	*Nelumbo nucifera*	Nymphaeaceae	Thamarai	Rhizome
64.	*Nymphaea alba*	Nymphaeaceae	Alli thamarai	Plant extract
65.	*Ocimum sanctum*	Lamiaceae	Thulasi	Leaf
66.	*Oroxylum indicum*	Bignoniaceae	Vangam	Root bark
67.	*Oryza sativa*	Poaceae	Nellu, arisi	Root
68.	*Papver somniferum*	Papaveraceae	Kasakasa	Fruit
69.	*Phaseolus aconotifolius*	Fabaceae	Tulkapyrai	Seed
70.	*Phaseolus munga*	Fabaceae	Patchipyaru	Seed
71.	*Phaseolus radiatus*	Fabaceae	Paunippayaru	Seed
72.	*Phaseolus vulgaris*	Fabaceae	Panipayar	Seed
73.	*Phyllanthus emblica*	Euphorbiaceae	Nelli	Fruit

Contd...

Table 7.1–Contd...

Sl.No.	Name of the Plant	Family	Local Name	Parts Used
74.	*Phyllanthus niruri*	Euphorbiaceae	Nelli	Whole plant
75.	*Picrorrhiza kurroa*	Scrophunariaceae	Katukurohini	Rhizome, Root
76.	*Pinus roxburghi*	Pinaceae	Shimaithevari	Bark
77.	*Plumbago zeylanica*	Plumbaginaceae	Kodiveli	Root
78.	*Pongamia glabra*	Fabaceae	Pungamaram	Root, Bark, Leaf
79.	*Polyalthia longifolia*	Annonaceae	Asogu, Nettilingam	Bark
80.	*Prunus amygdalus*	Rosaceae	Vadumai	Seed oil
81.	*Punica granatum*	Puniaceae	Madulai	
82.	*Pterocarpus marsupium*	Fabaceae	Vengai	Wood
83.	*Pterocarpus santalinus*	Fabaceae	Santhana Vengai	Seed
84.	*Rhizophora mangle*	Rhizophoraceae	Kandal	Leaf
85.	*Ricinus communis*	Euphorbiaceae	Amanakku	Leaf
86.	*Saccharum officinarum*	Gramineae	Karumbu	Stem
87.	*Salacia oblongata*	Hippocrateaceae	Ponkoranti	Root bark
88.	*Sesamum indicum*	Pedaliaceae	Ellu	Seed
89.	*Swertia chirata*	Gentianaceae	Nilavembu	Stalk
90.	*Syzigium cumini*	Myrtaceae	Naval	Seed
91.	*Terminalia cebula*	Combretaceae	Kudukkay	Fruit
92.	*Terminalia arjuna*	Combretaceae	Marudu	Fruit
93.	*Tribulus terrestris*	Zygophyllaceae	Neringil	Thorns and Fruit
94.	*Tinospora cardifolia*	Menispermaceae	Seenthil Kodi	Stalk, Stem
95.	*Trigonella foenumgraceum*	Fabaceae	Venthayam	Seed
96.	*Trichosanthus dioica*	Cucurbitaceae	Kommu Pudalai	Fruit
97.	*Vateria indica*	Dipterocarpaceae	Kundurukkam	Resin, Fruit
98.	*Woodfordia fruitcosa* Kurz	Lythraceae	Velakkai	Flower
99.	*Zingiber officinale*	Zingiberaceae	Inji	Rhizome
100.	*Zizyphus rugosa*	Rhamnaceae	Kattilandai	Bark

Finally, with multiplication, cultivation and selection, it is possible to create an immense number of plant varieties adapted to diverse condition that would eventually sustain our biodiversity and thus preserve a half of our massive habitat loss. If this is achieved in near future, we would go along way in sustaining medicinal plants.

References

Anonymous, 1978. *The Ayurvedic Formulary of Part I*, Edition 1. Ministry of Health and Family. Govt. of India, New Delhi.

Altieri, M. 1989. Rethinking Crop genetic resource conservation: A view from the south. *Conservation Biology*, 3: 77–79.

Beulah Jerlin, S., John De Britto, A., Balasingh, J., Swami doss, D.P. and Prakash, A.A. 2004. Medicinal flora of Koonthakulam Bird Sanctuary. *J. Ecobiol.*, 16(1): 7–16.

Chandola, H.M. 1999. Diabetes mellitus in Ayurveda. In: *National Seminar on Medicinal Plants and Traditional Medical Knowledge and Workshop on Traditional Knowledge and Management of Incurable Diseases like AIDS* Organized by Gujarat Ayurved University, Jamnagar.

Chandola, H.M. and Singh, G. 2001. Promotion of mental health of Ayurveda and yoga in diabetes mellitus. In: *Second International Seminar on Ayurveda,* Organized by Jamnagar Ayurved University, Jamnagar.

Cunningham, A.B. 1991. Indigenous knowledge and biodiversity. *Cultural Survival Quarterly,* 15: 4–8.

Farroqi, A.S., Sreeramu, B.S. and Srinivasappa, K.N. 2003. Cultivation of medicinal plants in South India. *Agrobios Newsletter,* 1: 21–22.

Jain, S.K. 1967. Ethanobotany, its scope and study. *Indu. Mus. Bull.,* 2(1): 39–43.

Kar, A.K., Upadhyay, B.N. and Ojha, D. 2001. A study on the prognosis of Prameha (Diabetes mellitus) on the basis of Insulin level of the body. In: *Second International Seminar on Ayurveda,* Organized by Jamnagar Ayurved University, Jamnagar.

Kirtikar, K.R. and Basu, D.D. 1933. *Indian Medicinal Plants,* 2nd ed. Lalit Mohan Basu Co., Allahabad, India.

Krishnan Marg, K.S. 1992. *The Useful Plants of India.* Publications and Information Directorate, CSIR, New Delhi.

Maheswari, J.K., Kunkel, G., Bhandari, M.M. and Duke, J.H. 1993 *Ethnobotany in India.* Scientific Publishers.

Mohanty, B. and Dwivedi, R.B. 2002. Antioxidant mechanism. Away of Biopurification in diabetes mellitus. In: *Second International Seminar on Ayurveda,* Organized by JamnagarAyurved University, Jamnagar.

Mundappa, A. and Ommen, S. 1988. Amruth 2:3 Med. Plan. Conservation Society, Bangalore, India.

Nadkarni, K.M. 1954. *Indian Materia Medica,* Vol. I and II, 3rd Ed. A.K. Nadkarni, Popular Book Depot, Bombay, India.

Saklani, Arvind and Jain, S.K. 1994. *Cross-cultural Ethnobotany,* Deep Publication, New Delhi.

Sharma, P.P. 1999. Importance of Dravyaguna Vijnana and list of drugs useful in Madhemeha, Amavata and AIDS. In: *National Seminar on Medicinal Plants and Traditional Medical Knowledge and Workshop on Traditional Knowledge and Management of Incurable Disease like AIDS,* Organized by Jamnagar Ayurved University, Jamnagar.

Shriranjini, S. and Acharya, G.S. 2001. Clinical evaluation of the hypoglycaemic effect of Asanad Kwatha in diabetes mellitus. In: *Second International Seminar on Ayurveda,* Organized by Jamnagar Ayurved University, Jamnagar.

Singh, D.K. and A.K. Sharma. 2001. Clinical evaluation of the hypoglycaemic effect of Asanad Kwatha in diabetes mellitus. In: *Second International Seminar on Ayurveda,* Organized by Jamnagar Ayurved University, Jamnagar.

Singh, J., Ashok Sharma and Susil Kumar. 1997. Inventorisation of herbal drugs used in the traditional and modern medicines. *J. Med. Aromatic Plant Sci.,* 19: 665.

Tumane, H.H., Mehta, P.J. and Chopra, K.K. 2001. Conservation of medicinal plants. In: *Second International Seminar on Ayurveda,* Organized by Jamnagar Ayurved University, Jamnagar.

Chapter 8

Epiphytic Mosses of Malshej Ghats (North Konkan) Maharashtra, India

B.L. Chaudhary and F.S. Bhagora*
Bryology Laboratory, Department of Botany, College of Science, M.L. Sukhadia University, Udaipur – 313 001, Rajasthan, India

ABSTRACT

Three epiphytic mosses have been collected from various parts of Malshej Ghats in different seasons. These are *Anomobryum auratum* (Mitt.) Jaeg, *Stereophyllum ligulatum* Jaeg and *Diaphanodon procambens* (C. Muell.) Ren.& Card.

Keywords: *Epiphyte mosses, Malshej Ghats, North Konkan.*

Introduction

The present study aims to enlist the underexplored moss flora of Malshej Ghats and to understand the distribution of mosses flora of this region. The study revealed that Malshej Ghats has a great variety of bryophytes and is a 'hot spot for them. Malshej Ghats falls in the Western Maharashtra and is on the way from Ahmednagar to Kalyan (state highway No. 2 of Maharashtra state). A major portion of the Ghats is included under the Thana district whereas a small portion comes under the Pune district. It is situated at an altitude of about 900m. Due to high rainfall and humidity it is the richest spot of angiosperm vegetation in general and bryophytes in particular. Trees and forest floor are densely covered with moss vegetation. It consists of mixed tree growth with *Ficus glomerata* Roxb,

* E-mail: fsbhagora@rediffmail.com; Fax: 091-294-2410300

as a predominant species in the most of area. The other tree vegetation are *Terminalia arjuna, Olea dioca* Roxb, *Eugenia jambolana* Lam, *Tectona grandis* Linn, *Ficus religiosa* Linn, *Boswellia serrata* Roxb.

Material and Methods

Field survey was conducted during the rainy season in 2004–2005. Mosses were collected, tagged, numbered and brought to the laboratory in small paper or plastic bags.

They were washed thoroughly and preserved wet and dry. The specimen were identified and submitted in Bryology Laboratory, Department of Botany, College of Science, M.L. Sukhadia University, Udaipur. Identification was done with the help of standard books (Chopra, 1975; Gangulle, 1969–1980, 1985; Chaudhary and Deora, 1993; Dhabade, 1993).

Results and Discussion

Anomobryum auratum (Mitt.) Jaeg. Ber. S. Gall. Naturw. Ges. 1873–74, 142, 1875

***Bryum auratum* Mitt; Musc. Ind. Or. 67, 1859 [Figure 8.1(A_1–A), Specimen No. MG-11]**

Plants densely tufted, shining green, julaceous, branched with catkin like branches up to 2 cm long densely radiculose below, branched by 2–3 equally julaceous sub floral innovations; leaves 1.5 mm long, cymbiform, ovate-elliptical, densely imbricate to give shoot a cylindrical appearance, apiculate to obtusely rounded, margin entire; costa slender, pale brown, ending well below apex; upper cells with thickened walls, linear, 10–20 × 1.7μ wide and 80–140 × 1.7μ long; basal cells thin walled hyaline, rhomboidal, hexagonal to subrectangular, 40–50 × 1.7μ wide and 60–150 × 1.7μ long. Middle cells 10–20 × 1.7μ wide 160–180 × 1.7μ long. Plant sterile.

Field Notes

Plant growing on the bark of *Ficus religiosa* Linn.

Distribution

East Nepal, Sikkim, Darjeeling, Bhutan, Mussoorie, Ranikhet, Kashmir, U.P., Mahabaleshwar, Nilgiri, Palni hills, Rajasthan, Gujarat, Sri Lanka, China, Japan, Australia, Pakistan.

Stereophyllum ligulatum Jaeg, In Ber. S. Gall Naturw. Ges. 1877–78: 277, 1880

[Figure 8.2(B_1–B_6), Specimen No. MG-3)]

Yellow green, corticolous, glossy plants. Branches creeping along bark, stem 1.5 cm long, complanate. Leaves patent to spreading (shrunk and appressed to stem, often with out-spread tips when dry), concave, asymmetrical, ovate with obtuse apex suddenly narrowing into a small tip, margin usually revolute at one side of base, almost smooth except some papillose tips of some border cells. Costa strong and single, covering about., half the leaf length. Leaf cells elongate, narrow rhomboid at tip (±7 × 4.77μ long and 4 × 4.77μ wide), some of these apical cells show papillose development of tips, cells towards border become shorter and more rhomboid; median leaf cells (15–18 × 4.77μ long and 3–4 × 4.77μ), cells tinged and larger along extreme base and specially differentiated at alar where the extreme base cells are large (8–150 × 4.77μ and 5–6 × 4.77μ wide), becoming shorter, quadrate and lattice-like above.

Field Notes

Plant growing on the bark of *Tectona grandis* Linn.

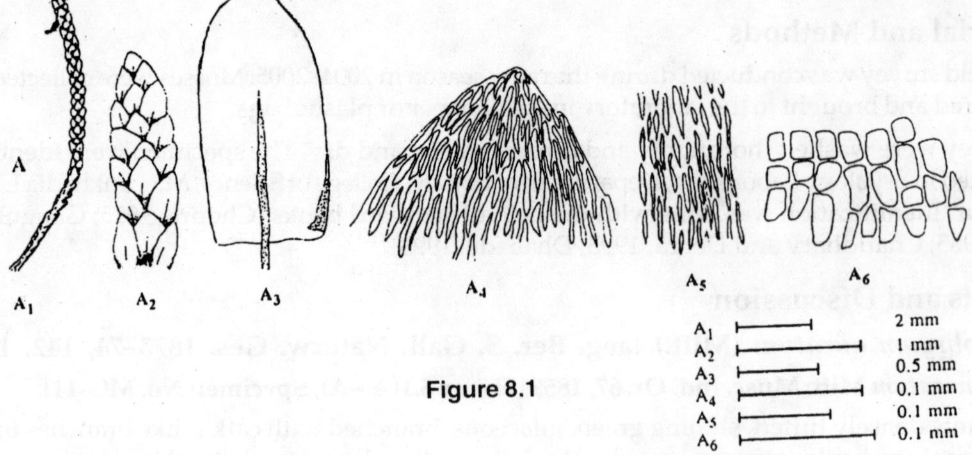

Figure 8.1

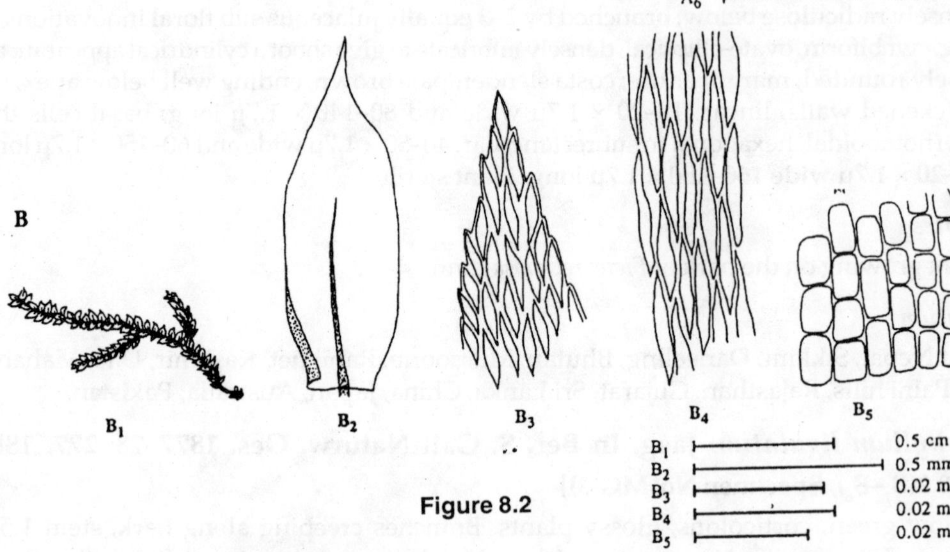

Figure 8.2

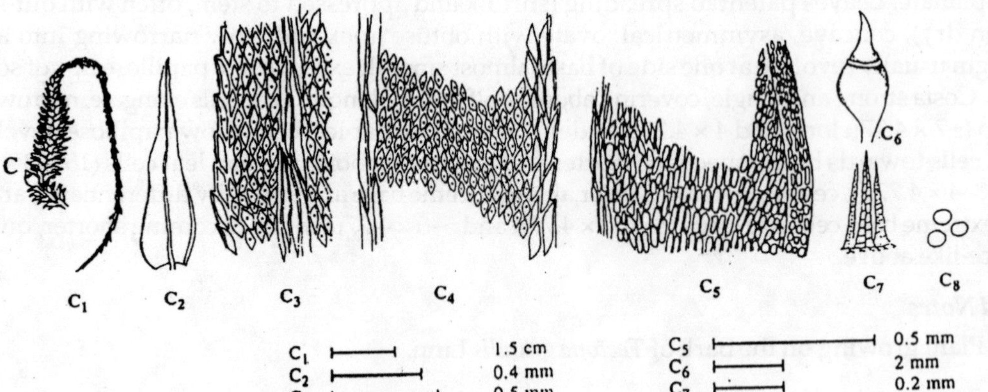

Figure 8.3

Distribution

Chhota Nagpur, Western Ghats, Palni hills, Rajasthan, Gujarat, endemic in peninsular-India.

Diaphanodon procumbens (C. Muell.) Renet Card.

[Figure 8.3(C_1–C_8), Specimen No. MG-8)]

Plants very slender, greenish brown, densely tufted. Primary stem creeping, filiform, Secondary stem pendulous, pinnate or bipinnately branched. Leaves closely appressed when serrate border. Nerve strong, ceasing below apex. Laminar cells oval to linear, more elongated towards the base. Alar cells quadrate to rectangular.

Sporophyte with oval capsule 900µ long and short seta, 2 mm long. Spores spherical to oval and 25–30µ diameter. This sporophyte is reported for first time In India. Generally plant is sterile every where in the world.

Field Notes

Plants growing on the bark of *Carissa carandas*.

Distribution

North West Himalayas, Eastern Himalayas (Kolkata), Mahableshwar, Ooty and also in South-East Asian Tropical Countries.

Acknowledgements

Authors are thankful to Dr. Virendra Nath and Dr. A.K. Asthana (N.B.R.I, Lucknow) for plant identification and to the UGC, New Delhi for financial assistance.

References

Chaudhary, B.L. and Deora, G.S., 1993. *Moss Flora of Rajasthan* (*India*). Himanshu Publication, New Delhi.

Chopra, R.S., 1975. *Taxonomy of Indian Mosses.* CSIR Publication, New Delhi, India.

Gangulee, H.C., 1969–1980. *Mosses of Eastern India and Adjoining Areas,* Vols. I, II and III. Eastend Printers: Calcutta, India.

Gangulee, H.C., 1985. *Handbook of Indian Mosses.* Amerind Publishing Co. Pvt. Ltd., New Delhi.

Dabhade, G.T., 1998. *Mosses of Khandala and Mahabaleshwar in the Western Ghats* (India). M/s Classic Graphics, Kalwa, Thane, India.

Chapter 9

Saccharification Studies by a Thermophilic Fungus *Sporotrichum thermophile* Isolated from Agriculture Waste of Bhopal

Rajhans Singh[1], Namita Singh[2], Anil Prakash[3] and Sarika Poonia[2]

[1]Maharana Partap Degree College, Kalor Road, Bhopal – 462 016

[2]Department of Biotechnology, Guru Jambheshwar University, Hisar – 125 001, Haryana

[3]Department of Biotechnology, Barkatullah University, Bhopal – 462 026, M.P., India

ABSTRACT

Thermophilic fungus has been reported from various habitat *viz.* municipal waste, natural compost, coal refuse, oil palm, and stored grains etc. Thermophilic fungus has been exploited for number of unique properties, like enzymes, antibiotic, amino acids production.

Cellulose is available in nature as plant residues. The cellulosic wastes cause environmental pollution, conservation of these wastes into useful products through microbes is in progress. Number of thermophilic fungi produces extra cellular enzymes, which diffuse and break the cellulose chain by saccharification.

In the present study attempt has been made to characterize and standardize the production of enzymes from *Sporotrichum thermophile* fungus isolated from agriculture waste. Following observations were recorded. High enzymatic activity was recorded between 12–15 day of incubation and a sharp decline in the activity after 15th days. 45° C temperatures shows better growth and production of enzyme and pH 6.8 showed better production of cellulose. Hence *S. thermophile* can be utilized for minimizing the environmental pollution by conservation of cellulosic waste into the useful products.

***Keywords**: Cellulose, Endo-β-glucanase, Exo-β-glucanase, Sporotrichum thermophile.*

Introduction

Cellulose is most abundantly produced biopolymer in terrestrial environments. Each year fixation of CO_2 yields more than 10^{11} tons of dry plant material all over world (Schlesinger, 1991) and almost half of this consist of cellulose (Eriksson *et al.*, 1990). Cellulose is one of the major components of municipal waste and agro wastes which cause potent sources of environmental pollution. This potentially valuable source of food energy is largely unavailable to monogastric animals because of resistance of cellulose to digestive enzymes (Coutts and Smith, 1976). Cellulose almost never occurs alone in nature but is usually associated with plant substances. This association affects its natural degradation. Bacteria and fungal degradation of cellulose occur exocellularly either in association with outer cell envelope layer or extracellularly. Their enzymes help in conversion of cellulose into simpler saccharides. Previously mesophilic microorganisms were known to produce cellulose enzyme (Coutts and Smith, 1976) but now-a-days thermophilic microorganisms have received considerable attention for production of more amount (Singh *et al.*, 1990; Wase *et al.*, 1985) and variety of enzymes. This is largely due to the fact that enzyme produced by thermophiles exhibit optimal activity at high temperature that their mesophilic counterparts (Maheshwari *et al.*, 2000; Feldman *et al.*, 1988; Grajek, 1987; Hagerdal *et al.*, 1980).

The thermophilic microorganisms have adopted themselves to a wide array and ecological niche such as comport, soil, birds nest, wood chips, seed grams, volcano and hot springs etc (Maheshwari *et al.*, 1987). The resistance to survive at high temperature has modified the metabolism of thermophiles in favour of production of enzymes and other secondary metabolites of industrial importance (Feldman *et al.*, 1988; Bhat and Maheshwari, 1987).

Thermophilic fungi have become an experimental system for the basic and applied research due to their unique properties (Maheshwari *et al.*, 1987, 2000). These include growth and temperature optima, habitat relationship, enzymatic abilities, degradation and deterioration characteristics as well as possible allergenic reactions. They produce important extracellular enzymes such as xylanase, amylase, protease, lipase and cellulase. The production of thermostable enzymes at elevated temperature by thermophilic fungi is of economic importance (Feldman *et al.*, 1988). Some of these have already been exploited commercially (Chaudhuri *et al.*, 1988).

The existence of cellulolytic microorganism is well documented. Several efforts have been made by many workers to develop preproductive cellulose mutants of thermophilic fungi (Hayashida and Mo, 1986; Sen and Chakarbarty, 1984). Cellulose is a prominent agrowaste in each country and creating too much environmental pollution. To minimize the environmental pollution these cellulosic materials should be converted into useful product. Among thermophilic fungi cellulose system has been examined exhaustively. However, *Sporotrichum thermophile* has been studied in the most detail as it synthesizes a complete set of enzymes necessary for the break down of crystalline cellulose. Therefore, *Sporotrichum thermophile* appeared to be a useful source of cellulases. Keeping this in view, *Sporotrichum thermophile* was selected to study the cellulolytic potential.

The present study was carried with the objective to identify different environmental conditions for optimum production of cellulose enzymes by locally isolated thermophilic fungi *S. thermophile.*

Materials and Methods

Organism and Culture Conditions

The organism, *Sporotrichum thermophile*, a thermophilic fungi used for the enzymatic saccharification of cellulose, was isolated from local agriculture waste (Bhopal, M.P., India). Culture of

Sporotrichum thermophile was maintained on basal medium (Pattersson *et al.*, 1963). Slants were kept at 4°C in cold room. Experiments were performed undersubmerged culture conditions.

Growth Parameters

The growth of *Sporotrichum thermophile* was estimated by measuring the extra cellular protein and mycelial protein. The extra cellular protein was measured by using Lowry's method (1951). Bovine Serum albumin was used as a standard mycelial protein was estimated by microbiuret method using Benedict's reagent (Herbert *et al.*, 1971). For studying growth behavior under different environmental conditions samples were taken after 3, 6, 9, 12, 15, 18 and 21 days incubation under different environmental conditions.

Enzymes Assay of Cellulose Enzymes

Cellulose activity was assayed viscometrically using Ostwald Fenske Viscometer (Reese *et al.*, 1950). The percentage loss in viscosity was calculated according to the following formula:

$$\%\ \text{loss in viscosity} = \frac{(ET_o - ET_t)}{(ET_o - ET_w)} \times 100$$

where,

ET_o: Efflux time at zero hour

ET_t: Efflux time at time t

ET_w: Efflux time of distilled water

Endo-β-glucanase (EC. 3.2.1.4)

The Endo-β-glucanase was assayed by the method of Sumner and Somer (1944) in terms of reducing sugar (R.S.) released from the substrate carboxyl methyl cellulose (CMC). 1 ml of cold CMC reagent was added to 1 ml of enzyme sample incubated at 50°C for 30 mins and then cooled to 4°C. Then amount of reducing sugar produced assayed by DNS reagent (Miller, 1959).

Exo-β-glucanase (E.C. 3.2.1.9.1)

The Exo-β-glucanase was assayed by the method suggested by Mendels and Weber (1969) by analyzing reducing sugars (R.S.) with Dinitrosalicyclic acid reagent 50 mg or cotton fibers in 1 mM citrate phosphate buffer pH-5 and 1 ml of crude enzyme preparation were incubated at 45°C for 24 hrs. Then reducing sugar was assayed by DNS reagent (Miller, 1959).

Results and Discussion

It is a well known fact that incubation period greatly influences the growth and production of the organism. To find optimum period of incubation cultures were kept for 3 to 15 days under submerged conditions. The optimum conditions of incubation for growth vary from organism to organism (Bhat and Maheshwari, 1987). The maximum growth in our experiment in terms of extracellular protein (105.4 μg/ml) and mycelial protein (98.2 μg/ml) was observed on 15th day of incubation after which the growth sharply declines (Figure 9.1). This finding was in contrast to Bhat and Maheshwari (1987) who reported maximum growth of organism on 4th day and Coutts and Smith (1976) reported that growth of organism increases with increase in incubation period.

The mycelial dry weight maximum was on 18th day of incubation (0.92 gm/70 ml) (Figure 9.1). It increases with increase in period of incubation and then remained approx stationary with culture age. Similar observation was reported by Bhat and Maheshwari (1987).

Bhat and Maheshwari (1987) reported that growth of *Trichoderma ressei* (mesophilic strain) was excellent in comparison to *S. thermophile* (thermophilic fungus). This was in contrast to our results that *S. thermophile* grow very well and to result of Maheshwari *et al.* (2000) who found that *S. thermophile* grew at five times the rate of *T. ressei* and to Tansey (1971) results.

Cellulolytic system mainly consists of three components, exoglucanase, endoglucanase and β-glucosidase that act in synergy to degrade cellulose (Ryu and Mandels, 1980; Grajek, 1987; Beguin, 1990; Gilbert and Hazlewood, 1993; Nidetzky and Steiner, 1993). Endoglucanase, exogluctnase and β-glucosidase may form ternary complex with substrate (Moloney, 1985). It is hypothesized that in *S. thermophile*, β-glucosidase physically associates with endo and exoglucanase to form a catalytically more active cellulose complex (Maheshwari *et al.*, 2000). The different level of celluloses produced by *S. thermophile* could be because of difference in induction by same substrate (Macris and Galiotou-Panayotou, 1986). Bhat and Maheshwari (1987) and Bagga *et al.* (1990) have reported that cellulases were not synthesized at the same time and same rate. Cellulose productivity also varies among strains of *S. thermophile* (Maheshwari *et al.*, 2000; Oberson *et al.*, 1992). The maximum production of endo-β-glucanase (40.6 μg R.S./ml) and exo-β-glucanase (31.2 μg R.S./ml) was on the same 12th day of incubation. These findings are not agreeable with the earlier concept of Bhat and Maheshwari (1987). Coutts and Smith (1976) reported that endo-β-glucanase appeared in early stages of growth. Similar observations were made by Neudoerffer and Smith (1970). However present study the production of both enzymes are on the same day. Production of endo and exoglucanase enzyme increased with increase in incubation period upto 12th day after which it declines. But the production of endo-β-glucanase is higher than the exoglucanase. This result was agreeable to some extent with Coutts and Smith (1976).

The maximum growth was on 15th day and maximum enzyme produced was on 12th day which is in actively growing period. This result was same to that of Maheshwari *et al.* (2000) that in *S. thermophile* endo and exoglucanase were typically formed during active growth. The percentage viscosity loss was maximum (67.5 per cent) on 12th day of incubation (Figure 9.4).

Growth and the production of enzymes by *Sporotrichum thermophile* was greatly affected by the hydrogen ion concentration of the synthetic medium. Exo-β-glucanase and endo-β-glucanase can tolerate a wide range of pH, the maximum production was at 6.5 pH (41.25 μg R.S./ml and 31.5 μg R.S./ml) (Figure 9.5). The percentage viscosity loss is maximum at pH 6.5 (69.5 per cent) and it declines to approx 31 per cent on moving from pH 6.5 to 8.5. Both exo and endoglucanase activity decline with increase in pH from 6.5. In contrast to our result many author reported that the cellulose production was observed at pH 6.0 by *Sporotrichum thermophile* (Bhat and Maheshwari, 1987). Singh *et al.* (1990) have also reported pH 6.0 as optimum for cellulase production by immobilized cells of *Sporotrichum thermophile*. Grajek (1987) has reported that the optimum pH for extracellular β-glucosidase production for his isolate of *Sporotrichum thermophile* from wood waste material was around 6.5 pH. In contrast to our results Coutts and Smith (1976) reported that his organism *Sporotrichum thermophile* produces exo and endoglucanases more efficiently at acidic pH. Similar results were observed by Rao (1977) that initial pH of 5.6 was found to be most favorable for obtaining high yield of cellulase.

For better growth of organism the basal medium was adjusted in the pH range from 3.5–8.5. The maximum growth was found when pH of the medium was adjusted to 6.5 prior to autoclaving

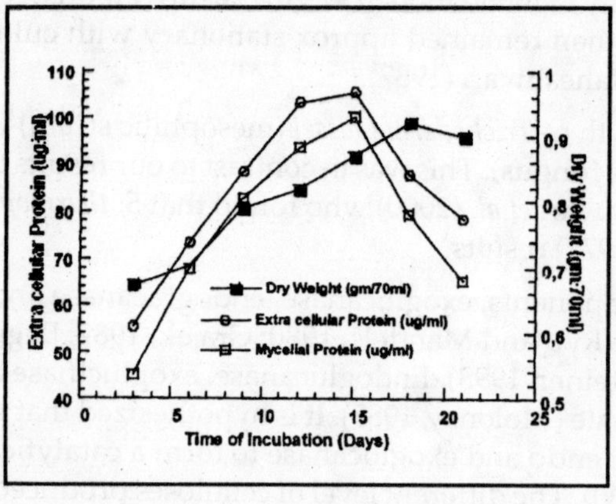

Figure 9.1

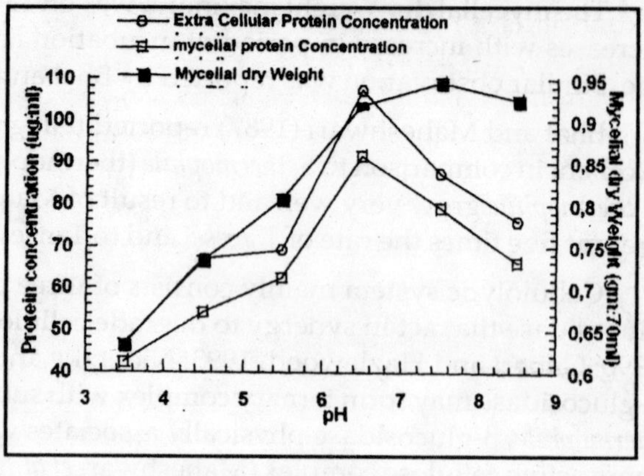

Figure 9.2

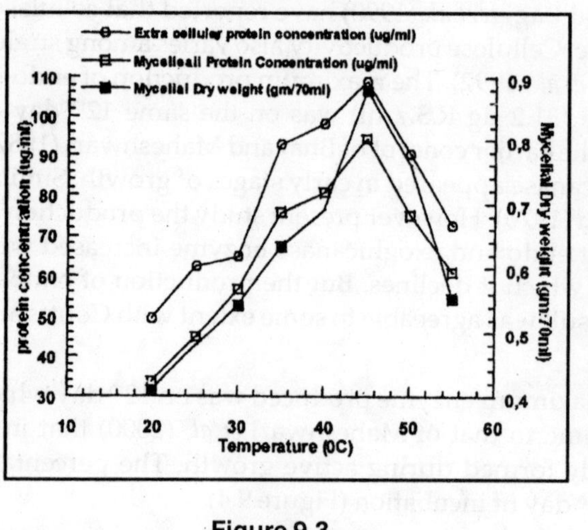

Figure 9.3

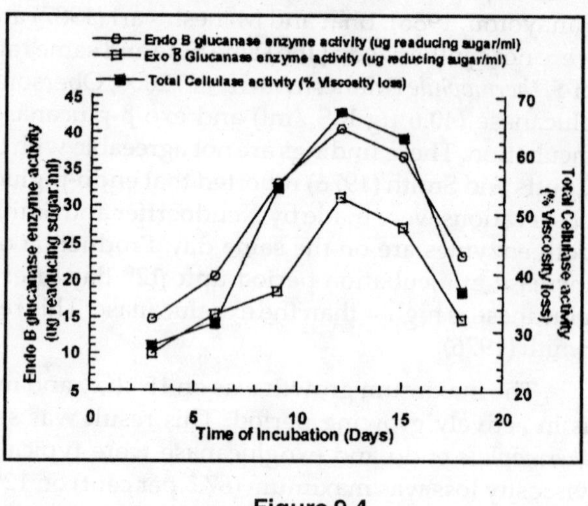

Figure 9.4

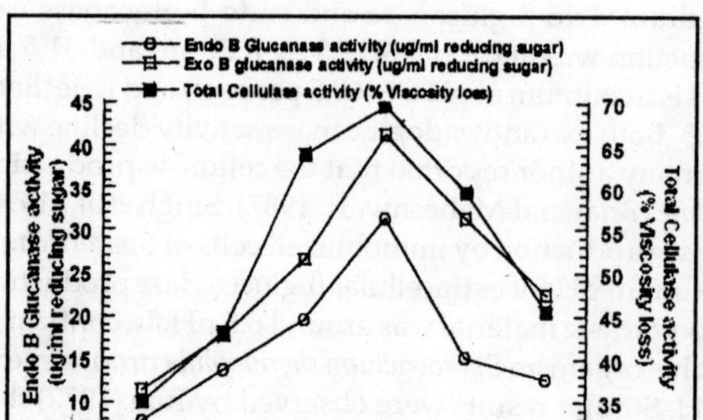

Figure 9.5

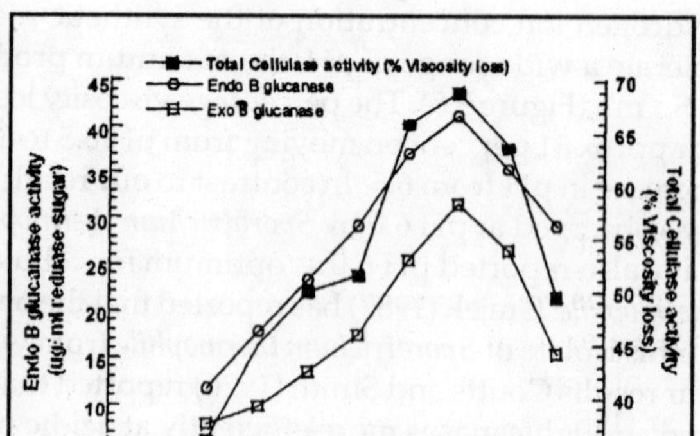

Figure 9.6

Figures 9.1–9.6

(extracellular protein concentration– 107.42 μg/ml and mycelial protein concentration–91.5 μg/ml) (Figure 9.2). pH 3.5–4.5 was found to be too acidic for growth of *S. thermophile.* However enhancement in pH from 6.5–8.5 show poor growth of organism. The maximum mycelial dry weight (0.9468 gm/70 ml) was found at 7.5 pH. This finding is in contrast with Coutts and Smith (1976), Bhatt and Maheshwari (1987), Maheshwari *et al.* (1987), Singh *et al.* (1990). Cossar and Canevascini (1986) reported that the growth of thermophilic fungus *Thermomyces lanuginosus* was at 6.5 pH. Optimum pH range for enzymatic activity was 4.6 to 5.2 and no activity was observed at 8.5 pH. Jensen and Olsen (1992) reported that the growth of the fungus at pH 6.5 was in accordance with our result. While about enzymatic activity the results are in contrast. Maheshwari *et al.* (2000) reported growth of thermophilic fungi at pH range 5.5–7.0.

From the results it is also apparent that percentage viscosity loss was also higher at 6.5 pH and low at acidic pH. From the above results it may be concluded that proper pH control is necessary to obtain high enzyme yield with the substrate.

Temperature plays an important role in production of enzymes. To optimize ideal temperature for growth and enzyme production flasks were kept at temperature ranging from 20–55°C. The production of cellulases by *Sporotrichum thermophile* was maximum at 45°C *i.e.* endo-β-glucanase, 41.25 μg R.S./ ml and exo-β-glucanase 31.5 μg R.S./ml. This temperature was good or both, fungal growth as well as higher yield of cellulases. In the case of mesophilic moulds the optimum temperature for cellulase is in the range 25–37°C (Rao, 1977; Johnson *et al.*, 1982; Wase *et al.*, 1985; Macris and Galiotou-Panayotou, 1986), but in the case of thermophiles an optimum of 45–60°C has earlier been reported (Prakash, 1984; Bhat and Maheshwari, 1987). Maheshwari *et al.* (2000) reported temperature optimum for *S. thermophile* at 45–50°C. Coutts and Smith (1976) suggested that greater enzyme production is at 45°C. Romanelli *et al.* (1975) suggest that maximum cellulose utilization was at 40°C. Grajek (1987) suggested that maximum glucosidase production by *S. thermophile* at 40°C.

From the technological point of view, the temperature effect on the enzymatic hydrolysis of cellulose would be of considerable interest. Thermophilic microorganisms have been reported to be more advantageous over the mesophilic microorganisms because thermophiles hydrolyze cellulose faster than mesophilic counterpart, (Coutts and Smith, 1976; Maheshwari *et al.*, 2000).

The growth of the organism always depends on tire temperature. Maheshwari *et al.* (2000) reported that for thermophiles the optimum temperature is 35°C. The effect of temperature on their growth has been evaluated taking into consideration either radial mycelial growth on solid media (Johri and Satyanarayana, 1986) or dry mycelial weight in liquid media (Emersson, 1968). Similar observation was also made in the present study that at low temperature the growth of *Sporotrichum thermophile* was low and with the increase in temperature the growth increases. The maximum extracellular protein concentration was 108.8 μg/ml and mycelial protein concentration 94.4 μg/ml at 45°C. The mycelial dry weight was also maximum (0.88 gm/70 ml) at 45°C (Figure 9.3).

Fergus and Amelung (1971), Satyanarayana and Johri (1981) have reported that thermophilic fungi differ in their ability to survive after exposure to temperature ranging from 60°C to 70°C to different length of time. However, in the present case *Sporotrichum thermophile* survive upto 45°C. Grajek (1987) reported best growth temperature was 43°C but maximum cellulose production at 40°C. Romanelli *et al.* (1975) reported that his strain of *Sporotrichum thermophile* grow at 40°C and degrade more cellulose. The results of Romanelli *et al.* (1975) was in contrast to our results because *Sporotrichum thermophile* in present case grown at 45°C and produced maximum enzyme at 45°C. Similar observation was also reported by Singh *et al.* (1990) and Coutts and Smith (1976). Therefore with the above discussion it may be concluded that temperature plays a very important role in growth as well as in the production of metabolites.

The results obtained in this experiment were very useful and encouraging, *Sporotrichum thermophile* is one of the most important organism which produce thermostable enzymes in nature.

References

Bagga, P.S., Sandhu, D.K. and Sharma, S., 1990. Purification and characterization of cellulolytic enzymes produced by *Aspergillus nidulans. Journal of Applied Bacteriology*, 68: 61–68.

Beguin, P., 1990. Molecular biology of cellulose degradation. *Ann. Rev. Microbiol.*, 44: 219–228.

Bhat, K.M. and Maheshwari, R., 1987. *Sporotrichum thermophile:* Growth, cellulose degradation, and cellulase activity. *Appl. Environ. Microbiol.*, 53(9): 2175–2182.

Chaudhuri, S., Thakur, I.S., Goel, R. and Johri, B.N., 1988. Purification and characterization of two thermostable xylanases from melanocarbus albomyces. *Biochem. Int.*, 17: 563–575.

Cossar, D. and Canevascini, G., 1986. Cellulose enzyme production during continuous culture growth of *Sporotricum* (*Chrysoporium*) *thermophile. Appl. Microbiology Biotechnology*, 24: 306–310.

Coutts, A.D. and Smith, R.E., 1976. Factors influencing the production of celluloses by *Sporotrichum thermophile. Appl. and Environ. Microbiol.*, 31: 819–825.

Emersson, R., 1968. Thermophiles. In: *The Fungi*, (Eds.) Ainsworth, G.C. and Sursman, A.S. Academic Press, New York, 3: 105–128.

Eriksson, K.E.L., Blanchette, R.A. and Ander, P., 1990. *Microbial and Enzymatic Degradation of Wood and Wood Components*. Springer-Verlag, Berlin, 407 pp.

Feldman, K.A., Lovett, J.S. and Tsao, G.T., 1988. Isolation of the cellulose enzymes from the thermophilic fungus *Thermoascus aurantiacus* and regulation of enzyme production. *Enzyme Microb. Technol.*, 10: 202–271.

Fergus, C.L. and Amelung, R.M., 1971. The heat resistance of some thermophilic fungi on mushroom compost. *Ibid*, 68: 675–679.

Gilbert, H.J. and Hazlewood, G.P., 1993. Bacterial celluloses and xylanases. *Journal of General Microbiology*, 139: 187–194.

Grajek, W., 1987. Comparative studies on the production of cellulases by thermophilic fungi in submerged and solid state fermentation. *Appl. Microbiol. Biotechnol.*, 26: 126–129.

Hagerdal, B., Ferchak, J.O. and Pye, E.K., 1930. Saccharification of cellulase by cellulolytic enzyme system of *Thermomonospora* sp. I. stability of cellulolytic activities with respect to time, temperature and pH. *Biotech. Bioeng.*, 22: 1515–1526.

Hayashida, S. and Mo, K., 1986. Production and characteristics of avicel-disintegrating endoglucanase from a protease-negative *Humicola grisea* var. *thermoidea* mutant. *Applied and Environ. Microbiol.*, 51(5): 1041–1046.

Herbert, D., Phipps, P.J. and Strange, R.E., 1971. Chemical analysis of microbial cells. In: *Method in Microbiology*, 5B: 209–344.

Jensen, B. and Olsen, J., 1992. Physio-chemical properties of a purified α-amylase from the thermophilic fungus *Thermamyces lanugenuses. Enzyme Microb. Technol.*, 14: 112–116.

Johnson, Eric A., Saka, J.M., Halliwell, G., Madia, A. and Demain, A.L., 1982. Saccharification of complex cellulosic substrate by the cellulose system from *Clostridium thermocelium. Appl. Environment Microbiol.*, 43(5): 1125–1132.

Johri, B.N. and Satyanarayana, T., 1986. Thermophilic moulds perspective in basic and applied research. *Indian Rev. Life Sci.*, 6: 75–100.

Lowry, O.H., Rosebrough, N.J., Farr, A.L. and Randall, R.J., 1951. Protein measurement with the folin phenol reagent. *Journal of Biological Chemistry*, 193: 265–275.

Macris, B.J. and Galiotou-Panayotou, M., 1986. Enhanced cellobiohydrolase production from *Aspergillus ustus* and *Trichoderma harzianum. Enzyme Microbiol Technology*, 8: 141–144.

Maheshwari, R., Bharadwaj, G. and Bhat, M.K., 2000. Thermophilic fungi: Their physiology and enzymes. *Microbiology and Molecular Biology Reviews*, 64(3): 461–488.

Maheshwari, R., Kamalam, P.T. and Balasubramanyam, P.V., 1987. The biogeography of thermophilic fungi. *Curr. Sci.*, 56: 151–155.

Mendels, M. and Weber, J., 1969. The production of cellulases. In: *Cellulases and their Applications.* Advances in chemistry series, (Ed.) F. Gould. American Chemical Society, Washington, 95: 391–413.

Miller, G.L., 1959. Use of dinitrosalicilic reagent for determination of reducing groups. *Anal. Chem.*, 31: 426–428.

Moloney, A.P., McCrae, S.I., Wood, T.M. and Coughlan, M.P., 1985. Isolation and characterization of 1,4-β-D-glucan glucanohydrolases of *Talaromyces emersonii. Biochem. J.*, 225: 365–374.

Neudoerffer, T.S. and Smith, R.E., 1970. An evaluation of fungal enzymes for the solubilization of wheat bean constituents. *Can J. Microbiol.*, 16: 139–146.

Nidetzky, B. and Steiner, W., 1993. A new approach for modeling cellulase cellulose adsorption and the kinetics of the enzymatic hydrolysis of microcystalline cellulose. *Biotechnology and Bioengineering*, 42: 469–479.

Oberson, J., Binz, T., Fracheboud, D. and Canevascini, G., 1992. Comparative investigation of cellulose-degrading enzyme systems produced by different strains of *Myceliophthora thermophila* (Apinis) v. Oorschot. *Enzyme. Microb. Technol.*, 14: 303– 312.

Pettersson, B., Cowling, E.B. and Porath, J., 1963. Studies on cellulolytic enzymes. I. Isolation of a low molecular weight cellulose from *Polypores versicolor. Biochem. Biphys. Acta*, 67: 1–8.

Prakash, A., 1984. Antagonistic attributes of thermophilic fungi and thermophilism. *Ph.D. Thesis*, Bhopal University, Bhopal.

Rao, R.R., 1977. Studies on cellulose production by an Aspergillus strain. *Indian Journal of Microbiology*, 17(3): 123–128.

Reese, E.T., Siu, R.G.H. and Levinson, H.S., 1950. The biological degradation of soluble cellulose derivatives and its relationship to the mechanism of cellulose hydrolysis. *Journal of Bacteriology*, 59: 485–497.

Romanelli, R.A., Houston, C.W. and Barnett, S.M., 1975. Studies on thermophilic cellulolytic fungi. *Appl. Microbiology*, 30: 276–281.

Ryu, D.D.Y. and Mandels, M., 1980. Cellulases: Biosynthesis and application. *Enzyme Microb. Technol.*, 2: 91–102.

Satyanarayana, T. and Johri, B.N., 1981. Lipolytic activity of thermophilic fungi of paddy straw compost. *Current Sci.*, 50: 682–688.

Schlesinger, W.H., 1991. *Biogeochemistry: An Analysis of Global Change*. Academic, San Diego, 443 pp.

Sen, S. and Chakraborthy, S.L., 1984. Isolation of cellulose hyperproductive mutants of *Myceliophthora thermophila*. In: *Proc. 7th In. Biotechnol. Symp.*, New Delhi, 2: 604–605.

Singh, A., Goel, R. and Johri, B.N., 1990. Production of cellulolytic enzymes by immobilized *Sporotrichum thermophile. Enzyme Microb. Technol.*, 12: 464–468.

Sumner, J.B. and Somer, G.F., 1944. *Laboratory Experiments in Biological Chemistry*. Academic Press, Inc., New York.

Tansey, M.R., 1971. Agar-diffusion assay of cellulolytic ability of thermophilic fungi. *Arch. Microbiol.*, 77: 1–11.

Wase, D.A. J., McManamey, W.J., Raymahasay, S. and Vaid, A.K., 1985. Comparison between cellulose production by *Aspergillus fumigatus* IMI 255901. *Enzyme Microb. Technol.*, 7: 225–229.

Chapter 10

Effect of Triazole on Growth and Chlorophyll Pigments in Radish

R. Sridharan,* A. Kishorekumar, R. Somasundaram, P. Manivannan, G.M. Alagu Lakshmanan, M. Gomathinayagam and R. Panneerselvam

Department of Botany, Annamalai University, Annamalainagar – 608 002, Tamil Nadu, India

ABSTRACT

The present study was conducted to investigate the effect of triazole compounds Triadimefon (10 mg L^{-1}) and Hexaconazole (5 mg L^{-1}) on radish. Each plants were treated with triadimefon and hexaconazole on 8th and 23rd DAS (Days After Sowing) in 30 cm plastic pot. The plants were harvested on 15th, 30th, 45th and 60th DAS for determining the effect of both triazoles on growth and chloroplast pigments. Both the triazole compounds increased the fresh and dry weight of tuber and pigments. However, both the treatments decreased the fresh and dry weight of shoot and leaf area.

***Keywords**: Triadimefon, Hexaconazole, Radish, Pigments, Growth.*

Introduction

Radish is one of the important edible vegetable tuber crop cultivated in India and south east countries. Triazole compounds are mostly used as the systemic fungicides, also have plant growth

* Corresponding Author.

regulating properties and effectively retard the shoot growth in monocotyledons and dicotyledons (Fletcher *et al.*, 2000). Triazole compounds like triadimefon and hexaconazole have growth regulating properties and induced many morphological changes like reduction in shoot elongation, stimulation of rooting, inhibition of gibberellin biosynthesis, increased chlorophyll content, alter carbohydrate status and increased cytokinin synthesis (Davis *et al.*, 1988; Isumi *et al.*, 1988).

Triazole compounds have been shown to improve the yield of many root crops such as carrot, sugar beet, and potato (Davis and Curry, 1991) Triazole compounds influences hormone balance, photosynthetic rate, enzyme activities, lipid peroxidation and yield components in various crop plants (Zhou *et al.*, 1993; Ye *et al.*, 1995; Zhou and Ye, 1996). Triazole inhibit cytochrome P-450 mediated oxidative dimethylation reaction, including those which are necessary for the synthesis of ergosterol and the conversion of kaurene to kaurenoic acid in the gibberellin biosynthetic pathway (Rademacher *et al.*, 1987). Paclobutrazol induces a variety of morphological and biochemical responses in plants including reduction in shoot elongation, stimulation of rooting, increased cytokinin synthesis and a transient increase in ABA content (Fletcher and Hofstra, 1988).

The objective of the present study is to estimate the effect of triadimefon and hexaconazole on the growth, fresh and dry weight of shoot and tuber and photosynthetic pigments in *Raphanus sativus.* (Cv. Pusa chetki) during the growth and maturation of tubers.

Materials and Methods

The seeds of *Raphanus sativus* L. Cv. (Pusa Chetki) were obtained from Mahyco-hybrid Seeds Co. Ltd, Maharashtra. The seeds were surface sterilized with 0.2 per cent $HgCl_2$ solution for 5 minutes with frequent shaking and thoroughly washed with tap water to remove $HgCl_2$. Two seeds were sown in each plastic pot of 30 cm diameter and 30 cm height containing 3 kg of soil mixture composed of red soil, sand and the farm yard manure (FYM) at 1 : 1 : 1 ratio. The seedlings were thinned to one per pot on 6th days after sowing. The pots were irrigated daily with tap water and care was taken to avoid leaching. The seedlings were treated with deionized water (control), 10 mg L^{-1} Triadimefon and 5 mg L^{-1} Hexaconazole solution alone per plant on 8th and 23rd days after sowing (DAS). Plants were harvested randomly on 15th, 30th, 45th and 60th DAS and separated into tuber and shoot and used for determining growth parameters. Total leaf area was determined using an electronic leaf area meter (Licor model L1-3100, Licoin USA). The dry weight was determined by drying the plant materials in an oven at 60°C until a constant dry weight was obtained. Chlorophyll were estimated by the method of Arnon (1949), carotenoid content was calculated using the formula of Kirk and Allen (1965). Analysis of variance (ANOVA) by the method outlined by Ridgman (1975).

Results and Discussion

Triazole treatment decreased the fresh weight of shoot significantly. Among the treatments triadimefon decreased it to a larger extent when compared to Hexaconazole (Table 10.1). Triazole treatments reduced the stem elongation, plant height and decreased the fresh weight of the shoot in bean (Asare-Boamah *et al.*, 1986), *Echingchola frumentacea* (Sankala *et al.*, 1992), grapevines (Hunter and Procter, 1992), citrus (Mehouach, *et al.*, 1996) *Prunus serotina* (Eliasson *et al.*, 1994) and Mango seedlings (Murti and Upreti 2005).

Triazole treatment decreased the dry weight of the shoot significantly in *Raphanus sativus*. Among the triazole treatment triadimefon decreased the shoot dry weight to a larger extent when compared to hexaconazole (Table 10.1). Triadimefon, triadimenol, paclobutrazol and S.3307 reduced the height and shoot dry weight in kentuchy blue grass (Kane and Smiley, 1983) and *Phaseolus vulgaris* (Fletcher *et al.*, 1986) and mango seedling (Murti and Upreti, 2005).

Table 10.1: Triadimefon and Hexaconazole Induced Variation in the Fresh Weight, Dry Weight and Leaf Area (Values are mean±SD of 3 replicates)

		Shoot				*Tuber*			
	DAS	*Control*	*TDM 10 mg L^{-1}*	*HEX 5 mg L^{-1}*	*LSD (P = 0.5)*	*Control*	*TDM 10 mg L^{-1}*	*HEX 5 mg L^{-1}*	*LSD (P = 0.5)*
Fresh weight	15	5.409	5.307	5.347	0.265	2.015	2.055	2.147	0.014
gram $plant^{-1}$ F.W	30	21.210	19.950	20.289	0.835	40.100	44.258	44.380	0.147
	45	34.526	30.180	31.880	1.235	72.950	89.510	90.090	1.834
	60	41.000	35.000	36.780	1.146	125.000	163.170	163.63	3.467
Dry weight	15	0.665	0.657	0.673	0.026	0.193	0.193	0.202	0.007
gram $plant^{-1}$ D.W	30	2.142	2.035	2.080	0.076	3.449	3.506	3.786	0.090
	45	3.384	2.979	3.124	0.123	5.676	6.623	6.675	0.128
	60	3.526	3.017	3.151	0.106	7.788	10.279	10.488	0.226
Leaf area	15	76.94	75.90	74.73	2.86				
cm^2 $plant^{-1}$	30	232.45	219.79	217.31	6.45				
	45	328.08	299.55	291.99	12.15				
	60	603.89	526.72	514.29	18.62				

Treatment with triazole compounds significantly increased the tuber weight of the radish. The fresh weight of the tuber increased with triazole treatments. Among the traizoles, hexaconazole treatment increased it to a higher level than triadimefon (Table 10.2). Triazole treated plants have exhibited thickened fleshy roots with increased root diameter and weight in apple and soybean (Steffens *et al.*, 1983; Sankhla *et al.*, 1986), Tomato (Berova *et al.*, 2000) and wheat (Berova *et al.*, 2002).

The dry weight of the tuber also increased by triazole treatment. Among the triazole, hexaconazole treatment significantly increased the dry weight of the tuber when compared to triadimefon (Table 10.1). The fresh and dry weight of the root were significantly higher in the triadimefon treated apple (Huang *et al.*, 1995) and mango seedlings (Murti and Upreti, 2005). Inhibited gibberellin biosynthesis and increased cytokinin and abscisic acid content induced by these triazoles might be the cause for decreased fresh and dry weight of shoot and increased fresh and dry weight of tuber growth in radish.

Both triadimefon and hexaconazole treatments inhibited the leaf area. Among these, hexaconazole treatment reduced the leaf area to a larger extent than triadimefon. Triadimefon reduced the leaf area in maize (Khalil *et al.*, 1990) and wheat (Gao *et al.*, 1988). Similar results were observed in tebuconazole treated *Barassica napus* (Child *et al.*, 1993).

Triadimefon and hexaconazole treatments increased the chlorophyll content to a larger extent when compared to control in radish. Among them hexaconazole treatment increased it to a larger level than triadimefon (Table 10.2). Increased chlorophyll content with triadimefon treatment was observed with uniconazole, LAB-150978, BAS-110 W treatments in *Cucumber* cotyledons (Thomas and Singh, 1995) and in paclobutrazol treated tomato leaves (Berova *et al.*, 2000).

The leaves of the triazole treated plants showed an increased carotenoid content when compared to control. Among these hexaconazole treatment showed higher level than triadimefon (Table 10.2). Higher level of carotenoid content was observed in triadimefon treated cowpea (Gopi *et al.*, 1999), uniconazole treated wheat (Fletcher and Hofstra, 1998), tomato (Senaratna *et al.*, 1988) and dichlorobutrazole treated *Zea mays* (Khalil *et al.*, 1990).

Table 10.2: Triadimefon and Hexaconazole Induced Variation in the Total Chlorophyll, Chlorophyll-a, Chlorophyll-b and Carotenoid (Values are mean±SD of 3 replicates)

	DAS	*Leaf*			
		Control	*TDM 10 mgL⁻¹*	*HEX 5 mgL⁻¹*	*LSD (P = 0.5)*
Total chlorophyll	15	0.154	0.191	0.180	0.008
mg g⁻¹.F.W	30	0.207	0.234	0.233	0.007
	45	0.307	0.330	0.335	0.012
	60	0.404	0.440	0.452	0.014
Chlorophyll-a	15	0.071	0.086	0.082	0.003
mg g⁻¹.F.W	30	0.090	0.104	0.105	0.005
	45	0.145	0.159	0.161	0.008
	60	0.215	0.242	0.248	0.012
Chlorophyll-b	15	0.082	0.104	0.098	0.003
mg g⁻¹.F.W	30	0.117	0.130	0.128	0.004
	45	0.162	0.171	0.174	0.006
	60	0.189	0.198	0.204	0.009
Carotenoid	15	0.370	0.390	0.393	0.007
mg g⁻¹.F.W	30	0.510	0.520	0.536	0.005
	45	0.626	0.682	0.684	0.009
	60	0.653	0.731	0.738	0.013

References

Arnon, D.I., 1949. Copper enzymes in isolated chloroplasts, polyphenol oxidase in *Beta vulgaris* L. *Plant Physiol.*, 24: 1–15.

Asare-Boamah, N.K., G. Hofstra, R.A. Fletcher and E.B. Dumbroff, 1986. Triadimefon protects bean plants from water stress through its effects on abscisic acid. *Plant Cell Physiol.*, 27: 383–390.

Berova, M., Z. Zlatev and N. Stoeva, 2000. Physiological response and yield of paclobutrazol treated tomato plants (*Lycopersicon esculentum* Mill). *Plant Growth Regul.*, 39: 117–123.

Bervova, M., Z. Zlatev and N. Stoeva, 2002. Effect of paclobutrazol on wheat seedlings under low temperature stress. *Bulg. J. Plant Physiol.*, 28(1–2): 75–84.

Berova, M. and Z. Zlatev, 2000. Physiological response and yield of paclobutrazol treated tomato plants (*Lycopersicon esculentum* Mill.). *Plant Growth Regul.*, 30: 117–123.

Child, R.D., D.E. Evans, J. Allen and G.M. Arnold, 1993. Growth responses in oil seed rape (*Brassica napus* L.) to combined applications of the triazole chemicals triapenthenol and tebuconazole interactions with gibberellin. *Plant Growth Regul.*, 13: 203–212.

Davis, T.D. and E.A. Curry, 1991. Chemical regulation of vegetative growth. *Critic. Rev. Plant Sci.*, 10: 151–188.

Davis, T.D., G.L. Steffens and N. Sankhla, 1988. Triazole plant growth regulators. In: *Hort. Rev.*, (Ed.) J. Janick. Timber Press, Portland, *Oregon*, 10: 63–105.

Eliasson, M.K., C.A. Beyl and P.A. Barker, 1994. *In vitro* responses and acclimatization of *prunus serotina* with Paclobutrazol. *J. Plant Growth Regul.*, USA, 13: 137–142.

Fletcher, R.A. and G. Hofstra, 1988. Triazoles as potential plant protectants. In: *Sterol Biosynthesis Inhibitors*, (Eds.) Berg, D. and M. Plempel. Ellis Horwood Ltd., Cambridge, England, pp. 321–331.

Fletcher, R.A., A. Gilley, N. Sankhla and T.M. Davis, 2000. Triazoles as plant growth regulators and stress protectants. *Hort. Rev.*, John Wiley and Sons Inc., 24: 56–138.

Fletcher, R.A., G. Hofstra and J. Gao, 1986. Comparative fungitoxic and plant growth regulating properties of triazole derivatives. *Plant Cell Physiol.*, 27: 367–371.

Gao, J., G. Hofstra and R.A. Fletcher, 1988. Anatomical changes induced by triazoles in wheat seedlings. *Can. J. Bot.*, 66: 1178–1185.

Gopi, R., B.M. Sujatha, S.N. Rajan, L. Karikalan and R. Panneerselvam, 1999. Effect of triadimefon in the sodium chloride stressed cowpea (*Vigna unguiculata*) seedlings. *Indian J. Agri. Sci.*, 69: 743–745.

Huang, W.D., T. Shen, S.H. Han and S. Liu, 1995. Influence of paclobutrazol on photosynthesis rate and dry matter partitioning in the apple tree. *Plant Nutr.*, 18: 901–910.

Hunter, D.M. and J.T.A. Procter, 1992. Paclobutrazol affects growth and fruit composition of potted grapevines. *Hort. Sci.*, 27: 319–321.

Izumi, K., S. Nakagawa, M. Kobayashi, H. Oshio, A. Sakurai and N. Takahashi, 1988. Levels of IAA, cytokinins, ABA and ethylene in rice plants as affected by a gibberellin biosynthesis inhibitor, uniconazole. *Plant Cell Physiol.*, 29: 97–104.

Kane, R.T. and R.W. Smiley, 1983. Plant growth regulating effects of systemic fungicides applied to kentucky blue grass. *Agron. J.*, 75: 469–473.

Khalil, I.A., E.I. Mercer and Z.X. Wang, 1990. Effect of triazole fungicides on the growth, chlorophyllase pigments and sterol biosynthesis of maize (*Zea mays* L.). *Plant Sci.*, 66: 21–28.

Kirk, J.T.O. and R.L. Allen, 1965. Dependence of chloroplast pigment synthesis on protein synthesis: Effect of acsidione. *Biochem. Biophys. Res. Commun.*, 21: 532–530.

Mehouachi, J., F.R. Tadeo, S. Zaragoza, E. Primo-Millo and M. Talon, 1996. Effects of gibberellic acid and paclobutrazol on growth and carbohydrate accumulation in shoots and roots of citrus rootstock seedlings. *Hort. Sci.*, 71: 747–754.

Murti, G.S.R. and K.K. Upreti, 2005. Paclobutrazol induced growth retardation of mango seedlings. *Indian Plant Physiol.*, 10: 9–13.

Rademacher, W., H. Fritsch, E. Graebe, H. Sauter and J. Jung, 1987. Tetcyclacis and triazole-type plant growth retardants: Their influence on the biosynthesis of gibberellins and other metabolic processes. *Pestic. Sci.*, 21: 241–252.

Ridgman, W.J., 1975. *Experimentation in Biology: An Introduction to Design and Analysis*. Thomson Litho Ltd., East Kilbridge, Scoland, pp. 81–100.

Sankhla, N., T.D. Davis, V.D. Jolley and A. Upadhyaya, 1986. Effect of paclobutrazol on the development of iron chlorosis in soybeans. *J. Plant Nutr.*, 9: 923–934.

Sankhla, N.A. Upadhyaya, T.D. Davis and D. Sankhla, 1992. Hydrogen peroxidase scavenging enzymes and antioxidants in *Echinochloa frumentacea* affected by triazole growth regulators. *Plant Growth Regul.*, 11: 441–442.

Senaratna, T., C.E. Mackay, B.D. McKersie and R.A. Fletcher, 1988. Uniconazole induced chilling tolerance in tomato and its relationship to antioxidant content. *J. Plant Physiol.*, 133: 56–61.

Steffens, G.L., S.Y. Wang, C.L. Steffens and T. Brennan, 1983. Influence of paclobutrazol (pp 333) on apple seedling growth and physiology. *Proc. Plant Growth Regul. Soc. Amer.*, 10: 195–205.

Thomas, R.M. and V.P. Singh, 1995. Effect of three triazole derivatives on mercury induced inhibition of chlorophyll and caroteniod accumulation in cucumber cotyledons. *Indian J. Plant Physiol.*, 38: 313–316.

Ye, Q.F., W.J. Zhou, W.F. Xi and J.Y. Fang, 1995. Effects of S-3307 on levels of endogenous (IAA, ABA and ZT) and some physiological of rape seedlings. *Acta. Agri. Zhejiang*, 7: 451–456.

Zhou, W.J., H.C. Shen, H.F. Xi and Q.F. Ye, 1993. Studies on the regulation mechanism of paclobutrazol to the growth of rape plant. *Acta Agri. Univ. Zhejian*, 19: 316–320.

Zhou, W.J. and Q.F. Ye, 1996. Physiological and yield effects of uniconazole on winter rape (*Brassica napus* L.). *J. Plant Growth Regul.*, 15: 69–73.

Chapter 11

The Role of Abiotic Factors in the Distributional Patterns of Acarina and Collembola in the Sub-tropical Forest Ecosystem of Manipur

Ch. Chitrapati and Th. Binoy Singh
Department of Life Sciences, Manipur University, Canchipur – 795 003, Imphal

ABSTRACT

The present investigation was conducted to determine the distributional pattern of Acarina and Collembola the two most dominant soil microarthopods group in the subtropical forest ecosystems of Manipur (24°45′N and 93°55′E). The soil samples were collected with the help of soil corer of 19.64 cm^2 in surface area and microarthopods were extracted by modified Tullgren funnel. Acarina and Collembola were recorded the most dominant group in both the study sites contributing 20–65 per cent to the total soil microarthopods. Maximum population density was recorded during rainy season and decline with the onset of winter season. The population density showed decreasing trend with the increase of soil depth in both the study sites. Among the four suborders of Acarina recorded, Cryptostigmata was the most dominant and Astigmata was the least. Nine genera each of Cryptostigmata and Collembola were identified. Microclimatic conditions in different layers of soil, primarily soil moisture content, soil temperature and soil organic carbon content exhibited a significant positive correlation ($P < 0.01$) with the changes in the population density of Acarina and Collembola. Sorensens Quotient of Community Similarity (Q/S) of oribatid and collembolan community analysed between the two study sites in different soil layers ranged from 25–90 per cent.

Keywords: *Soil microarthropods, Collembola, Acarina, Abiotic factors, Community similarity.*

Introduction

The soil is a physically and chemically diverse environment that is used by a wide variety of invertebrate animals and microorganisms. The species differ in the distribution of their activities in space, time and biological function. In the present study soil microarthropod would mean the microarthopods fauna in soil in general and Acarina and Collembola in particular. Soil microarthopods play a key role in regulation of microbial population, decomposition of organic matter and cycling of nutrients within the soils (Coleman and Crossley, 1996; Tian *et al.*, 1998). The abiotic factors are important in determining the soil faunal population pattern and species composition. Fluctuations in temperature and moisture are less secure than they are at and above the ground surface and the amplitude of these fluctuations decreases with depth in the soil (Geiger, 1959). Under Manipur climatic conditions except studies made by Singh *et al.* (1998), no work has been reported so far. Gradual soil degradation particularly due to slash and burnt method has led to the drastic changes in the population dynamics of the soil microarthopods and thereby either directly or indirectly affecting the soil ecosystem. With this objective the present study carried out in the sub-topical forest ecosystem of Manipur to know the distributional patterns of soil microarthopods under the different abiotic factors.

Materials and Methods

The present investigation was carried out from March 2000 to February 2001 in both the protected forest (Koirengei reserve forest) and degraded sites (Khonghampat area). A total of 324 soil samples were collected with the help of cylindrical soil corer of 19,64 cm^2 in surface area at three layers *i.e.* 0–5, 5–10 and 10–15 cm during the study period. The soil microarthropods were extracted using modified Tullgren funnel at a constant temperature of 40°C (±2°C) for 6–7 days depending on he moisture content of soil samples. Soil temperature, the soil moisture contents and the soil organic carbon (Walkley and Black method) were also analysed. Community similarity was analysed by the Coefficient of Similarity (CCS) of Sorensen (1948) also known as Quotient of Similarity (Q/S) as follows:

$$\frac{Q}{S} = \frac{2C}{S_1 + S_2} \times 100$$

where,

C: Number of species common to both the communities

S_1: Number of species in community 1 (protected site)

S_2: Number of species in community 2 (degraded site)

Results and Discussion

The total population density of Acarina and Collembola recorded higher in the protected forest site as compared to the degraded study site (Table 11.1) both annually and vertically which may be attributed to high soil moisture content, high organic carbon content (Figures 11..1 and 11.2) and close canopy in the study site 1. Vegetation itself does not directly influences Collembola but may affect them indirectly through microclimate, soil fertility, soil profile, microflora and soil porosity (Hagvar, 1982). The population density of Acarina exhibited highest aggregation of 38 per cent and 41 per cent in the uppermost soil layer *i.e.* at 0–5 cm in the protected forest and degraded sites respectively and least population density of Acarina was found at 10–15 cm soil layer contributing only 28 per cent in both the study sites. Collembola was also recorded maximum at the upper soil layer and decreased with depths. This may be attributed to adequate living space, favourable moisture conditions, aeration rates and rich accumulation of humus soil in the upper soil layer (Wallwork, 1970).

Table 11.1: Showing the Total Population Density of Acarina and Collembola in the Protected Forest and Degraded Ecosystem (Number±S.E.) × 10^2 m^{-2}

Soil Microarthropod	*Protected Forest*		*Degraded Ecosystem*
	Soil Layers (cm)	*Number±S.E.*	*Number±S.E.*
Acarina	0–5	718±36.05	495±19.20
	5–10	657±20.05	372±20.33
	10–15	530±34.97	339±25.69
	Total	**1905±91.07**	**1206±65.22**
Collembola	0–5	241±12.68	166±4.95
	5–10	190±3.32	148±3.8
	10–15	171±7.66	143±6.20
	Total	**602±23.66**	**457±14.15**

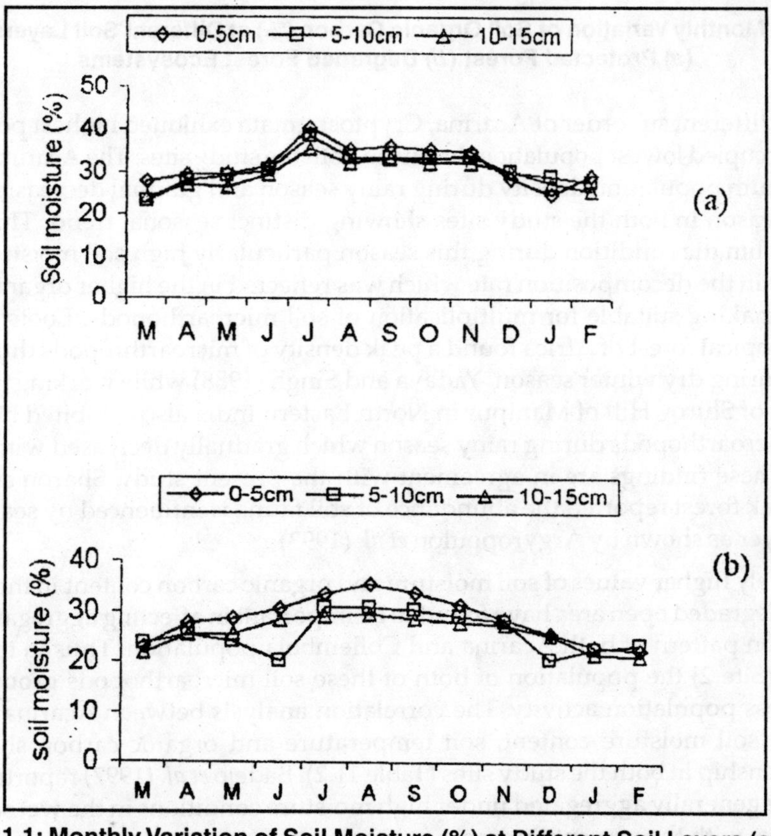

Figure 11.1: Monthly Variation of Soil Moisture (%) at Different Soil Layers (cm) in the (*a*) Protected Forest (*b*) Degraded Forest Ecosystems

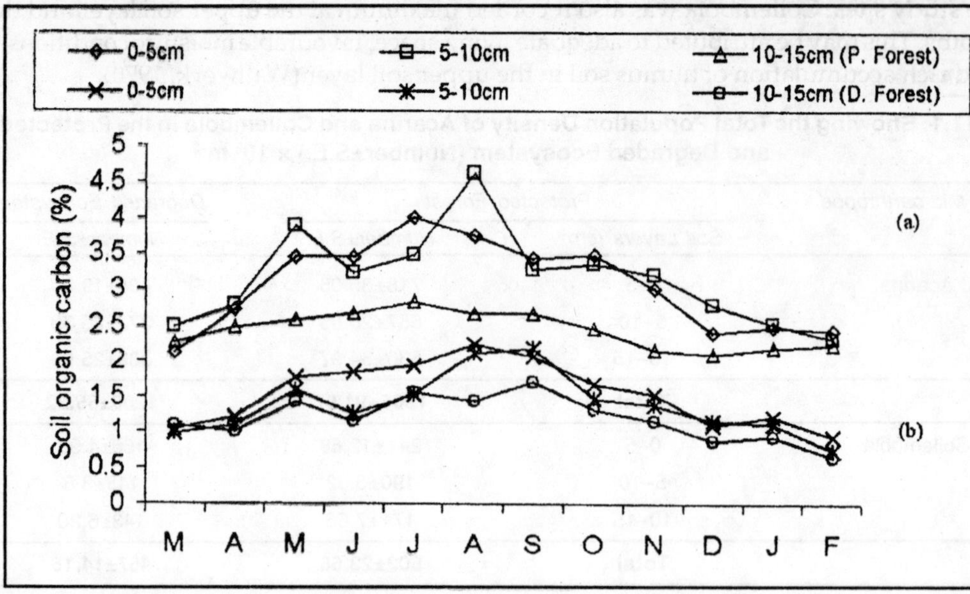

Figure 11.2: Monthly Variation of Soil Organic Carbon (%) at Different Soil Layers (cm) in the (*a*) Protected Forest (*b*) Degraded Forest Ecosystems

Among the different suborder of Acarina, Cryptostigmata exhibited highest population density and Astigmata occupied lowest population density in both the study sites. The Acarina and Collembola exhibited maximum population density during rainy season and gradual decreasing trend towards the dry winter season in both the study sites showing distinct seasonal trend. This may be due to congenial microclimatic condition during this season particularly high soil moisture content had a favourable affect in the decomposition rate which was reflected in the higher organic carbon content of the soil thus making suitable for multiplication of soil microarthopods. Loots and Ryke (1966) working in the tropical forest of Africa found a peak density of microarthropods during late summer and minimum during dry winter season. Yadava and Singh (1988) while working in the subtropical forest ecosystem of Shiroy Hill of Manipur in North Eastern India also exhibited higher population density of soil microarthopods during rainy season which gradually decreased with the approach of winter season. These findings are in agreement with the present study. Sharon *et al.* (2001) while working in an Oak forest reported the abundance of soil fauna is influenced by season and not only the upper leaf layer as shown by Argyropoulou *et al.* (1993).

Comparatively higher values of soil moisture and organic carbon content in the protected study site than that of degraded open area have found to be important in affecting aggregation behaviour as well as fluctuation pattern of both Acarina and Collembola population. Thus in the degraded and open study site (Site 2) the population of both of these soil microarthopods group resulted in the dispersed and less population activity. The correlation analysis between Acarina and Collembola population with soil moisture content, soil temperature and organic carbon showed a positive significant relationship in both the study sites (Table 11.2). Badejo *et al.* (1997) reported that Mites and Collembola were generally aggregated under high moisture conditions in the wet season which was also in accordance with the present finding.

Table 11.2: Correlationship Between Acarina and Collembola with Soil Moisture and Temperature in the Protected and Degraded Study Site

Variables	*Soil Layers (cm)*	*r*	*df*	*Y*	*t*	*p*
Acarina						
Protected Forest						
Soil moisture (%)	0–5	0.8660	10	–8.41+0.45x	5.4732	< 0.01
	5–10	0.6719	10	–6.81+0.39x	2.8670	> 0.05
	10–15	0.7040	10	–6.39+0.36x	3.32	< 0.01
Soil temperature (ºC)	0–5	0.8440	10	–1.03+0.35x	4.976	< 0.01
	5–10	0.807	10	–1.54+0.36x	4.315	< 0.01
	10–15	0.785	10	–11.94+5.48x	3.937	< 0.01
Soil organic carbon (%)	0–5	0.801	10	–3.13+2.93x	7.062	< 0.01
	5–10	0.860	10	–8.55+4.69x	5.318	< 0.01
	10–15	0.789	10	–8.08+14.12x	4.060	< 0.01
Degraded Ecosystem						
Soil moisture (%)	0–5	0.828	10	–3.79+0.27x	4.67	< 0.01
	5–10	0.673	10	–3.42+0.25x	2.875	< 0.01
	10–15	0.559	10	–2.34+0.20x	2.133	< 0.05
Soil temperature (ºC)	0–5	0.879	10	–0.49+0.23x	8.005	< 0.01
	5–10	0.915	10	–1.45+0.23x	7.197	< 0.01
	10–15	0.911	10	–1.38+0.21x	6.954	< 0.01
Soil organic carbon (%)	0–5	0.799	10	–0.005+2.68x	4.198	< 0.01
	5–10	0.500	10	1.71+0.93x	1.824	> 0.05
	10–15	0.649	10	0.12+2.50x	2.697	< 0.05
Collembola						
Protected Forest						
Soil moisture (%)	0–5	0.596	10	–2.81+0.15x	2.339	< 0.05
	5–10	0.514	10	–1.29+0.09x	1.897	> 0.05
	10–15	0.627	10	–1.91+0.11x	2.859	< 0.05
Soil temperature (ºC)	0–5	0.655	10	–0.64+0.14x	2.738	< 0.05
	5–10	0.628	10	–0.083+0.085	2.551	< 0.05
	10–15	0.627	10	–0.17+0.07x	2.543	< 0.05
Soil organic carbon (%)	0–5	0.678	10	–1.76+1.21x	2.914	< 0.05
	5–10	0.539	10	0.14+0.53x	2.022	> 0.05
	10–15	0.629	10	–2.19+1.48x	2.556	< 0.05
Degraded Ecosystem						
Soil moisture (%)	0–5	0.880	10	–1.35+0.09x	5.793	< 0.01
	5–10	0.858	10	–1.52+0.103x	5.291	< 0.01
	10–15	0.580	10	–0.55+0.066x	3.165	< 0.05
Soil temperature (ºC)	0–5	0.766	10	0.08+0.06x	3.765	< 0.01
	5–10	0.819	10	–0.15+0.68x	4.518	< 0.01
	10–15	0.576	10	0.17+0.05x	6.954	< 0.01
Soil organic carbon (%)	0–5	0.670	10	–0.12+0.92x	2.853	< 0.05
	5–10	0.711	10	0.25+0.72x	3.196	< 0.01
	10–15	0.440	10	0.61+0.49x	1.550	> 0.05

In the present investigation the community analysis have been confined to the individuals belonging to only nine genera each of Cryptostigmata (Acarina) and Collembola. Oribatid belonging to nine genera *viz.*, *Chaunoproctus* sp., *Scheloribates* sp., *Zetorchestes* sp., *Rhysotritia* sp., *Haplophorella* sp., *Oppia* sp., *Xylobates* sp., *Galumna* sp. and *Euphthiracus* sp. and nine genera of Collembola *viz.*, *Sminthurinus* sp., *Isotoma* sp., *Sinella* sp., *Folsomia* sp., *Lepidocyrtus* sp., *Onychiurus* sp., *Xenylla* sp. and *Tullbergia* sp. were analysed to show the similarity between the two study sites.

The value of Sorensen Quotient of Oribatid and Collembola showed high similarity between the two study sites (Table 11.3) which was regulated by different abiotic factors. Some genera such as *Chaunoproctus* sp., *Scheloribates* sp., *Oppia* sp. and *Xenylla* sp. were commonly available throughout the seasons and soil layers exhibited high similarities between the two study sites. Sharon *et al.* (2001) pointed out that every taxon has a geographical range which fluctuates through times and becomes discontinuous and though individuals were distributed within the range of various points at any given time, they may not be available at the same points. But Addison *et al.* (2003) reported that the community structure of Collembola in old-growth forests differed significantly from those of younger forests and could still be differentiated from those of mature forests 80–102 years after clearcutting. However, the results of the present study showed that Collembola and Acarina population tends to be similar in diversity and richness so long the microhabitat factors are similar and least difference in the elevation of the study site could not influence much in their distribution pattern. As it was revealed from the investigation in the protected site with congenial abiotic factors particularly soil moisture played important role in determining population distribution both horizontally as well as vertically. In the degraded site, population response, associated with clear cutting and fire, resulted in a low population density as compared to the protected study site. Both Collembola and Acarina abundance preferred humus rich soils as well as soil moisture content which was clearly indicated from the high Verhoef *et al.* (1988) and reported soil moisture as the most important environmental factor acting upon soil arthropods distribution. Thus the present study revealed that the microhabitat factors particularly soil moisture and soil organic matter content tended to respond more rather than the macrohabitat factors in the distribution pattern of Acarina and Collembola population both horizontally as well as vertically.

Table 11.3: Showing Community Similarity of Oribatid and Collembola in Different Seasons and Soil Layers

Soil Layer (cm)	*Season*	*Sorensen's Coefficient Q/S (%)*	
		Oribatid	*Collembola*
0–5	Summer	66.70	54.54
	Rainy	80.00	54.54
	Winter	72.72	66.67
5–10	Summer	54.54	60.00
	Rainy	66.70	60.00
	Winter	25.00	50.00
10–15	Summer	75.00	57.14
	Rainy	88.80	50.00
	Winter	40.00	28.57

References

Addison, J.A., Trofymow, J.A. and Marshall, V.G., 2003. Abundance, species diversity and community structure of Collembola in successional coastal temperate forests on Vancouver Island, Canada. *Applied Soil Ecology*, 24: 233–246.

Argyropoulou, M.D., M.D. Asikidis, G.D. Latrou and G.P. Stamou, 1993. Colonization patterns of decomposing litter in a maquis ecosystem. *European Journal of Soil Biology*, 29: 183–191.

Badejo, M.A., T.O. Obilade and B.A. Oblubakin, 1997. Spatial distribution and abundance of mites and springtails under different temperature and moisture regimes in a tropical rain forest floor. *Tropical Ecology*, 38(1): 31–38.

Coleman, D.C. and D.A. Crossley, 1996. *Fundamentals of Soil Ecology*. Academic Press, New York, USA.

Geiger, R., 1959. *The Climate Near the Ground*. Harvard University Press, Cambridge, Mass, USA.

Hagvar, S., 1982. Collembola in Norwegian coniferous forest soils: Relations to plant communities and soil fertility. *Pedobiologia*, 24: 255–296.

Loots, G.C. and P.A.J. Ryke, 1966. A comparative quantitative study of the microarthopods in different types of pasture soils. *Zool. Afr.*, 21:157–197.

Prabhoo, N.R., 1976. Soil microarthropods of virgin forest and adjoining tea fields in the Western Ghats in Kerala. *Oriental Insects*, 10(33): 435–442.

Sharon, R., G. Degani and M. Warburg, 2001. Comparing the soil macrofauna in two Oak wood forests: Does community structure differ under similar ambient conditions? *Pedobiologia*, 45: 355–366.

Singh, T.B. and P.S. Yadava, 1998. Seasonal fluctuations of oribatid mites in a subtropical forest ecosystem of Manipur, North Eastern India. *International Journal of Ecology and Environmental Sciences*, 24: 923–129.

Sorensen, T., 1948. A method of establishing groups of equal amplitude in plant sociology based on the similarity of species content and its application to analysis of the vegetation Danish commons. *Biol. Skr.*, 5: 1–34.

Tian, G., C.O. Adejuyigbe, G.O. Adeoye and B.T. Khang, 1998. Role of soil microarthopods in leaf decomposition and N release under various land-use practices in the humid tropics. *Pedobiologia*, 42: 33–42.

Verhoef, H.A., Prast, J.E. and Verweij, R.A., 1988. Relative importance of fungi and algae in the diet and nitrogen nutrition of *Orchesella cinta* (L.) and *Tomocerus minor* (Lubbock). *Funct. Ecol.*, 2: 195–201.

Wallwork, J.A., 1970. *Ecology of Soil Animals*. Mc-Graw Hill Publishing Company Limited.

Yadava, P.S. and E.J. Singh, 1988. Some aspects of ecology of oak forest in Shiroy Hills, Manipur (North-Eastern India). *International Journal of Ecology and Environmental Sciences*, 14: 111.

Chapter 12

Diversity of Phytoplankton in Mani Reservoir, Hosanagar, Karnataka

D.N. Veerendra[1], S. Manjappa[1] and E.T. Puttaiah[2]
[1]Department of Chemistry, UBDT Engineering College, Kuvempu University, Davanagere
[2]Department of Environmental Science, Kuvempu University, Shankaraghatta – 577 451

ABSTRACT

This article deals with the phytoplanktonic and physico-chemical composition of the Mani reservoir. A total of 34 species of phytoplankton were identified under 4 classes. Among these maximum density was recorded under Bacillariophyceae, Chlorophyceae Cyanophyceae and Euglenophyceae. Based on the results of the study it has been deduced that the reservoir is less in productivity.

Keywords: Mani reservoir, Phytoplankton, Productivity.

Introduction

Phytoplankton, the floating inconspicuous plant life plays a major role in the food chain of aquatic ecosystem by biosynthesis the organic matter and thus act as the primary producers of food on which other life forms depend. The status of the aquatic ecosystem is dependent on the abiotic propertied of water and biological diversity of the ecosystem.

Evaluation of the potentiality of an aquatic ecosystem is nothing but the estimation of the rate of its primary production, where it involves the primary fixation and its subsequent transfer of higher trophic levels. It is a well established fact that, the pollution of aquatic habitats is assessed by using algae as indictors (Patrick, 1973).

In view of the above thoughts, the present study was carried out which comprises the data on morphological features of the reservoir, physico-chemical characteristics of water and diversity of phytoplankton community. Based on this data an attempt has also been made to evaluate the trophic status of the reservoir.

Morphometric Features of the Reservoir

The Mani reservoir originates at near Mani village, located at 13°34′ to 13°45′ East and 74°52′ and 75°11′ North. It is situated about 95 km from Shimoga city.

Morphometry

1. Water spread area (sq km) : 163
2. Maximum height of the dam (mt) : 59
3. Maximum length of the dam (mt) : 585
4. Storage capacity (M cum) : 960

Materials and Methods

For the qualitative estimation of phytoplankton, water samples were collected with the help of plankton net of mesh size 50 microns during February 2004 to January 2005. Further these samples were fixed in 4 per cent formaldehyde and brought to the laboratory for identification was made with the help of available literature (Hegde and Bharathi, 1985; Vaishya and Adoni, 1992). Further distribution pattern of the phytoplankton recorded in Table 12.2.

The physico-chemical characteristics of water samples were analyzed from each tank at an interval of 30 days with help of black colored carboys can of 2 liters capacity. Key waters quality parameters like temperature, DO, free CO_2, BOD, pH, total hardness, total alkalinity, chloride, calcium, TDS, total acidity, sulphate and electrical conductivity was estimated by adopting the method given in APHA (1998).

Results and Discussion

The physico-chemical characteristics that are responsible for the abundance of phytoplankton. The temperature is one of the vital factors which control the abundance of phytoplankton in lentic ecosystem (Nazneen, 1980). In the present investigation temperature ranges between 25° C to 30° C favouring the growth of phytoplankton.

The data indicates that, the water in reservoir is slightly alkaline. Verma and Mohanthy (1995) are of the opinion that, higher pH values promote the growth of the algae and results in blooms. In the present investigation the pH value ranges from 6.5 to 8.5, Robert *et al.* (1974) suggested that, pH 5 to 8.5 is ideal for phytoplankton growth. The range in which the reservoir water under study also lies. Vashiya and Adoni (1992) have stated that, the alkaline pH favors the population of diatoms. The observations of the present investigation agrees with this.

Higher concentration of dissolved oxygen and low temperature favours the dominance of Chlorophyceae. In the present investigation dissolved oxygen ranges between 1.62 to 13.68 mg/l Venkteshwaralu (1969). Dhakar (1979) also, observed that, green alga prefer water with higher concentration of dissolved oxygen. In the present study members of the Chlorophyceae dominated the other groups. This is also agreement with the findings of the Khatri (1987).

Table 12.1: Monthly Variation of Physico-chemical Characteristics of Mani Reservoir, February 2004 to January 2005

Parameter	*Feb*	*Mar*	*Apr*	*May*	*Jun*	*July*	*Aug*	*Sept*	*Oct*	*Nov*	*Dec*	*Jan*	*Range*
AT	31	32	33	30.5	31	29	28.5	34	26.0	30	28	33	26–33
WT	27	27	30	29.5	28	28	26	28	25.0	29	26	27	25–30
pH	8.4	8.5	8.2	7.6	8.0	7.1	6.5	7.2	8.0	6.5	8.3	7.5	6.5–8.5
EC	120	129	138	92	135	76	68	62	57.0	63	66	63	57–138
TDS	76.8	82.5	88.32	58.88	89.1	48.64	43.52	39.68	36.48	37.12	42.24	40.32	37.12–89.1
Tur	20	17.2	20.01	17.1	19.1	20.1	21.1	19.1	15.1	10.9	24	32	17.1–21.1
D.O	8.10	9.97	8.10	1.62	7.29	6.08	3.24	2.43	5.67	16.2	13.78	5.39	1.62–13.78
Free CO_2	8.8	8.8	8.8	8.8	8.8	4.4	8.8	8.8	4.4	13.2	4.4	4.4	4.4–13.2
Cl_2	12.75	15.60	19.85	19.85	18.43	19.85	14.18	8.50	14.18	12.75	15.60	9.92	8.50–19.85
Ca^{++}	6.73	12.0	11.78	6.73	12.62	3.87	5.30	6.73	5.05	4.20	10.10	5.05	3.87–12.62
Mg^{++}	2.24	3.89	4.53	2.24	4.02	2.8	4.57	0.78	1.31	2.31	3.61	2.92	0.78–4.57
TH	32	46	48	26	48	21.33	32	20	40	18.0	20.40	24.6	18–48
BOD	3.64	3.49	3.64	4.06	3.65	3.06	5.4	3.2	0.3	6.09	6.08	1.09	0.3–6.09
Total alkalinity	40	50	50	40	50	40	20	20	40	10	40	40	20–50
PO_4	0.01	0.03	0.18	0.18	0.14	0.06	0.01	0.09	0.03	0.001	0.001	0.001	0.001–0.18
NO_3	0.21	0.11	0.12	0.11	0.14	0.21	0.26	0.18	0.13	0.17	0.16	0.18	0.11–0.26
SO_4	22.04	22.87	24.57	22.04	24.96	2.04	11.13	11.53	11.1	13.44	41.0	25.15	2.04–25.15
Na^+	2.4	3.1	3.0	3.2	2.8	3.2	2.4	3.1	2.4	1.4	3.2	2.8	1.4–3.2
K^+	1.6	2.3	1.9	2.4	1.1	1.4	1.6	2.0	1.6	1.2	1.6	1.4	1.1–2.0

Note: All the parameters are in mg/l except pH, Temperature (°C), Turbidity (NTU) and Electrical conductivity (μmhos/cm).

Table 12.2: Diversity of Phytoplankton in Mani Reservoir

Chlorophyceae	*Bacillariophyceae*	*Cyanophyceae*	*Euglenophyceae*
Ankistrodesmus acicularis	*Acnathes exilis*	*Apanocapsa stagnina*	*Euglena minuta*
Ankistrodesmus convolutes	*Diatoma vulgare*	*Gleocapsa minutus*	*Phacus caudatus*
Ankistrodesmus falcatus	*Gomphonema gracile*	*Merismopedia tenuissima*	*Phacus tortus*
Closterium lunula	*Melosira granulate*	*Microcystis aeruginosa*	
Cosmarium sp.	*Navicula digitotadiata*	*Nostoc* sp.	
Crucigenia quadrata	*Navicula punctata*	*Phormidium* sp.	
Crucigenia tetrapedia	*Pinnularia nobilis*	*Oscillatoria pseudogeminata*	
Pediastrum simples	*Rhopalofia gibba*		
Scenedesmus quadricuada	*Surirella biseriata*		
Selenastrum gracile	*Surirella ovata*		
Staurastrum arachne	*Synedra acus*		
Staurastrum sp.			
Ulothrix sp.			

Generally Calcium concentration in natural fresh water falls below 10 mg/l calcium concentration in the vicinity of lime stone ranges between 30–100 mg/l. Zafar (1964) has stated that, the calcium is one of the important elements influencing the distribution of Bacillariophyceae, in the present investigation calcium ranges between 3.67 to 12.68 mg/l, favoring the growth of Bacillariophyceae. This is agreement with the findings of Zafar.

Nitrate is an important source of nitrogen for controlling the occurrence of and abundance of phytoplankton. Nanadan and Patel (1992) have stated that, rainfall is responsible for increasing the nitrates in water. Higher concentration of nitrate is an indication of organic pollution. In the present study nitrate ranges between 0.1 to 0.26 mg/l. Based on the results the investigation concluded that, the water body is oligotrophic.

Phosphate is considered as one of the important nutrients limiting the growth of phytoplankton (Welch *et al.*, 1978). In the present study the phosphate ranges between 0.001 to 0.18 mg/l.

In all 34 species of phytoplankton belonging to four different classes of algae were recorded. The qualitative analysis showed that, Chlorophyceaee had 13 species, Bacillariophyceae 11 sp., Cyanophyceae 7 and Euglenophyceae 3 depicted in Table 12.2.

References

APHA, 1998. *Standard Methods for the Examination of Water and Wastewater*, 20th edition. Washington, D.C.

Dhakar, M.L., 1979. Studies in some aspects of the hydrobiology of Indrasagar tank (South Rajasthan). *Ph.D. Thesis*, University of Udaipur, Udipur.

Hegde, G.R. and S.G. Bharathi, 1985. Comparative phytoplankton ecology of fresh water ponds and lakes of Dharward, Karnataka State, India. In: *Proc. Nat. Symp. Pure and Appl. Limnology*. Bull. Bot. Soc. Sagar, 32: 24–29.

Khatri, T.C., 1987. Seasonal distribution of phytoplankton in Indukki reservoir of Kerala (India). *Environ. Ecol.*, 5: 71–73.

Nandan, S.N. and Patel, R.S., 1991. Ecological studies of algae. In: *Aquatic Ecology*, (Eds.) S.R. Mishra and D.N. Saksena. Ashish Publishing House, New Delhi, pp. 69–99.

Nanzeen, S., 1980. Influence of hydrological factors on seasonal abundance of phytoplankton in Kinjhar lake, Pakistan. *Int. Rev. Ges. Hydrobiol.*, 65(2): 269–282.

Vaishya, A.K. and Adoni, A.D., 1992. Phytoplanktonic seasonality and their relationships with physico-chemical properties in a hypereutrophic central Indian lake. *Proc Indian. Sci. Acad.*, 59(B): 153–160.

Venkateshwaralu, V., 1969. An ecological study of the algae of the river Mosi, Hyderabad (India) with special reference to water pollution. IV periodicity of some common species of algae. *Hydrobiol.*, 36: 45–65.

Verma, J. and Mohanty, R.C., 1995. Phytoplankton and its correlation with certain physico-chemical parameters of Danmukundpur pond. *Poll. Res.*, 14(2): 233–242.

Welch, E.B., Staurtevant, P. and Perkins, M.A., 1978. Dominance of phosphorus over nitrogen as the limiter of phytoplankton growth rate. *Hydrobiologia*, 57(3): 209–215.

Zafar, A.R., 1964. The ecology of algae in certain fish ponds of Hyderabad, India: Physico chemical complex, *Hydrobiolgia*, 23: 176–196.

Chapter 13

The Occurrences of the Heavy Metals from Three Reservoirs of Satara District (Maharashtra), India

Sandhya M. Pawar and Sanjay S. Sathe
Department of Biological Sciences, Padmabhushan Dr. Vasantraodada Patil Mahavidyalaya, Tasgaon, District Sangli – 416 312, India

ABSTRACT

The present work aims to study the level of heavy metals in different water reservoirs from Satara district and also to analyse other water quality parameters like pH, total hardness, electrical conductivity, free CO_2, dissolved oxygen, sodium, potassium, nitrates and phosphates. The analysis shows a low level of these metals below the permissible limits as the standards given by different health organizations.

Introduction

Heavy metals are one of the most hazards environmental pollution. Industrial effluents are discharged into sewage canals of cities causing serious pollution and health hazards (Baddesha and Rao 1986). The heavy metals like Pb, Cd, Cu and Zn are mostly absorbed and get accumulated in various parts of plants as free metals, which may adversely affect the plant growth and metabolism. (Barman and Lal, 1994). Some of the heavy metals are get contaminated into aquatic systems as result of natural activities (Weathering of soils and rocks from volcanic eruption) and from a variety of human activities involving the tanneries, mining, electroplating industries and various other industries producing pesticides and plants.

The heavy metals accumulate in the sediments due to adsorption process more over the ability to adsorb heavy metals is enhanced due to presence of organic particles. (Jackson, 1978) The deposition of heavy metals occurs in association with fine particles of silt and clay. They decrease root respiration and nutrient uptake by the plants, which inhibit cell mitosis in root meristmatic regions (Jameel, 2001).

The sub-metals such as Manganese, Iron, Copper and Zinc are essential micronutrients while others such as Mercury; Cadmium and lead are not required even in small amount by any organism. Virtually all metals including the essential metal micronutrients are toxic if exposure levels are sufficiently high. The Cadmium, Chromium, Copper, Lead, Mercury, Nickel and Zinc are particular relevance in fresh water. The increased circulation of toxic metals in recent times resulted in invariable build up of such toxic substances in the human food chain. Heavy metals are rapidly absorb to particulate materials as like Detritus, Plankton, suspended sediments and they are assimilated by living organisms. On a similar survey many workers, (Sharma and Konsal, 1986; Patel *et al.*, 2002) have been reported that industrial waste water contains high amount of elements, like Fe, Cu, Zn, Cd, Pb, Ni, Cr, Co, As and Hg which affect the soil, water and plant system. The concentration of metals dissolved in the water may give a highly misleading picture of the degree of metal pollution. Therefore the survey was undertaken to judge the suitability of different effluents in water for the drinking and irrigation.

Study Area

Kas Reservoir

It is situated 30 km away in the northern west part of Satara city. It is constructed at the origin of Urmodi River.

In 1882 the tank was constructed, at the time it's expenditure was Rs. 261000. The length of tank is 720 feet and depth of tank is 60 feet. The total area 7.03 sq km. From the tank 11 to 12 Lac gallon water is supplied to the Satara city. It is supplied by gravity force to city.

Kanher Reservoir

It is a artificial (man made) reservoir built on Venna River near Kanher. It is situated in the northern west part of Medha Tahsil of Satara Distrct. It is about 15 km away from city.

It's construction is began in 1976 and completed in 1988. It is about 1955 mts in length, 50.30 mts in height and has capacity of storage 286 million gallon of water.

Mahadare Reservoir

It is one of natural reservoir constructed at the foot of Yawateshwar by honorable Chh. Pratapsinh Maharaj to supply the water to the city.

It is situated in Mahadare village, 6 km away from city. The length of this tank is 260 feet, width is 257 feet and depth is 30 feet. It is constructed totally by stone. From this tank daily 5000 gallon water is supplied to the city.

Materials and Methods

The water samples were collected from all these three reservoirs in pre-sterilized plastic bottles according to standard method of APHA (1995) and by Manivasakam (1984). The samples were analyzed for different physico- chemical parameters as per method, described by Trivedy and Goel (1984). The traces of the heavy metals as Fe, Mn, Zn, Cu, Pb, Cd, Cr, Co and Ni were analyzed by Atomic Absorption Spectrophotometers.

Results and Discussion

The results of analysis of physico-chemical parameters are represented in Table 13.1 and characteristic of heavy metals of these reservoirs presented in Table 13.2. In last 25 years metals have received great attention because they provide a coded history of lake's environmental. (Pennington, 1981; Kiggey *et al.;* Jain and Salman, 1995 and Someshwara Rao *et al.*, 1997). Heavy metal is continuously realized into aquatic environment from natural process that is volcanic activity and weathering of rocks. Heavy metals as well as pesticides have a unique property and accumulation over a period of time along a food chain and very high levels can be accumulated in organisms from very low concentration in water and sediments.

Table 13.1: Annual Range of Some Physico-chemical Parameters in Kas, Kanher and Mahadare Reservoirs from March 2004 to April 2005

Sl.No.	*Parameters*	*Sites*		
		Water Reservoirs		
		Kas	*Kanher*	*Mahadare*
1.	pH	6.2–7.1	6.30–7.50	7.10–8.5
2.	Electrical conductivity	0.027–0.044	0.113–0.190	0.208–0.407
3.	Dissolved oxygen	3.15–8.51	3.00–7.49	4.05–8.10
4.	Free CO_2	2.64–12.0	4.4–10.2	5.5–16.4
5.	Total hardness	28–60	46–98	60–195
6.	Sodium	4.0–9.0	7.2–19.0	13.0–26.5
7.	Potassium	< 0.05–0.16	0.5–1.2	0.80–1.83
8.	Nitrates	2.4–8.0	10.0–19.0	15.0–24.0
9.	Phosphates	0.002–0.180	0.0002–0.280	0.013–0.220

All values are expressed in mg/Liter except pH and E.C.

Table 13.2: Annual Range of Heavy Metals in Kas, Kanher and Mahadare Reservoirs March 2004 to April 2005

Sl.No.	*Heavy Metals*		*Sites*		
			Water Reservoirs		
			Kas	*Kanher*	*Mahadare*
1.	Lead	ppm	0.004	BLD	0.003
2.	Chromium	ppm	0.002	0.004	0.008
3.	Cadmium	ppm	0.0128	0.0003	0.0200
4.	Arsenic	ppm	BLD	BLD	BLD
5.	Nickel	ppm	BLD	BLD	BLD
6.	Cyanide	ppm	BLD	BLD	BLD
7.	Iron	ppm	2.12	2.40	2.40
8.	Manganese	ppm	12	1.6	BLD
9.	Zinc	ppm	1.16	0.56	1.16
10.	Copper	ppm	0.028	0.011	0.016

BLD: Below Detectable Level.

The permissible limit presented by FAO (1992) for heavy metal is for Zn-2mg/L, Cu-0.2mg/L, Pb-0.04 to 0.5 mg/L, Cd-0.01 mg/L, Co-0.05mg/L, Cr-0.10 mg/L and for Fe-0.1 0 mg/L.

In the present investigation the heavy metals like Arsenic, Nickel and Cynide are not detected in these reservoirs. In drinking water Cadmium more, than 0.1 mg/L causes Bronchitis, Anemia, Renal stone formation, copper and zinc are found to be essential for proper metabolic functions where as other like Cd and Pb are highly toxic to aquatic organism and consequently to the consumers (Grepe *et al.*, 2000). Hypertension and Arteriosclerosis in animal human beings (Manivasakam, 1987; W.H.O., 1988). While water having 5 mg/L of chromium are toxic to animals (W.H.O., 1988). The Chromium values detected are 0.002 mg/L, 0.004 mg/L and 0.008 mg/L for Kas, Kanher and Mahadare reservoirs, respectively while Cadmium values recorded 0.0128 mg/L, 0.003 mg/L and 0.020 mg/L for these three reservoirs respectively. These values are with in the permissible limit of BIS (1991). Similar observations were recorded by Kaushik *et al.* (1997) for different reservoirs in Gwalior regions. The lead was found to be 0.003 to 0.004 mg/L. All the reported values were under the permissible limits (WHO Standard, 1983). The manganese ranged between 1.60 to 12 mg/L. The values of the Zinc reported between 0.56 to 1.16 mg/L. These values are under the permissible limits (USPH Standard, 1980). The Copper ranged between 0.011 to 0. 028 mg/L. All the values of it were beyond the permissible limits (WHO Standard, 1983). All these observations indicates that the water from all these reservoirs is highly potable and without any organic, pollution of heavy metals.

Acknowledgement

Authors are thankful to U.G.C. New Delhi and W.R.O. Pune for financial assistance to their minor research projects. Similarly thanks due to Principal Dr. R.V. Shejwal for encouragement and providing necessary facilities.

References

APHA, 1995. *American Health Association Standard Methods for Examination Water and Wastewater*, 19th Edition, New York.

BIS, 1991. *Bureau of Indian Standard Drinking Water Specification* (First revision), I.S. 10500.

Bruno, K., 1995. *Wastewater Treatment and Use in Agriculture,* (Ed.) Percal, M.B. Irrigation and Drainage Paper (4), FAO Rome, p. 13–19.

Grepe, M.S., Moreno, V.J. and Patat, M.L., 2000. Cadmium, zinc and copper accumulation in the squid III argentinus from southwest Atlantic ocean. *Mar. Biol.*, 136: 1039–1044.

Kaushik, S., Sabu, B.K., Lawania, R.K. and Tiwari, R.K., 1997. Occurrence of heavy metals in Lentic water of Gwalior Region. *Poll. Res.*, 16(4): 237–239.

Lloyd, R., 1961. *The Toxicity of Mixture of Zinc and Copper Sulphates Rainbow Trout.*

Manivasakam, N., 1984. *Physico-chemical Examination of Water Sewage and Industrial Effluent.* Pragati Prakashan, Meerut, U.P. p. 3–12.

Manivasakam, N., 1987. *Industrial Effluents*, Shakti publication, Coimbatore.

Narayan, T.N.N.S. and Madhyastha, M.N., 1985. *Heavy Metal Pollution in India: A Review Current Pollution Research in India.* Environment Pollution, Karad.

Patel, K.P., Pondya, R.R., Maliwal, G.L., Patel, K.C. and Ramani, V.P., 2002. Suitability of industrial effluences for irrigation around Bharuch and Ankleshwar industrial zone in Gujarat. *Pollution Res.*, Each (Accepted).

Sharma, V.K. and Kansal, B.D., 1986. Heavy metal contamination of soils and plants with sewage irrigation, *Pollution Res.* Each, 4: 86–91.

Trivedi, R.K. and Goel, P.K., 1984. *Chemical and Biological Method for Water Pollution Studies.* Environmental Publish, Karad, India.

WHO, 1988. *Guidelines for Drinking Water Quality,* Volume 2, WHO, Geneva.

Chapter 14

Effect of Manures and Fertilizers on Growth and Yield of Cucumber in Inceptisol

I.R. Bagwan[1], S.D. Rane[1], T.K. Tiwari[2] and S.N. Bhende[1]

[1]Department of Soil Science and Agricultural Chemistry,

[2]Department of Agronomy,

Mahatma Phule Krishi Vidyapeeth, Rahuri, Dist Ahmednagar (M.S.) India

ABSTRACT

A field experiment was carried out during summer 2003–04 to study the effect of manures and fertilizers on growth and yield of cucumber in Inceptisol. Integrated use of manures and fertilizers significantly increased the length of vine, number of internodes per vine and yield contributing characters.

Significantly the highest fruit yield (15.79 t ha^{-1}) was recorded with application of recommended dose of fertilizers (100 : 50 : 50 kg NPK ha^{-1}) + BDM Bio-organic manure at 2.5 t ha^{-1}.

Keywords: Cucumber, Yield, Manures and fertilizers, Bioorganic manure.

Introduction

Cucumber is one of the important vegetable crops in summer season as it replenishes the water loss from human body and helps to avoid dehydration problem when consumed as a raw as it contains about 90.6 per cent moisture. The continuous and indiscriminate use of inorganic fertilizers leads to decrease in nutrient uptake and adversely affect the quality of vegetables (Agrawal, 2003). The growing awareness about organic vegetables free of chemical residues has lead to integrated approach of nutrient management in cucumber. The information on integrated use of bioorganic manures and fertilizers in cucumber is very meagre which necessitates conducting this experiment.

Materials and Methods

The field experiment was conducted in Inceptisol at All India Coordinated Vegetable Improvement Project, MPKV, Rahuri, Dist. Ahmednagar (M.S.). The experiment was laid out in randomised block design with eight treatments and three replications and the details of which are furnished in Table 14.2. The soil was low in available nitrogen and phosphorus (153.66 and 13.50 kg ha^{-1}) and very high in potash (358.70 kg ha^{-1}) with alkaline reaction (pH 8.22). The crop was dibbled in furrows opened at 1.5 m distance in triangular fashion at 100 × 25 cm spacing on 25th February 2003.

Farm yard manure (FYM) and BDM bioorganic manure were used as the organic sources of nutrients and their composition is presented in Table 14.1 whereas commercial mixed fertilizer Vijay (17 : 17 : 17) and urea were used as inorganic sources of nutrients. The recommended dose of fertilizers applied for cucumber was 100 : 50 : 50 kg NPK ha^{-1}, respectively. The organic manures were applied during land preparation whereas for inorganic nutrient management whole quantity of P and K and 50 per cent N were applied at the time of sowing as basal dose and remaining 50 per cent N was top dressed after 30th day of sowing. The remaining field operations were carried out uniformly as per the recommendations.

Table 14.1: Composition of Organic Manures

Composition	*FYM*	*BDM Bio-organic Manure*
pH (1 : 10)	8.33	6.90
EC (1 : 10 $dSmo^{-1}$)	18.23	7.17
Organic carbon (per cent)	23.49	8.70
Nitrogen (per cent)	0.63	0.45
Phosphorus (per cent)	0.52	0.34
Potassium (per cent)	1.24	0.21

Results and Discussion

Growth Characters

Application of 100 per cent recommended dose of N, P and K through inorganic sources alongwith BDM-bioorganic manure @ 2.5 t ha^{-1} recorded significantly more length of vine (189.8 cm) and number of internodes per vine (46.1) than rest of the treatments except for length of vine as recorded with application of 50 per cent recommended dose of N, P and K through inorganic sources alongwith BDM-bioorganic manure @ 5 t ha^{-1} (187.6 cm) which was at par with it. The bioorganic manure might have increased the availability of the nutrients, which as a result would have enhanced the length of vine and number of internodes per vine. This confirms the findings of Nirmala and Vadivel (2000). It was further noticed that the length of vine and number of internodes per vine were significantly the lowest in absolute control where neither manure nor fertilizer application was made.

Table 14.2: Effect of Manures and Fertilizers on Growth and Yield of Cucumber in Inceptisol

Treatments	*Length of Vine (cm)*	*No. of Internodes per Vine*	*Days Required for 50% Flowering*	*Number of Fruits per Vine*	*Average Fruit Weight (g)*	*Yield* ($t h^{-1}$)
Absolute control	90.3	19.33	51.13	4.2	100.3	9.2
NPK 100 per cent as recommended dose	176.3	41.67	46.00	5.0	120.3	13.6
NPK 50 per cent as recommended dose	162.8	32.60	48.47	4.6	100.6	10.3
BDM-Bioorganic manure @ 5 t ha^{-1}	163.7	35.53	48.67	4.8	110.0	11.3
BDM-Bioorganic manure @ 2.5 t ha^{-1}	148.5	27.60	50.27	4.3	100.6	9.4
NPK 100 per cent + BDM-Bioorganic manure @ 2.5 t ha^{-1}	189.8	46.06	45.73	6.0	120.0	15.7
NPK 50 per cent + BDM-Bioorganic manure @ 5 t ha^{-1}	187.6	41.47	47.13	5.5	110.3	14.4
FYM @ 5 t ha^{-1}	163.0	33.73	48.33	4.6	101.0	11.1
SE±	0.72	0.53	0.54	0.05	2.51	0.27
CD (P = 0.05)	2.20	1.61	1.63	0.14	7.62	0.83
General mean	160.3	34.75	48.22	4.9	107.9	11.9

Yield

Significantly the less number of days (45.73) were required for 50 per cent flowering in cucumber with application of 100 per cent recommended dose of N, P and K through inorganic sources alongwith BDM-bioorganic manure @ 2.5 t ha^{-1}. It was further noticed that application of either 50 per cent recommended dose of N, P and K through inorganic sources alongwith BDM-bioorganic manure @ 5 t ha^{-1} or 100 per cent recommended N, P and K through, inorganic fertilizers required statistically the same number of days for 50 per cent flowering. This might be due to more availability of chemically fixed nutrients especially phosphorus and its uptake by cucumber crop which as a result might have enhanced the flowering of cucumber by 5 days as compared to absolute control without fertilizer application. Also, the combined application of bioorganic manure with fertilizers might have added micronutrients, increased the microbial population and physical condition of the soil. Similar observations were recorded by Subbarao and Ravishankar (2000). The days required for 50 per cent flowering were significantly more in absolute control and application of 2.5 t ha^{-1} BDM-Bioorganic manure.

Application of 100 per cent recommended dose of N, P and K through inorganic sources alongwith BDM-bioorganic manure @ 2.5 t ha^{-1} recorded significantly the highest number of fruits per vine (5.5), average fruit weight (120.0 g) and fruit yield (15.7 t ha^{-1}) owing to more availability and uptake of nutrients. The chemically fixed forms of nutrients were made available by bioorganic manures through liberation of organic acids and the organic manure on decomposition also might have added some nutrients. This confirms the findings of Warade *et al.* (1995) and Nirmala and Vadivel (2000). Absolute control without fertilizer or manurial application and application of 2.5 t ha^{-1} BDM-Bioorganic manure recorded significantly less number of fruits per vine, average fruit weight and fruit yield as compared to other nutrient management practices.

Thus, it can be concluded that integrated nutrient management through combined application of 100 : 50 : 50 kg N, P and K ha^{-1} through inorganic fertilizers with BDM-bioorganic manure @ 2.5 t ha^{-1} is superior in enhancing the growth and fruit yield of cucumber in Inceptisol.

References

Agrawal, A.K., 2003. Role of organic enrichers in management of soil salinity. *Agrobios*, 2: 21–23.

Nirmala, R. and Vadivel, E., 2000. Effect of Organic manure and biofertilizers on growth and productivity of cucumber. *South Indian Hort.*, 47: 252–254.

Subbarao, T.S.S. and Ravishankar, C., 2000. Effect of organic manures on growth and yield of brinjal. *South Indian Hort.*, 49: 288–289.

Warade, S.D., Desale, S.B., and Shinde, K.G., 1995. Effect of organic, inorganic and biofertilizers on yield of onion bulb cv. B-780. *J. Maharashtra Agric. Univ.*, 20: 467–468.

Chapter 15

Silver Toxicity to *Anabaena doliolum* Bharadwaja

Praveen K. Chandel*, Dilip K. Rathore and Krishna Chandra Sonie
Department of Botany, University College of Science, Mohanlal Sukhadia University, Udaipur – 313 001, Rajasthan, India

ABSTRACT

Cyanobacteria have tremendous capacity of nitrogen fixation, as well as, capacity of adaptability of high order. Detergents and heavy metals are discharged from various industrial processes, modern agricultural practices, acid mine drainage and human wastes in aquatic habitat. Sustenance of life in such contaminated water bodies requires proper management and restoration by way of nutrient and metal removal. *Anabaena doliolum* can play an efficient role in this kind of activity. It has exhibited the survival potential up to 0.75 ppm of silver concentration. So, its growth therein will help in removal and gleaning of this particular metal ion. Thus, cyanobacteria are valuable tool for bioassay of metal toxicity and the tolerant forms can be used for water self-purification processes in natural water bodies.

***Keywords**: Heavy metals, Gleaning, Water self-purification and Toxicity.*

Introduction

Cyanobacteria are fascinating group of photosynthetic prokaryotes, endowed with a tremendous capacity of nitrogen fixation and its distribution in ecologically diverse habitats proves their capacity

---

* E-mail: cruzshiphar@yahoo.co.in

of adaptability of high order. They constantly face stressful environment due to presence of detergents and heavy metals in aquatic habitat which are discharged from various industrial processes, modem agricultural practices, acid mine drainage and human wastes.

Metal ions may mark their presence in trace or ultra trace amounts and may be acting as micro or macronutrients. Some metals are considered as heavy metals input into aquatic environment by continued anthropogenic activities. They possess a potential threat to the structure and functioning of the ecosystem. Metal can be accumulated by aquatic, organisms, they enter the food chain leading to man and thus become a human health problem (Rachlin *et al.*, 1982). The toxicity of heavy metal to eukaryotic algae has been widely acknowledged (Rai *et al.*, 1981; Stokes, 1983; Whitton, 1984). Cyanobacteria have, however, been little explored in this regard (Stratton and Corke, 1979a,b; Singh and Yadav, 1983, 1984; Rai and Raizada, 1985). Heavy metal toxicity is influenced by presence of various group due to their metal binding properties (Rai *et al.*, 1981; Rai and Raizada, 1985; Raizada and Rai, 1985).

Many works related to cyanobacterial interaction with heavy metal provide us toxicity sequence of heavy metal Ag > Ni > Cr > Pb (Rai *et al.*, 1990). Silver is widely distributed in nature but in micro amounts, silver is also universally present in soils and plants. It is well known as less toxic element for higher organisms but probably the most toxic element for microorganism.

Exposure of *Anabaena doliolum*, a pale blue-green filamentous, heterocystous cyanobacterium with barrel shaped cells at different concentration of silver has resulted in reduced growth and heterocyst production. This also has led to decrease in the content of chlorophyll a, carbohydrates, proteins and amino acids.

Materials and Methods

A. doliolum is selected to find out the silver toxicity and physiological responses in terms of chlorophyll-a, total carbohydrate, protein and amino acids. *A. doliolum* was cultured in nitrogen-free Allen and Amon's (1955) medium.

Growth was measured using Systronics UV-VIS spectrophotometer–117 by taking optical density of pigments at 630, 645 and 665 nm and calculated as follows (Kaushik, 1987).

Total Growth (Chl.a) = $C_a \cdot V_a / V_c$ mg/l

where,

Ca: Amount of Chl.a

V_a: Volume of 80 per cent acetone

V_c: Volume of culture concentrate

$(C_a = 11.6\ OD_{665} - 1.31\ OD_{645} - 0.14\ OD_{630})$

and the growth rate percentage was calculated using formula given below.

Growth rate percentage = $dx/dt \cdot 1/x \cdot 100$

where,

dx: Change in growth (final – initial growth)

dt: Time (days)

x: Initial growth

Specific growth rate was calculated at 663 nm using Systronics UV-VIS Spectrophotometer–117. Protein was estimated by Lowry's method (Lowry *et al.*, 1951).

Carbohydrate was estimated by Acid hydrolysis Anthrone reaction method (Plummer, 1979) after the growth period of 18 days.

Heterocyst frequency is expressed as number of heterocysts per hundred vegetative cells. Observations were taken from at least 10–15 fields and an average of 10 filaments were taken as per requirement of the experiment.

Results and Discussion

Various physiological and biochemical parameters were analyzed for the interaction of *A. doliolum* with silver.

In growth measurements, specific growth rate at 0.05 ppm has shown stimulation over control and later a gradual decrease was observed with increasing concentration (Table 15.1). Similar trend was also observed for growth rate percentage (Figure 15.1).

Table 15.1: Effect of Different Concentrations of Silver on Specific Growth Rate

Concentration of Silver (ppm)	*Specific Growth Rate (ΔA_{663})*	
	Mean±SD	*Inhibition (%)*
0.00	0.541±0.002	–
0.05	0.584±0.002	–7.95
0.10	0.443±0.005	18.11
0.15	0.428±0.003	20.89
0.20	0.427±0.003	21.07
0.25	0.411±0.003	24.03
0.30	0.331±0.001	38.82
0.35	0.328±0.002	39.37
0.40	0.298±0.004	44.92
0.45	0.295±0.005	45.47
0.50	0.250±0.005	53.79
0.55	0.243±0.005	55.08
0.60	0.221±0.002	59.15
0.65	0.171±0.003	68.39
0.70	0.166±0.004	69.32
0.75	0.144±0.006	73.38

Data with negative sign show stimulation over control.

At various concentrations of silver, initially total carbohydrate content increased up to 0.10 ppm (3.4089 g/l) and then a regular decrease was seen. The maximum value for total protein content was observed in control (4.4618 g/l) and thereafter, it decreased gradually. Amino acid content of the test organism showed the peak value at 0.05 ppm (1.0194 g/l) and then a declined towards the toxic level (Figure 15.2). pH of supernatant obtained after centrifugation of cultured teat organism showed slight variation in relation to different concentrations of silver (Figure 15.3).

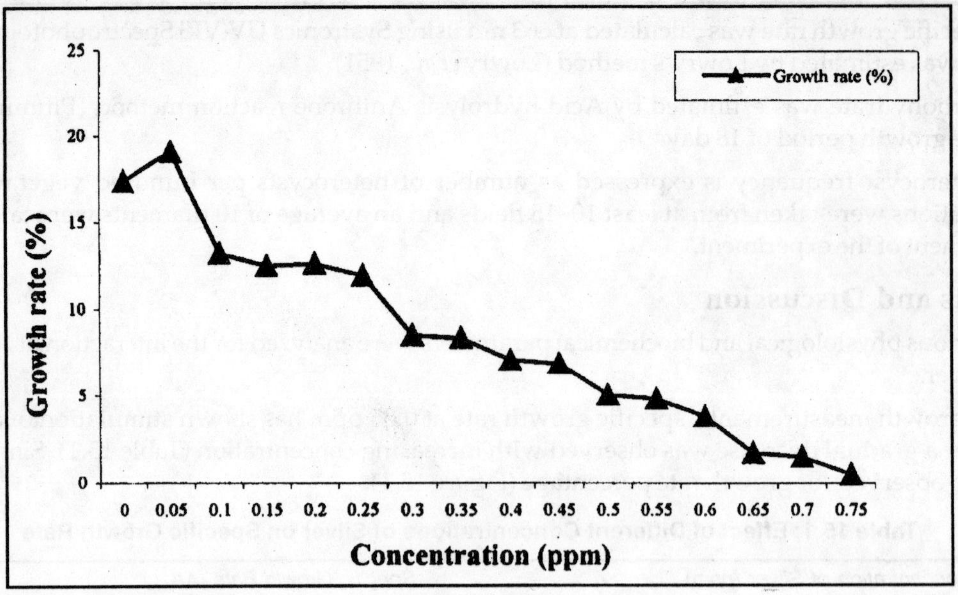

Figure 15.1: Growth Rate Percentage of *Anabaena doliolum* at Various Silver Concentration

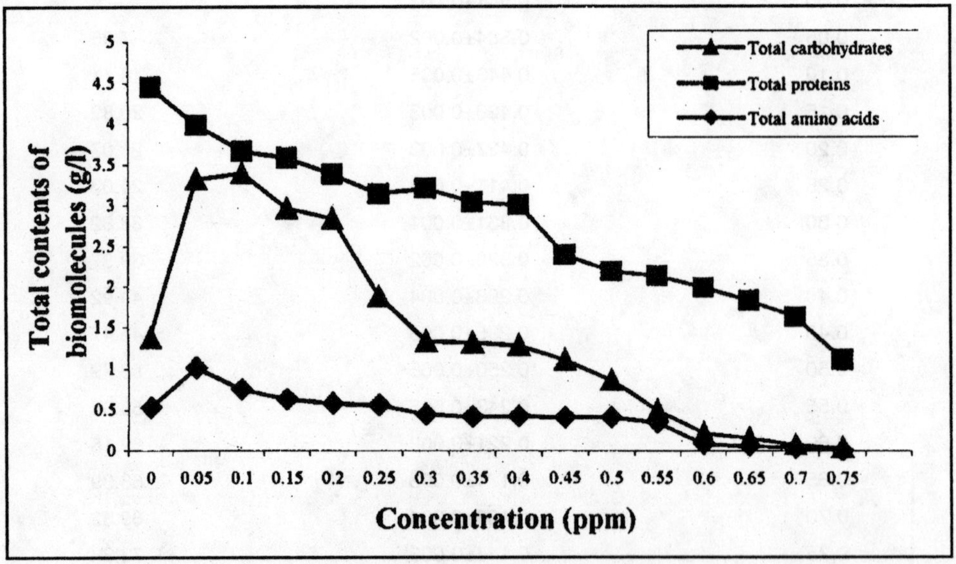

Figure 15.2: Total Content of Carbohydrates, Proteins and Amino Acids of *Anabaena doliolum* at Different Concentrations Silver

Heterocyst frequency was maximum at control (0.00 ppm) and decreased regularly up to 0.65 ppm. At 0.70 ppm and 0.75 ppm, no significant frequency (*i.e.*, nearly zero) was observed which might be due to silver toxicity. In contrast, heterocyst dedifferentiation showed an increase up to 0.75 ppm (Figure 15.4).

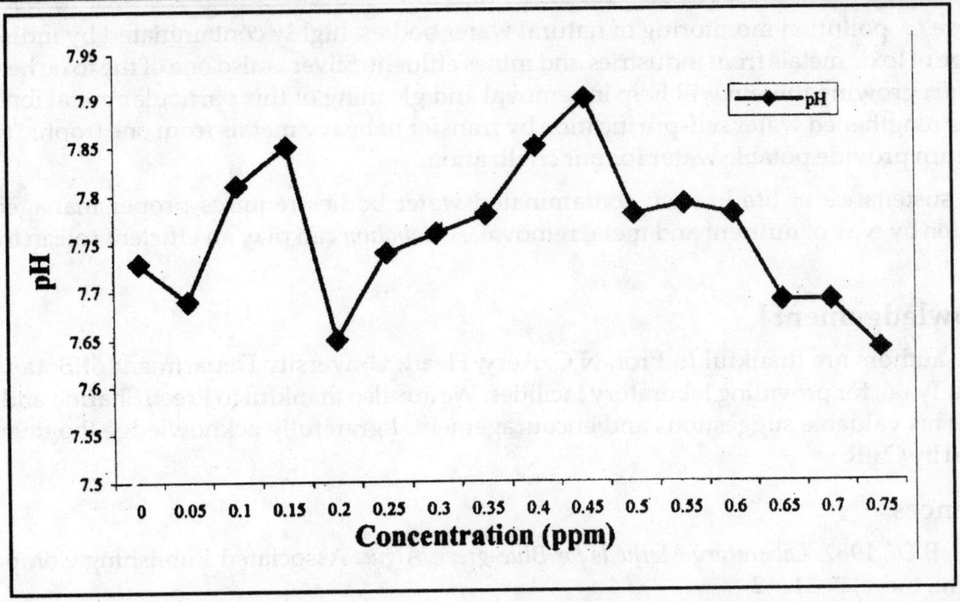

Figure 15.3: pH of the Medium after Incubation Period at Different Concentrations of Silver Ion

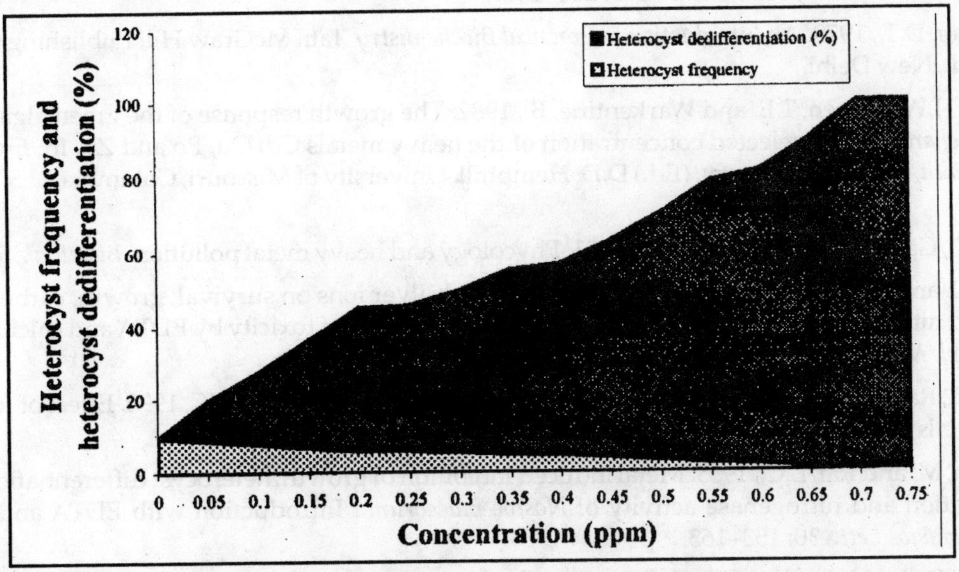

Figure 15.4: Heterocyst Frequency and Heterocyst Dedifferentiation of *Anabaena doliolum* at Different Concentrations of Silver

Conclusion

A. doliolum exhibited the survival potential up to 0.75 ppm of silver concentration. So, the present study and finding will help towards the major environmental health problems confronting both developed and developing countries related to municipal water supply and proper facilities for disposal of sewage *i.e.* pollution monitoring of natural water bodies, highly contaminated by indiscriminate discharge of toxic metals from industries and mines effluent. Silver is also one of the toxic heavy metal. *A. doliolum* growing therein will help in removal and gleaning of this particular metal ion. This will further strengthened water self-purification by transfer of heavy metals from one trophic to another, thus in turn provide potable water for our civilization.

So, sustenance of life in such contaminated water bodies requires proper management and restoration by way of nutrient and metal removal. *A. doliolum* can play an efficient role in this kind of activity.

Acknowledgement

The authors are thankful to Prof. N.C. Aery, Head, University Department of Botany, and Dr. Sandhya Tyagi, for providing laboratory facilities. We are also thankful to Preeti Sharma and Premlata Vikal for his valuable suggestions and encouragement. I gratefully acknowledge the financial help provided by CSIR.

References

Kaushik, B.D., 1987. *Laboratory Methods for Blue-green Algae*. Associated Publishing Company, New Delhi, India, p. 31–62.

Lowry, O.H., Rosebrough, N.J., Farr, A.L. and Randall, R.J., 1951. Protein measurement with the folin-phenol reagent. *J Biol. Chem.*, 193: 265–275.

Plummer, D.T., 1979. *An Introduction to Practical Biochemistry.* Tata McGraw Hill Publishing Company Ltd., New Delhi.

Rachlin, J.W., Jensen, T.E. and Warkentine, B., 1982. The growth response of the green alga (*Chlorella saccharophila*) to selected concentration of the heavy metals Cd, Cu, Pb and Zn. In: *Environment Health, XVI: A Symposium,* (Ed.) D.D. Hemphill. University of Missouri, Columbia, U.S.A, pp. 45–54.

Rai, L.C., Gaur, J.P. and Kumar, H.D., 1981. Phycology and heavy metal pollution. *Biol. Rev.*, 56: 99–151.

Rai, L.C. and Raizada, M., 1985. Effect of nickel and silver ions on survival, growth, carbon fixation and nitrogenase activity of *Nostoc muscorum:* Regulation of toxicity by EDTA and calcium. *J. Gen. Appl. Microbiol.*, 31: 329–337.

Rai, L.C., Raizada, M., Mallick, N., Husaini, Y., Singh, A.K. and Dubey, S.K., 1990. Effect of four heavy metals on the biology of *Nostoc muscorum. Bio. Metals,* 2: 229–234.

Raizada, M. and Rai, L.C., 1985. Metal induced inhibition of growth, heterocyst differentiation, carbon fixation and nitrogenase activity of *Nostoc muscorum* : Introduction with EDTA and calcium. *Microbios Lett.*, 30: 153–163.

Singh, S.P. and Yadav,V., 1983. Cadmium induced inhibition of nitrate uptake in *Anacystis nidulans*: Interaction with copper and other divalent cation. *J. Gen. Appl. Microbiol.*, 29: 297–304.

Singh, S.P. and Yadav,V., 1984. Cadmium induced inhibition of ammonium and phosphate uptake in *Anacystis nidulans*: Interaction with other divalent cation. *J. Gen. Appl. Microbiol.*, 30: 79–86.

Stokes, P.M., 1983. Response of freshwater algae to metals. In: *Progress in Phycological Research,* (Eds.) F.E. Round and V.J. Chapman. Elsevier, Amsterdam, 2: 87–111.

Stratton, G.W. and Corke, C.T., 1979a. Effect of cadmium ion on the growth, photosynthesis and nitrogenase activity of *Anabaena inaequalis. Chemosphere*, 5: 277–282.

Stratton, G.W. and Corke, C.T., 1979b. Effect of mercuric ion on the growth, photosynthesis and nitrogenase activity of *Anabaena inaequalis. Environ. Microbiol.*, 38: 537–543.

Whitton, B.A., 1984. Algae as monitors of heavy metals in freshwaters. In: *Algae as Ecological Indicators*, (Ed.) L.E. Shubert. Academic Press, London, pp. 257–280.

Chapter 16

Limnobiotic Study of Kanhan River at Mouda, Near Nagpur (Maharashtra)

R.R. Khapekar*, P.R. Chaudhari and S.R. Wate
National Environmental Engineering Research Institute (NEERI), Nehru Marg, Nagpur – 440 020 (India)

ABSTRACT

Limnobiotic investigation was conducted for qualitative assessment of Kanhan River at Mouda Town near Nagpur city of Maharashtra state during March 2004 to February 2005. The study was concentrated on three sampling sites of Kanhan River namely Site 1: Kesalpur, Site 2: Kiranapur and Site 3: Sawangi (confluence point of polluted Nag River and Kanhan River). The present investigation aimed to study the physico-chemical and biological features of 3 sites of Kanhan River. For the investigation of water quality, total seventeen physico-chemical parameters were analyzed. It was found that there is gradual increase in pollution level in river water quality from Site-1 to Site-3 as evident from increase in levels of C.O.D., T.D.S., chlorides and sulphate and decrease in D.O. concentration. The highest degrees of pollution were found at Site-3. The results revealed that water quality at Site-3 was more polluted due to addition of sewage and various industrial effluents of Nagpur city into Kanhan River through polluted Nag River.

Biological studies revealed an increase in species diversity and population density of members belonging to Chlorophyceae, Bacillariophyceae (phytoplanktons) and insects (macro invertebrates) from Site-1 to Site-3. During the investigation period it is observed that few clear water species recorded at Site-1 were disappeared and certain pollution tolerant species of phytoplanktons and zooplanktons were appeared at Site-3. The present investigation indicate progressive pollution levels in the river present near urban environment due to unscrupulous discharge of wastewater.

Keywords: *Kanhan River, Water pollution, Physico-chemical parameters, Biological parameters, Urban area.*

* E-mail: rkhapekar@rediffmail.com dr.khapekar@gmail.com

Introduction

Among all the other natural resources, water is the most precious one, and life is not possible without water. Although the global freshwater in our planet is about 1500 million cubic kms (Penman, 1970) but most of it is useless for us as it contains too much of salt. Total amount of estimated available freshwater on our planet is only 84.4 million cubic kms. In recent years, rapid population growth increasing living standard, wide spheres of human activities and industrialization have resulted in greater demand of good quality water, while on other hand pollution of water resources is also increasing steadily (Murugesan and Sukumaran, 1999).

India is a country having various landforms and rivers. There are fourteen major rivers in India, which share about 83 per cent of the total drainage basin and contribute 85 per cent of total surface flow Nilay (Chaudhary, 1983). The total annual flow in all the rivers is estimated to be 18,18,100 million m^3 according to the Central Water and Power Commission (Kayastha, 1989). Enormous anthropogenic activities make all this water resources unfit for any purpose. Eutrophication of water bodies is rapidly increasing due to rapid increase in quality and quantity of sewage discharge, which in turn enhance algal proliferation. A vast literature exists on the limnological studies of algae. Some of the important contributors are Phillipose (1960), Zafar (1964), Venkateswarlu (1969), Munawar (1970), Kant and Kachroo (1971), Kumar (1977), Khan and Zutschi (1980).

The present investigation was carried out by exploring the various sites of Kanhan River of Mouda Town near Nagpur City in Maharashtra. The present investigation includes the study of physico-chemical and biological characteristics of different sites of Kanhan River.

Materials and Methods

Three sampling sites, namely S-1 (Keslapur); S-2 (Kiranapur) and S-3 (Sawangi) were selected for present study. The area near Kanhan River is predominantly agricultural interspersed by various kinds of grasses, herbs, shrubs and trees. The 3 sites of Kanhan River were selected with purpose to collect water samples for physico-chemical analysis and biological parameters at seasonal intervals during March 2004 to February 2005.

Analysis of physico-chemical and biological parameters were made after following the work of Edmondson (1959), Needham and Needham (1962), APHA (1992). For physico-chemical analysis water samples were collected in 1 lit polyethylene bottle during morning hours. For biological analysis, samples were collected in 100 ml polyethylene bottles. Fifty liter of water was passed through zooplankton net to collect the concentrated water sample for zooplankton analysis. Water samples directly collected for phytoplankton was preserved in Lugol's Iodine solution and zooplankton samples were preserved in 5 per cent formaldehyde solution at the sampling sites. Samples were brought to the laboratory for microscopic examination. Identification of organisms was done by following the work of Ward and Whipple (1959), Mellanby (1963), Tonapi (1980), Gaur *et al.* (2000).

Results and Discussion

Physico-chemical River Water Quality

Seasonal physico-chemical river water quality at Site-1, 2 and 3 are given in Tables 16.1–16.3 respectively. The river water temperature at Site-3 remained comparatively higher (31.50°C–34.0°C) than Site-1 (30.50°C–33.50°C) and Site-2 (30.5°C–33.0°C) through out the period of investigation. Site-3 is a confluence point of Nag and Kanhan River and Nag River contains effluents from various areas of Nagpur City including domestic and industrial effluents. Klien (1956), Jolly and Chapman (1966),

Venkateswarlu and Jayanti (1968) and Munawar (1970 a) also made similar observations and explained heat is generated when chemicals react with each other and discharged into water and thus bringing about an increase in water temperature.

Table 16.1: Seasonal Physico-Chemical River Water Quality at Site-1

Sl.No.	*Parameters*	*Seasons*		
		Summer	*Monsoon*	*Winter*
1.	Temperature	33.00	32	30.5
2.	pH	7.22	7.20	7.50
3.	TS	290	410	350
4.	TDS	165	320	255
5.	TSS	125	90	95
6.	DO	3.1	3.8	3.4
7.	Free CO_2	–	–	–
8.	Total alkalinity	134	131	152
9.	Carbonate	7	8	10
10.	Bicarbonate	45	31	30
11.	Total hardness	32	42	37
12.	Calcium hardness	3.727	4.008	2.926
13.	Magnesium hardness	33.116	37.992	29.236
14.	Chloride	25.72	22.92	19.72
15.	Sulphate	70.10	81.81	80.70
16.	Phosphate	0.132	0.147	0.283
17.	COD	138.36	153.53	115.312

All values in mg/l except temperature and pH.

The recorded pH values were comparatively lower at Site-1 (7.20–7.50) and Site-2 (7.00–7.20) and higher at Site-3 (7.00–8.50). Chandra *et al.* (1981) also reported an increase in pH values with addition of effluents from various industries. Similarly, Site-2 showed higher values of total alkalinity (143 mg/l–167 mg/l) than at Site-1 (131 mg/l–152 mg/l) and Site-2 (115 mg/l–127 mg/l). Chandra *et al.* (1981) and Badrinath *et al.* (1987) also recorded higher total alkalinity where a different factory effluent was discharged in water.

Total solid content at different river sites were estimated in the range of 290–350 mg/l, 205–255 mg/l and 430–560 mg/l respectively at Site-1, Site-2 and Site-3. During present investigation it was observed that the highest values of TS, TDS and TSS were recorded at Site-3. It might be due to higher percentage of suspended as well as other impurities present at this particular site. It imparts slight turbidity to the water. The similar observations were also recorded by Bhosle *et al.* (1994).

DO contents at Site-1 remained high (3.1 mg/l–3.8 mg/l) as compared to Site-2 (3.0 mg/l-3.2 mg/l). Whereas, at Site-3, the DO values 3 (1.5 mg/l–2.0 mg/l) were comparatively low during the period of investigation. Addition of sewage and industrial effluent might be responsible for low value of DO. A similar impact of effluents from fertilizer factory on the DO contents in water has been observed by Woodward (1984) and Mathuthu *et al.* (1993).

Table 16.2: Seasonal Physico-Chemical River Water Quality at Site-2

Sl.No.	*Parameters*	*Seasons*		
		Summer	*Monsoon*	*Winter*
1.	Temperature	33.0	32	30.5
2.	pH	7.20	7.10	7.00
3.	TS	205	320	255
4.	TDS	95	245	170
5.	TSS	80	70	68
6.	DO	3.0	3.2	3.1
7.	Free CO_2	–	–	–
8.	Total alkalinity	127	115	121
9.	Carbonate	19	24	27
10.	Bicarbonate	109	91	90
11.	Total hardness	36	44	40
12.	Calcium hardness	6.012	6.412	5.112
13.	Magnesium hardness	35.432	37.588	31.182
14.	Chloride	35.62	33.99	31.19
15.	Sulphate	78.62	87.22	86.87
16.	Phosphate	0.148	0.188	0.211
17.	COD	157.36	163.42	161.53

All values in mg/l except temperature and pH.

Table 16.3: Seasonal Physico-Chemical River Water Quality at Site-3

Sl.No.	*Parameters*	*Seasons*		
		Summer	*Monsoon*	*Winter*
1.	Temperature	34.0	32	31.5
2.	pH	8.50	7.00	7.20
3.	TS	430	560	490
4.	TDS	290	450	360
5.	TSS	140	110	130
6.	DO	1.8	2.0	1.5
7.	Free CO_2	–	–	–
8.	Total alkalinity	167	143	156
9.	Carbonate	26	28	29
10.	Bicarbonate	112	115	110
11.	Total hardness	49	64	58
12.	Calcium hardness	8.126	9.619	6.828
13.	Magnesium hardness	49.861	54.381	42.738
14.	Chloride	60.10	56.98	51.10
15.	Sulphate	99.13	90.48	83.48
16.	Phosphate	2.160	3.166	3.837
17.	COD	182.10	194.16	142.61

All values in mg/l except temperature and pH.

During the whole investigation period, CO_2, was absent at all the three sites. The absence of free CO_2 might be due to cause of pollution and photosynthetic activities of algae (Sreenivasan, 1971). Similarly nil free CO_2 values in most of the month were also recorded by Nandkar, (1979) and Shankhadarwar (2002), which resembles with the present study.

During the entire study period, the bicarbonate values were higher than carbonate values. It was also found that the Site-3 has always higher values of carbonate (19 mg/l–27 mg/l) and bicarbonate (90 mg/l–109 mg/l). The higher values of carbonate and bicarbonate might be due to high pollutional load at Site-3. The present observation is coinciding with the results of Shankhadarwar, (2002) and Khapekar, (2005).

Total hardness, calcium hardness, magnesium hardness, chloride and sulphate were found to be increased at Site-3 because Site-3 is a confluence point of Nag and Kanhan. River and it contains effluents from various industries indicating presence of salts of these ions. Similar observations were made by Wetzel (1975), Chandra *et al.* (1991) and Chapman and Kimstach (1992).

Concentration of chloride were found to vary within 19.72 mg/l to 25.72 mg/l, 31.19 mg/l–35.62 mg/l and 51.10–60.10 respectively at Site-1, Site-2 and at Site-3. During the whole study period, concentrations of chloride were high at Site-3. The higher concentration of chloride at Site-3 is because of addition of sewage and industrial effluents. Chloride increases degree of eutrophication (Sinha, 1986). Klein (1957) found a direct correlation between chloride concentration and pollution load. Similar observations were also recorded by Chatterjee *et al.* (2002).

The values of sulphate ranged within 70.10 mg/l to 81.81 mg/l, 78.62 mg/l to 87.22 mg/l and 83.48 mg/l to 99.13 mg/l respectively at Site-1, Site-2 and Site-3. Higher values were always recorded at Site-3. Higher values can be attributed to addition of sewage and industrial effluents into the river. The present observation finds support with the work of Kannan (2000) and Chatterjee *et al.* (2002).

Phosphate content of river water were observed comparatively higher at Site-3 (2.160 mg/l–3.8373 mg/l) than Site-1 (0.132–0.285 mg/l) and Site-2 (0.148–0.211 mg/l). Similar observations were recorded by De Smet and Evans (1972), Michael (1984), Das (1972), Dogra and Kaw (1989), Ouzounis *et al.* (1989), Gopal (1990), Chapman and Kimstach (1992), Meybeck *et al.* (1992) and Mathuthu *et al.* (1993).

The COD concentrations of Site-1 (115.31 mg/l–153.55 mg/l) and Site-2 (157.30 mg/l–163.42 mg/l) were comparatively lower as compared to the Site-3 (142.61 mg/l–194.16 mg/l). The COD content of water increased near pollution input point. Hence, during the whole investigation period, highest values of physico-chemical parameters were found at Site-3. The results of present investigation were coincides with the work of Vijay Kumar and Sagwal (2000), Asif Khan *et al.* (2002) and Shankadarwar (2002).

Biological River Water Quality

The observations on distribution of phytoplanktons and zooplanktons population at three sampling sites of Kanhan River are given in Tables 16.4 and 16.5. In all, 20 phytoplankton species belonging to Chlorophyceae, Cyanophyceae and Bacillariophyceae were observed in Kanhan River. In terms of population density, Bacillariophyceae was found to be the most dominant group at all 3 sampling sites. Species diversity and population density of phytoplankton were found to be higher at Site-3 where effluents from Nag river is poured in Kanhan river. The effluents of Nag River contains various industrial sewage and domestic wastes which enrich water with a wide variety of nutrients required for growth of various phytoplankton.

Table 16.4: Distribution of Phytoplanktons at Three Sites

Sl.No.	Species/Genera	Sampling Sites		
		Site-1	Site-2	Site-3
I	**Chlorophyceae**			
1.	*Spirogyra* sp.	+	+	+
2.	*Pithophora* sp.	+	+	+
3.	*Stigeochlonium* sp.	+	+	+
4.	*Pediastrum* sp.	+	+	+
5.	*Hydrodictyon* sp.	+	+	+
6.	*Closterium* sp.	–	–	+
7.	*Scenedesmus* sp.	–	–	+
8.	*Cosmarium* sp.	+	–	+
9.	*Oedogonium* sp.	–	–	+
10.	*Mougeotia* sp.	+	+	–
11.	*Vaucheria* sp.	+	+	–
II	**Cyaophyceae**			
1.	*Oscillatoria* sp.	–	+	+
2.	*Lyngbya* sp.	+	+	+
3.	*Phormidium* sp.	–	–	+
III	**Baciliariophyceae**			
1.	*Melosira* sp.	+	+	+
2.	*Synedra* sp.	+	+	+
3.	*Navicula* sp.	+	+	+
4.	*Pinnularia* sp.	+	+	+
5.	*Fragilaria* sp.	+	+	+
6.	*Gomphonema* sp.	–	–	+

Table 16.5: Distribution of Zooplanktons at Three Sites

Sl.No.	Species/Genera	Sampling Sites		
		Site-1	Site-2	Site-3
I	**Protozoa**			
1.	*Euglena* sp.	+	+	+
2.	*Difflugia* sp.	+	+	+
3.	*Ceratium* sp.	+	+	+
4.	*Chilomonas* sp.	+	+	–
5.	*Euglypha* sp.	+	+	+
6.	*Paramecium* sp.	+	+	+

Contd...

Table 16.5–Contd...

Sl.No.	Species/Genera	Sampling Sites		
		Site-1	Site-2	Site-3
II	**Rotifera**			
1.	*Monostyla* sp.	+	+	–
2.	*Philodina* sp.	+	+	–
3.	*Trichotoria* sp.	+	+	+
4.	*Hexartha* sp.	+	+	–
5.	*Notholca* sp.	+	+	+
6.	*Neptunis* sp.	+	+	–
7.	*Gastropus* sp.	–	+	+
8.	*Asplanchna* sp.	+	+	–
III	**Crustacea**			
(*a*)	**Copepoda**			
1.	*Cyclops* sp.	+	+	+
2.	*Mesocyclops* sp.	+	+	+
3.	*Diaptomus* sp.	+	+	+
(*b*)	**Cladocera**			
1.	*Daphnia* sp.	+	+	+
2.	*Chydorus* sp.	+	+	+
3.	*Bythotrephes* sp.	–	–	+
4.	*Moina* sp.	+	+	–
5.	*Simocephalus* sp.	+	+	+
6.	*Eurycercus* sp.	–	–	+
IV	**Insecta**			
1.	*Chironomus* sp.	+	+	+
2.	*Culex* sp.	+	+	+
3.	*Hydrophilis* sp.	+	+	+
4.	*Eristalis* sp.	–	–	+
5.	*Baetis* sp.	–	–	+
6.	*Corixa* sp.	+	+	+
7.	*Tipula* sp.	–	–	+
8.	*Tabanus* sp.	–	–	+
9.	*Perla* sp.	–	–	+
10.	*Tanypus* sp.	–	–	+
11.	*Ephemera* sp.	+	+	–

In the present investigation among Chlorophyceans, *Mougoetia* sp. and *Vaucheria* sp. were found only in clean water, similar observations were also recorded by Sladeck (1961), David and Ray (1966), Bulusu *et al.* (1967) and Gunale and Balkrishnan (1981), *Closterium* sp., *Scenedesmus* sp. and *Oedogonium* sp. were observed only at Site-3 where water is contaminated with different types of wastes. Whereas, *Spirogyra* sp., *Pithophora* sp., *Stigeoclonium* and *Pediasterum* sp. were observed at all the three sites. All these species have been reported in a wide variety of polluted waters by various authors (David and Ray, 1966; Jolly and Chapman 1966; Venkateswarlu and Jayanti 1968; Singh *et al.*, 1969, Munawar 1970b; Verma *et al.*, 1978; Govindan and Sudaresan 1979; Zutschi *et al.*, 1980 and Gunale and Balkrishnan 1981).

Among Cyanophyceans *Oscillatoria* sp. was observed at Site-2 and Site-3, but profuse growth of this species was found at site-3 indicating the higher degree of pollution. The same species have been observed in clean as well as in polluted waters by Kolkwitz and Marsson (1908), Sladeck (1961), David and Ray (1966), Jolly and Chapman (1966), Govindan and Sundareean (1979), Zutshi *et al.* (1980) and Paramasivam and Sreenivasan (1981). *Lyngbya* was found at all the three sites. *Lyngbya* was found in clean water by Verma and Shukla (1970) and Paramsivam and Sreenivasan (1981). In the present investigation, *Lyngbya* sp. was also reported in polluted water. This might be due of mixing of clean water with polluted water. *Phormidium* sp. was observed only at Site-3 and was conspicuous by its absence at other sites. Thus this species as well as *Oscillatoria* can be served as the indicator of polluted water.

Among Bacillariophyceans, *Melosira* sp., *Synedra* sp., *Navicula* sp. *Fragilaria* sp. and *Pinnularia* sp. were reported at all the three sites where as *Gomphonema* sp. were reported only at Site-3. Among all observed species of Bacillariophyceae, *Gomphonema* may be labeled as a pollution indicator species because it was observed only at Site-3.

In all, 34 zooplanktons species belonging to Protozoa, Rotifera, Crustacea, Cladocera and Insecta were recorded from Kanhan River. Amongst all the reported protozoans *i.e. Euglena* sp., *Difflugia* sp. *Ceratium* sp., *Chilomonas* sp., *Euglypha* sp., and *Paramecium* sp. only *Chilomonas* sp. was not reported at Site-3. Species diversity and population density of zooplanktons were more or less similar at all 3 sites. It was found that growth of aquatic vegetation and amount of nutrients were favorable for growth of photosynthetic protozoans. These findings were in accordance with some other workers *viz.* Zafar (1959a). Phillipose (1960), Venkateswarlu (1969) and Munawar (1970b). Amongst the rotifers, the presence of *Trichotoria* sp. and *Notholca* sp. was common at all the three sites. *Monostyla* sp., *Philodina* sp., *Hexartha* sp., *Asplanchna* sp. and *Neptunis* sp. were observed in clean water at Site-1 and Site-2. Among this only *Neptunis* sp. were earlier reported to be tolerant species (Algarsamy *et al.*, 1967, Bilgrami and Datta Munshi, 1979; Vasisht and Sra, 1979; Sladecek, 1983; Goel and Chavan, 1991). *Trichotoria* sp., *Notholca* sp. and *Gastropus* sp. were found at Site-3. These species were reported to be pollution tollerant species. (Arora, 1966b; Krishnamoorthi and Visvesvara, 1966; De Smet and Evans, 1972; Sandhu *et al.*, 1948).

In the present investigation, in all 9 species of crustaceans were found at Site-1, Site-2 and Site-3. Out of these, 6 species were common to all the three sites. *Cyclops* sp., *Mesocyclops* sp., and *Diaptomus* sp. were the representative of Copepoda. Among cladocerans, *Daphnia* sp., *Chydorus* sp. and *Simocephalus* sp. were present at all the sites, *Moina* sp. were found only at Site-1 which has no direct source of pollution and *Bythotrephes* sp. and *Eurycercus* sp. were observed only at Site-3. Krishnamoorthi and Visvesvara (1966), Sitaramaih (1966), Liberaman (1970), Verma *et al.* (1978, 1984) and Batish and Kumari (1986) also reported *Moina* to be a clean water species. Rest of the crustaceans are reported to be tolerant species (Dextar, 1959; Mallanby, 1963; Jolly and Chapman, 1966; Das, 1989). Hence, in this investigation Site-3 is characterized by disappearance of a clean water species and appearance of two pollution tolerant species.

At Site-1 and Site-2 only 5 insects species were found whereas at Site-3, 10 species were found. The insects discovered at Site-1 and Site-2 were *Chironomus* sp. (larvae), *Culex* sp. (larvae), *Hydrophilis* sp., *Corixa* sp. and *Ephemera* sp. In present study, *Ephemera* sp. was found only in clean water at Site-1 and Site-2. Similar type of observation was also reported by Verma and Shukla (1969), Gaufin (1974), Badala and Singh (1981) and Das (1989). Site-3 is characterized by the presence of *Eristalis* sp., *Baetis* sp., *Tabanus* sp., *Perula* sp. and *Tanypus* sp. All these species of insects are pollution tolerant species as also reported by Daviet and Ray (1966), Gaufin (1974) and Das (1989).

Conclusion

During the investigation period, it was observed that the water quality at confluence point of Nag and Kanhan river *i.e.* Site-3 (Sawangi) was grossly polluted because of the addition of all domestic wastewater and different industrial effluents from Nagpur City into the Kanhan river. Physico-chemical investigation showed that Site-3 is highly polluted as compared to Site-1 and Site-2. Biological evaluation of water quality also supports this observation. The pollution tolerants groups of phytoplanktons (Cyanophyceae) and zooplanktons (Insecta) were more dominant at Site-3. The clean water indicator species *viz. Mougetia, Vaucheria* (green algae), *Lyngbya* (blue-green algae), *Chilomonas* (protozoa), *Monosyla, Phyilodina, Hexartha, Alsplanchna* (rotifera), *Moina* (crustacea), *Culex, Corixa, Ephemera, Hydrophilas* (insecta) were reported from Site-1 and Site-2 which are relatively unpolluted region of Kanhan River. While pollution indicators species *viz. Closterium, Scenedesmus, Oedogonium* (green algae), *Phormidium* (blue-green algae), *Gomphonema* (diatom), *Euglena, Parameacium* (protozoa), *Trichotoria, Notholca, Gastropus* (rotifera), *Bythtrephes, Eurycercus* (crustaca), *Eristalis, Baetis, Tabanus, Perula, Tanypus* (insecta), were recorded from Site-3 of Kanhan River which is highly polluted due to confluence of polluted Nag River.

Acknowledgement

The authors wish to thank Director, NEERI, Nagpur for providing facilities to carryout this work.

References

Algarsamy, S.R. Abdulappa, M.K. and Bopardikar, M.V., 1967. *High Rate Deep Stabilization Pond, Part II: Studies on the Faunal Assemblage of the System.*

Arora, H.C., 1966 b. Rotifers as indicators of trophic nature of environments. *Hydrobiologia*, 27(1–2): 146–159.

APHA, 1992. *Standard Methods for the Examination of Water and Wastewater*, 18th Ed. American Public Health Association, AWWA, WPCF, Washington, D.C.

Asif Khan, Untoo, S.A. and Parveen, S., 2002. Limnology of a reservoir receiving effluents from a thermal power plant. In: *Ecology of Polluted Water*, (Ed.) Arvind Kumar. A.P.H. Publishing Corp., New Delhi.

Badrinath, S.K., Kaul, S.N. and Gadkari, S.K., 1987. Fertilizer industries, an environmental appraisal. *T.J.E.P.*, 7(3): 201–208.

Batish, S.K. and Kumari, P., 1986. Effect of physico-chemical factors on the seasonal abundance of cladocera in typical pond at village of Raaba, Ludhiana. *Indian. J. Ecol.*, 13(1): 146–151.

Bhosle, A.B. Agarkar, S.V. and Ambore, N.E., 1994. A brief pollution survey of godavari river water at Nanded. In: *Nat. Sympo. on Eco-Environmental Impact and Organism Response.*

Bilgrami, K.S. and Datta Munshi, J.S., 1979. Limnological survey and impact of human activities on river Ganges (Barauni to Farakka range). *Tech. Rep. Man and Bios. Prog. Project–5,* UNESCO, Bhagalpur University, Bhagalpur.

Bulusu, K.R., Arora, H.L. and Aboo, K.M., 1967. Certain observations on self-purification of Khan river and its effect on Kshipra river. *Indian. Environ. Hlth.,* 9: 275–295.

Chandra, K., Singh, B., Shrivastava, G.N. and Meharotra, S.N., 1981. Pollution from wastes of industries manufacturing nitrogenous fertilizer: A case study from river Ganga near Allahabad. *Int. Symp. Water Res. Cons. Poll. and Abatement,* (Ed.) Chituranashi, V.R., Roorkee.

Chapman, D. and Kimstach, V., 1992. The selection of water quality variables. In: *Water Quality Assessment,* (Ed.) D. Chapman. Chapman and Hall, London, New York, Tokyo.

Chatterjee, Chinmoy and Raziuddin, M., 2002. Determination of water quality index (WQI) of a degraded river in Asansol industrial area (W.B.). *Ecology of Polluted Water,* Vol. II, (Ed.) Arvind Kumar. APH Publishing Corporation, New Delhi.

Chaudhari, N., 1982. *Water and Air Quality Control: The Indian Context.* Central Board for the Prevention and Control of Water Pollution, New Delhi, India.

Das, S.M., 1989. *Handbook of Limnology and Water Pollution.* South Asian Publishers Pvt. Ltd., New Delhi.

De Smet, W.H.O. and Evans, F.M.J.C., 1972. A hydrobiological study of the polluted river Lieve (Ghent, Belgium). *Hydrobiologia,* 31(1): 91–154.

Dexter, R.W., 1959. *Anosteraca*: 558–571. Quoted from Ward. H. B. and Whipple, G.C., 1959.

Dogra, S. and Kaw, J.L., 1989. Industrial effluents and aquatic ecosystem: A toxicological over view. In: *Management of Aquatic Ecosystem,* (Eds.) Agrawal, V.P., Desai, B.N. and Bidi, S.A.M. Soc. of Biosci., Muzaffarnagar, India, p. 271–279.

Edmondson, W.T., 1959. *Freshwater Biology.* John Wiley and Sons, Inc., New York, 1248 pp.

Gaur, R.K., Asif, A. Khan, Sultanat Parveen and Sayeed, A. Untoo, 2001. Sediment quality characteristics of a leach ate reservoir receiving effluents from a thermal power plant. *J. Ecophysial. Occup. Hlth.,* 1: 16–178.

Goel, P.K. and Chavan, V.R., 1991. Limnological study of a polluted fishery tank at Kolhapur. In: *Proc. Nat. Symp. Planning for Environ. Sus.,* New Delhi.

Gopal, B., 1990. Indian subcontinent and the aquatic habitats. In: *Ecology and Management of Aquatic Vegetation in the Indian Subcontinent,* Kluwer Academic Publ., Netherlands, p. 7–28.

Govindan, V.S. and Sudaresan, B.B., 1979. Seasonal succession in algal flora in polluted region at Adyar river. *Indian J. Environ. Hlth.,* 21(2): 131–142.

Gunale, V.R. and Balkrishnan, M.S., 1981. Biomonitoring of eutrophication in the Pavana, Mula and Mutha rivers flowing through Poona. *Indian. Environ. Hlth.,* 23(4): 316–321, *Hydrobiologia,* 27: 501–514.

Jolly, V.H. and Chapman, V.A., 1966. A preliminary biological study of the effects of pollution on farmers Creek and Cox's river, New South Wales. *Hydrobiologia,* 27: 160–187.

Kannan, K., 1991. *Fundamentals of Environmental Pollution.* S. Chand and Co. Ltd., New Delhi.

Kant, S. and Kachroo, P., 1974. Limnology studies in Kashmir lakes IV seasonal dynamics of phytoplankton in the Dal and Nagin. In: *Proc. Indian Nat. So. Aca.*, 40: 77–99.

Khapekar, R.R., 2005. Ecophysiological studies on algae of thermal power station. *Ph.D. Thesis*, R.T.M., Nagpur University, Nagpur.

Khatavkar, S.D., Kulkarni, A.Y. and Goel, P.K., 1989. Limnological study on two lentic freshwater bodies at Kolhapur with special reference to pollution. *J.E.P.*, 9(3): 198–203.

Klein, L., 1956. The problems of river pollution in industrial area. *Ray. Sec. Prov. Hlth. F.*, 76: 348–357.

Klein, L., 1957. *Aspects of River Pollution*. Butterworths Scientific Publication, London.

Kolkwitz, R. and Marsson, M., 1908. Oekologider Pflanzlichen Sprolien. *Ber. Bot. Ges.*, 26: 505–519.

Krishnamoorthi, K.P. and Visvesvara, G., 1996. Hydrobiological studies in Gandhisagar (Jammu Tank). *Seasonal Variations in Plankton* (1961–62).

Lieberman, M.E., 1970. The response of *Maina brachiata* (Jurine) 1820 to biological oxygen demand and light. *Hydrobiologia*, 36(1): 9–16.

Mathuthu, A.S., Zaranyika, F.M. and Jannalagadde, S.B., 1993. Monitoring of water quality in upper Mukuvisi river in Harare, Zimbabwe. *Environ. Int.*, 19(1): 51–61.

Mellanby, H., 1963. *Animal Life in Freshwater*, 6th ed. Chapman and Hall Ltd., London.

Meybeck, M., Friedrich, G., Thomas, R. and Chapman, D., 1992. Rivers. In: *Water Quality Assessments*, (Ed.) D. Chapman. Chapman and Hall, London, New York, Tokyo, p. 239–316.

Michael, P., 1984. *Ecological Methods for Field and Laboratory Investigations*. Tata McGraw Hill. Corp. Ltd., New Delhi.

Munawar, M., 1970. Limnological studies on freshwater ponds of Hyderabad, I: Biotope. *Hydrobiologia*, 35: 127–162.

Munawar, M., 1970b. Limnological studies on freshwater ponds of Hyderabad, II: Biocenose. *Hydrobiologia*, 36(1): 105–128.

Munawar, M., 1970. Limnology on freshwater ponds of Hyderabad, India: The biotops. *Hydrobiologia*, 35C 17, IR 7–162.

Muragesan, A.G. and Sukumaran, N., 1999. Impact of urbanization and industrialization on the river Tamirabarani, the life line of Tirunelveli and Thoothukudi districts. *Proc. Sem. Env. Prob.*, p. 1–6.

Nandkar, P.B. and Marathe, K.V., 1983. Estimation of the oxygen produced by dominant algae at six polluted streams. *J. Indian Bot. Soc.*, 62: 245–252.

Needham, J.G. and Needham, P.R., 1962. *A Guide to the Study of Freshwater Biology*. Holden Day Inc., Francisco, 108 pp.

Ouzounis, K.G., Zachariadis, G.A. and Straits, J.A., 1989. Preliminary environmental studies of New Karvali Bay (Kaval Gulf. North Greece). *Toxicol. Environ. Chem.*, 20(21): 227–232.

Paramasivam, M. and Sreenivasan, A., 1981. Changes in algal flora due to pollution in Cauvery river. *Indian J. Environ Hlth.*, 23(3): 222–238.

Penman, H.L., 1970. *The Water Cycle Scientific American*, 233(3): 99.

Philipose, M.R., 1960. Freshwater phytoplankton of Inland fisheries. *Proc. Sym. Algology*, ICAR, p. 272–291.

Sandhu, J.S., Sehgal, N.S. and Toor, N.S., 1984. Rotifer species diversity in manured experimental fish tanks. *Indian J. Ecoi.*, 3(2): 303–308.

Shankhadarwar, S.D., 2002. Eco-physiological studies on toxic wastewater algae. *Ph.D Thesis*, R.T.M., Nagpur University, Nagpur.

Singh, V.P., Saxena, P.N., Tiwari, A., Lonsane, B.K., Khan, M.A. and Arora, I., 1969. Algal flora of sewage Uttar Pradesh in relation to physico-chemical variables. *Environment. Hlth.*, 11: 208–219.

Sinha, A.K., Shrivastava, R.K., Diwivedi, A. and Pande, D.P., 1991. Sediment characteristics of river Ganga at Shuklaganj (Unnao), Gagasso (Rae Bareli) and Phakhamau (Allahabad). *Proc. Nat. Conf. on Ag. Science in India,* New Delhi.

Sinha, M.P., 1986. Limnobiotic study on trophic status of a polluted freshwater reservoir of coal field area. *Poll. Res.*, 5(1): 13–17.

Sitarmaiah, P., 1960. Studies on the ecology of a freshwater pond community. *Hydrobiologia,* 27: 529–547.

Sladeck, J., 1961. A guide to limno-saprobial organisms. *Sc. Proc. Inst. Tech. Progr. Technol. Water,* 7: 543–562.

Tonapi, G.R., 1980. *Freshwater Animals of India: An Ecological Approach*. Oxford and IBH Publication Co., New Delhi.

Trivedy, R.K. and Goel, R.K., 1986. *Chemical and Biological Methods for Water Pollution Studies*. Environ. Publis., Karad, India.

Vasisht, N.S. and Sra, G.S., 1979. *Proc. Symp. Environ. Biol.*, p. 429–440. Quoted from Saksena, D.N. and Mishra, S.R., 1990.

Venkateshwarlu, V., 1969. An ecological study of river Moosi, Hyderabad (India) with special reference to water pollution, I: Physico-chemical complexes. *Hydrobiologia,* 32: 352–363.

Venkateshwarlu, Y., 1969. An ecological study of the algae of the river Moosi. Hyderabad (India) with special reference to water pollution, I: Physico-chemical complexes. *Hydrobiologia,* 33: 117–143.

Ventakeswarlu, V. and Jayanti, R.V., 1968. Hydrobiological studies of river Sabarmati to evaluate quality. *Hydroblologia,* 31: 442–449.

Verma, S.R. and Shukla, G.R., 1969. Pollution in a perennial stream, Khala by the sugar factory effluent near Laksar, Shaharanpur, *India. Environ. Hlth.*, 11: 145–162.

Verma, S.R. and Shukla, G.R., 1969. Pollution in perennial stream, Khala, by the sugar factory effluent near Laksar, Shaharanpur, India. *Environ. Hlth.*, 11: 145–162.

Verma, S.R. and Shukla, G.R., 1970. The physico-chemical conditions of Kamala Nehru tank, Muzaffaranagar in relation to the biological productivity. *Environ. Hlth.*, 12: 110–128.

Verma, S.R. Tyagi, A.K. and Dalela, R.C., 1978. Physico-chemical and biological characteristics of Kadarabad drain in U.P. *Indian J. Environment. Hlth.*, 20(1): 1–13.

Verma, S.R., Sharma, P., Tyagi, A., Rani, S., Gupta, A.K. and Dalela, R.C., 1984. *Limnologica* (Berlin), 15: 69–133. Quoted from Saksena, D.N. and Mishra, S.R., 1990.

Vijay Kumar and Sagwal, O.P., 2000. Recent studies on soil and water pollution in some parts of India: A Review. *Agric. Rey.*, 21(3): 186–192.

Ward, H.B. and Whipple, G.C., 1959. *Freshwater Biology*, 2nd ed. John Wiley and Sons. Inc., New York, London.

Wetzel, R.G., 1975. *Limnology*. W.B. Saunders, Co., Philadelphia, London and Toranto.

Woodward, G.M., 1984. Pollution control in Hamber estuary. *Water Pollution Control*, 83(1): 82–90.

Zafar, R., 1964. On the ecology of algae in certain fish ponds of Hyderabad. India, Physico-chemical complex. *Hydrobiologia*, 23: 179–195.

Zutshi, D.P., Subla, B.A., Khan, M.A. and Wagaňeo, A., 1980. Comparative limnology of nine lakes of J&K Himalayas. *Hydrobiologia*, 72: 101–112.

Chapter 17

Assessment of Groundwater Quality of Roorkee Town with Special Reference to Seasonal Variations in Some Physico-chemical Properties

Deepali, Kamal K. Gangwar and B.D. Joshi
Department of Zoology and Environmental Sciences, Gurukul Kangri University, Haridwar – 249 404, India

ABSTRACT

Present study deals with the seasonal variations in some physico-chemical parameters of groundwater at five selected locations of Roorkee town observed during about 12 months period of study. The aim of the present study is to assess the suitability of drinking water at all the selected locations. The parameters included in this study are pH, electric conductivity, total solids, alkalinity, total hardness, sulphates, phosphates, chloride, sodium, potassium, calcium and magnesium. All the parameters were compared with standard values which are recommended by Bureau of Indian Standard (BIS) and World Health Organization (WHO).

Keywords: Groundwater quality, Seasonal variations, Roorkee town, Standard values.

Introduction

About 80 per cent of water pollution in developing countries like India is caused by domestic waste which adversely affects groundwater quality in the form of sewages and solid wastes. Sewer leakage cause a variety of deleterious changes in the quality of water as indicated by the BOD, COD, nitrate, organic chemicals and bacterial values of the groundwater (Todd *et al.*, 1980; Nathonson, 1986).Groundwater pollution due to agriculture sources is mainly derived from fertilizers, animal wastes and crop waste disposals as pointed out by Kakkar (1986).

Studies have already been carried out in different cities of India (Balakrishnan and Karuppusamy, 2005; Chaudhary *et al.*, 2005; Prajapati, and Mathur, 2005; Rani *et al.*, 2004; Paul and Mishra, 2004; Kaushik *et al.*, 2002; 2004; Sohani *et al.*, 2001; PCRI, 1995, etc.). In the present investigation an attempt has been made to evaluate the quality of groundwater at five selected sites i.e., Irrigation Colony, Durga Chowk, Civil Lines, Chaw Mandi and IIT Campus of Roorkee with respect to some selected physico chemical parameters.

Materials and Methods

Roorkee town is part of Haridwar district at 29°51′ North to South latitude and 77°53′ East to West East longitude and at the elevation of 274 MSL. It covers an area approximate 811.0 hectares and population is 97,064 (according to census, 2001).

The groundwater samples were collected monthly, round the year, from five selected sites *i.e.*, Irrigation Colony, Durga Chowk, Civil Lines, Chaw Mandi and IIT Campus. The study was carried out from February 2005 to January 2006. The data, as given in Table 17.1, are pooled for each season. The methods for all parameters were followed vide APHA (1998) and Trivedi and Goal, (1986).

Results and Discussion

Roorkee is an important commercial and institutional town of Haridwar district. Groundwater is the major source of water for domestic use in the city. The parameters for the groundwater quality of Roorkee town are depicted in Table 17.1.

The pH of water was found to vary from 7.0 to 8.5 for all the stations round the year. The highest and lowest mean values were found 7.77 and 7.87 for pre monsoon season at Chaw Mandi and Civil Lines, 7.3 and 7.17 for monsoon season at Chaw Mandi and Durga Chowk and 7.8 and 7.6 for post monsoon season at Irrigation Colony, Civil Lines, Chaw Mandi and IIT Campus, respectively. The pH values were within the permissible limits prescribed by WHO (1984) and ISI (1991).

Balakrishnan and Karuppusamy (2005) have reported minimum and maximum pH range, 6.8 to 7.3, respectively. Chaudhary *et al.* (2005) have reported higher pH value in pre and post monsoon season but slightly lower in monsoon season. The pH values were fluctuated between 7.2 and 7.8. Sohani *et al*, (2001) have reported the minimum and maximum pH value during pre monsoon, 6.8 to 8.0 and in post monsoon 7.1 to 8.1, respectively. On the other hand slightly acidic values of pH in groundwater were observed by Murugesan *et al.* (2004), in some samples of east-west region of Tamil Nadu. Similar observation has been made by Rani, *et al.* (2004) and Prajapati and Mathur (2005).

The electric conductivity varied from 130.0 μmho/cm to 746.0 μmho/cm with highest and lowest mean values 740.05 μmho/cm and 159.02 at IIT Campus and Chaw Mandi, during pre monsoon season. Similarly, 719.05 μmho/cm, and 137.85 μmho/cm and 728.82 μmho/cm and 142.1 μmho/cm during monsoon and post monsoon season, respectively, at IIT Campus and Chaw Mandi. Electric conductivity for all the samples were according to the permissible limits of WHO (1984), for drinking water. Paul and Mishra (2004), observed electric conductivity between the range of 97.0 to 1378.0 μmho/cm. Sohani *et al.* (2001), observed minimum and maximum electric conductivity values, 100.0 to 1170.0, μmho/cm in pre monsoon and 820.0 to 1550.0 μmho/cm in post monsoon season. Rani *et al.* (2004), also reported similar observations. On the other hand, slightly higher values were reported by Kaushik *et al*, (2004).

Table 17.1: Seasonal Variations in Physico-chemical Parameters of Groundwater *viz.*, Irrigation Colony, Durga Chowk, Civil Lines, Chaw Mandi and IIT Campus of Roorkee Town. All values are ranges, mean±SD of four observations each, during February, 2005 to January, 2006.

Sl.No.	*Parameters*	*Seasons*	*Irrigation Colony*	*Durga Chowk*	*Civil Line*	*Chaw Mandi*	*IIT Campus*
1.	pH	PRM	7.85±0.18	7.85±0.24	7.87±0.33	7.77±0.22	7.82±0.20
			(7.6–8.1)	(7.60–8.1)	(7.5–8.2)	(7.5–8.5)	(7.6–8.0)
		M	7.24±0.17	7.17±0.17	7.25±0.12	7.3±0.14	7.2±0.14
			(7.0–7.4)	(7.0–7.4)	(7.1–7.4)	(7.1–7.4)	(7.1–7.4)
		POM	7.8±0.15	7.67±0.25	7.8±0.22	7.8±0.24	7.6±0.24
			(7.6–8.0)	(7.4–8.0)	(7.6–8.1)	(7.6–8.1)	(7.3–7.8)
2.	E.C	PRM	282.79±5.73	467.6±16.11	698.75±3.77	159.02±5.12	740.05±5.47
			(278.0–291.0)	(456.5–491.5)	(693.8–703.0)	(152.6–164.5)	(732.8–746.0)
		M	266.45±13.92	451.65±3.61	684.17±4.91	137.85±8.25	719.0±7.23
			(247.0–278.0)	(448.5–456.8)	(680.0–691.0)	(130.0–148.0)	(711.0–728.0)
		POM	262.37±7.71	475.95±12.27	688.37±7.37	142.1±7.74	728.82±4.38
			(252.0–268.5)	(466.0–491.2)	(681.8–696.3)	(136.4–152.8)	(723.0–732.8)
3.	Total Solids	PRM	257.5±21.72	267.87±28.31	246.75±26.88	396.37±19.73	239.25±13.43
			(225.5–274.0)	(226.0–288.0)	(208.0–269.5)	(374.5–420.5)	(220.0–251.0)
		M	305.25±12.46	330.85±42.84	297.25±29.94	532.37±50.10	327.25±34.24
			(288.5–318.0)	(302.5–394.0)	(266.0–337.0)	(460.0–571.0)	(293.5–373.5)
		POM	274.75±19.70	271.75±20.73	258.5±18.49	459.12±52.49	279.12±22.35
			(249.5–295.0)	(245.5–293.0)	(237.0–279.5)	(402.0–516.0)	(250.5–302.5)
4.	Alkalinity	PRM	233.0±23.13	217.5±14.20	189.0±11.01	211.25±7.93	256.0±10.42
			(210.0–265.0)	(198.0–229.0)	(176.0– 200.0)	(201.0–218.0)	(244.0–267.0)
		M	171.0±11.60	178.25±9.56	161.5±7.18	186.5±4.50	237.0±1.82
			(156.0–182.0)	(165.0–187.0)	(152.0–168.0)	(180.0–190.0)	(235.0–239.0)
		POM	235.0±17.06	209.5±9.81	183.05±9.94	202.75±9.17	255.75±7.93
			(216.0–255.0)	(198.0–221.0)	(172.0–195.0)	(192.0–213.0)	(246.0–265.0)
5.	Total Hardness	PRM	142.26±12.01	162.2±10.01	224.17±7.03	444.40±13.81	331.97±11.83
			(130.55–156.0)	(148.4–172.2)	(216.55–230.30)	(426.0–457.2)	(317.0–342.1)
		M	134.92±5.41	178.62±20.75	199.48±26.16	438.22±4.29	322.45±7.39
			(130.0–142.6)	(156.2–197.5)	(175.5–222.8)	(432.4–440.0)	(316.0–333.0)
		POM	149.17±2.48	140.34±40.23	201.99–14.93	445.82±5.70	331.35±6.07
			(147.2–152.6)	(104.4–197.88)	(187.5–221.60)	(439.5–452.7)	(324.8–339.4)
6.	Sulphate	PRM	33.72±4.40	28.74±4.43	35.15±4.15	35.74±4.54	39.93±3.23
			(30.0–39.74)	(23.80–34.34)	(29.80–39.60)	(31.60–41.65)	(36.84–44.26)
		M	19.75±3.72	17.58±1.72	26.69±2.06	28.87±3.05	34.84±0.86
			(18.0–23.50)	(15.55–19.32)	(23.88–28.66)	(24.68–31.69)	(32.77–35.80)
		POM	32.44±4.81	27.90±4.52	33.26±3.36	38.12±3.03	40.36±2.86
			(27.33–37.80)	(23.30–33.84)	(29.90–37.70)	(33.84–40.85)	(37.42–43.80)

Contd...

Table 17.1–Contd...

Sl.No.	Parameters	Seasons	Irrigation Colony	Durga Chowk	Civil Line	Chaw Mandi	IIT Campus
7.	Phosphate	PRM	0.216±0.072	0.240±0.088	0.210±0.070	0.215±0.071	0.317±0.086
			(0.131–0.301)	(0.130–0.340)	(0.155–0.280)	(0.130–0.295)	(0.215–0.415)
		M	0.401±0.042	0.470±0.043	0.380±0.010	0.397±0.053	0.526±0.050
			(0.346–0.450)	(0.420–0.520)	(0.365–0.400)	(0.325–0.455)	(0.470–0.580)
		POM	0.200±0.127	0.190±0.105	0.150±0.100	0.230±0.113	0.317±0.122
			(0.075–0.360)	(0.080–0.330)	(0.040–0.285)	(0.100–0.365)	(0.165–0.440)
8.	Chloride	PRM	13.28±3.34	26.67±4.87	12.70±2.88	28.72±10.85	44.51±12.29
			(9.55–17.35)	(21.30–32.10)	(9.20–16.23)	(16.80–41.81)	(31.33–59.67)
		M	17.68±2.55	33.60±1.89	17.50±2.08	48.25±7.89	67.46±8.13
			(14.20–19.85)	(31.20–35.80)	(14.80–19.50)	(38.90–56.70)	(56.81–75.66)
		POM	10.85±2.71	22.82±4.26	10.80±1.78	21.82±8.01	36.90±10.07
			(8.50–13.62)	(18.70–27.60)	8.80–13.00)	(13.95–32.30)	(26.23–48.96)
9.	Sodium	PRM	15.37±1.70	19.97±1.33	25.25±0.69	22.37±0.87	22.65±0.88
			(13.60–17.40)	(18.4–21.6)	(24.60–25.90)	(21.30–23.40)	(21.40–23.50)
		M	6.95±5.56	11.12±4.46	12.17±5.78	17.35±1.48	13.47±4.91
			(3.50–15.20)	(8.40–17.80)	(8.30–20.60)	(15.90–19.40)	(10.40–20.80)
		POM	9.57±2.72	14.62±3.09	17.25±4.28	1.9.5±1.29	15.86±2.99
			(6.50–12.25)	(11.20–18.30)	(12.80–22.20)	(18.20–20.80)	(12.30–19.35)
10.	Potassium	PRM	7.77±0.68	12.07±1.18	8.12±0.68	15.72±1.06	11.42±1.39
			(7.10–8.65)	(10.10–12.70)	(7.40–9.0)	(14.28–16.85)	(10.40–13.45)
		M	3.77±2.50	7.17±2.02	5.07±2.23	9.81±3.33	6.60±2.70
			(2.10–7.45)	(5.30–9.80)	(3.60–8.40)	(7.90–14.80)	(4.80–10.60)
		POM	5.39±0.96	8.15±1.34	5.67±1.14	11.37±1.88	8.28±1.61
			(4.66–6.80)	(6.30–9.40)	(4.20–6.80)	(9.40–13.40)	(6.05–9.80)
11.	Calcium	PRM	50.09±7.98	66.72±9.51	66.72±10.60	80.80±9.17	72.56±12.59
			(40.30–58.92)	(54.34–76.47)	(53.10–77.67)	(70.34–90.70)	(59.92–87.12)
		M	35.02±5.32	43.27±6.33	42.54±7.91	69.18±8.27	53.52±6.19
			(28.65–40.85)	(36.65–49.52)	(32.34–51.20)	(60.55–79.20)	(46.67–61.49)
		POM	55.17±6.48	70.30±9.30	73.87±10.51	90.76±7.80	82.08–12.23
			(46.90–61.11)	(59.32–80.92)	(62.49–86.30)	(82.66–100.3)	(68.64–96.65)
12.	Magnesium	PRM	28.78±3.14	33.82±2.69	25.99±1.86	44.45±2.33	37.49±3.85
			(24.84–31.42)	(29.84–35.55)	(23.86–27.60)	(41.86–47.28)	(32.33–40.66)
		M	15.92±3.42	22.38±3.17	17.38±2.92	34.56±0.60	25.00±2.39
			(12.50–20.66)	(20.12–26.80)	(15.20–21.70)	(33.90–38.44)	(22.7–27.65)
		POM	21.64±4.51	30.41±3.77	2.1.07±3.23	40.24±2.34	35.21–3.45
			(16.68–26.80)	(26.84–33.20)	(17.48–24.82)	(37.80–43.42)	(30.5–37.9)

Total solids ranged from 208.0 mg/l to 516.0 mg/l. The maximum and minimum mean values for pre monsoon season was observed 396.37 mg/l and 239.25 mg/l at Chaw Mandi and IIT Campus, 532.37 mg/l and 297.25 mg/l for monsoon season at Chaw Mandi and Civil Line, respectively. While, during the post monsoon season the maximum and minimum values fluctuated between 459.12 mg/l and 258.5 mg/l at Chaw Mandi and Civil Line. Whereas, total solids ranges between 122.0 to 5652.0 mg/l in pre monsoon, 115.0 to 5642.0 mg/l in monsoon and 120.0 to 5640.0 mg/l in post monsoon, were reported by Mishra and Patel (2001).

The alkalinity ranged between 152.0 mg/l to 267.0 mg/l. The highest and lowest mean values were recorded 256.0 mg/l and 189.0 mg/l, 237.0 mg/l and 161.5 mg/l and 255.75 mg/l and 183.25 mg/l at IIT Campus and Civil Lines for pre-monsoon and post monsoon season, respectively. Balakrishnan and Karuppusamy (2005), reported alkalinity range 50.0 to 180.0 mg/l. Similarly, Mishra and Patel (2001), reported alkalinity range between 54.0 to 106.0 mg/l.

Sohani *et al.* (2001), have reported minimum and maximum alkalinity ranges in pre monsoon nil to 520.0 mg/l and in post monsoon 200.0 to 600.0 mg/l. Chaudhary *et al.* (2005), observed a high alkalinity of groundwater during winter and pre monsoon due to slow leaching of carbonates and bicarbonates during and after rainy season. They have reported alkalinity from 200.0 to 377.0 mg/l.

The total hardness values fluctuated between 104.4 mg/l to 457.2 mg/l. The maximum and minimum mean value recorded 444.40 mg/l and 142.26 mg/l, 438.22 mg/l and 134.92 mg/l at Chaw Mandi and Irrigation Colony for pre monsoon and monsoon season, respectively. While the highest and lowest values recorded 445.82 mg/l and 140.34 mg/l at Chaw Mandi and Durga Chowk for post monsoon season. Sohani *et al.* (2001), reported the values of hardness in pre monsoon 344.0 to 1388.0 mg/l and in post monsoon 252.0 to 680.0 mg/l. Mishra and Patel (2001), observed alkalinity values in pre monsoon season from 27.0 to 110.0 mg/l in monsoon season from 29.0 to 114.0 mg/l and in post monsoon from 25.88 to 170.0 mg/l. Similar results were also observed by Kaushik *et al.* (2002, 2004); Paul and Mishra, (2004); Rani *et al.* (2004); Barbar and Kaplay, (1999); Chaudhary *et al.* (2005).

During the period of investigation the sulphates ranged between 15.55mg/l to 44.26 mg/l. The maximum mean values of sulphates recorded 39.93 mg/l, 34.84 mg/l, and 40.36 mg/l, while, the minimum mean values recorded 28.74 mg/l, 17.58 mg/l and 2.7.90 mg/l, at IIT Campus and Durga Chowk, respectively, for pre monsoon, monsoon and post monsoon season. The values of sulphates were well below the limits of WHO (1984). Murugesan *et al.* (2004), have reported sulphate between the range of 10.0 to 60.8 mg/l in monsoon and post monsoon season. On the other hand, Sohani *et al.* (2001), have observed very high range of Sulphate in pre monsoon, (510.26 to 4214.76 mg/l) and in post monsoon (641.94 to 13052.7 mg/l). The higher values of sulphates were also reported by Kaushik *et al.* (2002).

The phosphates ranged from 0.040 mg/l and 0.580 mg/l. The minimum mean values noted 0.210 mg/l, 0.380 mg/l and 0.150 mg/l at Civil Lines for pre monsoon, monsoon and post monsoon season, respectively. The maximum values recorded 0.317 mg/l, 0.526 mg/l and 0.317 mg/l at IIT Campus, during pre monsoon, monsoon and post monsoon season, respectively. Phosphate values were found below the standard limit prescribed by WHO (1984). While, Abraham and Beegum (2001), have been observed that phosphate values fluctuate from 0.03 to 0.66 mg/l.

Chloride values varied from 8.50 mg/l to 75.66 mg/l. The highest mean values recorded 44.51 mg/l, 67.46 mg/l and 36.90 mg/l and lowest 12.70 mg/l, 17.50 mg/l, and 10.80 mg/l at IIT Campus and Civil Line for pre monsoon, monsoon and post monsoon seasons, respectively.

While, the higher values of chloride were observed by Kaushik *et al.* (2002); Chaudhary *et al.* (2005), reported much higher chloride content of groundwater samples than the permissible limits. They have observed that the chloride contents in pre monsoon ranged between 150.0 to 310.0 mg/l, in monsoon 124.0 to 177.0 mg/l and in post monsoon 162.0 to 361.0 mg/l. High chloride contents indicate heavy pollution. While, chloride contents in lower quantities (13.5 to 35.9 mg/l) were recorded by Balakrishnan and Kartuppusamy (2005). Rani *et al.* (2004), observed chloride concentration in pre monsoon period between 25.2 to 36.2 mg/l, in monsoon period between 21.5 to 32.5 mg/l and in post monsoon between 13.5 to 35.9 mg/l.

Sodium values ranged between 3.50 mg/l to 25.90 mg/l. The maximum and minimum mean values were recorded 25.25 mg/l and 15.37 mg/l for pre monsoon season at Civil Line and Irrigation Colony, respectively. While the highest values of sodium recorded 17.35 mg/l and 19.5 mg/l and lowest 6.95 mg/l and 9.57 mg/l at Chaw Mandi and Irrigation Colony, for monsoon and post monsoon season, respectively. Similar results were observed by Babar and Kaplay (1999). However, much higher values were recorded by Kaushik *et al.* (2002, 2004). Chaudhary *et al.* (2005), have also observed the higher values of sodium in pre monsoon and post monsoon but slightly lower values in monsoon season. Percolation of rain water during and after monsoon through sodium rich surface layers of soil can lead to enrichment of groundwater sodium level.

Potassium contents varied from 2.10 mg/l to 16.85 mg/l. The minimum values recorded 7.77 mg/l, 3.77 mg/l and 5.39 mg/l and maximum 15.72 mg/l, 9.8.1 mg/l and 11.37 mg/l at Irrigation Colony and Chow Mandi during pre monsoon, monsoon and post monsoon season respectively. Chaudhary *et al.* (2005), observed potassium values in pre monsoon from 39.67 to 161.0 mg(l, in monsoon season from 13.50 to 138.0 mg/l and in post monsoon from 25.11 to 146.0 mg/l. Similar observations have been made by Babar and Kaplay, (1999), but slightly higher values of potassium were observed by Kaushik *et al.* (2002, 2004).

The calcium ranged between 28.65 mg/l to 100.3 mg/l with highest and lowest mean values 80.80 mg/l and 50.09 mg/l, 69.18 mg/l and 35.02 mg/l and 90.76 mg/l and 55.17 mg/l, during the pre monsoon, monsoon and post monsoon season, for Chow Mandi and Irrigation Colony, respectively. All the values are with in permissible limits of WHO (1984), standards.

According to Sohani *et al.* (2001), calcium range varied between 86.57 to 291.78 mg/l in pre monsoon and 67.33 to 142.68 mg/l in post monsoon. Similarly, during winter season calcium were found between 30.0 to 35.7 mg/l by PCRI (1995). Mishra and Patel (2001), have been reported 17.60 to 25.0 mg/l calcium in pre monsoon and in. rainy season 23.0 to 32.0 mg/l calcium and in post monsoon 20.70 to 50.4 mg/l calcium. Chaudhary *et al.* (2005), observed high calcium values in pre monsoon from 123.0 to 239.0 mg/l, in monsoon from 96.87 to 180.0 mg/l and in post monsoon value varied from 157.0 to 212.0 mg/l.

Magnesium values ranged from 12.50 mg/l to 47.28 mg/l. The maximum and minimum mean values observed 44.45 mg/l and 25,99 mg/l at Chaw Mandi and Civil Line respectively, during pre monsoon season. During the monsoon the maximum and minimum values recorded 34.56 mg/l and 15.92 mg/l at Chaw Mandi and Irrigation Colony and for post monsoon season 40.24 mg/l and 21.07 mg/l at Chaw Mandi and Civil Line, respectively. The maximum permissible limit of Magnesium as prescribed by WHO (1984), is 150.0 mg/l. All the magnesium values in the present study were much below the standard permissible limit. Sohani *et al.* (2001), observed magnesium range in pre monsoon from 31.19 to.161.78 mg/l and in post monsoon 81.11–86.73 mg/l, which are slightly higher value than the standard limits.

PCRI (1995), have reported magnesium value in winter season between 9.3 to 12.7 mg/l. Higher magnesium concentration were observed by Balakrishanan and Karuppusamy (2005), Mishra and Patel (2001) reported lower concentration of magnesium in all the seasons. Chaudhary *et al.* (2005) observed that the magnesium values varied from 42.61 to 88.82 mg/l in pre monsoon, 31.81 to 71.54 mg/l in monsoon and 39.31 to 99.88 mg/l in post monsoon season, respectively.

References

APHA, AWWA and WPCF, 1995. In: *Standard Methods for the Examination of Water and Wastewater,* 20th edn. American Public Health Association, Washington, D.C., New York.

Abraham, B.T. and Sajeela Beegum, T.M., 2001. Comparison of water quality of well and borewell of Moothakunnam: Malian Kara area with USPHS standards. *Indian J. Environ. and Ecoplan.*, 5(3): 593–596.

Babar, M.D. and Kaplay, R.D., 1999. Groundwater quality around Pingalgad Nala in Parbhani district, Maharashtra, India. *Ecol. Env. and Cons.*, 5(2): 141–143.

Balakrishnan, V. and Karuppusamy, S., 2005. Physico-chemical characteristic of drinking water samples of Palani, Tamil Nadu. *J. Ecotoxicol. Environ. Monit.*, 15(3): 235–238.

Chaudhary, S. Anuradha and Sastry, K.V., 2005. Groundwater quality in Faridabad: An industrial town of Haryana. *J. Eco-toxicol. Environ. Monit.*, 15(3): 263–271.

ISI, 1991. *Indian Standard Specification for Drinking Water.* Bureau of Indian Institute, New Delhi, ISI: 10500.

Kakkar, Y.P., 1986. Groundwater pollution. *Groundwater News*, 5(11–12): 51–56.

Kaushik, A., Kumar, K., Kanchan, Taruna and Sharma, H.R., 2002. Water quality index and suitability assessment of urban groundwater of Hisar and Panipat in Haryana. *J. Environ. Biol.*, 23(2): 325–333.

Kaushik, A., Kumar, K., Sharma, I.S. and Sharma, H.R., 2004. Groundwater quality assessment in different land-use areas of Faridabad and Rohtak cities of Haryana using deviation index. *J. Environ. Biol.*, 25(2): 173–180.

Mishra, P.C. and Patel, R.K., 2001. Study of the pollution load in the drinking water of Rairangpur: A small tribal dominated town of North Orissa. *Indian J. Environ. and Ecoplan.*, 5(2): 293–298.

Murugesan, S., Dhamodhar Kumar, S., Rajan, S. and Chandrika, D., 2004. Comparative study of groundwater resources of east to west region of Chennai, Tamil Nadu. *Nature Environment and Pollution Technology*, 3(4): 495–499.

Nathonson, Jerry A., 1986. *Basic Environmental Technology*. John Wiley and Sons, pp. 479.

Mishra, A.K., 2004. An assessment of pollution of water used for domestic purposes in Lumding Town, Assam, India. *Nature Environment and Pollution Technology*, 3(4): 531–534.

PCRI (Pollution Control Research Institute), Hardwar, 1995. Environmental impact assessment for Anpara Thermal Power Station proposed extension, stage –CC2x500 MW), Anpara, Uttar Pradesh.

Prajapati, P. and Mathur, R., 2005. Statistical studies on the groundwater at the rural areas of Sheopurkalan, Madhya Pradesh. *J. Ecotoxicol. Environ. Monit.*, 15(1): 47–54.

Rani, Rekha, Gupta, B.K. and Shrivastava, K.B.L., 2004. Studies on water quality assessment in Satna city (MP): Seasonal' parametric. *Nature Environment and Pollution Technology*, 3(4): 563–565.

Sohani, D., Pande, S. and Srivastava. V.S., 2001. Groundwater quality at tribal town Nandurban (Maharashtra). *Indian J. Environ and Ecoplan.*, 5(2): 475–479.

Todd, D.K., 1980. *Groundwater Hydrology*, 2nd Edition. John Wiley and Sons, Inc., pp. 535.

Trivedi, R.K. and Goel, P.K., 1986. *Chemical and Biological Methods for the Water Pollution Studies.* Environment Publication, Karad.

WHO, 1984. *Guidelines for Drinking Water Quality Vol. 1: Recommendations.* World Health Organization, Geneva, p. 1–130.

Chapter 18

Studies on Sugar Mill Effluents: The Physico-chemical Characteristics and Concentration of Heavy Metals

Mukul Chaturvedi and H.B. Singh
School of Chemical Sciences, St. John's College, Agra – 282 002, India

ABSTRACT

Present work deals with the studies of physico-chemical parameters and heavy metal contents of Chhata Sugar Mill (District Mathura, India) effluents by using different analytical methods. The parameters like colour, pH, electrical conductivity, COD, total dissolved solid, total hardness, Ca^{2+}, Mg^{2+}, sulphate, chloride, fluoride, alkalinity, temperature and concentration of heavy metals (like Cr, Mn, Ni, Zn, Cd, Fe, Cu, Pb and Hg) of the effluent collected from the various sites between the exit point at the Mill and discharge point in a natural drain in different months have been determined and explained.

Keywords: Sugar mill, Heavy metals, Toxic metals, Physico-chemical parameters.

Introduction

Pollution of the ecosystem by man is a world wide phenomenon with increase in population as well as industrial activity (Manas, 1976; Kundra and Purthy, 1979). As there are close links in the ecological and technical fields, the environmental protection installations require regular maintenance in the industries. Further, the personnel training must periodically be imparted to make the personnel aware of environmental issues. But unfortunately proper attention is not paid to and, therefore, the industrial effluents are posing toxicological hazards to the environment (Matkar and Gangotri, 2003; Matkar, 1997; Beynt, 1984; Shrinivas *et al.*, 1984).

Furthermore, the effluents discharged from different industries have different chemical compositions, and thus there can not be a single common method for their treatment and safe disposal. Therefore, before disposing off into environment or deciding on the methods of appropriate utilisation in agriculture the chemical composition of these effluents should be thoroughly studied and investigated.

The sugar industry contributes to the general economic development of a country, including the intensification of agriculture, start of the general industrialisation of rural areas and job creating in agriculture and manufacture, all of which attracts the potential work force in the surrounding area. But the sugar industry alone produces a huge amount of wastewater. These effluents contribute a number of chemical pollutants such as carbonates, bicarbonates, nitrites, nitrates in addition to suspended solids, dissolved solids and scores of water toxic materials. Thus the analysis of sugar mill effluents is essential so that the required measures could be taken to refrain from the possible hazards thereby.

However, a careful survey of the literature reveals that only a scanty work (Jameel *et al.*, 2003; Arora *et al.*, 1974; Rao *et al.*, 1981; Matkar and Gangotri, 2002) has been done so far on the studies of the sugar mill effluents.

Chhata sugar mill is located in Mathura district on National Highway No. 2 and 20 km far off Mathura towards New Delhi. The sugar mill is playing an important role in the enhancement of the prosperity of the area around. But the effluents from sugar mill are posing threat to the ecology of the surrounding area and may be even more severe in the near future. It was, therefore, decided warthcarrying the studies on the effluents discharged by the said sugar mill in Mathura.

Experimental

The three different points on the drain *viz.*, Point-1 (the exit in the premises), Point-2 (½ km far from Point-1) and Point-3 (½ km far from Point-2) were selected for sample collection of sugar, mill effluents in different months *viz.*, December 2005, January and February 2006.

The physico-chemical analysis of sugar mill effluents carried out as per the standard methods employed of analysis of water, wastewater and the industrial effluents (Manivaskam, 1996; APHA, 1995; Trivedi and Goel, 1996; USEPA, 1995). pH was determined by Digital pH Meter (Bio-Craft NDC 736) and electrical conductivity by Electrical Digital Conductivity Meter (Bio-Craft NDC 736), sulphate was determined by gravimetric method, whereas DO has been measured by Winkler's Method. Fluoride contents have been determined by Ion Selectivity Meter (Thermo Orion Model 720).

The samples were analyzed far heavy metal contents by the Atomic Absorption Spectrophotometer (Perkin-Elmer Analyst 100).

Results and Discussion

Physico-chemical Parameters

The Physico-Chemical Characteristics of sugar mill effluents have been enlisted in Table 18.1

Table 18.1: Physico-chemical Analysis and Concentrations of Heavy Metals of Chhata Sugar Mill Effluents at Different Time Intervals (All concentrations are reported in ppm (mg/L) except pH, Temperature in (°C) and Electrical Conductivity in mho/cm)

Sl.No.	*Parameters*	*December, 2005*			*January, 2006*			*February, 2006*		
		Point-1	*Point-2*	*Point-3*	*Point-1*	*Point-2*	*Point-3*	*Point-1*	*Point-2*	*Point-3*
1.	Colour	Yellow	Light Yellow	Very Light Yellow	Yellow	Light Yellow	Very Light Yellow	Yellow	Light Yellow	Very Light Yellow
2.	Odour	Sweet	Light Sweet	Light Alcoholic	Sweet	Light Sweet	Light Alcoholic	Sweet	Light Sweet	Light Alcoholic
3.	Temperature	40°C	25°C	18°C	35°C	23°C	17°C	30°C	18°C	13°C
4.	pH	3.86	4.00	5.20	3.90	4.50	5.25	4.60	5.70	5.90
5.	E.C.	6.10	5.85	5.00	4.89	4.59	4.41	5.50	4.91	4.45
6.	C.O.D.	460	300	80	380	200	60	440	350	140
7.	T.D.S.	11.60	9.60	7.60	.14.00	10.00	6.00	24.80	16.80	8.80
8.	T.H.	502	440	400	.490	442	390	500	4S2	402
9.	Ca^{2+}	197.60	172.40	144.60	185.80	.164.00	153.90	195.10	170.70	164.80
10.	M^{2+}	74.20	62.20	62.30	74.22	67.80	57.60	74.39	75.96	57.88
11.	D.O.	2.63	4.66	7.09	2.70	4.50	7.20	2.83	4.66	7.20
12.	Alkalinity	402	370	280	410	360	270	418	390	298
13.	Cl^-	460.94	283.65	172.28	425.48	304.90	212.74	418.31	283.60	248.15
14.	F^-	3.34	2.17	2.00	1.63	1.71	1.68	1.88	1.24	1.91
15.	SO_4^{2-}	21.60	16.60	9.20	22.60	18.90	11.10	23.80	21.30	14.80
16.	Ni^{2+}	0.042	0.013	0.010	0.011	0.012	0.022	0.016	0.020	0.022
17.	Zn^{2+}	1.014	0.430	0.055	0.539	0.346	0.628	0.039	1.653	0.383
18.	Pb^{2+}	0.263	0.144	0.218	0.283	0.197	0.243	0.312	0.208	0.304
19.	Cd^{2+}	0.038	0.044	0.045	0.054	0.056	0.061	0.061	0.064	0.060
20.	CU^{2+}	1.183	0.788	0.029	1.178	1.014	1.571	0.035	0.380	0.013
21.	Mn^{2+}	1.423	0.725	0.102	1.019	0.327	0.667	0.319	2.062	0.171
22.	Fe^{2+}	22.930	21.480	2.474	15.310	12.860	15.610	5.253	24.54	2.517
23.	Cr^{3+}	0.420	0.380	0.493	0.470	0.472	0.397	0.561	0.542	0.670
24.	Hg^{2+}	00.00	00.00	00.00	00.00	00.00	00.00	00.00	00.00	00.00

It has been observed that the effluents are yellow in colour and intensity decreases from Point-1 to Point-3. The temperature is lowest at Point-3 and increases appreciably upto Point-1. The effluents' odour of mill found sweet to alcoholic from Point-1 to Point-3. With increase in the distance pH values are increased perhaps due to the continuous increase in amount of alcohol by the slow fermentation of the sugar contents within the effluents. These values are higher than the standard limits recommended by ISI and WHO (Trivedi and Gurudeep, 1995).

COD found is of the order 60–460 mg/l. The values are higher than the standard limits of industrial effluents as adviced by ISI. The concentration of Ca^{+2} and Mg^{+2} ions are 144.60–197.60 mg/l and 57.60–75.96 mg/l respectively. The total hardness ranges between 390–502 mg/l. Alkalinity values found varying over 270–418 mg/l. Cl^- ion concentration was obtained 9.20–23.80 mg/l. Fluoride contents range over 1.63–3.34 mg/l. The quantity of SO_4^{-2} ion has been found to be 9.20–23.80 mg/l.

Heavy (Toxic) Metals

The concentration values of heavy metals enlisted in Table 18.1 indicate that 0.042 mg/l, 1.653 mg/l, 0.144 mg/l and 0.064 mg/l are the highest concentrations of Ni, Zn, Pb and Cd in the month of December 2005, February 2006, December 2005 and February 2006 respectively. It was observed that Cu concentration value (1.57 mg/l) was higher in the month of January 2006. Concentrations of Mn and Fe exhibit variation from 0.102 to 2.062 mg/l and 2.74 to 24.54 mg/l respectively. Cr Concentration values have been found from 0.380 to 0.670 mg/l. However, Hg was almost nil in sugar mill effluents.

Thus the above data of physico-chemical parameters and concentrations of toxic metals are higher than the acceptable limits 17–19. Therefore, the discharge of the mill effluents as such will badly affect the environment in the surrounding area (Khursheed, 1998).

Acknowledgement

The authors are highly thankful to the Principal, St. John's College, Agra for providing the necessary facilities to carry out the present studies.

References

American Public Health Association (APHA), 1995. *Standard Methods for Examination of Water and Wastewater,* 18th edition. Washington, USA.

Anand, M., 1987. Impact of sugar policy. *J. Bhartiya Sugar Dir.*, 1: 213–221.

Arora, H.C., Routh. T., Chattopadhyay, S.N. and Sharma, U.P., 1974. *Indian. J. Env. Health,* 16(3): 233–246.

Beynt, A.M., 1984. Biological testing of industrial effluents: Problems of aquatic toxicity, biotesting and water quality management. In: *Proc. 45 B-USSR Symposium* p. 222–235.

Borale, D.D. and Patil, P.R., 2004. *Poll. Res.*, 23(1): 83–86.

Jameel, A., Abdul and Sirajuddin, J., 2003. *Indian J. of Env. Prot.*, 23(6): 663–660.

Jesudass, I ., Aklla L., 1996. The physico-chemical characteristics of sugar factory effluent. *Indian J. Env. Protection,* p. 808–810.

Khurshid, S., Basheer and Abdul, 1998. Effect of waste disposal on water quality in parts of Cochin, Kerala. *Indian J. Environ. Health,* 40(1): 45.

Kundra, R. and Purthy, S., 1979. Sugar industry and pollution. In: *Ecology and Development,* (Eds.) Deshbandu R. and Bharadwaj, V. Indian Env. Soc., New Delhi, 178–190.

Manas, C., 1976. Water pollution in textile, leather and paper industry water. *Ind. Chem. Man.*, 14(3): 13–14.

Manivaskam, N., 1996. *Physico-chemical Examination of Water Sewage and Industrial Effluents.* Pragati Prakashan, Meerut.

Matkar, L.S. and Gangotri, M.S., 2002. *J. of Industrial Pollution Control,* 18(2): 139–144.

Matkar, L.S. and Gangotri, M.S., 2003. *Poll. Res.*, 22(2): 269–276.

Matkar, L.S., 1997. Effect of sugar industrial effluents on the freshwater carb, *Barytelphusa guerini* (H. Milne Edwards) (Decapota : Potamida). *Ph.D. Thesis*, University of Pune.

Rao, V.V.S., Kotinandaraman, V., Rao, M. and Sastry, C.A., 1981. *Indian J. Env. Health*, 23(4): 288–299.

Shrinivas, P., Bangarij, B. and Rao, P., 1984. Toxicity of textile industry effluent on some freshwater animal. *Indian J. Environ. Health*, 26(1): 57–59.

Trivedi, P.R. and Gurudeep, Raj, 1997. Vol. 11, Akashdeep Publishing House.

Trivedi, R.K. and Goel, R.K., 1996. *Chemical and Biological Methods for Water Pollution Studies*. Environmental Publication, Karad.

U.S.EPA, 1995. *Methods of Chemical Analysis of Water and Wastes*. Washington D.C., U.S.A.

Chapter 19

Antimicrobial Activity of Terpenoidal Compound Isolated from *Sphaeranthus indicus* against *Bacillus subtilis*

L.N. Dubey[1], S.K. Bansal[1] and Brijesh Sahu[2]
[1]Pest Control and Ayurvedic Drug Research Laboratory,
[2]Department of Botany,
S.S.L. Jain Postgraduate College, Vidisha – 464 001, India

ABSTRACT

Sphaeranthus indicus Linn. (Family Compositae) called mundi in Hindi is a common scandent Indian indigenous herb. Its leaves eaten as a pot herb. The extract shows the presence of Terpenoidal Compound with dose dependent inhibitory activity against *Bacillus subtilis*, a bacteria causing flacherie of 2nd and 3rd instar larvae of silk moth *Bombyx mori*. Minimum inhibitory concentration was found to be highest (0.10 mg) for purified fraction-Sf_2. The identification of the compound on spectral analysis gave a Bicyclic Sesquiterpene lactone purified fraction.

***Keywords**: Antimicrobial activity, Sphaeranthus indicus, Minimum inhibitory, Concentration, Flacherie.*

Introduction

Herbal medicine has been improved in developing countries as an alternative solution to health problems and costs of pharmaceutical products.

Plants used for traditional medicine contain a wide range of substance that are used to treat chronic as well as infectious diseases.

Survey of certain Pharmacologically active phyto constituents in wild plants to be used in folk medicine in Madhya Pradesh and adjoining Uttar Pradesh have been carried out by Dixit and Pandey (1984); Sinha and Dogra (1985); Saxena *et al.* (1992).

In the recent year Agrawal and Rangari (2002) have reported antibacterial activity of essential oils of *Strobilanthus* plant. Koli *et al.* (2002), have also reported antimicrobial activity of isoflavone glycoside. Ahmad *et al.* (1998) and Perumal Samy *et al.* (1998) have been reported antimicrobial properties of some Indian medicinal plants. Chaudhary *et al.* (2002) have reported causal organism of flacherie in the silk worm *Antheraea assama* Ww: Isolation, Characterization and its Inhibition by garlic extract.

Looking to the pertaining literature and in view of the necessity to search the suitable antibacterial drugs from plants, the present investigation was proposed to be investigated against *Bacillus subtilis*, a causative agent of flacherie of silkmoth *Bombyx mori.*

Plant Material

The plants *Sphaeranthus indicus* were collected from a village of Raisen district of M.P. in India during Winter season and were dried in shade. After identification in the Botany department, a Voucher Specimen was procured in the herbarium record of Pest Control and Ayurvedic Drug Research Lab. at Vidisha (M.P.) India at Sl.No. 2.

The dried plants were powdered (40 to 60 mesh) and extracted for 148 h. with acetone in Soxhlet apparatus. The solvent was removed by Vacuum distillation and the extract was evoporated to dryness under reduced pressure the 70 per cent methanolic extract was evaporated to dryness in vacuum evaporator to get greenish crude material. This was made in different Concentration for bioassay test against *Bacillus subtilis.*

Purification and Isolation of Crude for Biologically Active Principle

The isolation of only 90 per cent methanolic crude extract was carried out using glass column and TLC methods. Total 3 fractions were isolated and tested separately. The most biologically active fraction Sf_2 was used for spectral analysis work which was done at CDRI Lucknow by sending purified samples for IR, UV, HNMR, CNMR and Mass spectrum.

The spectral analysis of the compound was carried out for IR bads at 3473.4 cm^{-1} (OH), 1646 cm^{-1} and 1462.5 cm^{-1} (showed O and B unsaturated bilactonic group containing exocyclic double bond). UV maximum at 242 nm confirmed double bond in the same ring. Mass spectrum display 154 and other fragments ions at m/z 101, 107, 120, 136, 154, 176, 228, 254, 289 which supports molecular formula $C_{15}H_2O_3$. The HNMR spectrum showed 2, methyl group at 1.0 (3H). On the basic of this, a Terpenoidal compound (Bicyclic Sesquiterpene Lacton) was isolated from the whole plant of *Sphaeranthus indicus.*

Experimental Bioassays

Bacillus subtilis ATCC 6633 bacterial strain was used for the present study. The bacterial strain procured from MTCC, Chandigarh.

The microorganism suspensions used for inoculation was prepared at 10^6 Cf_u/ml by diluting fresh culture.

Agar liquid nutrient medium was used for diluting the microorganism suspension and for two fold dilution of the extract. The bacterial strain kept in small petridishes of Agar media. Concentration of plant compounds were mixed by microsyringe and the mortality was observed at different intervals the treated petridishes were placed at 36°C for 16–20 h. and then again observed mortality. Inhibition zones of the extracts formed on the medium were measured with a transparent rules in millimeters.

Results and Discussion

Table 19.1 reports the percentage yield of crude drug in soxhlet at different solvents maximus yield was noticed in water extract. This shows the solutions gave polar compounds. Table 19.2 shows the TLC fractions of the crude drug obtained from the plant along with Rf value which shows the presence of terpenoidal compound. Table 19.3 shows the different concentration of the purified fraction were applied on the petridishes containing bacterial culture. After 24 h. inhibition of growth was measured and maximum inhibitory concentration was recorded.

Table 19.1: Percentage Yield of the Crude Extracts by Soxhletion of *Sphaeranthus indicus*

Sl.No.	*Name of the Solvent Used*	*Wt. of Powder (g)*	*Volume of Solvent (ml)*	*Wt. of Extract Obtained (gm)*	*Yield (per cent)*
1.	n-Hexane	125	700	5.22	3.36
2.	Petroleum ether	125	700	5.40	324
3.	Chloroform	125	700	2.11	1.68
4.	90% Methanol	125	700	5.30	2.65
5.	Water	200	700	12.32	7.66

Table 19.2: TLC of Crude Extract of *Sphaeranthus indicus*

Sl.No.	*Name of the Fraction*	*Colour of the Fraction in*			*R_f Value*
		Visual Light	*UV Light*	*Iodine Chamber*	
1.	Sf_1	Green	Red	Red	2/9 = 0.22
2.	Sf_2	Brown	Green	Green	8/9 = 0.89
3.	Sf_3	Black	Dark Green	Black	4/9 = 0.44

Solvent system used in TLC

Methylene dichloride	:	Chloroform	:	Ethyl acetate	:	n Propanol
45	:	45	:	4.5	:	4.5

Table 19.3: MIC Value of the Different Fractions of Terpenoidal Compound of *Sphaeranthus indicus* Against *Bacillus subtilis*

Fraction of Terpenoidal Fraction	*Microorganism Minimum Inhibitory Concentration*
Sf_1	0.73
Sf_2	0.10
Sf_3	0.31

In conclusion, the present study shows that the extracts examined have variable antimicrobial activities.

References

Agrawal, R.B. and Rangari, V.D., 2002. Antibacterial and antifungal activity of essential oil of *Strobilanthus ixiocephala. Indian Drug*, 39(1): 36–38.

Ahmad, I., Mehmood, I. and Mohammad, F., 1998. Screening of some Indian medicinal plants for their antimicrobial properties. *Journal of Ethnophormacology*, 62(2): 182–193.

Chaudhary, A., Guha, A., Yadav, A., Umni, B.G. and Roy, M.K., 2002. Causal organism of flacherie in the Silkworm *Antheraea assarna* Ww: Isolation, characterization and its inhibition by garlic extract. *Phytotherapy Research*, 16(51): 89–90.

Dixit, R.S. and Pandey, H.C., 1984. Plants used as folk medicine in Jhansi and Lalitpur Section of Bundelkhand (U.P.). *Int. J. Crude Drug Res.*, 22: 47–51.

Koli, M.C., Choudhary, Rashmi, Kumar Saurabh, Thakur Suman, Shukla Deepti and Saxena, R.C., 2002. As isoflavone glycoside from the stem of *Euphorbia hirta* (Linn) As antimalarial compound. *Asian Journal of Chemistry*, 14(3–4): 1673–1677.

Perumal Samy, R., Ignacimuthu, S. and Sen, A., 1998. Screening of 34 Indian medicinal plants for antibacterial properties. *Journal of Ethnopharmacology*, 62(2): 173–182.

Saxena, R.C., Dixit, O.P. and Padma, S., 1992. Laboratory assessment of Indigenous plant extract for anti-juvenile hormone activity in *Culex fasiatus. Indian J. Med Res.*; 98: 204–206.

Sinha, S.K.P. and Dogra, J.M.V., 1985. A survey of the plants of Bhagalpur and Santhal pargana for saponin, flavonoids and alkaloids. *Int. J. Crude Drug Res.*, 23: 77–86.

Chapter 20

Determination of Physico-chemical and Microbiological Analysis of Nambul River at Imphal City, Manipur

W. Vidyarani[1], H. Bisheswar Singh[2] and M. Gonchandra Sharma[3]

[1]Department of Life Sciences, Manipur University, Manipur – 795 003

[2]Department of Horticulture and Soil Conservation, Manipur – 795 001

[3]Manipur Pollution Control Board, Lamphel, Manipur – 795 004

ABSTRACT

The physico-chemical and microbiological parameters of water samples from both upstream and downstream of the Imphal City were analysed during different years before and after onset of rain. Most of the parameters were within the acceptable range of drinking water standard. However pH was lower than the normal limit and alkalinity decreased through years. BOD and COD was much lesser in the upstream than the downstream. Microbiological study suggest untreated domestic and household disposal to the water body causing heavy pollution to water.

Keywords: Pollution, pH, Alkalinity, Coliform bacteria.

Introduction

Water Pollution is one of the most serious problem to humankind. It is an established fact that water quality is closely related to the surrounding environment and prevalent land use (APHA, 1992). Nambul River is one of the major river in Manipur (23.80°N to 25.68°N Latitude and 93.03°E to 94.78°E longitude) with 62.70 kms length, originates from Kangchup Hill range in the western side at an elevation of 1830 m above mean sea level. The river flows through the thickly populated area of the city and ultimately discharges into the Loktak lake, the biggest freshwater lake in the North-Eastern Region. The potentially polluted stretch of the river is within the Imphal Municipality area for a length of about 1.45 kms and its tributary Naga Nala for a length of about 1 km. The people inhabiting around this river uses the partially treated water for bathing and washing purposes. The present study attempts to analyze the physico-chemical property and microbiological aspect of this river water at three selected points where maximum pollution is expected during May (when the water body is low prior to monsoon) and June (just after heavy rain when the water level is high) of 1994, 1998 and 2003 respectively and the results were compared with the standard value for drinking water given by WHO (1984), Ministry of Works and Housing(1975) and I.S. 2966 : 1982.

Materials and Methods

The water samples were collected in the morning hours from three different Stations:

Station 1: Naoremthong–2.5 km upstream from main market

Station 2: Hump Bridge–0 km from main market

Station 3: Keisamthong–1.5 km downstream from main market.

Parameters like temperature, pH, was measured on the spot. Other parameters were analyzed in the laboratory by the adopted methods of APHA (1992).

Results and Discussion

The results for the analysis of water samples collected from the three stations are shown in Tables 20.2–20.4. Table 20.1 shows the standard of drinking water recommended by WHO, Ministry of Works and Housing and I.S 2966: 1982.

Table 20.1: Drinking Water Standards

Sl.No.	*Parameters*	*WHO, 1984* *Maximum Permissible*	*Ministry of Works and Housing 1975 Acceptable*	*I.S. 2966 : 1982 Conventional Treatment Followed Disinfection*
1.	Temp. °C			
2.	pH	7.0–8.5	7.0–8.5	6.5–8.5
3.	Turbidity (NTU)	25	2.5	–
4.	Alkalinity (mg/l)	120	–	–
5.	Conductivity (μ mho/cm)	–	–	–
6.	Dissolved oxygen (mg/l)	6	–	4
7.	BOD (mg/l)	–	–	3
8.	COD (mg/l) 10	–	–	–
9.	CO_2 (mg/l)	–	–	–
10.	Hardness (mg/l)	–	–	–
11.	Nitrogen (mg/l)	–	–	–
12.	Chloride (mg/l)	250	200	600
13.	Phosphate (mg/l)	–	–	–

Contd...

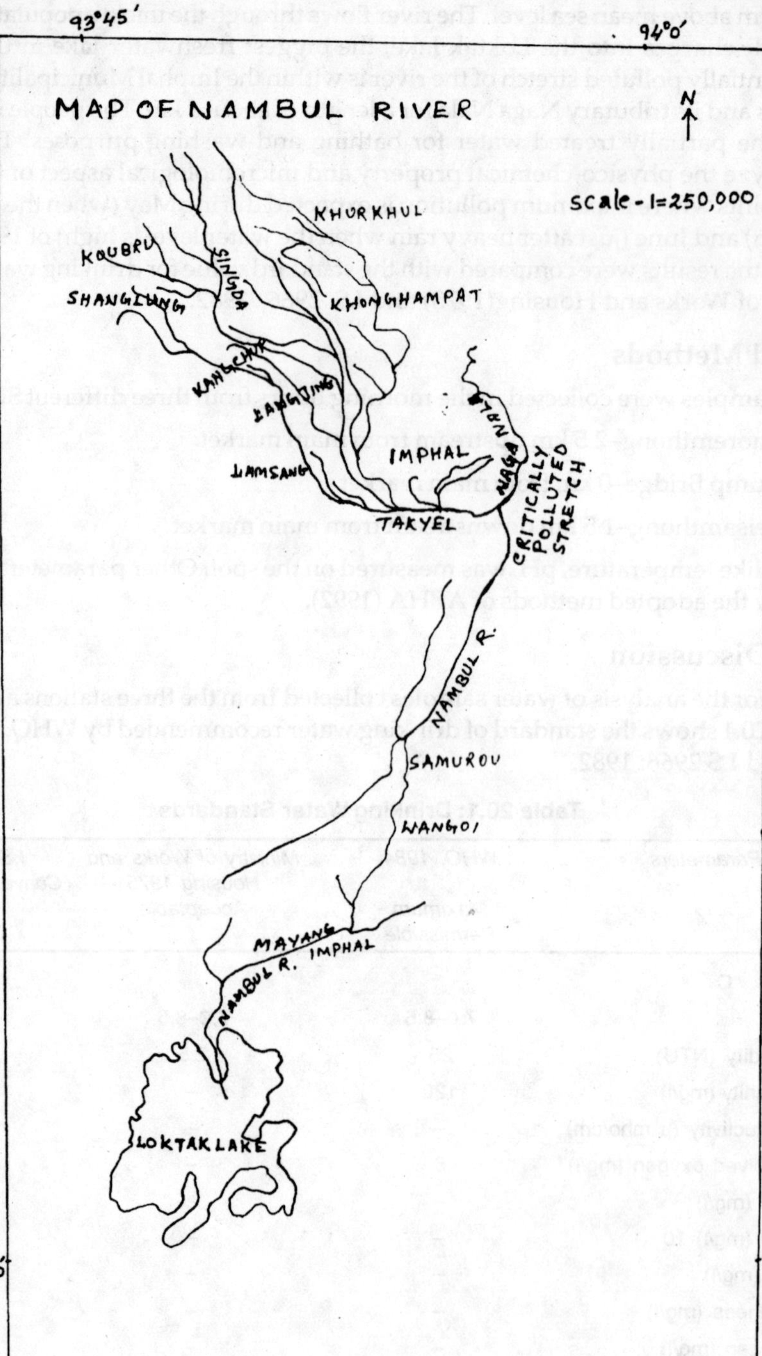

Figure 20.1: Map of Nambul River

Table 20.1–Contd...

Sl.No.	Parameters	WHO, 1984 Maximum Permissible	Ministry of Works and Housing 1975 Acceptable	I.S. 2966 : 1982 Conventional Treatment Followed Disinfection
14.	Sulphate (mg/l)	200	200	400
15.	Sodium (ppm)	–	–	–
16.	Potassium (ppm)	5–10	–	–
17.	Calcium (mg/l)	75	–	–
18.	Magnesium (mg/l)	150	200	–
19.	Total suspended solids	500	500	1500
20.	SPC/ml	–	–	–
21.	Faecal coliform count MPN/100 ml.	< 4	–	50

Table 20.2: Physico-chemical and Microbiological Analysis of Nambul at Naoremthong During May and June

Sl.No.	Parameters	May			June		
		1994	1998	2003	1994	1998	2003
1.	Temp °C	18.0	25	26	25.0	27.0	25.0
2.	pH	6.5	7.2	7.4	6.9	7.1	7.1
3.	Turbidity (NTU)	48.0	98.0	110.0	160	103	110
4.	Alkalinity (mg/l)	40.0	30.0	35.0	67	32	35
5.	Conductivity (μMho/cm)	128.7	113.0	147.0	53	120	91
6.	Dissolved oxygen (mg/l)	4.83	6.2	7.08	11.25	7.0	7.8
7.	BOD (mg/l)	4.06	2.0	1.21	2.82	1.8	2.43
8.	COD (mg/l)	12.19	4.1	3.7	7.1	6.3	6.8
9.	CO_2 (mg/l)	20.12	20.0	28.0	7.2	23.0	26.0
10.	Hardness (mg/l)	53.0	20.0	27.0	48	18	17
11.	Nitrogen (mg/l)	0.48	0.4	0.53	0.32	0.45	0.58
12.	Chloride (mg/l)	11.36	15.0	17.6	9.6	17.0	10.7
13.	Phosphate (mg/l)	0.18	0.10	0.13	0.09	0.09	0.12
14.	Sulphate (mg/l)	57.0	48	60	43.6	43.0	60.0
15.	Sodium (ppm)	19.0	13.0	15.0	32.0	20.0	65.0
16.	Potassium (ppm)	8.0	7.0	10.0	13.0	16.0	16.0
17.	Calcium (mg/l)	6.41	16.0	16.0	13.9	16.0	13.0
18.	Magnesium (mg/l)	0.01	4.0	11.0	9.6	2.0	4
19.	Total suspended solids	8.0	80.0	96.35	150.0	82.0	86.0
20.	SPC/ml	15,500	3500	6,000	4,200	3,700	6,500
21.	Faecal coliform count MPN/100 ml.	8,800	110	200	360	130	250

Table 20.3: Physico-chemical and Microbiological Analysis of Nambul at Hump Bridge During May and June

Sl.No.	Parameters	May			June		
		1994	1998	2003	1994	1998	2003
1.	Temp °C	18.5	25.2	26.1	24.5	26.9	25.2
2.	pH	6.9	7.9	7.7	7.1	7.7	7.2
3.	Turbidity (NTU)	68.0.	110.0	113.0	240	112	109
4.	Alkalinity (mg/l)	70.0	42.0	40.0	79.9	36	37.5
5.	Conductivity (μMho/cm)	351.10	120.0	153.0	39.6	126	130
6.	Dissolved oxygen (mg/l)	2.2	3.0	3.85	8.84	3.2	5.28
7.	BOD (mg/l)	80.4	38.0	40.0	8.04	42	53.24
8.	COD (mg/l)	149.9	67.0	90.0	12,8	64.0	98.3
9.	CO_2 (mg/l)	30.8	28.0	30.0	5.9	25.0	32.0
10.	Hardness (mg/l)	94.0	28.0	32.0	59	15	18
11.	Nitrogen (mg/l)	0.49	0.45	0.60	0.39	0.42	0.49
12.	Chloride (mg/l)	31.24	21.0	28.0	19.3	23.0	23.2
13.	Phosphate (mg/l)	1.96	53	0.96	0.39	0.48	0.5
14.	Sulphate (mg/l)	68.0	53	62	42.9	56.0	62.0
15.	Sodium (ppm)	33.0	14.0	17.0	30.1	24.0	67.0
16.	Potassium (ppm)	24.0	9.0	13.0	19.9	18.0	20.0
17.	Calcium (mg/l)	16.83	20.0	20.0	19.6	12.0	14.0
18.	Magnesium (mg/l)	12.67	8.0	12.0	13.1	3.0	4
19.	Total Suspended Solids	27.1	78.0	51.70	200.0	88.0	30.0
20.	SPC/ml	4,75,000	4700	8000	45000	5200	7900
21.	Faecal coliform count MPN/100ml.	16,866	450	580	900	400	600

Temperature ranged from 18°C to 27°C. During 1994 and 1998 temperature in May was lower than June but found to be higher in 2003 which can influence the property of the water body. The pH ranges from 6.5 to 7.9 and is lesser than the recommended value which may increase the corrosive activity on surfaces. Turbidity increases as the water volume increased after rain which ranged between 48.0 to 306 and was much higher than the prescribed range showing pollution. Alkalinity decreases as the year passed however the recorded value was found within the limits (WHO). Conductivity was also found to be decreased when the water volume increases after rain. Parameters like carbon dioxide content, nitrogen and sulphate were found more or less similar in all the tested samples during the study period. During June the dissolved oxygen content increases. BOD value was lower in the upstream showing less pollution but in the downstream its value increases much higher than the recommended value. Values of hardness, chloride, phosphate and magnesium decreased as the water volume increased in June. The estimated value of potassium was lower in 1994 but during 1998 and 2003 the values were higher showing more pollution due to disposals. Calcium varied from 6.41 to 21 and was found to be within acceptable range. Total suspended values ranged between 8 and 206 which is within acceptable range. Bacterial count was fairly high showing the degree of pollution. The data on faecal coliform count also suggest that nature of pollution which is much higher than the acceptable value. The value was higher during May and found highest in 1994.

Table 20.4: Physico-chemical and Microbiological Analysis of Nambul at Keisamthong During May and June

Sl.No.	*Parameters*	*May*			*June*		
		1994	*1998*	*2003*	*1994*	*1998*	*2003*
1.	Temp. °C	19.0	25.5	26.3	24.6	26.8	25.2
2.	pH	6.9	7.7	7.6	7.3	7.4	7.2
3.	Turbidity (NTU)	62.0	97.0	114.0	306	117	109
4.	Alkalinity (mg/l)	75.0.	44.5	47.5	83.6	37	42.5
5.	Conductivity (μMho/cm)	374.46	117.0	149.0	44.2	123	159
6.	Dissolved oxygen (mg/l)	1.9	3.2	3.04	9.24	3.6	5.67
7.	BOD (mg/l)	76.38	32.0	43.0	9.26	40	44.0
8.	COD (mg/l)	146.7	70.0	98.6	11.2	72.0	84.0
9.	CO_2 (mg/l)	35.2	26.0	20.6	6.3	24.0	30.0
10.	Hardness (mg/l)	54.0	24.0	34.0	54	20	21
11.	Nitrogen (mg/l)	0.50	0.47	0.57	0.36	0.43	0.52
12.	Chloride (mg/l)	28.40	18.0	23.6	20.3	19.0	19.5
13.	Phosphate (mg/!)	1.03	0.44	0.50	0.46	0.45	0.56
14.	Sulphate (mg/l)	54.0	47	56	41.3	49.0	56.0
15.	Sodium (ppm)	32.0	17.0	19.0	29.6	23.0	59.0
16.	Potassium (ppm)	26.0	12.0	15.0	18.6	17.0	22.0
17.	Calcium (mg/l)	18.43	18.0	21.0	17.6	16.0	15.0
18.	Magnesium (mg/l)	1.95	6.0	13.0	12.4	4.0	6
19.	Total Suspended Solids	26.9	83.0	69.15	206.0	87.0	51.0
20.	SPC/ml	3,60,000	4500	7,500	60,000	4500	7800
21.	Faecal Coliform count MPN/100 ml.	14,666	480	620	1260	420	750

Manipur is not an industrial state and moreover the course of the river do not receive bulk volume of out flowing rice field water. Thus the main pollutants added to the river system comes from domestic waste/refuge, sewage, animal excreta and hospital waste. As per the standard norms, 0.4 to 0.5 kg of waste per capita gets generated daily. Approximately 85 tonnes of waste is being generated from the city of Imphal everyday which contributes towards the degradation of the river water quality. Thus it is suggested that proper waste management should be taken up before evacuating into the water body during rain.

References

APHA, AWWA, WEF, 1992. *Standard Methods for the Examination of Water and Wastewater*, 18th edition. American Public Health Association Washington D.C.

Devi Jayalakshmi, O., Belagali, S.L, Ramaswamy, S.N., Dass, R.S. and Janardhana, G.R., 2005. Investigation on water quality characteristics in selected areas of Karnataka state, India. *Ind. J. Environ. Ecoplan.*, 10: 405–408.

Sharma, G.C., 1993. A preliminary pollution study of Imphal City, Manipur, Paper presented at Workshop on SAVE IMPHAL" organised by Environment Protection Society, Manipur.

Sharma, M.G. and Singh, T.H., 1988 Limnological studies of major water bodies of Manipur Report submitted to UGC, Government of India.

Singh, R.P., 1982. *Geography of Manipur*. National Book Trust, India, p. 1–185.

Trivedy and Goel, 1984. *Chemical and Biological Methods for Water Pollution Studies*. Environmental Publication, Karad (India), p. 1–251.

WHO, 1984. *Guidelines for Drinking Water Quality*. WHO Geneva, Vol. 1 and 2: Recommend, WHO, Geneva.

Chapter 21

Influence of Diet Composition and Starvation on the Growth of the Fingerlings of the Grass Carp *Ctenopharyngodon idella*

R. Sivakami[1], R. Ramya[1], M. Suganya[1], M. Marieeswari[1] and G. Premkishore[2]

[1]P.G. and Research Department of Microbiology, Kurinji College of Arts and Science, Trichirappalli – 620 002, Tamil Nadu

[2]Alan Feldman Public School and Junior College, Kazhzkuttam, Tirvandrum, Kerala

ABSTRACT

With the exploding human population, all the third world countries especially India is faced with a precarious situation to feed its people with a protein rich diet. Aquaculture, specifically the production of fish, comes as a relief as fish is rich in protein. However, one of the serious factors that confront an aquaculturist is feed. Today, most of the feeds consist of fish meal. Hence it was thought worthwhile to attempt a study on fish growth using a low cost commercial feed in the grass carp, *Ctenopharyngodon idella*. In addition, a study was also attempted to assess the ability of the fish to withstand starvation and hence growth during star vation was also studied. Results indicate that there was a significant difference ($p < 0.05$) in growth between feeding and starvation. Biochemical changes in the selected tissues of fish showed differences in their composition during starvation. Liver appeared to be the most affected tissue indicating that the energy stored in the liver is utilized first to overcome the stress caused during starvation.

Introduction

A clear understanding of the nutritional requirements of fish and feeding schedules is a must to develop a suitable cost effective balanced diet in fish culture (Naik *et al.*, 2000). The quantity and quality of feed given greatly influences the fish growth and thus the production. In order to assess the ecological success of a fish in a particular habitat, a knowledge of its efficiency to convert the food material into its body substance is of much importance (Kinne, 1960, 70, 71; Brett, 1970). For this, the growth and conversion in a fish has to be observed under different food items (Ritakumari and Nair, 1979).

Aquaculture production in India has been increasing rapidly and this has resulted in an increased demand for a variety of formulated feeds. A major share in expense in intensive fish culture is accounted for by supplementary feeds. The commonly used conventional feed in India is a mixture of rice bran and oil cake in equal propositions. However, it has been found that this conventional feed is nutritionally imbalanced to achieve fast growth of fish., Further, considerable increases in the cost of ingredients of conventional feed in these days have necessitated to look for alternate protein sources. Under these circumstances, it was thought worthwhile to attempt the use of a commercially available low cost animal feed commonly used for growing aquarium fishes in the present study. This food was tested in the fingerlings of the grass carp, *Ctenophayngodon idella*. In addition, starvation studies have also been attempted with a view to understand the ability of the fish to withstand unfavourable conditions.

Materials and Methods

Fingerlings of the grass carp. *Ctenopharyngodon idella* were collected from the Fish Farmers Development Agency, Trivandrum. In the laboratory, the fishes were stocked in glass aquarium tanks of 25L capacity. The fishes were acclimated to the lab conditions using well water for a fortnight before the start of the experiment. The fishes during this period were fed daily with the experimental diet.

For the experiment, a total of 20 fishes of almost similar size and weight were chosen. Only active and healthy fishes were taken. Prior to the experiment, the fishes were starved for a period of 72 hours in order to evacuate their gut contents completely (Windell, 1966). The fishes were then placed individually in glass containers containing the same well water with which they were acclimated. Before introducing the fish, the fish was blotted with a filter, paper and weighed. As the process of weighing was quick, the fish were not subjected to undue stress. Simultaneously, 3 fishes were weighed the same way and the muscle and liver dissected out and estimated for its protein, carbohydrate and lipid levels. This was used as the control.

The food was left in the experimental tank for 6 hours a day as the exposure of food for a duration of 6 to 8 hours a day was reported to be congenial for maximum consumption (Mercybai, 1970). The remaining of the food were collected after the exposure, blotted with a filter paper and weighed. The faeces released by the fish were also collected at regular intervals and weighed after blotting with a filter paper. In the present study, fed at the rate of 5 per cent of their initial body weight throughout the experiment.

After a period of 14 days feeding, the fish was starved for a period of 24 hours and then weighed individually. Five fishes were then sacrificed, its liver and muscle dissected out and estimated for its protein, carbohydrate and lipid levels.

The remaining fishes were then subjected to starvation to assess the effect of starvation on growth. At the end of the 14th day, the fishes were again weighed. In addition, specific tissues like the liver and muscle were dissected from 5 fishes, similar tissues pooled and estimated for its protein, carbohydrate and lipid contents. While protein content was analysed after Lowry *et al.* (1951), carbohydrate was estimated after Roe (1955) and lipids after Folch *et al.* (1975).

Some physicochemical variables of the water which have an effect on the growth of fishes were also analysed. While water temperature was measured using a mercury thermometer, pH was measured using a pH meter (Elico). Dissolved oxygen, free carbondioxide, phenolphthalein and methyl orange alkalinities were analysed by following APHA (1989).

Results

The water quality variables of the media in which the experimental fishes were kept is presented in Table 21.1.The proximate composition of the commercial feed is provided in Table 21.2. The growth performance of the fingerlings of the grass carp while feeding and during starvation are provided in Table 21.3. As evident from this table, from an initial weight of 958.95 mg, the fish increased in weight to 991.35 mg thus recording a net gain of 32.40 mg during the period under feeding. The food conversion ratio at the end of the experiment was 4.45 with the survival percentage being 95 per cent. The growth of fingerlings after a period of 14 days starvation recorded a net decline in growth of 8.97 mg and the percentage survival during this period was 86.6 per cent.

Table 21.1: Physico-chemical Variables of Water

Sl.No.	*Variable*	*Range*
1.	Water Temperature (°C)	30.1–31.8
2.	pH	7.1–8.0
3.	Free carbon dioxide (mg/l)	0–10
4.	Dissolved oxygen (mg/l)	3.8–5.3
5.	Total alkalinity (mg/l)	200–250

Table 21.2: Proximate Composition of Experimental Feed

Sl.No.	*Variable*	*Content (per cent)*
1.	Protein	32
2.	Moisture	10
3.	Lipid	03
4.	Crude fibre	05

Table 21.3: Growth Performance of *C. idella* after Feeding and Starvation

Variable	*After Feeding*	*After Starvation*
Initial weight (mg)	958.95±2.7	990.26±2.18
Final weight (mg)	91.35±2.63	981.29±3.6
Weight gain (mg)	32.6±4.47	8.97±2.86
Relative growth rate (per cent)	3.61	–0.97
Survival (per cent)	95	86.8
Food Conversion Ratio	4.45±0.44	Nil

± indicates Standard deviation.

$$\text{Relative Growth Rate} = \frac{\text{Final weight} - \text{Initial weight}}{\text{Initial weight}} \times 100$$

$$\text{Food Conversion Ratio} = \frac{\text{Total food taken}}{\text{Total weight gain}}$$

The biochemical composition of the muscle and liver tissues after feeding and starvation are presented in Table 21.4. After feeding and starvation, the protein, carbohydrate and lipid contents showed changes in both the liver and muscle. While biochemical changes after starvation revealed no major changes in the muscle tissue, the liver tissue appeared to show more drastic changes, especially with regard to the carbohydrate and lipid levels than with the protein level.

Table 21.4: Biochemical Composition of Selected Tissues of the Grass Carp After Feeding and Starvation

Variable (per cent)	*Control*		*After Feeding*		*After Starvation*	
	Muscle	*Liver*	*Muscle*	*Liver*	*Muscle*	*Liver*
Protein	55.5	22.4	59.4	24.0	62.2	35.6
Carbohydrate	31.0	56.0	32.4	60.3	50.1	52.8
Lipid	13.5	21.6	8.2	15.7	7.7	11.6

Discussion

Information regarding the relationship between feed ratio and growth is important in ensuring minimal feed loss and economical production cost (Hashim *et al.*, 1999). In addition to the nutrient content in the artificial feed, feeding rate, water temperature and fish age are also considered to be the important factors affecting fish growth (Brett, 1979).

As far as the growth in terms of weight after feeding was concerned, the present study could record a growth of 32.4 mg. Comparing this with a related work done by Leena (2001) using an animal feed in the same fish, reveals almost comparable results. However, a comparison of the growth rate using a plant feed (Rani, 2001), revealed a lower growth rate in the present study. The differences in the growth rate obtained using different feeds in the same animal could be attributed to the composition of suitable organic and inorganic substances in different proportions in the various feeds.

When the fishes were exposed to starvation and then assessed for growth, it was found that after the starvation period, the fishes registered a loss in growth in terms of weight by 8.7mg. Comparing this with the growth obtained during feeding reveals a significant difference ($p < 0.05$).

Experiments with the present study indicated an FCR of 4.45. It is a well-known fact that a lower FCR value indicates abetter utilization of the feed while higher levels indicate poor utilization. Nevertheless, it is also common knowledge that FCR levels tend to be high during the initial days and the levels gradually decrease as the number of days increases. Hence, it is only natural that a high FCR level was obtained in the present study as it was only of a short duration and that there is every possibility for the FCR to decrease as days progressed. Nevertheless, an accurate measurement of the FCR can be made only after a prolonged study.

Biochemical studies were done to understand the changes in the protein, carbohydrate and lipid levels in the liver and muscle tissues after feeding. Results indicate that subtle changes have occurred in these tissues clearly indicating the role of feed in these changes. Biochemical changes after starvation revealed a decrease in the carbohydrate and lipid contents in both the muscle and liver tissues. However, among the two, liver appeared to be the most affected. The decrease in carbohydrates and lipid contents is due to their utilization under stress conditions. Carbohydrates represent the principal and immediate energy precursors of animals when exposed to stress conditions with the protein being the spare energy source during chronic conditions of stress. The reduction in carbohydrate content in the tissues may be due to the rapid degradation of stored glycogen or decreased rate of glycogenesis or glyconeogenesis or both (Sivakami *et al.*, 1995). Such a condition was also noticed by Jayachandran and Chockalingam (1986) in *Channa puntatus*, Premkishore and Chandran (1995) in freshwater turtle and Sivakami *et al.*, (1994) in *Mystus vittatus*. These observations are in line with the present study.

It does not, however, mean that all substances are simultaneously utilized. If the principal and immediate energy source is depleted, then the other sources of energy exhibit a proportional depletion as it is well known that the metabolism of these substances is inter linked through the metabolic pathway, the Tricarboxylic cycle.

References

APHA, 1989. *Standard Methods for the Examination of Water and Wastewater*, 17th edition. American Public Health Association, Washington, DC.

Brett, J.R., 1970. Temperature in animals: Fishes. In: *Marine Ecology, Vol. 1: Environmental Factors*, (Ed.) O. Kinne. Wiley Inter Science, London, 38–57 pp.

Brett, J.R., 1979. Environmental factors and growth. In: *Fish Physiology, Vol. 8: Bioenergetics and Growth*, (Eds.) Hoar, W.S., Randall, D.J and Brett, J.R. Academic Press, London, 599–675 pp.

Folch, J., Less, M. and Stoars-Stanely, A.H., 1957. A simple method for determination of total lipid from animal tissues. *J. Biol. Chem.*, 226: 497–508.

Hashim, R., Chong, A. and Ali, A., 1999. The effect of feeding rate on water quality and growth of red *Tilapia* fry cultured in portable canvas tanks. *J. Biosci.*, 10: 39–46.

Jayachandran, K.P.S. and Chockalingam, S., 1986. Effects of tannery effluent on the respiratory parameters and biochemical constituents in the air breathing fish, *Channa punctatus*. *Comp. Physiol. Ecol.*, 12: 197–200.

Kinne, O., 1960. Growth, food intake and food conversion in an europlastic fish exposed to different temperatures and salinities. *Physiol. Zool.*, 33: 288–317.

Kinne, O., 1970. Temperature in animals: Invertebrates. In: *Marine Ecology, Vol. 1: Environmental Factors*. Wiley Inter Science, London, 290–350 pp.

Kinne, O., 1971. *Marine Ecology, Vol. 2: Environmental Factors*. Wiley Inter Science, New York, 683–1244 pp.

Leena, S.N., 2001. A comparison of the growth of the finglerlings of the grass carp *Ctenopharyngodon idella* using an animal feed and a plant feed. *M.Sc. Dissertation*, University of Kerala, India.

Lowry, O.H., Rosebrough, N.J., Farr, A.L. and Randall, R.J., 1951. Protein measurement with folin phenol reagent. *J. Biol. Chem.*, 195: 265–275.

Mercybai, V., 1970. Studies on food requirements and rate of metaolism of a tropical fish. *Ph.D. Thesis*, Madurai Kamaraj University, Tamil Nadu, India.

Naik, M.K., Reddy, H.R.V. and Annappaswamy, T.S., 2000. Influence of diet composition, starvation and feeding frequency on gastric evacuation rates in catla, *Catla catla* (Hamilton) fingerlings. *Ind. J. Anim. Sci.*, 70: 1090–1093.

Premkishore, G. and Chandran, M.R., 1995. Effect of starvation on the tissue chemistry of the freshwater turtle, *M. trijuga trijuga* of Tamil Nadu, India. *Zoos Print*, 10: 44–45.

Rani, S.N., 2001. On the use of aquatic weeds (*Hydrilla* and *Pistia*) as feed for the growth of the grass carp, *Ctenopharyngodon idella. M.Sc. Dissertation*, University of Kerala, India.

Roe, J.H., 1955. The determination of sugar in blood and spinal fluid with anthrone reagent. *J. Biol. Chem.*, 212: 335–343.

Ritakumari, S.D. and Nair, B.N., 1979. Effect of feeding rate on the growth, conversion efficiency and chemical composition of *Noemacheilus triangularies* (Day) and *Lepidocephalus thermalis* (Cuv. and Val.). *J. Anim. Morphol. Physiol.*, 26: 296–306.

Sivakami, R., Premkishore, G. and Chandran, M.R., 1994. Sublethal effects of chromium on the feeding energetics and growth in the freshwater cat fish, *Mystus vittatus. J. Freshwater Biol.*, 6: 165–175.

Sivakami, R., Premkishore, G. and Chandran, M.R., 1995. Sublethal effects of mercury on the feeding energetics and body composition in the freshwater cat fish, *Mystus vittatus. J. Aqua. Trop.*, 10: 109–117.

Windell, J.T., 1966. Rate of digestion in the blue gill sunfish. Invest. *Indiana Lakes and Streams*, 7: 185–214.

Chapter 22

Response of Greengram (*Vigna radiata* L. Wilczek) to Raw and Biological Treated Sugar Mill Effluent

L. Baskaran, P. Sundaramoorthy, K. Sankar Ganesh, K. Jayakumar and P. Vijayarengan

Division of Environmental Biology, Department of Botany, Annamalai University, Annamalainagar – 608 002, Tamil Nadu, India

ABSTRACT

Sugar mills play a major role in polluting the water bodies by discharging a large amount of wastewater as effluent. The sugar mill effluent analysis showed that raw effluent contains the higher amount of suspended solids, dissolved solids, biological oxygen demand (BOD), chemical oxygen demand (COD) and considerable amount of chlorides, sulphates, nitrates calcium and magnesium are also present in the effluent. The presence of these chemicals in large quantities in the effluent not only affect the germination and growth of crop plants but also deteriorate the soil properties when it is used for irrigation, so it was decided to give biological treatment by growing the aquatic macrophytes such as *Eichhornia*, *Pistia* and *Ludwigia* in sugar mill effluent upto 7 days. So the present research work has been carried out to find out the effect of raw and biological treated sugar mill effluent on greengram (*Vigna radiata* L.). The various concentrations (10, 25, 50, 75 and 100 per cent) of both raw and biologically treated sugar mill effluent were used for germination studies. The various morphological parameters such as germination percentage, shoot length, root length, fresh and dry weights were taken into consideration. All these parameters increased upto 25 per cent in biological treated effluent and upto 10 per cent raw in effluent.

Keywords: *Greengram, Sugar mill effluents, Biological treatment, Photosynthetic pigment.*

Introduction

Today, water resources have been the most exploited of the natural systems, most of our water bodies is seriously polluted due to rapid population growth, industrial proliferations, urbanizations, increasing living standards and wide spheres of human activities. Groundwater, rivers, sea, lakes, ponds are streams and finding it more and more difficult to escape from pollution. Many rivers of the world receive heavy flux due to sewage industrial effluents. The waste consisting of substances varying from simple nutrient to highly toxic hazardous chemicals water is necessary for irrigation to increase the productivity in agriculture. Most of the Indian rivers are already polluted by industrial effluents, different kinds of industries play a major role for polluting the water bodies, like sugar industries, Sago factory tannery, textile, dairy industries etc. Among them sugar industries produced large amount of wastewater as effluent. The effluent contains more amount of physico-chemical elements and toxic in nature. They are harmfully affect to agricultural areas, when it is used for irrigation, so we want to take some treatment to that kinds of effluents then only will make a good source of environment in that way. Biological treatment is an eco-friendly technique, evolved to remove toxic pollutants from solutions. It has been gained importance in decade over the old conventional chemicals which generated secondary pollutants in the environment (Patterson, 1978). Biological treatment involves the use of aquatic plants for the removal of contaminants such as organic, inorganic and heavy metals from aqueous solution (Banuelos *et al.*, 1997). The role of macrophytes as bio-filter in water bodies to cleaning the toxic contents from the water. Presence of pollutants in water body accumulate the aquatic plants and help in reducing concentration of metallic contaminants (Parashar *et al.*, 2003). Water hyacinth, an aquatic weed has the fastest growing capacity in the world. The water hyacinth can be used for the treatment of sugar mill effluent and the treated wastewater can also be successfully used for the irrigation. Number of works are already made the biological treatment on industrial effluents (Singaram, 1995; Sundaramoorthy, 2003; Rogers and Davis *et al.*, 1994; Parashar *et al.*, 2003). So the present investigation deals with the response of greengram to raw and biological treated sugars mill effluent.

Materials and Methods

Sugar Mill Effluent Sample

The effluent sample was collected from N.P.K.R. Ramasamy Sugar Mill Ltd., Thalainayar, Nagapattinam District of Tamil Nadu. The collected effluent sample was stored in plastic containers and they are used to biological treatment. The Raw and Biological treated effluents were analysed for its various physico-chemicals parameters as per APHA (1992).

Three types of aquatic plants were used for biological treatment to sugar mill effluent. The aquatic plants such as *Eichhornia, Pistia* and *Ludwigia* were collected from freshwater ponds. The experimental crop greengram was collected from Tamil Nadu Rice Research Station Aduthurai, Tamil Nadu. The seeds were used for germination.

Five litres of sugar mill effluent was taken in plastic container. The same aged aquatic plants (1 kg of Biomass) such as *Eichhornia, Pistia* and *Ludwigia* were allowed to grow on the effluent. The experiment was conducted with three different plants in separate containers. Plants were allowed to grow on effluent upto 13 days. After seven day the treated effluent was taken from the plastic container and various physico-chemical parameters were analysed. The different concentrations of this biological treated effluent were prepared and used for germination studies.

Seed Germination

Germination studies were conducted in with greengram seeds irrigated with biologically treated sugar effluent. The morphological parameters such as germination percentage, seeding length and dry weight were measured and recorded on 7th day old seedlings.

Results and Discussion

The physico-chemical properties of raw and biological treated sugar mill effluent are presented in Table 22.1. It reveals that there was a considerable amount of chloride, total dissolved solids, BOD, COD, calcium, magnesium, sodium, potassium, iron and sulphate. The results are in agreement with previous workers of Chandrasekar *et al.* (1998), Kumar (1999), Thamizhiniyan (2000), Ramakrishnan *et al.* (2001) and Boral and Patil (2004). The variation in physico-chemical properties may be due to the process involved raw materials used in sugar separation. At the same time the biological treated sugar mill effluent reduces considerable amounts of pollutant. The same findings were also noted in various effluent treatments (Gopal Sharma, 1981 ; Gupta, 1982; Rogers and Davis *et al.*, 1972). In the process three types of aquatic macrophytes such as *Eichhornia, Pistia, Ludwigia* were used. All the three plants reduced the pollutants in the sugar mill effluent. Among them the *Eichhornia* treated effluent reduce considerable amount of pollutants when compared with all the other treatment the aquatic vascular plants possess a tremendous capacity absorbing nutrients and other substances from the wastewater.

Table 22.1: Physico-chemical Analysis of Raw and Biological Treated Sugar Mill Effluent

Sl.No.	*Physico-chemical Parameters*	*Name of the Aquatic Macrophytes*			
		Raw Effluent	*Eichhornia*	*Pistia*	*Ludwigia*
1.	pH	4.55	6.6	5.5	5.3
2.	Electrical conductivity	2.23	2.42	2.38	2.32
3.	Chloride	26	32	28	28
4.	Total dissolved solids	1480	1508	1492	1488
5.	BOD	1783	1448	1672	1542
6.	COD	2359	1983	2150	2218
7.	Total hardness as $CaCO_3$	245	228	240	238
8.	Calcium as $CaCO_3$	91	80	86	87
9.	Magnesium as $MgCO_3$	224	210	206	213
10.	Sodium	40	32	36	36
11.	Potassium	0	0	0	0
12.	Iron	42	35	38	38
13.	Sulphate	246	236	238	240

All parameters except colour, odour, pH, EC and temperature are expressed in mg/L.

Table 22.2 shows that the untreated sugar mill effluent effect on seed germination, seedling growth, fresh and dry weights of greengram. All the parameters are increased at 10 per cent effluent concentration. At the same time the higher concentrations of all the parameters are decreased. The same findings were noted earlier reports, Goel and Kulkami (1994); Chandrasekar *et al.* (1998); Thamizhiniyan *et al.* (2000); Kumar (2000); Ramakrishnan *et al.* (2001); Krishna and Leelavathi (2002).

The increase in growth might be due to reduction in level of toxic metabolites by dilution and better utilization of nutrients present in the effluent (Kannan, 2000).

Table 22.2: Effect of Different Concentrations of Sugar Mill Effluent on Shoot Length, Root Length, Fresh and Dry Weights of Greengram (*Vigna radiata* L.) Seedlings

Effluent Concentrations (%)	*Germination Percentage*	*Shoot Length (cm/seedling)*	*Root Length (cm/seedling)*	*Fresh Weight (g/seedling)*	*Dry Weight (g/seedling)*
Control	96±4.80	15.3±0.765	7.2±0.360	0.53±0.026	0.29±0.0145
10	100±5.0	17.4±0.870	9.6±0.48	0.63±0.031	0.35±0.017
25	85±4.25	13.2±0.660	5.3±0.265	0.50±0.025	0.22±0.011
50	72±3.60	11.9±0.595	6.1±0.305	0.48±0.024	0.10±0.005
75	66±3.30	11.0±0.550	5.0±0.250	0.39±0.019	0.10±0.005
100	45±2.25	8.6±0.430	2.2±0.11	0.20±0.012	0.10±0.005

±: Standard deviation.

The reduction in germination at higher concentration may also be due to the excess amount of minerals and nutrients in the effluent (Kumar, 2000) the untreated sugar mill acidic in nature and contains a considerable amount various forms of pollutants the effluent has high BOD and COD values. However the biological treated sugar mill effluent showed a reduction of 30 per cent BOD (Subramani *et al.*, 1999). The germination studies of green gram grown under different concentrations of biologically treated sugar mill effluent. All the parameters are increased upto 25 per cent effluent (Table 22.3) concentration. It is in agreement with the findings of Subramani *et al.* (1999); Thamizhiniyan (1995). This was due to the reduction in drastic effects of the effluent by biological treatment to a considerable extent (Subramani *et al.*, 1999).

Table 22.3: Effect of Biologically Treated Sugar Mill Effluent on Germination Studies of Greengram (*Vigna radiata* L.)

Effluent Concentrations (%)	*Eichhornia Treated Effluent*			*Pistia Treated Effluent*			*Ludwigia Treated Effluent*		
	G %	*SLG*	*SDW*	*G %*	*SLG*	*SDW*	*G %*	*SLG*	*SDW*
Control	96 ±4.80	23.3 ±1.165	0.82 ±0.041	96 ±4.80	23.3 ±1.165	0.82 ±0.041	96 ±4.80	23.3 ±1.165	0.82 ±0.041
10	100 ±5.0	28.3 ±1.415	1.10 ±0.055	100 ±5.0	28.2 ±1.41	1.08 ±0.054	100 ±5.0	27.8 ±1.390	0.98 ±0.049
25	100 ±5.0	29.8 ±1.490	1.28 ±0.064	100 ±5.0	29.5 ±1.475	1.16 ±0.058	100 ±5.0	28.5 ±1.425	1.12 ±0.056
50	90 ±4.50	22.2 ±1.110	0.78 ±0.039	88 ±4.40	21.0 ±1.05	0.73 ±0.036	86 ±4.25	20.2 ±1.01	0.70 ±0.035
75	72 ±3.60	18.5 ±0.925	0.66 ±0.033	70 ±3.50	18.2 ±0.910	0.56 ±0.028	70 ±3.50	17.8 ±0.890	0.52 ±0.026
100	60 ±3.0	12.6 ±0.630	0.48 ±0.024	58 ±2.90	12.0 ±0.60	0.42 ±0.021	56 ±2.80	11.2 ±0.56	0.40 ±0.020

±: Standard deviation; G%: Germination percentage; SLG: Seedling length; SDW: Seedling dry weight.

References

APHA, AWWA and WPCF, 1992. *Standard Method for the Examination of Water and Wastewater,* 17th Edition. Washington, DC.

Banuelos, G.S., Ajwa, H.A., Wu, B., Mackey, L., Cook, C., Akohoue, S. and Zambruzuski, S., 1997. Evaluation of different plant species used for phytoremediation of high soil selenium. *J. Environ. Qual.*, 26: 639–646.

Borale, D.D. and Patil, P.R., 2004. Studies on physico-chemical parameters and concentration of heavy metals in sugar industry. *Poll. Res.*, 23: 83–86.

Chandrasekar, W., Subramani, A. and Saravanan, S., 1998. Fffect of sugar mill effluent on germination and early seedling growth of blackgram (*Vigna mungo* L.) Hepper var. ADT 3. *J. Indust. Poll. Control,* 14: 73–78.

Goel, P.K. and Kulkarni, S.M., 1994. Effects of sugar factory waste on germination of gram seeds (*Cicer arietinum* L.). *J. Environ. Poll.*, 1: 35–43.

Gopal, B. and Sharma, K.P., 1981. Water hyacinth, *Eichhornia crasipes* the most trouble some weed of the world. Hindasia Publishers, New Delhi.

Gupta, G.C., 1982. Use of water hyacinth wastewater treatments: A literature view. *Environ. Hlth.*, 43: 980–982.

Kannan, J., 2001. Effect of distillery effluents on crop plants. *Ad. Plant Sci.*, 14: 127–132.

Krishna, K. and Leelavathi, S., 2002. Toxicity of sugar factory effluent to germination, vigour index and chlorophyll content of paddy. *Nat. Environ. Poll. Technol.*, 1: 249–253.

Kumar, A., 2000. Carbonaceous sugar mill effluent on germination of *Hordeum vulgare* 1865. *Ad. Plant Sci.*, 2: 49–52.

Parashar, V., Dubey, A., Bajpai, A. and Mishra, S.M., 2003. Biomagnification of heavy metals in macrophytes of a tropical wetland. *Poll. Res.*, 22: 569–576.

Patterson, W.J., 1978. *Wastewater Treatment Technology,* 3rd Ed., Ann Arbor Science Publishers, Ann Arbor, pp. 265.

Ramakrishnan, K., Thamizhiniyan, P., Sundaramoorthy, P. and Senthilkumar, R.S., 2001. Effect of sugar mill raw and yeast treated effluent on seed germination and seedling growth of paddy (*Oryza sativa* L. ADT 36). *Asian J. Microbiol. Biotech. Env. Sci.*, 3: 37–40.

Rogers, H.H. and Davis, D.E., 1972. Nutrient removal by water hyacinth. *Weed Sci.*, 20: 423–528.

Singaram, P., 1995. Effect of industrial and sewage effluents on soil and crop production. In: *Agricultural Inputs and Environment,* (Ed.) S.P. Palaniappan. Scientific Publishers, Jodhpur, pp. 357–388.

Subramani, A., Sundaramoorthy, P., Saravanan, S., Selvaraj, M. and Lakshmanachary, A.S., 1999. Impact of biologically treated distillery effluent on growth behaviour of greengram [*Vigna radiata* (L.) Wilczek]. *J Indust. Poll. Control,* 75: 281–286.

Sundaramoorthy, P., 2003. Physical, chemical and biological treatments to textile dyeing industry effluent. *INSA Report,* 89.

Sundaramoorthy, P., Lakshmi, S. and Rajesh, M., 2003. Screening of paddy cultivars for tolerance to sugar mill effluent irrigation. *J. Annamalai Univ., Part B, Science*, p. 53–58.

Thamizhiniyan, P., 1995. Effect of sugar mill raw effluent and biologically treated effluent on germination, growth and biochemical changes of groundnut (*Arachis hypogaea* L. VRI 3) and paddy (*Oryza sativa* L. var. TKM 9). *Ph.D. Thesis*, Annamalai University.

Thamizhiniyan, P., Sundaramoorthy, P. and Lakshmanachary, A.S., 2000. Effect of germination, growth and pigment contents of groundnut (*Arachis hypogaea* L.) and paddy (*Oryza sativa* L.). *Indian J. Appl. Pure Biol.*, 15: 151–155.

Chapter 23

Water Quality Studies on Freshwater Lakes of Kolar Taluk, Kolar, Karnataka

H. Krishna Ram, M. Ramachandra Mohan, Chithanagiri and N. Latha

Water Quality Research Laboratory, Department of Zoology, Bangalore University, Bangalore – 560 056, Karnataka, India

ABSTRACT

Water is one of the five elements required for the very existence of life. Freshwaters most basic and important benefit is health of peoples. Ecosystem and societies, but tragedy of the present century in that human negligence is depleting this vital source indispensable for the very existence of human race on earth. The immediate and future well being of all countries depends on their abilities to manage water resources more effectively. In lake one the water will not be fit for domestic purpose, only for irrigation purposes and other three lakes were used for domestic and irrigational purpose. The study reveals that all the parameters are varied in nature.

Keywords: *Water quality, Nutrients, Kolar.*

Introduction

Water is considered as one of the most basic and important element for human civilization our dependence on freshwater resources has accelerated in the last century due to rapid world population growth and economic development (Postel, 1992). As a result, freshwater resources have deteriorated both in quantity and quality in many areas of the world, lakes providing the majority of easily accessible freshwater resources, have been most influenced. Many lakes in the world suffer environmental problems such man made eutrophication, chemical contamination etc. consequently causing ecosystem and human health damage.

Freshwater is Man's most precious needs. Since the beginning of civilization man has tried to settle at the banks of great rivers which supplied freshwater not only for the basic physiological needs of man but also for various socio-economic requirement. The freshwater present on the surface of the earth is put to multifarious uses, it is used for drinking, domestic and municipal uses, agricultural irrigation, industries navigation, recreation etc. after being used in the above mentioned human activities the water flowing out becomes contaminated. Some of the extensive contributions on Limnological studies of freshwater bodies in India are of Zutshi and Vas (1978). Harshey *et al.* (1982), Kaul Sidharth (1983), Zafar (1964), Munawar (1970), Seenayya (1971) and Rao (1991). This study therefore attempts to evaluate the quality of lentic waters of Kolar Taluk in South of Karnataka State with respect to domestic and irrigational purposes.

Materials and Methods

Study Site

The study was conducted in Kolar District. It is one of the districts falling a madian area of Karnataka having a geographical area of about 8240.00 sq km with the population of about 25 lakhs. It is a predominantly on agricultural district. Where in the major profession of the rural mass is agriculture the average annual rainfall of,the district, is between 700 to 750 mm. In this study we are selected Kolar lakes of Kolar Dist.

Kolar Amanikere–1

It is situated on the Kolar–Bangalore Road (NH4) to the East of Kolar city. It is a natural lake with a total area of one and half square kilometers and an average depth 2.20 m. Now the lake turned out to be unfit for human consumption and for agricultural and industrial purposes. It receives enormous amount of domestic sewage. At the very look the water appears greenish with periodic algal blooms of throughout the year.

Tambihally Lake–2

It is located 15 km far from district head quarters of Kolar. The lake is man made rain fed and ofterly used by people for cattle drinking and cloths washing the catchment area is 0.64 sq.km and water spread area is 5.76 hectares. It receives enormous amount of Agricultural wastes.

Harati Lake–3

It is located 20 km far from district head quarters of Kolar. The lake is man made fed by rain. The lake is run off materials from surrounding agricultural plots. The lake used by people for cattle drinking and washing. The catchment area is 0.77 sq.km and water spread area is 1.28 hectares.

Muduvadi Lake–4

It is located 15 km far from district headquarters of Kolar. The lake is man made, rain fed, and ofterly used by people for agricultural purposes. The catchment area is 14.74 sq.km and water spread area is 266.25 hectares.

Physio-chemical Analysis

Water samples were collected during the period January 2005 to December 2005, in the polythene bottles before 10 am for analysis of following standard methods. Trivedy and Goal (1984) and Atomic absorption sepectrophotometery (AAS).

Results and Discussion

The physio-chemical parameters of an aquatic ecosystem is an essential component which drastically effects the flora and fauna of it.

pH

pH represents the effective concentration (activity) of hydrogen ions (H^+) in water. In the present study of lakes the pH of the lake-1, Maximum', pH was recorded 8.3 in site-7 and minimum pH was recorded 7.0 in site-1. In lake-2, the maximum pH was recorded 8.4 in site-7 and minimum pH was recorded 7.1 in site-2 (Table 23.1). In lake-3, the maximum pH was recorded 7.8 in site-2 and minimum pH was recorded 7.1 in site-7. Inlake-4 the maximum pH was recorded 8.1 in site-5 and minimum pH was recorded 7.3 in sites-1, 6, 7 (Table 23.2). pH gives an idea to the type and intensity of pollution (Verma *et al.*, 1984 and Mishra and Saksena, 1991) and pH is one of the most important single factor which influences aquatic production (Swingle., 1967). The life process and growth of microorganisms is influenced by H-ion concentration of the water a greater extent (Shastree, 1991) observed that pH was not very significant with reference to other physico-chemical as well as biological features.

Table 23.1: Physio-chemical Characteristics of Kolar Amanikere and Tambihally Lake 2005

Parameters	*Kolar Amanikere*							*Tambihally Lake*						
	KS1	*KS2*	*KS3*	*KS4*	*KS5*	*KS6*	*KS7*	*TS1*	*TS2*	*TS3*	*TS4*	*TS5*	*TS6*	*TS7*
pH	7	7.5	7.8	7.9	7.7	7.6	8.3	7.3	7.1	7.8	7.5	7.3	7.5	8.4
Conductivity	2200	1800	2500	1010	1500	1640	1840	1980	2200	2500	4700	3000	1580	1320
Hardness	564	512	920	324	520	516	496	560	320	780	1840	932	416	292
TDS	1470	1050	1510	570	870	940	1070	1140	1280	1460	2620	1750	910	758
Carbonate (CO_3)	24	16	9	10	8	6	2	0	14	8	7	12	5	14
Bicarbonate (HCO_3)	552	534	356	343	339	362	368	421	578	306	279	181	397	196
Calcium (Ca)	208	164	228	75	126	126	114	200	93	263	547	270	142	77
Magnesium (Mg)	41	24	88	34	53	50	53	15	22	29	118	64	15	25
Sodium (Na)	176	170	142	79	98	134	188	197	357	208	222	250	173	170
Potassium (K)	23	4	9	5	5	8	8	3	0	7	14	16	0	2
Iron (Fe)	0.20	0	0	0.10	0.05	0.40	0.03	0.04	0.30	0.00	0.30	0.03	0.2	0.15
Sulphates (SO_4)	340	80	100	56	123	130	152	72	67	126	117	145	60	50
Chloride (Cl)	364	263	144	170	232	269	3i6	398	330	608	1414	120	283	294
Nitrate (NO_3)	10	35	27	33	25	11	22	13	72	20	7	73	7	40
Floride (F)	0.40	1.00	0.60	0.50	0.75	1.00	0.95	1.30	0.80	1.25	0.90	1.00	1.6	1.6

All the parameters expressed in mg/ltr except pH, conductivity, micro mhos.

Conductivity

Conductivity is a measure of how well water can conduct an electrical current. In the present investigation the lake-1, Maximum conductivity was recorded 2500 micro mhos in site-3 and minimum conductivity was recorded 1010 micro mhos in site-4. In lake-2, the maximum conductivity was recorded 4700 micro mhos in site-4 and minimum conductivity was recorded 1320 micro mhos in site-7 (Table 23.1). In lake-3, the maximum conductivity was recorded 2700 micro mhos in site-6 and minimum conductivity was recorded 580 micro mhos in site-1. In lake-4 the maximum conductivity was recorded 1240 micro mhos in site-1 and minimum conductivity was recorded 650 micro, mhos in site-4, (Table 23.2). Conductivity increases with increasing amount and mobility of ions these ions, which come from the breakdown of compounds, conduct electricity because they are negatively and positively charged when dissolved in water (CDPHE-WQCD) the increase concentration of salts because of discharge of domestic effluents and organic matter from the near by residential areas. Which bring along with in the lake. High levels of conductivity reflects on the pollution status as well as trophic levels of the aquatic body (Anitha *et al.*, 2005).

Table 23.2: Physio-chemical Characteristics of Harati Lake and Muduvadi Lake–2005

Parameters	*Hariti Lake*							*Muduvadi Lake*						
	HS1	*HS2*	*HS3*	*HS4*	*HS5*	*HS6*	*HS7*	*MS1*	*MS2*	*MS3*	*MS4*	*MS5*	*MS6*	*MS7*
pH	7.2	7.8	7.4	7.2	7.3	7.7	7.1	7.3	8.0	7.4	7.8	8.1	7.3	7.3
Conductivity	580	1000	880	1090	2000	2700	2000	1240	740	890	650	860	790	730
Hardness	200	280	244	284	588	816	640	300	240	236	212	348	304	296
TDS	350	590	520	680	1190	1610	1180	720	415	530	380	495	450	410
Carbonate (CO_3)	10	0	9	0	0	0	0	22	0	9	19	14	0	19
Bicarbonate (HCO_3)	196	377	225	270	441	252	372	304	204	225	240	240	270	220
Calcium (Ca)	51	84	60	80	198	208	197	86	56	62	54	88	54	80
Magnesium (Mg)	18	16	23	21	53	60	37	21	24	20	19	32	42	24
Sodium (Na)	40	85	89	141	128	257	163	145	68	95	52	31	40	33
Potassium (K)	2	7	3	15	8	7	3	3	5	2	2	9	2	4
Iron (Fe)	0.5	0.00	0.00	0.20	0.10	0.20	0.0	0.2	0.0	0.0	0.0	0.1	0.0	0.1
Sulphates (SO_4)	26	67	58	50	144	116	97	53	29	66	31	58	32	18
Chloride (Cl)	41	76	123	171	282	614	370	196	59	123	44	104	87	89
Nitrate (NO_3)	35	38	37	40	127	195	95	9	9	6	9	15	22	12
Floride (F)	2.00	1.10	1.70	0.50	1.00	1.00	1.30	2.00	2.40	1.80	1.80	0.40	1.60	0.6

All the parameters expressed in mg/ltr except pH, conductivity, micro mhos.

Hardness

Hardness is measure of polyvalent cations (ions with a charge greater than +1) in water. Hardness generally represents the concentration of calcium (Ca^{2+}) and magnesium (Mg^{2+}) ions because these are the most common polyvalent cations. In the present study lake-1, the maximum hardness was recorded 920 mg/L. in site 3 and minimum hardness was recorded 324 mg/L in site 4. In lake-2 the maximum hardness was recorded 1840 mg/L in site-4 and minimum hardness was recorded 292 mg/L in site-7 (Table 23.1). In lake-3 the maximum hardness was recorded 816 mg/L in site-6 and minimum.hardness was recorded 200 mg/L in site-1. In lake-4 the maximum hardness was recorded 348 mg/L in site-5 and minimum was recorded 212 mg/L in site-4 (Table 23.2). Hardness often employed as indicator of water quality depends on the concentration of carbonate and bicarbonate salts of calcium and magnesium (temporary hardness) or sulphate chloride or other anions of mineral acids (permanent hardness) (Anitha, *et al.*, 2005). The higher values depended upon geographical formation of the study area and also available minerals. The hardness exhibited positive association with calcium, magnesium chloride and sulphates.

TDS

Total dissolved solids as a measure of the amount of material dissolved in water. The present study lake-1, the maximum TDS was recorded 1510 mg/L in site-3 and minimum TDS was recorded 570 mg/L in site-4. In lake-2 the maximum. TDS was recorded 2620 mg/L in site-4 and minimum TDS was recorded 758 mg/L in site-7 (Table 23.1). In lake-3 the maximum TDS was recorded 1610 mg/L in site-6 and minimum TDS was recorded 350 mg/L in site-1. In lake-4 the maximum TDS was recorded 720 mg/L in site-1 and minimum TDS was recorded 380 mg/L in site-4 (Table 23.2). The changes in TDS concentrations can be harmful because the density of the water determines the flow of water into and out of an organism's cells (Mitchell and Stapp, 1992). However, TDS concentrations are too high or too low. The growth of many aquatic life can be limited and death may occur water with high TDS often has a bad taste and/or higher water hardness, and could result in a laxative effect.

Carbonate (CO_3)

In the present study the lake-1, the maximum CO_3 was recorded 24 mg/L in site-1 and minimum CO_3 was recorded 2 mg/L in site-7. In lake-2 the maximum CO_3 was recorded 14.0 mg/L in site-2 and 7 and minimum CO_3 was recorded in 0.0 mg/L in site-1 (Table 23.1). In lake-3 the maximum CO_3 was recorded 10 mg/L in site-1 and minimum CO_3 was recorded 0.0 in sites-2, 4, 5, 6, 7, Lake-4 the maximum CO_3 was recorded 22 mg/L in site-1 and minimum CO_3 was recorded 0.0 mg/L in sites-2 and 6 (Table 23.2). Carbonate alkalinity (CO_3) is a measure of the buffering capacity of water or the capacity of bases to neutralize acids, measuring carbonate alkalinity is important in determining a lakes ability and neutralize acidic pollution from rainfall or wastewater. Carbonates all added to a water system if the water passes through soil and rock that contain carbonate minerals. Carbonate ions in water can remove toxic metals (such as lead, arsenic and cadmium).

Bicarbonate (HCO_3)

In the present study lake-1 the maximum HCO_3 was recorded 552 mg/L in site-1 and minimum HCO_3 was recorded 339 mg/L in site-5. In lake-2 the maximum HCo3 was recorded 578 mg/L in site-2 and minimum HCO_3 was recorded 181 mg/L in site-5 (Table 23.1). In lake 3 the maximum HCO_3 was recorded 441 mg/L in site-5 and minimum HCO_3 was recorded 225 mg/L in site-3. In lake-4 the maximum HCO_3 was recorded 270 mg/L in site-6 and minimum HCO_3 was recorded 204 mg/L in site-2 (Tables 23.1 and 23.2). Bicarbonate alkalinity (HCO_3) in a measure of the buffering capacity of water. The capacity of bases to neutralize acids. Bicarbonates are added to a. water system if the water passes through soil and rock that contain Bicarbonate minerals. Bicarbonate ions in water can remove toxic metals (such as lead, arsonic, and cadmium) by precipitating the metals out of solution. Generally above pH 8.3, alkalinity is mostly in the form of carbonate (CO_3) and below 8.3 alkalinity is present mostly as bicarbonate (HCO_3).

Calcium (Ca)

In the present study the lake-1, maximum Ca was recorded 228 mg/L in site-3 and minimum Ca was recorded 75 mg/L in site-4. In lake-2 the maximum Ca was recorded 547 mg/L in site-4 and minimum Ca was recorded 77 mg/L in site-7 (Table 23.1). In Lake-3 the maximum Ca was recorded 208 mg/L in site-6 and minimum Ca was recorded 51 mg/L in site-1. In lake-4 the maximum Ca was recorded 88 mg/L in site-5 and minimum Ca was recorded 54 mg/L in sites-4 and 6 (Table 23.2). This element is an important component of the carbonic buffer system and also cycles through biotic and a biotic component of the eco system. Calcium was 3rd in the order of equivalent ionic dominance in the lakes (Anitha, *et al.*, 2005). Calcium as a main cations factors causing water hardness in nature water and originate from natural process *i.e.* on a dissolvent of mineral containing Ca and other sources can be industrial wastes and sometime agricultural waste. It is non-toxic.

Magnesium (Mg)

In the present study the lake-1, maximum Mg was recorded 88 mg/L in site-3 and minimum Mg was recorded 24 mg/L in site-2, In lake-2 the maximum Mg was recorded 118 mg/L in site-4 and minimum Mg was recorded 15 mg/L in site-6 (Table 23.1). In Lake-3 the maximum Mg was recorded 60 mg/L in site-6 and minimum Mg was recorded 16 mg/L in site-2. In lake-4 the maximum Mg was recorded 42 mg/L in site-6 and minimum Mg was recorded 19 mg/L in site-4 (Table 23.2). Magnesium as main cation are factors causing water hardness. In natural water Mg originate from natural process that is as dissolvent of mineral containing Mg and other sources can be industrial waste and some times agriculture waste Mg is non-toxic.

Sodium (Na)

In the present study the lake-1, maximum Na was recorded 188.mg/L in site-7 and minimum Na was recorded 79 mg/L in site-4. In lake-2 the maximum Na was recorded 357 mg/L in site-2 and minimum Na was recorded 170 mg/L in site-7 (Table 23.1). In Lake-3 the maximum Na was recorded 257 mg/L in site-6 and minimum Na was recorded 40 mg/L in site-1. In lake-4 the maximum Na was recorded 145 mg/L in site-1 and minimum Na was recorded 31 mg/L in site-5 (Table 23.2). Sodium is main cation in natural water in freshwater such as lake water. Sodium in collected from natural minerals dissolved in water. In low concentration sodium in drinking water is a mineral necessary for electrolyzing blood balance in the human bodies.

Potassium (K)

In the present study the lake-1, maximum K was recorded 23 mg/L in site-1 and minimum K was recorded 4 mg/L in site-2. In lake-2 the maximum K was recorded 16 mg/L in site-5 and minimum K was recorded. 0.0 mg/L in site-6 (Table 23.1). In Lake-3 the maximum K was recorded 15 mg/L in site-4 and minimum K was recorded 2 mg/L in site-1. In lake-4 the maximum K was recorded 9 mg/L in site-5 and minimum k was recorded 2 mg/L in sites-3, 4 and 6 (Table 23.2). Potassium is main cation in natural water in freshwater such as lake water. Potassium occurs in natural water salty through found in small amount. It plays a vital role in metabolism offish environment and is a important macro nutrients.

Iron (Fe)

In the present study the lake-1, maximum Fe was recorded 0.40 mg/L in site-6 and minimum Fe was recorded 0.0 mg/L in sites-2 and 3. In lake-2 the maximum Fe was recorded 0.30 mg/L in site-2 and 4 and minimum Fe was recorded 0.03 mg/L in site-5 (Table 23.1). In lake-3 the maximum Fe was recorded 0.5 mg/L in site-1 and minimum Fe was recorded 0.0 mg/L in sites-2, 3 and 7. In lake-4 the maximum Fe was recorded 0.20 mg/L in site-1 and minimum Fe was recorded 0.0 mg/L in sites-2, 3, 4 and 6 (Table 23.2). In aquatic environments metals have been termed as conservative pollutants because once added to the environment. They prevail forever. Further the seriousness and persistence of heavy metals in water are compounded by the fact that generally they are water soluble non degradable and strongly bounded to many biochemical especially poly peptides and proteins besides, other materials which loses electron rich functional groups such as sulfohydryl, amino and imidasole. Being stable compounds like heavy metals are non-reality removed by oxidation, precipitation other processes and affects the activities of the animals (Anitha, *et al.*, 2005).

Sulphates (SO_4)

In the present study the lake-1, maximum SO_4 was recorded 340 mg/L in site-1 and minimum SO_4 was recorded 56 mg/L in site-4. In lake-2 the maximum SO_4 was recorded 145 mg/L in site-5 and minimum SO_4 was recorded 50 mg/L in site-7 (Table 23.1). In Lake-3 the maximum SO_4 was recorded 144 mg/L in site-5 and minimum SO_4 was recorded 26 mg/L in site-1. In lake-4 the maximum SO_4 was recorded 66 mg/L in site-3 and minimum SO_4 was recorded 18 mg/L in site-7 (Table 23.2). Sulphur exists is a number of oxidation status from the most oxidized sulphate to the most reduced sulphide. The biological reduction of sulphur can take place in both aerobic and anaerobic conditions. A high concentration of sulphates stimulates the action of sulphate reducing bacteria. Which produce hydrogen sulphide a gas highly toxic to fish life. The high concentration of sulphate associated with the depletion of oxygen led to anoxic or anaerobic conditions in the lake waters.

Chloride (Cl)

In the present study the lake-1, maximum Cl was recorded. 364 mg/L in site-1 and minimum Cl was recorded 144 mg/L in site-3. In lake-2 the maximum Cl was recorded 1414 mg/L in site-4 and minimum Cl was recorded 120 mg/L in site-5 (Table 23.1). In Lake-3 the maximum Cl was recorded 614 mg/L in site-6 and minimum Cl was recorded 41 mg/L in site-1. In lake-4 the maximum Cl was recorded 196 mg/L in site-1 and minimum Cl was recorded 44 mg/L in site-4 (Table 23.2). This element is an important component of the carbonic buffer system and also cycles through biotic and a biotic component of ecosystem the chloride values is maximum in bottom layer and this may be due to washing down of organic matter from the surrounding catchment area (Datta *et al.*, 1984).

Nitrates (NO_3)

In the present study the lake-1, maximum NO_3 was recorded 35 mg/L in site-2 and minimum No_3 was recorded 10 mg/L in site-1. In lake-2 the maximum NO_3 was recorded 73 mg/L in site-5 and minimum No_3 was recorded 7 mg/L in site-4 and 6 (Table 23.1). In Lake-3 the maximum NO_3 was recorded 195 mg/L in site-6 and minimum NO_3 was recorded 35 mg/L in site-1. In lake-4 the maximum NO_3 was recorded 22 mg/L in site-6 and minimum No_3 was recorded 9 mg/L in sites-1, 2 and 4 (Table 23.2). It is one of the most, important elements in aquatic systems, as it helps in synthesis and maintenance of proteins and productivity of the water. Nitrate concentration of the water bodies all influenced with the geochemical conditions organic load the rate of their mineralization, the rainfall is to be responside for increasing the amount of nitrate in water (Prasad and Sexana, 1980). According to (Wetzel, 1983) nitrate is generated by heterotrophic microbes as a primary and product of decomposition of organic matter either directly from protein or from the organic compound. The presence of excessive nitrate in water is due to man made domestic activities and fertilizers from fields. Nitrate represent high load all cause of eutrophication of the aquatic body (Zutshi and Khan, 1988). Nitrates stimulate the growth of plankton and water needs that provide food for fish. It is may increase the fish population. However if argue grow too wildly, oxygen levels will be reduced and fish will die (USEPA, 1986) and (Knepp and Arkin, 1973).

Fluoride (F)

In the present study the lake-1, maximum F was recorded mg/L in site-2 and 6 and minimum F was recorded 0.40 mg/L in site-1. In lake-2 the maximum F was recorded 1.60 mg/L in site-6 and 7 and minimum F was recorded 0.80 mg/L in site-2 (Table 23.1). In Lake-3 the maximum F was recorded 2 mg/L in site-1 and minimum F was recorded 0.50 mg/L in site-4. In lake-4 the maximum F was recorded 2.40 mg/L in site-2 and minimum F was recorded 0.40 mg/L in site-5 (Table 23.2). Fluoride is the highest member of the halogen family is the most electronegative among all chemicals elements (Hodge and Smith, 1965) In ions inhabit a variety of enzymes often by forming complexes with magnesium ions and other metals ions (Nas, 1971a). In low concentration fluoride prevents dental caries and higher concentration of fluoride also causes respiratory failure fall of blood pressure and general paralysis.

References

Anitha, G., Chandrashekar, S.V.A. and M.S. Kodarkar, 2005. Limnological, studies on Mir Alam lake, Hyderabad. *Poll. Res.*, 24(3): 681–687.

Colorado, 2002. Department of Public Health and Environment and Water Quality Control Division (DPHE– WQCD). The basic standards and. methodologies for surface water. 5CCR, 1002–31.

Datta, N.C., Choudhary, A. and S. Choudhary, 1984. Annual rhythm of some hydrobiological parameters and their interaction in a brackish water impoundment of West Bengal, India. *Camp. Physiol. Ecol.*, 9(2): 149–154.

Harshey, D.K., Pakl, S.G. and D.F. Singh, 1982. Limnological studies on a trophical freshwater tank of Jabalpur, India. I. Abiotic factors. *Geobios*, 1(2): 98–102.

Hodge, H.C. and Smith, F.A., 1965. *Fluorine Chemistry*, Vol. 4, Academic Press, New Delhi.

Kaul Sidharth, 1983. Limnological studies of an aquatic body, Lalpari Talab at Rajkot. In: *Proc. Indian Nat. Sci. Acad.*, B45: 564–570.

Knepp and Arkin, 1973. Ammonia Toxicity Levels and Nitrate Tolerance of, Channel Catfish. *The Progressive Fish Culturist*, 35: 221.

Mishra, S.R. and Saksena, D.N., 1991 Pollutional ecology with reference to physico-chemical characteristics of morar (Kalpi) river, Gwalior (M.P.). *Current Trends in Limnology*, 1: 159–184.

Mitchell and Stapp, 1992. *Field Manual for Water Quality Monitoring*.

Munawar, M., 1970. Limnological studies on freshwater ponds of Hyderabad–II. The biocenose. *Hydrobiologia*, 36(1): 105–128.

NAS, 1971a. *Biologic Effects of Atmospheric Pollutants: Fluorides*. Washington, D.C. National Academy of Science. National Research Council. Committee on Biological Effects of Atmospheric Pollutants, pp. 239.

Postel, S., 1992. *Last Oasis: Facing Water Scarcity*. W.W. Norton and Company, Inc.

Rao, V.S., 1991. An Ecological Study of three freshwater ponds of Hyderabad, India, the environment. *Hydrobiol.*, 35(1): 213–223.

Seenayya, G. and A.R. Zafar, 1979. An Ecological Study of the Mir Alam Lake. Hyderabad, India. *Ind. J. Biol.*, 2(2): 214–220.

Shastree, N.K., Islam, M.S., Pathak, S, and M. Afsan, 1991. Studies on the physico-chemical dimensions of the lentic hydrospere of Ravindra Sarovar (Gaya). *Current Trends in Limnology*, 1: 133–153.

Swingle, H.S., 1967. *Standardisation of Chemical Analysis for Water and Muds*. FAO. Fish Rep., 4(4): 397–421.

USEPA, 1986. *Quality Criteria for Water*. U.S. Environmental Protection Agency. EPA #440/5–86–001.

Verma, S.R., Sharma, P., Tyagi. A., Rani, Gupta, A.K. and R.C. Dalela, 1984. Pollution and sporshic status of Eastern Kalinadi. *Limnologica (Barlin)*, 15(1): 69–133.

Zafer, A.R., 1964. On the Ecology of algae in certain fish ponds of Hyderabad, India. I. Physio-chemical complexes. *Hydrobiologia*, 23: 179–195.

Zutshi, D.P. and Vass, K.K., 1978. Limnological Studies on Dal Lake: Chemical features. *Indian J. Ecol.*, 5(1): 90–97.

Chapter 24

Effect of Crop Stand Establishment Methods and Split Application of N and K on Crop Density, Flowering and Post Harvest Fertility Status of CoRH2 Rice Hybrid

R. Gobi, S. Ramesh, B.J. Pandian and B. Chandrasekaran
Department of Agronomy, Tamil Nadu Agricultural University, Coimbatore – 641 003, Tamil Nadu, India

ABSTRACT

Field experiment was conducted at Agricultural college and Research Institute, establishment methods and split application of N and K on the plant population, days to 50 per cent flowering and post harvest nutrient status of the rice hybrid CoRH2. The experiment was conducted in factorial randomized block design replicated thrice with two establishment methods, two plant densities and three methods of split application of N and K were compared. Significant interaction was noticed between plant population and application of nutrients in different splits. Line planting with 50 hills m^{-2} combined with five splits of N and four splits of K(P_2S_3) registered significantly higher plant population as compared to other treatments. Stand establishment methods and split application of N and K significantly influenced the days needed for completion of 50 per cent flowering. Post harvest soil available N, P and K status and organic carbon content of the crop were significantly influenced by the stand establishment methods and split application of N and K.

Keywords: *Establishment methods, Hybrid rice, N and K splits.*

Introduction

The development of hybrid rice technology and the adoption of hybrid varieties to Indian environment offers one approach to solve the problem of matching food supply to expected demand. To meet the current and future demand for rice, exploitation of hybrid vigour has become a practical tool with the release of several promising rice (*Oryza sativa* L) hybrids (Surekha *et al.*,1999). After the successful performance of hybrid rice in China, few hybrids in India have been developed and released to enhance the yield in the context of pleauting trend in yield with conventional high yielding variety (Hari Om *et al.*, 2000). Economically viable, easily adoptable and efficiency package of practices are more crucial to have maximum resource utilization and productivity in hybrid rice (Rajarathinam and Balasubramaniyan, 1999).

Growing hybrid rice is a complex process and especially agronomic management of hybrid rice differs from that of conventional varieties because of the heterosis. Developing a package of optimum production practices for hybrid rice cultivation is essential to explore the full hybrid potential. Optimum plant population with appropriate establishment techniques are the viable package of practices to enhance the productivity of hybrid rice. Fertilizer nitrogen (N) is one of the costliest and the most essential nutrient element limiting the rice yield. Adequate nutrient supply is crucial in realizing the full potential of hybrid rice. Potassium is a major nutrient, accounts for a greater share of total nutrient removal by rice crop. Split application of N and K plays vital role in enhancing their use efficiency. In this context, the present investigation was undertaken to study the effect of different crop establishment methods and split application of N and K on the plant population, days to 50 per cent flowering and fertility status of CoRH2 rice hybrid under Tambiraparani command area.

Materials and Methods

Field experiment was conducted at Agricultural college and Research Institute Killikulam, during late *Pishanam* season (November–April) to study the effect of establishment methods and split application of N and K on the plant population, flowering and post harvest fertility status of hybrid rice CoRH2. The soil of the experimental field was sandy clay loam with a pH of 6.8. Regarding fertility status, the soil was classified as low in available nitrogen, high in available Phosphorus and medium in available potassium content. The hybrid CoRH2 is a medium duration hybrid rice (125 days) developed by three line breeding system using cytoplasmic male sterile line, maintainer line and restorer line. The grain is medium slender and white. The experiment was laid out in factorial randomized block design and replicated thrice.

The details of treatments are given below:

A: Establishment Methods

P_1: Line planting at 25 × 10 cm (40 hill m^{-2})

P_2: Line planting at 20 × 10 cm (50 hill m^{-2})

P_3: Seedling broadcasting with 40 hills m^{-2}

P_4: Seedling broadcasting with 50 hills m^{-2}

B. Split Application of N and K

S_1: Three splits of N at 50 per cent B, 25 per cent AT, 25 per cent PI and two splits of K at 50 per cent B, 50 per cent PI

S_2: Four splits of N at 50 per cent B, 16.5 per cent AT, 16.5 per cent PI, 16.5 per cent H and three splits of K at 50 per cent B, 25 per cent PI, 25 per cent F

S_3: Five splits of N at 50 per cent B, 12.5 per cent AT, 12.5 per cent PI, 12.5 per cent H, 12.5 per cent F and four splits of K at 50 per cent B, 25 per cent PI, 12.5 per cent H, 12.5 per cent F.

(B: Basal, AT: Active tillering, PI: Panicle initiation, H: Heading and F: Flowering).

Under line planting seedlings of 25 days old were planted with single seedling hill-1 by adopting spacing as per the treatments. Seedlings were broadcasted as per the treatments over the field by maintaining thin film of water. The recommended dose of 200 : 50 : 50 kg of NPKha^{-1} was applied to CoRH2 rice hybrid in the form of urea, superphosphate and potash respectively. Superphosphate was applied in full as basal and potash and nitrogenous fertilizers as muriate of potash and urea respectively were applied in different splits as per the treatments.

Plant Population

Number of plants in four randomly selected spot using 0.25 m^2 quadrat were counted in each plot at tillering stage and the mean value was expressed in number m^{-2}.

Days to 50 per cent Flowering

Number of days from sowing the seeds until the date when approximately 50 per cent of the plants flowered were counted and recorded.

Soil Analysis

Soil samples from each plot in surface layer (0–15 cm) were collected at post harvest stage and shade dried, powdered and sieved through 2 mm sieve for the analysis of the following nutrients (Table 24.1).

Table 24.1: Analysis of Nutrients

Nutrients	*Method*	*Reference*
Available N	Alkaline permanganate	Subbiah and Asija (1956)
Available P	Olsen's (Calorimetry)	Olsen *et al.* (1954)
Available K	N-N Ammonium acetate extract	Stanford and English (1949)
Organic carbon	Chromic acid wet digestion	Walkey and Black (1934)

Results and Discussion

Plant Population

Rice plants require optimum space for the maximum exploitation of their inherent yield potential. The competition in a community of rice plant has greatly influenced by the morphology of the variety and population density. Stand establishments and split application of N and K treatments showed significant influence on the plant population (Table 24.2). Line planting with 50 hill m^{-2} (P_2) ensured significantly higher plant population of 47.80 number m^{-2} and it was on par with seedling broadcasting with 50 hills m^{-1}. Among the split application of N and K, five splits of N and four splits of K (S_3) produced significantly higher population of 43.70 number m^{-2}. Significant interaction was noticed between plant population and application of nutrients, where P_2S_3 recorded higher plant population (48.80 no.m^{-2}).

Table 24.2: Effect of Stand Establishment Methods and Split Application of N and K on Plant Population (no.m^{-2}) of CoRH2 Rice Hybrid

Treatment	*Plant Population*			
	S_1	S_2	S_3	*Mean*
P_1	38.20	38.40	38.60	38.40
P_2	47.20	47.60	48.80	47.86
P_3	36.60	37.80	39.20	37.86
P_4	47.40	47.80	48.20	47.80
Mean	42.35	42.90	43.70	42.98
	SEd	*CD (0.005)*		
P	0.21707	0.45019		
S	0.18799	0.38987		
PS	0.37598	0.77975		

Days to 50 per cent Flowering

Stand establishment methods and split application of N and K significantly influenced the days needed for completion of flowering (Table 24.3). Line planting with 50 hills m^{-2}(P_2), seedling broadcasting with 50 hills m^{-2}(P_4) and seedling broadcasting with 40 hills m^{-2}(P_3) were on par with days to 50 per cent flowering took 85.29, 85.32 and 85.31 days for flowering respectively. Among split application of N and K, five splits of N and four splits of K(S_3) took 85.39 days for 50 per cent flowering. Interaction effect was significant during the crop season, where P_2S_3, P_3S_2, P_4S_3, P_4S_1 and P_3S_3 were statistically on par with days to 50 per cent flowering. Split application of N and K extended after panicle initiation stage to meet the plant requirement of N and K led the higher values of growth characters and days to 50 per cent flowering. Similar findings were observed by Biswas (1976) and Senthilvel and Palaniappan (1985).

Table 24.3: Influence of Stand Establishment Methods and Split Application of N and K 50 per cent flowering of CoRH2 Rice Hybrid

Treatments	S_1	S_2	S_3	*Mean*
P_1	85.40	85.21	85.32	85.19
P_2	85.29	85.08	85.50	85.29
P_3	85.13	85.49	85.32	85.31
P_4	85.40	85.14	85.43	85.32
Mean	85.21	85.23	85.39	
	SEd	*CD (0.005)*		
P	0.03930	0.08151		
S	0.03404	0.07059		
PS	0.06808	0.14118		

Post Harvest Nutrient Status

Available N, P and K status and organic carbon content of the soil after crop harvest were significantly influenced by the stand establishment methods and split application of N and K (Table 24.4). Among the stand establishment methods, higher soil available N (201.51 kg ha^{-1}) and P (18.16 kg ha^{-1}) was recorded with line planting 40 hills m^{-1}(P_1), but higher soil available K (246.32 kg ha^{-1}) and organic carbon (0.57 per cent) was registered in seedling broadcasting with 50 hill m^{-2} (P_4). Regarding the split application of N and K, the higher soil available N (208.64 kg ha^{-1}), P (18.33 kg ha^{-1}), K (252.41 kg ha^{-1}) and organic carbon (0.59 per cent) was noticed in three splits of N and two splits of K (S_1). Significant interaction effect was observed between treatments for soil available nutrients. The higher availability of N was recorded in P_1S_1 which was on par with P_3S_1 and P_1S_1. P and K and organic carbon values were higher with P_2S_1 which was on par with P_2S_1 and P_4S_1. The lower available nutrients and organic carbon content under S_3 and S_2 might be due to increased uptake of nutrients.

Table 24.4: Post Harvest Soil Fertility Status (kg ha^{-1}) of CoRH2 Rice Hybrid as Influenced by Crop Stand Establishment Methods and Split Application of N and K

Treatments	*Available N*	*Available P*	*Available K*
P_1	201.51	18.16	246.19
P_2	201.21	18.08	245.60
P_3	200.94	17.55	245.74
P_4	201.10	18.06	246.32
SEd	0.1579	0.1705	0.1909
CD (0.05)	0.3275	0.3537	0.3961
S_1	209.03	18.83	252.41
S_2	198.56	17.86	244.53
S_3	195.99	17.19	240.95
SEd	0.1367	0.1477	0.1654
CD (0.05)	0.2836	0.3063	0.3308
PS			
SEd	0.2735	0.2954	0.3308
CD (0.05)	0.5672	0.6127	0.6860

References

Biswas, C.R., 1976. Effect of split application of potash on the efficiency of nitrogen utilization by paddy in saline soils in West Bengal. *Indian Potash J.*, 1(2): 33–36.

Hari Om, S., Katyak, K. and Dhiman, S.D., 2000. Response of two rice (*Oryza sativa*) hybrids to graded levels of nitrogen. *Indian J. Agric. Sci.*, 70(3): 140–142.

Olsen, S.R., Cole, C.L., Watanabe, P.S. and Den, D.A., 1954. Estimation of available phosphorus in soils by extraction with sodium bicarbonate. U.S.D.A. Circ. No. 939.

Rajarathinam, P. and Balasubramaniyan, P., 1999.Optimum plant population, seedling density and N levels for medium duration hybrid rice (*Oryza sativa*). *Indian J. Agric. Sci.*, 69(8): 583–585.

Senthilvel, T. and Palaniappan, S.P., 1985. Effect of K top dressing through NK granules on yield and nutrient uptake of rice under lowland conditions. *Journal of Potassium Research*, 1(3): 166–173.

Stanford, S. and English, L., 1949. Use of flame photometer in rapid soil tests for K and Ca. *Agron. J.*, 41: 446–447.

Subbiah, B.V. and Asija, G.L., 1956. A rapid procedure for estimation of available nitrogen in the soil. *Current Science.*, 25: 259–260.

Surekha, K., Narayana Reddy, M., Kumar, R.M. and Vijayakumar, C.H.M., 1999. Effect of nitrogen sources and timing of yield and nutrient uptake of hybrid rice. *Indian J. Agric. Sci.*, 69(7): 447–481.

Walkley, A. and Black, C.A., 1934. An examination of methods for determining organic carbon and nitrogen in soils. *J. Agric. Sci.*, 25: 598–609.

Chapter 25

Physico-chemical Water Quality of River Beas

Surinder Deswal[1]* and Pankaj Chandna[2]

[1]Assistant Professor, Civil Engineering Department,
[2]Assistant Professor, Mechanical Engineering Department
National Institute of Technology, Kurukshetra

ABSTRACT

The physico-chemical analysis of water quality of River Beas is presented in this study. Water quality of Beas River is deteriorating, particularly in the tourist-season, due to the disposal of untreated sewage/wastewater into this river and its tributaries from the cities/towns located on its bank. The various parameters examined are temperature, colour, pH, electrical conductivity, total dissolved solids (TDS), total suspended solids (TSS), total solids, dissolved oxygen (DO), chloride, total alkalinity and total acidity. The findings of the study revealed that River Beas is experiencing pollution during the tourist-season (April to September). The disposal of untreated wastewater into the watercourse may render the river water unsuitable for sustenance and propagation of aquatic life and unfit for human consumption in due course of time if suitable wastewater disposal and treatment measures are not taken at the earliest.

***Keywords**: River Beas, Water quality, Dissolved oxygen, TDS, TSS.*

Introduction

The Beas is one of the important snow-fed perennial rivers of Indus River System. The river was known as "Arjiki" or "Vipasa" to the ancient Indians and the "Hyphasis" or "Hypanis" or "Bipasis"

* Corresponding Author: E-mail: sdeswal@nitkkr.ac.in.

to the Ancient Greeks. It forms the beautiful world famous valleys of Kullu and Kangra in Himachal Pradesh. But ironically, its one of the source, known as Beas Rishi, is an insignificant looking igloo like structure on the right of Rohtang Pass in Pir Panjal range. To the South of this source lies another source known as Beas Kund. Both these mountain streams meet at Palchan village, nine kilometres North of Manali to form the river Beas. From Manali, after covering hundreds of kilometres through the hills, the river embraces the river Satluj at Hari Ka Patan in Ferozpore district of Punjab before flowing into Pakistan. The total length of this river is 460 km. It flows for 256 km. in Himachal Pradesh, along the National Highway-21 (from Palchan to Mandi) and then along National Highway-20 (from Mandi onwards).

River Beas bears the testimony of historical events which do not find mention in the recorded history. It has played a significant role in the development of a peculiar hill culture which pervades the life of hill people living in the towns and surrounding villages since ages. The river also serves as drinking water source to many towns on and around its bank, but due to non-availability of proper sewerage/disposal system, the wastewater of the towns/cities is being discharged into Beas River and its tributaries. The important settlements on the bank of Beas river along National Highway 21 are Manali, Raison, Kullu, Bhunter, Aut, Pandoh and Mandi. All these hill-stations, Manali and Kullu in particular, attract lacs of tourists every year during the tourist-season (April to September). As a prerequisite for better hospitality, these hill-stations have large number of hotels/restaurants/resorts/ etc. to meet the needs of the tourists. However, the wastewater of these towns and other settlements is polluting the Beas River and deteriorating its water quality, particularly during the tourist-season in summers.

The water quality of rivers, lakes/water tanks, seas/oceans or groundwater is measured by taking the physico-chemical and/or biological parameters (Gupta *et al.*, 2004; Ingole and Kadam, 2003; Khadse *et al.*, 2005; Saha *et al.*, 2003; Singh and Singh, 2003; Unnisa and Khalilullah, 2004). In the present study, the water quality of the Beas River was analyzed during the tourist-season in a stretch of about 130 kilometres (from Manali up-stream to Mandi down-stream), in terms of important physico-chemical parameters, for better understanding of extent and nature of pollution during the tourist-season.

Materials and Methods

Water samples were collected on a bright sunny day of 21st May, 2006 from 07:40 AM to 03:10 PM in two-litres pre-sterilized plastic bottles along the stretch of 130 kilometres of the Beas river from Manali up-stream (u/s) to Mandi down-stream (d/s). The sample stations are marked as 1, 2, 3, 4, 5, 6, 7, 8, 9, 10, 11, 12, 13 and 14 as shown in Figure 25.1. It is important to mention that Pandoh Reservoir, which is a part of the Beas-Sutlej Link Project for diverting the Beas water into the Sutlej basin, is located at about 1.5 kilometres up-stream of Pandoh town. As a result of this the discharge/flow in the down-stream of the Beas River reduces.

The sample preservation for physico-chemical analysis was done as per the standard method prescribed in ISI-2888 (1968). The temperature (T), electrical conductivity (EC), pH and colour were measured at the sampling locations itself. Temperature was recorded bya digital thermometer having an accuracy of 0.1°C. Electrical conductivity (in μmhos/cm) and pH were recorded by battery-powered portable conductivity meter and pH-meter respectively; while colour (in Hazen) was measured by a portable BPH Lovibond Nessleriser instrument that employ coloured glass disks which stimulate the various colour standards. Samples for dissolved oxygen (DO) were collected separately and fixed at site. DO was estimated by modified Winkler method? All the other parameters were determined as per

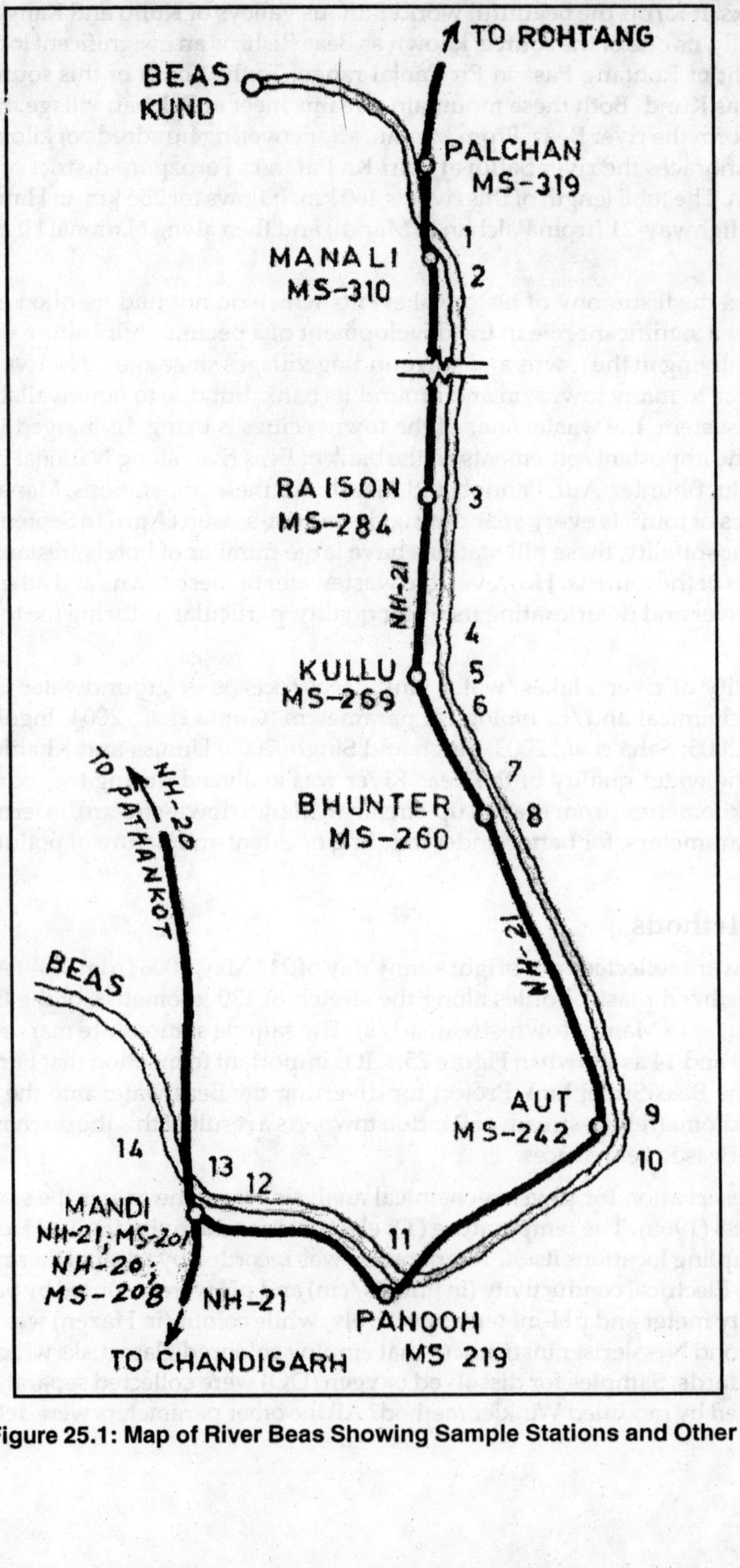

Figure 25.1: Map of River Beas Showing Sample Stations and Other Details

standard methods (APHA, 1998). The physico-chemical parameters measured in this study were temperature, electrical conductivity, pH, colour, dissolved oxygen (DO), total acidity, total alkalinity, total solids (TS), total suspended solids (TSS), total dissolved solids (TDS) and chloride. All the chemicals used in the chemical examination were of analytical grade.

Results and Discussion

The results of the physico-chemical analysis of water samples are shown in Table 25.1 The Mile Stones (MS) of cities/towns on the bank of Beas River and the distance of sample stations from Beas Kund (the origin/source of Beas River) are also given in Table 25.1 so as to locate and identify the sample stations.

In the present study, the colour of all the water samples taken along the length of Beas River was less than 5 Hazen units. The pH values were found in the range of 6.20–7.15. The change might be due to decomposition of organic matter. The lowest pH (=6.2) was observed at sample station 11. This may be due to disposal of waste from the Pandoh town in suddenly decreased flow rate of the river (due to the presence of Pandoh Reservoir at the up-stream of the river). It is evident from the present study that pH values were within or slightly lower than the permissible limit of 6.5–8.5 for "Use class-A" according to Central Pollution Control Board standards (CPCB, 1995), but depending only upon the pH value the quality of water cannot be judged (Garg *et al.*, 1990).

EC measurement is an excellent indicator of total dissolved solids, which is a measure of salinity that affects the taste of water. The gradual increase in the concentration of TDS was observed from station 1 to 14. This is due to the discharge of untreated wastewater into the river from the cities/towns situated on the banks of River Beas. The concentration of TS increases from station 1 to station 8, then suddenly drops at station 9 and 10, and then gradually increases. Similarly, there is a gradual increase of TSS from station 1 to 8 and then TSS value suddenly drops at station 9 and 10 (*i.e.* at Aut up-stream and Aut down-stream); however, after station 10 the concentration decreases. It seems that high turbulence of river water due to topography of the area from station 1 (Manali up-stream) to station 8 (Bhunter down-stream) plays a significant role in keeping the solids in suspension. After this, most of the suspended solids settled down at Pandoh Reservoir (just before station 11), and also the turbulence in river decreased. The variations of parameters TDS and TSS at sample stations 1 to 14 along the length of Beas River are shown in Figure 25.2.

The study also revealed that there was no alarming DO deficit along the length of the river indicating that re-oxygenation rate from the atmosphere higher than de-oxygenation rate due to decomposition. At all the sample stations (except station 2, where DO=4.7 mg/L), the DO value was observed more than 6 mg/L. Ellis *et al.* (1946) stated that the minimum DO for maintaining fish in a healthy condition is 5.0 mg/L at 20°C, the critical value being 3.0 mg/L. The same limits have also been mentioned by Dimiek and Merryfield (1947).

In surface water the concentration of chloride is due to sewage and many of the soluble salts found in soil. The chloride concentration was found to be in the range of 9.2–17.1 mg/L. The variations of DO, chloride, alkalinity and acidity at different sample stations along the watercourse are shown in Figure 25.3.

In brief, the results of physico-chemical analysis of River Beas, along a stretch of about 130 kms. from Manali up-stream to Mandi down-stream, revealed that:

1. The pH values (6.2–7.15) were within or slightly lower than the permissible limit of 6.5–8.5 for "Use class-A" (*i.e.* the source is designated best use as drinking water source without

Table 25.1: Physico-chemical Parameters of Beas River

Sample Station	Sample Site	Distance of Sample Station from Bease Kund (Kms)	Temperature (°C)	Colour (Hazen)	pH	Electrical Conductivity (µmhose/cm)	Total Dissolved Solids (mg/l)	Total Suspended Solids (mg/l)	Total Solids (mg/l)	Dissolved Oxygen (mg/l)	Total Acidity (mg/l, as $CaCO_3$)	Total Alkalinity (mg/l, as $CaCO_3$)	Chloride (mg/l)
1.	Manali u/s	14.5	10.6	<5	8.2	122.9	62.1	172.7	234.8	10.43	8.9	19.2	8.6
2.	Manali d/s	18.3	11.1	<5	6.41	334.3	167.9	433.3	601.2	4.7	27	48	15.6
3.	Raison	49	11.7	<5	7.1	200.3	117.5	400	517.5	8.35	18.4	40	9.6
4.	Kullu u/s	53.2	12	<5	6.6	333.2	166.6	466.7	633.3	6.75	16	38.4	9.8
5.	Kullu city	56.2	12.5	<5	6.62	328.2	162.3	553.4	715.7	6.18	20.4	41.2	10.4
6.	Kullu d/s	62	13.2	<5	7.04	373.8	189.8	566.7	756.5	7.34	16.8	40.4	10.4
7.	Bhunter u/s	64	14.3	<5	7.05	448.5	200.3	667.3	867.6	7.37	17.1	39.6	10
8.	Bhunter d/s	69.1	16	<5	6.86	456.7	233.3	790.7	1024	7.12	22.7	43.7	10.9
9.	Aut u/s	82.5	13.6	5	6.45	759.9	394.6	266.7	661.3	7.92	15.6	30.4	8.8
10.	Aut d/s	86.7	13.9	5	6.92	400	199.8	167.2	367	7.8	12.8	28	6.8
11.	Pandoh	108	19	5	6.2	856.3	433.3	366.7.	800	6.93	29.2	33.2	9.2
12.	Mandi u/s	122.2	20.2	<5	6.73	761.8	398.3	200	598.3	7.67	39.6	51.2	13.4
13.	Mandi city	129	21.5	<5	6.86	1036.3	534.2	100	634.2	7.69	23.6	48.4	17.1
14.	Mandi d/s	130.8	22.5	<5	7.15	995.2	513.3	166.7	680	6.53	49:2	52.4	16.4

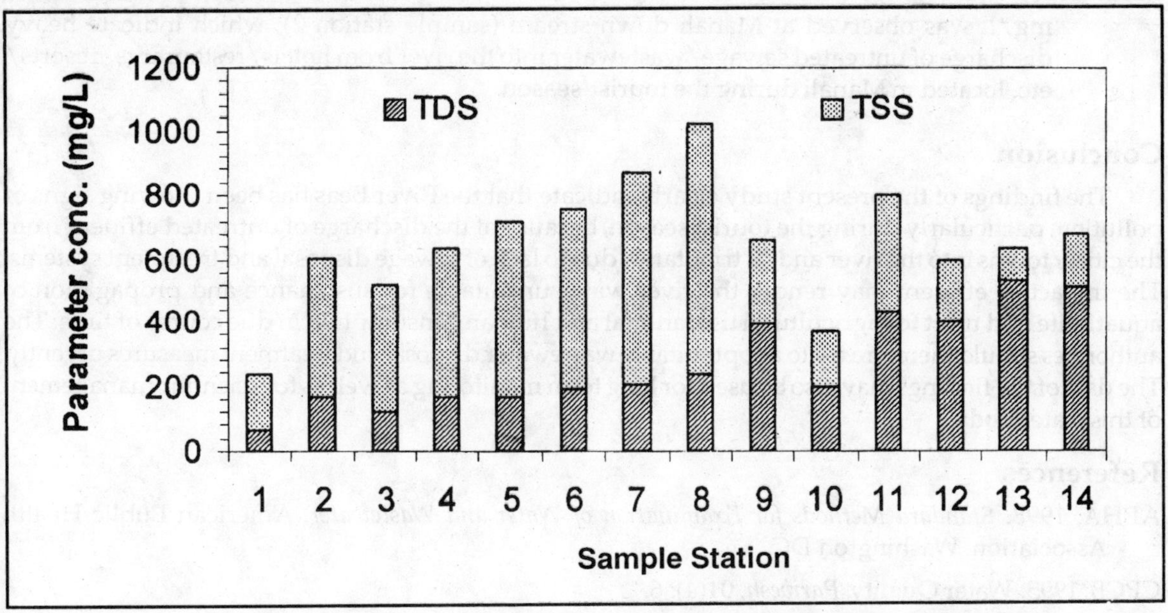

Figure 25.2: Variation of TDS and TSS Along the Length of the Beas River

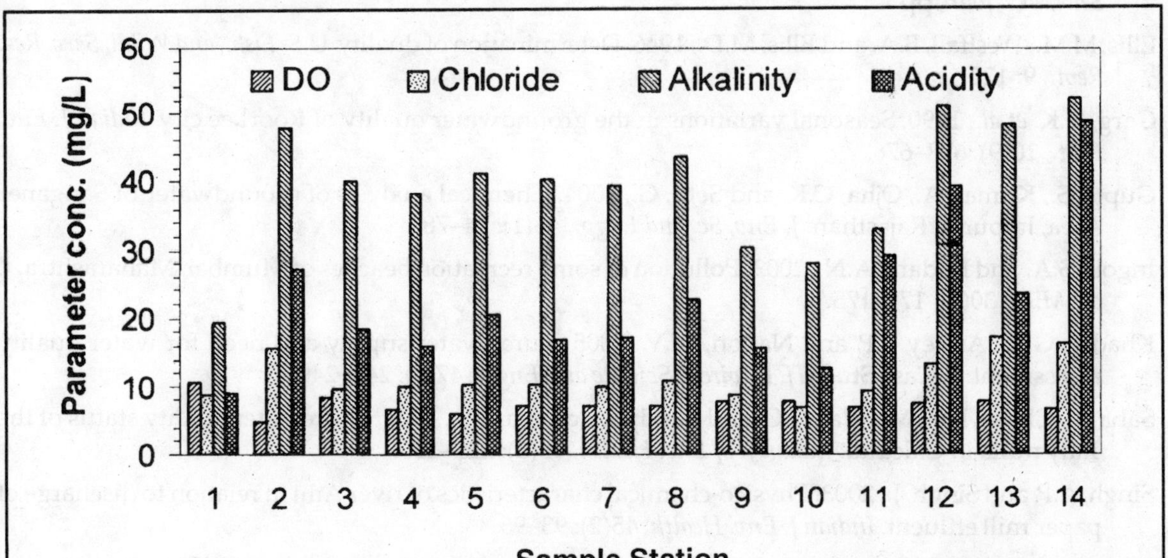

Figure 25.3: Parameter Variation Along the Length of the Beas River

conventional treatment but after disinfections) according to Central Pollution Control Board standards (Khadse *et al.*, 2005).

2. Also the DO of the river water samples (except at sample site 2) was more than 6 mg/L, the desirable limit for "Use class-A" of freshwaters (Khadse *et al.*, 2005). The lowest DO of 4.7

mg/L was observed at Manali down-stream (sample station 2), which indicate heavy discharge of untreated sewage/wastewater into the river from hotels/restaurants/resorts/ etc. located in Manali during the tourist-season.

Conclusion

The findings of the present study clearly indicate that the River Beas has been showing signs of pollution, particularly during the tourist season, because of the discharge of untreated effluent from the cities/towns into the river and its tributaries due to lack of sewage disposal and treatment systems. The impact of effluent may render the river water unsuitable for sustenance and propagation of aquatic life and unfit for agricultural use, animal and human consumption in due course of time. The authorities should therefore try to adopt suitable wastewater disposal and treatment measures urgently. The data of the findings may also be used for long term monitoring as well as for scientific management of this water body.

References

APHA, 1998. *Standard Methods for Examination of Water and Wastewater*. American Public Health Association, Washington DC.

CPCB, 1995. Water Quality. *Parivesh*, 01(4): 6.

Dimiek, R.F. and Merryfield, E., 1947. The fishes of the Willamette river system in relation to pollution. *Sew. Wks. Jour.*, pp. 49.

Ellis, M.M., Westfall, B.A. and Ellis, M.D., 1946. Determination of duality. *U.S. Fish., and Wildl. Serv. Res. Rept.*, 9: 122.

Garg, D.K. *et al.*, 1990. Seasonal variations in the groundwater quality of Roorkee city. *Indian J. Env. Prot.*, 10(9): 673–676.

Gupta, S., Kumar, A., Ojha, C.K. and Seth, G., 2004. Chemical analysis of groundwater of Sanganer Area, Jaipur in Rajasthan. *J. Env. Sc. and Engg.*, 46(1): 74–78.

Ingole, S.A. and Kadam, A.N., 2003. Pollution of some recreation beaches of Mumbai, Maharashtra. *J. of IAEM*, 30(3): 172–175.

Khadse, G.K., Andey, S.P. and Nanoti, M.V., 2005. Rural water supply and need for water quality assessment: A Case Study. *J. Environ. Science and Engg.*, 47(3): 242–249.

Saha, T., Ghose, P.B., Mandal, C.C. and Bandyopadhaya, T.S., 2003. Wastewater quality status of the Tally nulla in Calcutta. *Indian J. of Env. Health*, 45(3): 221–226.

Singh, A.P. and Singh, J., 2003. Physico-chemical characteristics of river Ami in relation to discharge of paper mill effluent. *Indian J. Env. Health*, 45(2): 93–96.

Unnisa, S.A. and Khalilullah, M., 2004. Impact of industrial pollution on ground and surface water quality in the Kattedan industrial area. *J. IAEM*, 31(1): 77–80.

Chapter 26

Erosion Resistant Surface Coatings by Varying Current Density and Time in Electroforming Techniques

S. Ramachandran[1]*, T.M. Sridhar[2] and J. Abbas Mohaideen[3]**

[1]Assistant Professor, Department of Mechanical Engineering, Sathyabama Deemed University, Old Mamallapuram Road, Jeppiaar Nagar, Chennai – 600 019

[2]Department of Biomedical Engineering, Sathyabama Deemed University, Old Mamallapuram Road, Jeppiaar Nagar, Chennai – 600 019

[3]Maamallan Institute of Technology, Sriperumpudur, Kanchipuram District, India

ABSTRACT

Electroforming is the production and reproduction of articles by plating sufficient thickness of metal such as chromium, nickel or copper over the given mandrel and subsequently the deposit (electroformed deposit) is separated from the mandrel to act as a separate entity. This techniqùe is used for surface coating on mild steel pipes. The effect of coating variables like, bath current density and forming time on the electroforming of copper and nickel on mild steel pipes are investigated. By varying the time duration of electroforming and the concentration of the electroplate, the thickness of the coating can be varied. Applied current density and forming time critically influence the eletroforming process in forming on mild steel pipes. Electroforming can be used as a successful technique to form coatings of chromium, nickel and copper at the optimum parameters obtained with a coating thickness of 8 μm. The optimum time was found to be 40 minutes and a thickness of 8 μm above which it is not economical to increase the coating thickness.

Keywords: *Electroforming, Mild steel pipes, Current density, Nickel.*

* *E-mail: aishram_2006@yahoo.com, Phone: 91-44-24661909, Mobile: 9444001904; **E-mail: tmsridhar@gmail.com.

Introduction

Electroforming is a highly specialized process for fabricating a metal part by electrodeposition in a plating bath over a base form or mandrel which is subsequently removed. The advantage of the electroforming process is that it faithfully reproduces the form or mandrel exactly, to within one micron, without the shrinkage and distortion associated with other metal forming techniques such as casting, stamping or drawing. And, since the mandrel is machined as an outside surface, close dimensional tolerances and high.surface finishes can be held and maintained on complex interior configurations (Harty *et al.*, 1981; Hart and Watson, 2000). The main advantage of electroforming is that it is an atomic scale process assuring replication fidelity that is unmatched by any other technology.

The principles of the electroforming process are demonstrated in Figure 26.1. This shows a suitable metal being electrodeposited onto the surface of a conical mandrel the term mandrel is used to describe the former onto which an electroform is grown). The positively charged electroformed metal source (anode) at the left is broken down (ionized) in the copper electrolyte solution and is attracted to the negative charged mandrel (cathode). When the deposit has attained a sufficient thickness it is then separated from the mandrel to become metallic product with a totally independent existence. Build-up is achieved over all mandrel surfaces at an approximate deposition rate of 0.001" per hour.

Electroforming is by far the most accurate method of manufacturing metallic articles. Both the dimensional tolerances that can be achieved and the ability of the process to reproduce surface detail with absolute fidelity are unmatched by any other metal-fabrication processes. Electroforming technique is used for the production and reproduction of a variety of engineering components, like parts for nuclear and space applications, moulds and dies for plastic, glass and zinc-die cast manufacturers, precision electronic parts, high temperature pressure vessels, turbine blades and nozzles and thrust chambers. Various structural properties can be controlled during electroforming (McGeough *et al.*, 2001; Sun *et al.*, 1997; Lamb, 1968).

Over the years, electroforming has progressed to using nickel rather than copper as the primary material. But neither copper nor nickel possesses the tensile strength and elasticity that high-grade

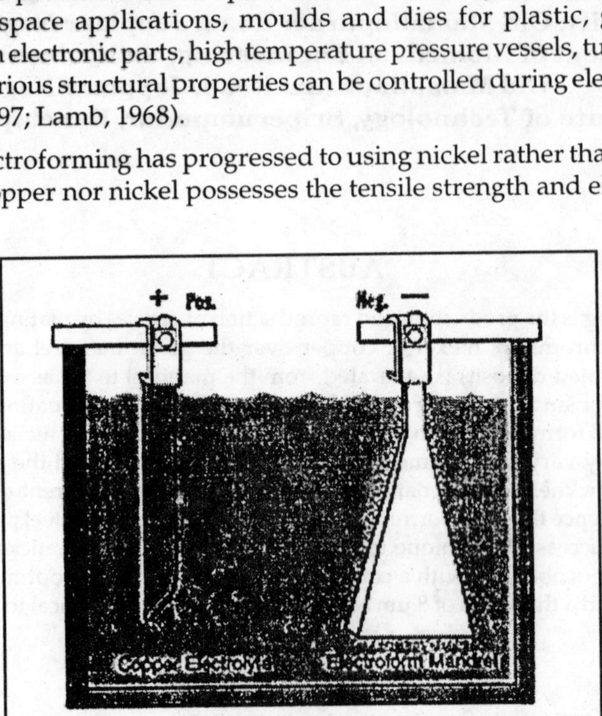

Figure 26.1: The Electroforming Process

steels can offer, and both are insufficiently hard and wear resistant. Therefore, electroformed articles in applications requiring high mechanical properties have found only limited use. Historically, most electroforms required post machining because the deposited layer was neither smooth nor uniform enough in thickness. This increased the cost of electroforms and hindered competitiveness of electroforming as a manufacturing method.

The aim of electroforming is to manufacture metallic articles and to a large extent, articles and coatings with different properties can be obtained by properly selecting and control the conditions of electrodeposition (Sun *et al.*, 1997). Bright nickel coating has been widely used in the area of decoration because of its sound appearance and protective performance (Sun *et al.*, 1997). Also, bright deposit is useful for the electroforming of articles because it is always attended by the improvement of other properties, such as levelling, grain refined.

Electroforming should be thought of as a basic manufacturing process when considering alternatives best suited for making any particular item. Electroforming is a real problem solver in these instances. Electroforming is used in manufacturing pipes which are subjected to high temperature fluids. The combustion takes place in the combustion chambers and very high temperature fluid passes through the inner passage of pipe and erosion of pipe takes place. The erosion in the pipe changes the velocity of the passing fluid and hence the performance of the engine is affected. Coatings on inner surface of the pipes will not be uniformly distributed in other coating techniques or metal working processes. Some nozzles and thrust chambers contain cooling passages on their walls and it is extremely difficult to make coatings by other metal working processes. These basic metal forming processes include casting, forging, stamping, deep drawing, machining and fabricating may serve well for most applications, but can be pressed beyond their limits when requirements call for extreme tolerances, complexity or light weight. Moreover in these techniques, many components have to be furnace brazed together to form the passages and the tooling costs required to complete the coating is very high and prohibitive. On the other hand, this can be electroformed in one piece conveniently, economically and to exact specification. Since the thickness of the layer electroformed is thin, it is not easily peeled off from the parent metal. Electroformed parts have their own distinct properties and dimensional tolerances are very good (Lamb, 1968; Ricq *et al.*, 2001; Barkalow *et al.*, 1980).

In the present work electroformed coatings are applied on the inner surface of mild steel pipe by copper, nickel and chromium on mild steel pipes from their respective baths. The effect of coating variables like applied current density and forming time on the electroforming of mild steel pipes are investigated.

Experimental Setup

The electroforming was carried out from three different bath solutions of chromium, nickel and copper baths. Chromium bath consist of chromic acid (250 g/l), H_2SO_4 2.5 g/l, pH 3 and graphite anode was used. Electroforming of nickel was performed with electrolyte in the solution containing $NiSO_4.7H_2O$ 240 g/l, $NiCl_2.6H_2O$ 20 g/l and boric acid 20 g/l and pH 4.5. All solutions were prepared by freshly distilled water. Nickel strip was used as the anode. A mild steel cylindrical pipe mandrel was used as the cathode (Figure 26.2, 100 × 50 mm) and the deposit area was formed on the inner surface area. The current density was varied from 2–12 A/dm^2. Before deposition, the mandrel was mechanically polished up to 1000 grit SiC papers, degreased with organic solvent, rinsed with ethanol. After the cylinder mandrel had been rotating in the electrolyte for several minutes, current was gradually increased to the desired magnitude. At the end of the process, the cathode was immediately withdrawn, washed with distilled water, dried and then subjected to further characterization measurements.

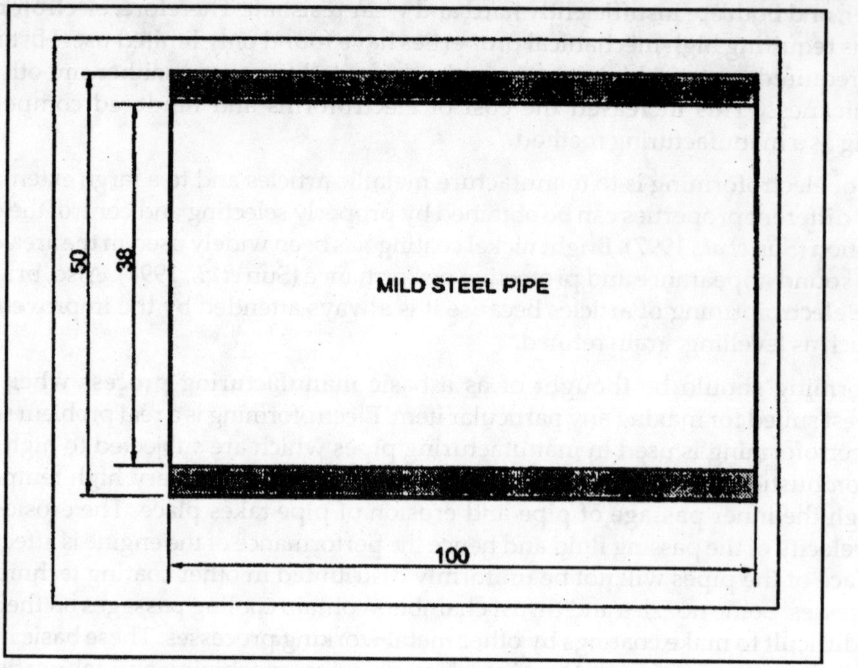

Figure 26.2: Dimensions of the Mild Steel Pipe Samples

Electroforming of copper was performed in the same methodology and parameters as nickel deposition but with electrolyte in the solution containing copper sulphate 50 g/l, Rochelle salt 70 g/l and ammonium hydroxide was used to raise the pH to 8.

Results and Discussion

Effect of Variation of Current Density on Electroforming

Chromium Forming

The choice of materials is limited not only by the normal constraints that apply to electrodeposition in general but also by the additional requirement for low internal deposit stress. Thus, a metal such as chromium, which would have many desirable properties as an electroform, particularly in relation to its hardness, cannot be used because of the extremely high internal tensile stresses generated in the electrodeposited metal.

Various types of materials are coated on the inner surface of the pipe using electroforming technique. The thickness of the coating depends on the bath composition, concentration of the electrolyte, pH of electrolyte, current density and forming time. Figure 26.3 shows the changes in coating thickness obtained on electroforming of chromium on mild steel pipes with varying current densities.

The parameters are summarized in Table 26.1. The coating thickness increases linearly with the increase in current density and varies from 4 to 8 µm. The forming was carried out at a constant time of 30 minutes and a temperature of 324 K.

Table 26.1: Electroforming Parameters of Chromium on Mild Steel Pipes at Various Applied Current Densities

Sl.No.	*Bath Temperature K*	*Rotational Speed of the Cathode (rpm)*	*Forming Time (min.)*	*Current Density (A/dm²)*	*Thickness of the Coating (mμ)*
1.	324	15	30	2	4.32
2.	324	15	30	4	5.60
3.	324	15	30	6	6.76
4.	324	15	30	8	7.24
5.	324	15	30	10	8.38

Deposit Thickness

If an electroplated coating is to fulfill its function it is essential that the adhesion between the deposit and the substrate be good. Electroforming, however, requires exactly the opposite. It is vital that there is little or no adhesion between the deposit and the mandrel so that the product can be separated without damage when the forming process is complete (Ricq *et al.*, 2001). Whereas in electroplating the pretreatment processes are selected to maximize deposit adhesion, electro-forming pretreatments are specifically chosen to reduce the adhesion between the deposit and the mandrel to the minimum required to prevent premature separation.

Electroplated deposits used as coatings in order to confer corrosion resistance, wear resistance, or simply aesthetic appeal are generally 1- to 30-micrometers thick. Electroformed products, however, need sufficient mechanical strength to be able to exist independently of the mandrel onto which they are deposited. They are, therefore, normally much thicker, ranging from foils of 10-micron thickness to mold faces that can be up to 5-millimeters thick (Lamb, 1968). The forming obtained above a current

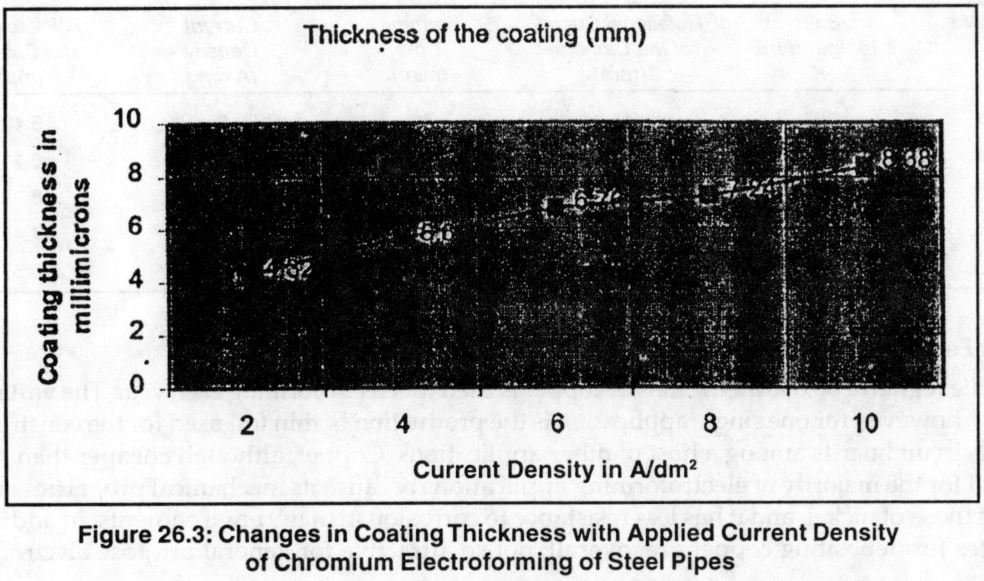

Figure 26.3: Changes in Coating Thickness with Applied Current Density of Chromium Electroforming of Steel Pipes

density 10 A/dm^2 indicate a change in the appearance of the samples that could affect its strength. There was also no considerable increase in the thickness with increase in current density above 10 A/dm^2. It is not economical when we increase the thickness more than 8 μm? Moreover, it saves material consumption and time without sacrificing the strength. This indicates that the optimum range of current density for obtaining chromium forming on mild steel pipes ranges from 4 to 10 A/dm^2.

Nickel Forming

Nickel can be considered the workhorse of the electroforming industry. Although a larger tonnage of copper than nickel is deposited, it is only used for a very limited range of products. It is probable that 95 per cent or more of the products manufactured by electroforming are produced in nickel. This is due to a favorable combination of properties, both of the electrodeposit and of the deposition processes. Electroformed nickel shows excellent strength, toughness, and ductility and is extremely resistant to corrosion in a wide range of environments. In addition it can readily be deposited in a low-stress condition, and its hardness and strength can be varied over a wide range by selecting appropriate deposition conditions (Barkalow *et al.*, 1980; Weil and Paquin, 1960). The present bath composition is highly reliable and relatively easy to control, being both chemically and electrochemically stable and environmentally friendly. The operating allows a considerable amount of control to be exercised over the properties of the deposits obtained.

Figure 26.4 shows the changes in coating thickness obtained on electroforming of nickel on mild steel pipes with varying current densities and the parameters are summarized in Table 26.2. Unlike chromium, the coating thickness increases linearly with the increase in current density up to 8 μm and then decreases gradually. The maximum coating thickness of 8 μm was achieved at a forming current density of 6 A/dm^2. This indicates that the optimum range of current density for obtaining the maximum thickness of 8 μm by nickel forming on mild steel pipes is 6 A A/dm^2 while range varies from 4 to 10 A A/dm^2.

Table 26.2: Electroforming Parameters of Nickel on Mild Steel Pipes at Various Applied Current Densities

Sl.No.	*Bath Temperature K*	*Rotational Speed of the Cathode (rpm)*	*Forming Time (min.)*	*Current Density (A/dm^2)*	*Thickness of the Coating (mμ)*
1.	323	15	30	2	5.42
2.	323	15	30	4	6.11
3.	323	15	30	6	8.33
4.	323	15	30	8	7.45
5.	323	15	30	10	6.57

Copper Forming

In the region of 80,000 metric tons of copper is used for electroforming each year. The vast majority of this is, however, for one single application is the production of thin foil used for the construction of printed circuit boards among a host of other applications. Copper, although cheaper than nickel, is not used for the majority of electroforming applications because its mechanical properties are not as good as those of nickel, and it has less resistance to corrosion in many environments. In addition, the processes for depositing copper are, overall, not so attractive for general purpose electroforming.

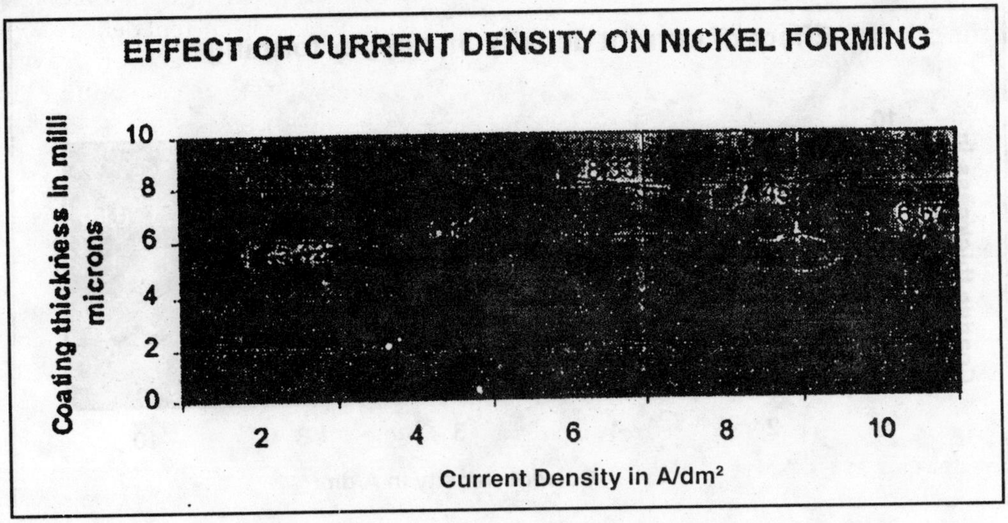

Figure 26.4: Changes in Coating Thickness with Applied Current Density of Nickel Electroforming of Steel Pipes

Though the copper deposits from the baths give smooth surfaces, they contain impurities that reduce the electrical conductivity and cause outgassing when heated to a high temperature or at electrical discharge (Tajiri *et al.*, 2001).

The bath has wide current density ranging from 2 to 10 A/dm^2, which makes this electroforming suitable for complex configuration components. Here, current density means both cathodic current density at normal cycle times and anodic current density for reverse-cycle times on treating surface, and they are set equal. The deposits produced by this electroforming show fine grains, excellent mechanical properties and low impurity content less than 35 ppm or so (Tajiri *et al.*, 1999).

The electroforming on the inner surface of the cavity body was performed by immersing the cavity vertically into the deep tank. The electrode is designed to have maximized surface and an appropriate distance from the cavity for successful deposition. The surface area of the electrode would be roughly equal to the treating surface area. The electroforming was performed by immersing., the cavity vertically into the deep tank. A compressed air jet from the small nozzles of the manifold attached to the bottom of the cavity generated an uprising solution stream through the space between the cavity and the electrode during the electroforming.

Electroforming on the edges of the cavity was performed by immersing the cavity horizontally into the wide tank having electrodes at both sides. The electrodes facing the treating surfaces offered uniform deposition on the edges. An electroformed copper layer of 1 mm or more was successfully built up on the cavity showing even surfaces with no harmful defects in appearance.

Figure 26.5 shows the changes in coating thickness obtained on electroforming of copper on mild steel pipes with varying current densities and the parameters are summarized in Table 26.3. The coating thickness increases linearly with the increase in current density up to 8 μm and then decreases gradually. The maximum coating thickness of 8 μm was achieved at a forming current density of 8 A/dm^2. This indicates that the optimum range of current density for obtaining the maximum thickness of 8 μm by copper forming on mild steel pipes is 8 A/dm^2 while range varies from 4 to 10 A/dm^2.

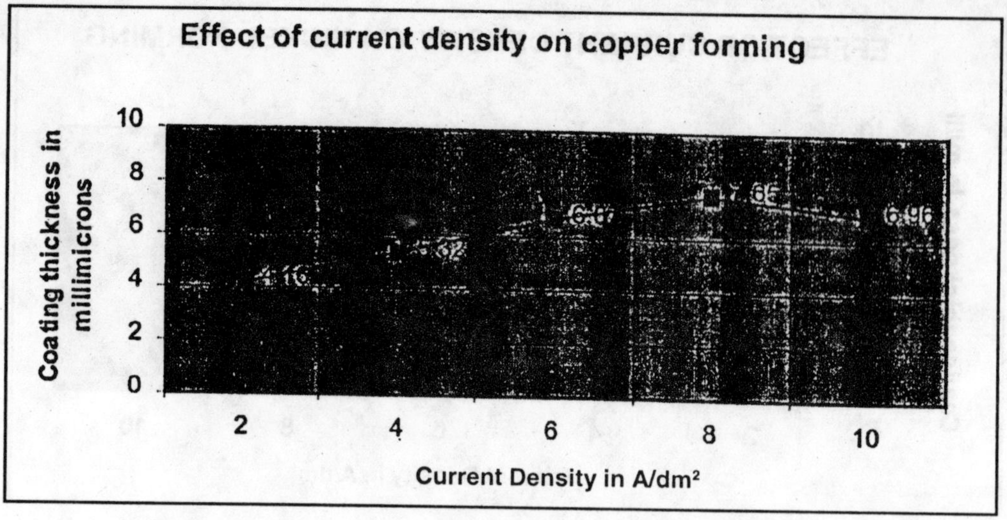

Figure 26.5: Changes in Coating Thickness with Applied Current Density of Copper Electroforming of Steel Pipes

Table 26.3: Electroforming Parameters of Copper on Mild Steel Pipes at Various Applied Current Densities

Sl.No.	Bath Temperature K	Rotational Speed of the Cathode (rpm)	Forming Time (min.)	Current Density (A/dm^2)	Thickness of the Coating (mμ)
1.	328	1.5	30	2	4.16
2.	328	15	30	4	5.32
3.	328	15	30	6	6.67
4.	328	15	30	8	7.65
5.	328	15	30	10	6.96

Effect of Variation of Time on Electroforming

Chromium Forming

Table 26.4 shows the changes in coating thickness obtained on electroforming of chromium on mild steel pipes with varying time at an applied current density of 10 A/dm^2. The coating thickness increases with the increase in time up to 30 minutes up to 8 μm and then decreases gradually with increase in time. The maximum coating thickness of 8 μm was achieved at a forming time of 30 minutes. This indicates that the optimum range of current density for obtaining the maximum thickness of 8 μm by chromium forming on mild steel pipes is 8 A/dm^2 and can be obtained over a time range of 20 to 40 minutes.

Nickel Forming

Table 26.5 shows the changes in coating thickness obtained on electroforming of nickel on mild steel pipes with varying time at an applied current density of 6 A/dm^2. The current density of 6 A/dm^2

was chosen as the optimum coating thickness was obtained when forming was carried out at varying current densities. The coating thickness increases linearly with the increase in time up to 40 minutes *i.e.* up to 8 μm and then decreases gradually with increase in time. Above 50 minutes no significant increase in forming thickness was observed. The maximum coating thickness of 8 μm was achieved at a forming time of 40 minutes. This indicates that the optimum range of current density for obtaining the maximum thickness of 8 μm by chromium forming on mild steel pipes is 8 A/dm² and can be obtained over a time range of 30 to 50 minutes.

Table 26.4: Electroforming Parameters of Chromium on Mild Steel Pipes at Various Forming Time Ranges

Sl.No.	*Bath Temperature K*	*Rotational Speed of the Cathode (rpm)*	*Current Density (A/dm²)*	*Forming Time (min.)*	*Thickness of the Coating (mμ)*
1.	324	15	10	10	7.92
2.	324	15	10	20	8.02
3.	324	15	10	30	8.38
4.	324	15	10	40	8.11
5.	324	15	10	50	7.68

Table 26.5: Electroforming Parameters of Nickel on Mild Steel Pipes at Various Forming Time Ranges

Sl.No.	*Bath Temperature K*	*Rotational Speed of the Cathode (rpm)*	*Current Density (A/dm²)*	*Forming Time (min.)*	*Thickness of the Coating (mμ)*
1.	323	15	6	10	7.62
2.	323	15	6	20	7.95
3.	323	15	6	30	8.33
4.	323	15	6	40	8.56
5.	323	15	6	50	8.21

Copper Forming

Table 26.6 shows the changes in coating thickness obtained on electroforming of capper on mild steel pipes with varying time at an applied current density of 8 A/dm². The current density of 8 A/dm² was chosen as the optimum coating thickness was obtained when forming was carried out at varying current densities. The coating thickness increases linearly with the increase in time up to 40 minutes *i.e.*, up to 8 μm and then decreases gradually very marginally with time. Above 50 minutes no significant increase in forming thickness was observed. The maximum coating thickness of 8 μm was achieved at a forming time of 40 minutes. This indicates that the optimum range of current density for obtaining the maximum thickness of 8 μm by chromium forming on mild steel pipes is 8 A and can be obtained over a time range of 30 to 50 minutes.

Conclusions

Applied current density and forming time critically influence the electroforming process in forming on mild steel pipes. Electroforming can be used as a successful technique to form coatings of chromium, nickel and copper at the optimum parameters obtained with a coating thickness of 8 μm. For chromium forming the parameters are forming time of 30 minutes, current density of 10 A/dm^2 to obtain a coating thickness of 8 μm. In case of nickel the parameters are forming time of 40 minutes and current density of 6 A/dm^2. The parameters for copper forming are a time of 40 min, current density of 8 A/dm^2.

Table 26.6: Electroforming Parameters of Copper on Mild Steel Pipes at Various Forming Time Ranges

Sl.No.	Bath Temperature K	Rotational Speed of the Cathode (rpm)	Current Density (A/dm^2)	Forming Time (min.)	Thickness of the Coating (mμ)
1.	328	15	8	10	6.53
2.	328	15	8	20	6.86
3.	328	15	8	30	7.70
4.	328	15	8	40	8.00
5.	328	15	8	50	7.95

References

Barkalow, R.H., Goebel, J.A. and Petti, F.S., 1980. High temperature erosion-corrosion by high velocity (200 m/s) particles prepared by Pratt and Whitney Aircraft Group for Electric Power Research Institute, Palo Alto, CA., 94304; EPRI CS–1448, Project 979–4.

Hart, T. and Watson, A., 2000. Electroforming. *Met. Finish*, 98: 388–99.

Harty, S.F., Mcgeough, J.A. and Tulloch, R.M., 1981. A review of the electroforming of iron and iron-nickel alloy. *Surf Tech.*, 12: 39–55.

Lamb, V.A., 1968. Plating and coating methods: Electroplating, electroforming and electroless deposition. In: *Techniques of Materials Preparation and Handling*, Part 3, (Ed.) R.G. Bundshah. Interscience Publishers, New York.

McGeough, J.A., Leu, M.C., Rajurkar, K.P., De Silva, A.K.M. and Liu, Q., 2001. Electrofornling process and application to micro/macro manufacturing. *Ann CIRP*, 50: 499–514.

Ricq, L., Lallemand, F., Gigandet, M.P. and Pagetti, J., 2001. Influence of sodium saccharin on the electrodeposition and characterization of CoFe magnetic film. *Surf Coat. Tech.*, 138: 278–283.

Sun, L., Changsong, D., Dianlong, W. and Xinguo, H., 1997. Electroforming of nickel and partially stabilized zirconia (Ni + PSZ) gradient coating. *Surf Coat. Tech.*, 91: 131–135.

Tajiri, K. and Imamura, T., 1999. Controllable copper electroforming for mechanical properties from acid copper sulfate bath, Abstracts of the 195th Meeting of the Electrochemical Society, No. 102.

Tajiri, K., Nakamura, T., Kabeya, Z., Yamanaka, Y., Naito, F., Kato, T. and Takasaki, E., 2001. Development of an electroformed copper lining for accelerator components. *Electrochimica Acta*, 47: 143–148.

Weil, R. and Paquin, R., 1960. The relationship between brightness and structure in electroplated nickel. *J. Electrochem. Soc.*, 107: 87–91.

Chapter 27

Phytoremediation of Aqueous Dye Solution Using Blue Devil (*Eichhornia crassipes*)

M. Vasanthy, A. Geetha, M. Jeganathan and M. Bhuvaneswari
P.G. and Research Department of Environmental Sciences, Government Arts College, Ariyalur – 621 713, Tamil Nadu, India

ABSTRACT

The present study deals with the treatability of aqueous azo dye solutions containing Majanta HE. Different concentrations of dye solution (5, 10, 15, 20 and 25 ppm) were prepared and young saplings of *Eichhornia crassipes* were introduced and were exposed to sunlight. The growth of the plant and per cent of colour reduction have got increased day by day. This plant has been found to be effective and economical in the removal of colour. The time taken for 95 per cent colour removal from 5 ppm dye solution was 144 hrs (6 days). The increase in biomass was noticed with all concentrations. The highest per cent removal of colour from 25 ppm dye solution was 70 per cent after 144 hrs.

Thus the study revealed the fact that *Eichhornia crassipes* has absorbed the dye effectively and new saplings has emerged with in 144 hrs in all the concentrations of aqueous dye solutions.

Keywords: *Phytoremediation, Majanta HB, Eichhornia crassipes, Sunlight.*

Introduction

The industrial landscape in India is going through an extraordinary change over the last 10 years. For more than two years, the Indian economy has not exactly been in a robust state of health.

One such industry of rapid development is the dyeing industry. The dyeing industry effluents are highly coloured and toxic to some or other organisms and causes the destruction of the exposed communities It also affects the groundwater quality (Asfour *et al.*, 1985)

The industrial effluent poses environmental and health problems affecting mankind and nature. The textile effluent is intensely coloured and removal of colour is a must, before discharging it into the rivers or sewers. The coloured effluent percolates in deserts too and contaminates the groundwater. As the natural degradation of coloured effluent is very slow, the colour removal by different treatment techniques becomes a must (Daga *et al.*, 2001)

Phytoremediation is the term that refers to the usage of plants for cleaning up contaminants in soil, groundwater, surface water and air. The use of such technique can be a non-polluting and cost effective way to remove or stabilize toxic chemicals that might otherwise be leached out of the soil by rain to contaminate near by watercourses.

Phytoremediation practices are undertaken with water hyacinths (*Eichhornia crassipes*), poplar tress (*Papulus* spp), forage kochia (*Kochia* spp), alfalfa (*Medicago sativa*), Kentucky bluegrass (*Poa pratensis*), Scirpus spp, coontail (*Ceratophyllum demersvm* L.) Amercian pondweed (*Potamogeton nodosus*) and the emergent common arrowhead (*Sagittaria latifolia*) amongst others. (www.phytorem.com).

Typically, researchers look for suitable phytoremediation properties among, both cultivated and wild varieties of plants. If suitable wild species are not available, researchers can try to improve the effectiveness of phytoremediation by introducing different genetic varieties.

A major barrier to the implementation of phytoremediation is that it is new and not fully developed. There is little regulatory experience with phytoremediation and it has to be considered on a site by site basis. Furthermore, the intrinsic characteristics of phytoremediation limit the size of the niche that it occupies in the site remediation market.

Momentum for the use of phytoremediation as a cleanup technique is building, particularly in application niches where other technologies are less suitable or do not exist. There will also likely be combined applications of bioremediation and phytoremediation.

The plant used for phytoremediation in the present study, is the water hyacinth (*Eichhornia crassipes*). It is a floating water plant which grows mainly in the rivers, canals of the world. Water hyacinth removes both nutrients and organic materials directly from water through their extensive root system, which can increase at a phenomenal rate of 15 per cent of their surface area per day, producing at least 20 tonnes net weight per hectare per day. Based on measured growth rates it is estimated that one hectare of water hyacinth could remove the nitrogen and phosphorous wastes discharged by 800 people per day. (Rai, 2001)

Water hyacinth has also shown its capacity to accumulate heavy metals. It has been found to absorb and concentrate heavy metals such as Pb, Cd, Hg and Ni (Wolverton and Mc Donald, 1975 and O keefee *et al.*, 1984).

As the ecological indicator it is found to be a cost effective when compared with the regular treatment techniques. Thus, this method gains much importance on economical point of view.

Hence a study was conducted for the phytoremediation of aqueous Majanta HB dyes using the ecological indicator.

Materials and Methods

Estimation of Dye Concentration

The wavelength of maximum absorbance (λ max) and the dye concentration were determined using systronics UV–Vis spectrophotometer Model No. 118. The wavelength of maximum absorbance (λ max) was used as the monitoring wavelength.

Adsorbate Used

The dye used for this study is Majanta HB

Commercial Name: Majanta HB

λ max: 407 nm

Preparation of Stock Dye Solution

Accurately 1000 mg of Majanta HB dye was weighed and dissolved in 1000 ml of water to get 1000 ppm dye solution.

Preparation of Working Solution

Appropriate amount of stock solutions were diluted and were made upto 1000 ml to get 5, 10, 15, 20 and 25 ppm of dye solutions.

Batch Type Experiments

Effect of Time and Initial Dye Concentration on the Removal of Colour from Aqueous Dye Solution

The stock culture of the plant was maintained in the laboratory pond at ambient temperature. About 1 m^3 containers with 5 litre of dye solution were prepared and the weighed plants were introduced. Sample solutions were withdrawn daily and the colour reduction was recorded.

Results and Discussion

Batch Type Experiments

Effect of Time and Initial Dye Concentration on the Removal of Dye from Aqueous Dye Solution (Majanta HB)

The results of the experiment have been tabulated in Table 27.1. The batch type experiments have clearly shown that the colour imparted by 5 ppm Majanta HB solution colour has been removed by 95 per cent after 144 hrs and from the 10 ppm solution 88 per cent colour has been removed after 144 hrs. The maximum of 83 per cent colour removal has been reported after 144 hrs in 15 ppm aqueous dye solution. At the same time, the maximum dye removal from the other 2 dye concentration that is 20 and 25 ppm has been reported to be 80 and 71 per cent.

Table 27.1: Percentage of Colour Reduction from Aqueous Dye Solution Using *Eichhornia crassipes*

Concentration (ppm)	*Percentage (%) Dye Removal*					
	24 hrs	*48 hrs*	*72 hrs*	*96 hrs*	*120 hrs*	*144 hrs*
5	65.4	66.1	76.4	79.5	83.5	95
10	56.9	63.1	64.6	71.5	78.5	87.7
15	8.8	55.2	64.7	66.2	77.2	83.1
20	7.3	52.7	54.7	62.7	72.6	80
25	6.9	49.7	54.1	59.8	67.9	71.7

Similarly Orozoco, 1979 has reported 96 per cent colour removal from the textile solution using industry wastewater. A maximum of 82.4 per cent colour removal from aqueous has been reported by Shalini Menon (2000) with Procion Orange M2R using *Pseudomonas* spp., and *Aspergillus flavus.* Jothimani and Bhaskaran (2003) have reported that the *Trametes* spp., has shown 53 per cent colour reduction from the dye effluent and the percentage of colour reduction has increased up to 70.3 per cent after 4 days due to addition of glucose 0.2 per cent and urea 0.05 per cent.

Thus the colour removal obtained using water hyacinth is comparatively higher and offers good potential for treatment of the textile industrial wastewaters. During the period of study luxurious growth of the aquatic plant water hyacinth has been noticed which shows that the plant is tolerant to the toxic condition and is capable of performing the treatment of water. After treatment, these plant materials could be subjected to composting.

Even sugar industry effluent have been treated by Sinha and Sinha (1969) and removal of BOD by 94 per cent, turbidity by 97 per cent and odour by 96 per cent during a time period of 7 days. Thus a number of industrial waste such as dairy waste, sugar industry waste, tannery waste, electroplating industry waste, paper and pulp mill waste, textile industry waste including dyeing industry waste have been treated using water hyacinth (Trivedy, 1998).

Thus, its clear that aquatic plant namely *Eichhronia crassipes* could be effectively used for the wastewater treatment thereby making the recycling and reuse of the effluent possible.

References

Asfour, H.M., M.M. Nassar, O.A. Fadali and M.S. Elgelundi, 1985. Colour removal from textile effluent using hard wood sawdust as an adsorbent. *J. Chem. Tech. Biotechnol.*, 35: 28–35.

Daga, K., Loonker, S., Rathore, S.J. and Katyal, S., 2001. *Solution by Polymeric Resin.* Department of Chemistry, J.N.V. University, Jodhpur, Rajasthan, India.

Jothimani, P. and A. Bhaskaran, 2003. Influence of enrichment sources on fungal decolourisation of dye effluent under shaking condition. *J. Ecotoxicol. Environ. Monit.*, 12(2): 89–95.

Keeffee, O., Hary, D.H., Anjanee, J.K. and Rao, R., 1984. Cadmium uptake by water hyacinth, Effects of solution factors. *Environmental Poll.*, 34: 133–147.

Menon, Shalini, 2000. Sorption removal of procion orange M2R from aqueous solutions using *Pseudomonas* sp., S5 and *Aspergillus flavus. M.Sc. Dissertation*, Bharathiar University, Coimbatore.

Orozoco, A., 1979. Remocion decoloury transpiration *Jacintos deaqua. Tech. Invest. Tratam Medio Ambiente,* 1: 15–30.

Rai, G.D., 2001. *Non-conventional Energy Sources*. Khanna Publishers, Delhi.

Sinha, S.N. and Sinha, L.P., 1969. Studies on the use of water hyacinth in oxidation ponds digested sugar wastes and effluents of septic tanks. *Ind. J. Env. Hlth.*, 11(3): 197–207.

Trivedy, R.K., 1998. *Advances in Wastewater Treatment Technologies*. Global Science, Aligarh.

Wolverton, B.C. and McDonald, R.C., 1975. Water hyacinth for upgrading sewage lagoons to meet advanced waste treatment standards Part–II. NASA, Tech X–72720, October, pp. 25.

www.phytorem.com.

Chapter 28

Some Medicinal Important Plant Gum Consumed by Monkeys for Self Medication

Shivesh Pratap Singh
Department of Zoology, Government P.G. College, Satna – 485 001, M.P.

ABSTRACT

The rhesus monkey (*Macaca mulatta*) and Hanuman langur (*Presbytis entellus*) are the most common monkeys of Chitrakoot area. They feed on a large number of medicinal plants as well as on plants gum, directly or indirectly for their survival. During present investigation 16 gum producing plants have been identified, which are exploited during their feeding by monkeys. The maximum gum consumption was observed during April when 8 plants gum were eaten by monkeys. Gum feeding was seen slightly higher in pregnant female monkey before and after birth compared to other females.

Keywords: *Medicinal plants, Gum, Self medication.*

Introduction

Gum is a group of amorphous, acidic polysaccharides of complex and variable structural composition which are soluble in water. Several authors have recorded feeding of plant exudates as an important category in the natural diet of nonhuman primates. Gum may well have more importance as a mineral source than as a source of carbohydrate (Bearder and Martin, 1980).

Apart from leaves, flowers and fruits, monkeys were found to feed on gum. The gum was eaten throughout the year but was particularly prevalent in the diet during the dry months with maximum consumption in March; there is a fall in the monsoon and early winter months from June to December. Two distinct peaks one in November and another in March are evident. Gum feeding was observed in only few plant species of the study area. They are *Prosopis juliflora, Acacia nilotica* and *Anogeissus pendula*. During January–March, *P. juliflora* exudes a gum. The present study was undertaken with the objectives of identifying gum producing plants, consumed by monkeys for their self medication in Chitrakoot area.

Materials and Methods

The present studies were conducted for a period of one year in Chitrakoot forest and Kamathnath temple sites (lies between latitutes 80–52′ North and longitudes 25–10′). The field observation on gum feeding nature were made by spending six hours daily for four two days (Sunday.) in every month throughout the year. The care was taken to monitor the activities concerning feeding in between an hour before Sunrise till the monkeys turned motionless on the places of their roosting a little after Sunset.

Results and Discussion

During winter and hot days monkeys eats gum in Chitrakoot area. There were 16 plant species, which produces gum in Chitrakoot area. Out of them gum of 14 tree species and one shrub species were consumed. The gum of two species *Cassia fistula* and *Terminalia belerica* were avoided by monkeys. The maximum gum consumption was observed during April when 8 plants gum were eaten by monkeys, where as in rainy days monkeys were not observed to take gum because in this season gum are more in liquid form. Mostly gum feeding was observed during November to February in winter months.

Table 28.1: Gum Producing Plants Found in Study Area and their Consumption by Monkeys in Different Seasons

Sl.No.	*Gum Producing Plants*	*Consumption Months*
1.	*Acacia catechu*	November to April
2.	*Acacia nilotica*	December to March
3.	*Acacia leucophloea*	December to April
4.	*Anogeissus pendula*	December to April
5.	*Anogeissus latifolia*	December to June
6.	*Azadirachta indica*	November to April
7.	*Boswellia serrata*	December to March
8.	*Bridelia squamosa*	December to April
9.	*Butea monosperma*	November to April
10.	*Buchanania lanzan*	October to March
11.	*Cassia fistula*	Not eaten
12.	*Lannea coromandilica*	November to April
13.	*Mangifera indica*	January to April
14.	*Sterculia urens*	December to April
15.	*Shorea robusta*	December to March
16.	*Terminalia belerica*	Not eaten

In small-bodies primates (such as Prosimians and Callitachids), the ecological and evolutionary significance of gum feeding has been considered in a number of recent publications (Martin, 1972; Petter, 1978; Hladik, 1979; Bearder and Martin, 1980 and Harcourt, 1986). Jolly (1985) reported gum feeding in several New World monkeys. At Kanha National Park, in Madhya Pradesh, langurs were observed to spend 1 per cent of their feeding time in gum-feeding (Newton, 1985). However, the details of physical and analytical process of gum-feeding and which animals participated has not been recorded in Newton's study.

There are considerable changes in the use of gum between the hot wet summer and the cold dry winter (Harcourt, 1986). Likewise, langurs also consume maximum gum during dry months. The possible explanations for this behaviour are:

1. Which are the months when gum production is highest in plants.
2. Gum can provide a valuable source of carbohydrate and trace element.
3. In summer gum tends to be soft and presumably easy to ingest, whereas it become much harder and more crystalline in winter (Bearder and Martin, 1980).

In rhesus macaques of northern India contains 36 per cent fruits and seeds, 9 per cent flowers, 34 per cent leaves and rest gum, buds, grass, root, bark and insects (Makwana, 1979). Langurs around Jodhpur exploit five species gum, as compared to 10 in Kanha (Newton, 1985). Chemical composition indicates that gums provide a valuable source of carbohydrates at least during the harsh period of the year. In addition the presence of other trace contents could be an important factors giving rise to gum-eating throughout the year.

Acknowledgement

Author wishes to acknowledge University Grant Commission, Bhopal (M.P) for providing financial support.

References

Bearder, S.K. and Martin, R.D., 1980. *Acacia* gum and its use by lesser bush babies, *Galago senegalensis* (Primates : Lorisidae). *Int. J. Primatology*, 1: 102–128.

Harcourt, C., 1986. Seasonal variation in the diet of South African Galagos. *Int. J. Primatology*, 7: 491–506.

Hladik, C.M., 1979. Diet and ecology of Prosimians. In: *The Study of Prosimian Behavior*, (Eds.) Doyal, G.A. and Martin, R.D. Academic Press, New York, pp. 307–357.

Makwana, S.C., 1979. Field ecology and behaviour of the rhesus macaque *Macaca mulatta*. II. Food, feeding and drinking in Dehradun forests. *Indian. J. Forestry*, Dehradun, 2(2): 242–253.

Martin, R.D., 1972. Adaptive radiation and behaviour of Malagasy lemurs *Phil. Trans. R. Soc.*, London, 264: 295–352.

Newton, P.N., 1985. The behavioural ecology of forest Hanuman langurs (*Presbytis entellus*). *Int. J. Primatology*, 8: 199–232.

Petter, J.J., 1978. Ecological and physiological adaptations of five sympatric nocturnal lemurs to seasonal variation in food production. In: *Recent Advances in Primatology*, (Eds.) Chivers, D.J and Herbert, J. Academic Press, London, 1: 211–223.

Chapter 29

Combined Economic-Environmental Dispatch Using Adaptive Simulated Annealing Technique

R.K. Santhi* and S. Subramanian
Department of Electrical Engineering, Annamalai University

ABSTRACT

This article describes a methodology for economic dispatch task to minimize the cost including the environmental objectives. The harmful ecological effects caused by emission of the particulate gaseous pollutants can be reduced by adequate distribution of load between the power system plants. However, this leads to an increase in the operating cost of the plants. An efficient adaptive annealing algorithm been proposed for emission minimum cost with controlled emission dispatches. The applicability of the algorithm has been demonstrated on a three generator and six generator system and the results are compared with a standard dispatch technique. The results clearly exhibit that the proposed solution method is valid for real-time applications.

Keywords: Simulated annealing, Reannealing, Emission.

Introduction

Thermal power generating station is one of the most important sources of carbon dioxide (CO_2), sulphur dioxide (SO_2) and nitrogen oxides (NO_x) which create atmospheric pollution. Economic

* E-mail: rksanthieee@yahoo.co.in

dispatch is allocation of amount of power to be generated by different plants in the system on an optimum economy basis, without violation of operating constraints. Some of the states in India experience severe power shortage, where to meet out this problem generation of hydro plants sources are limited. So to meet out the peak load condition, generation of fossil based generating units is the choice, which claim that electricity generation imposes environmental costs on society that are not reflected in the prices paid for electricity (Koomey, 1990; Bernow and Biewald, 1991). The fossil fuel burning imposes not only burdens on the environment through emission to air of SO_2, NO_x, CO_2 and particulates but also pollutes water, land use, etc. Operating fossil fired power plants impose emission to air during normal power plant operation. During start up, up-ramping periods the amount of emission to air is significant. Now-a-day's several analysts incorporate environment constraint along with the electric energy dispatch. The power dispatch to meet the customer demand for electricity may be met as economically as possible. But minimizing the fuel cost and emission level simultaneously are in conflict. They are usually in a trade-off relationship and often incommensurable. Therefore, the minimum cost can no longer be the goal of operation, if society is to have a clean atmosphere. Srikrishana and Palanichamy (1991) proposed a method for combined emission and economic dispatch using a price penalty factor developed a technique to add emission constraints to the standard classical economic dispatch problem. Based on emission, shadow prices (Eto and Hilcke, 1990) describe various emission dispatching strategies and solution procedure. The classical approaches need to introduce a compromise factor in order to decide on an optimal solution results in complicated problem formulation and computational inefficiency.

When the power system size increases, the solution procedure further complicates because of several constraints and objectives to be solved. Recently several heuristic approaches such as Fuzzy, SA are developed (Hota *et al.*, 1999; Basu, 2005) to optimize emission and economic dispatch problems. The author (Baskar *et al.*, 2002) has discussed Hybrid GA technique for the multi-objective combined emission and economic dispatch problem. Genetic algorithm is a powerful technique and gives global optimization but introduces nonlinearities and discontinuities to appear in the solution space. SA technique has the advantage to give global or near global solution for optimization problem but has the disadvantage of high convergence time.

An improved SA technique is thus suggested to statistically find the best global fit for a nonlinear constrained non-convex characteristic over a multidimensional space. Moreover, this algorithm allows rescaling of the annealing time permitting adaptation to changing sensitive in the multidimensional parameter space. This annealing schedule is faster than fast cauchy annealing and Boltzmann annealing, because annealing schedule exponentially decreases in annealing-time step.

In this article, multi objective formulation of the combined economic-environmental load dispatch (CEELD) of thermal generation is established. The two objectives are cost and emission level to be minimized. Thermal units with non smooth fuel cost and emission level function with transmission losses. The pollutant nitrogen oxide (NO_x) considered because its control is a significant issue at the global level.

Nomenclature

- n: Number of thermal units
- F_t: Total fuel cost (Rs/hr)
- P_i: Power output of the ith generating unit (MW)
- P_D: Total system power demand (MW)
- P_l: Total transmission losses (MW)

E_i: NO_x emission of thermal unit (Kg/hr)
P_{imin}: Minimum power output of ith unit
P_{imax}: Maximum power output of ith unit
a_i, b_i, c_i: Fuel cost coefficients of ith unit
d_i, e_i, f_i: NO_x emission coefficients of ith unit
Ψ_i: Both the fuel cost and NO_x emission output of thermal unit i
w: Single point in space
W: State apace
P_{acc}: Probability of acceptance
T_{gen}: Generating temperature
T_{acc}: Acceptance temperature
K_a, K_i: Annealing time indexes
J(w): cost function
S_i: Sensitivity factor
P_{unif}: Uniform random variable

Problem Formulation

The economic dispatch problem is defined as to minimize in

$$F_t = \sum_{i=1}^{n}(a_i P_i^2 + b_i P_i + c_i) \text{ Rs/hr}$$

subject to

System operating constraint

$$P_{imin} \le P_i \le P_{imax} \quad \text{......(1)}$$

System demand constraint

$$\sum_{i=1}^{n} P_i = P_D + P_L \quad \text{......(2)}$$

Minimum NO_x emission dispatch problem is defined as to minimize

$$E_i = \sum_{i=1}^{n}(d_i p_i^{\,2} + e_i p_i + f_i) \text{ Kg/hr}$$

subject to operating and demand constraint (1) and (2)

Combined economic and emission dispatch problem is defined as to minimize

$$\psi_i = \sum_{i=1}^{n} F_i + \sum_{i=1}^{n} E_i \text{ Rs/hr}$$

$$\psi_i = \sum_{i=1}^{n}(a_i P_i^2 + b_i P_i + c_i) + \sum_{i=1}^{n}(d_i P_i^{\,2} + e_i P_i + f_i) \text{ Rs/hr}$$

subject to system operating constraints demand constraint (1) and (2) and emission constraint subject to

$$\sum_{i=1}^{n} E_i \le Nox_i$$

The proposed adaptive simulated annealing (Ingber, 1995) is a powerful tool to find the best compromising solution in multi objective problems. The multi objective optimization is concerned with generation. Since cost and emission are the objectives, when cost is reduced the other objective emission is increased. The proposed simulated annealing a powerful optimization technique minimizes such complicated multi objective problem.

The SA procedure thus simulates the slow process of cooling of molten metal to achieve the minimum function value in a minimization problem. The cooling phenomenon is simulated by controlling a temperature like parameter introduced with the concept of the Boltzmann distribution and during the search process, the annealing schedule reduces in a logarithmic way. The proposed algorithm based on SA searches the minimum function values the annealing schedule temperature is rescaled. Otherwise the annealing schedule for temperature decreasing exponentially in annealing-time K and is related as $T = To \exp(-CK)^{1/M}$, where To is the initial temperature, C is user defined value, K annealing time step and M is the considered multi objectives. This re-annealing process also allows automatic adaptation to changing sensitivities in the multidimensional parameter space. The adaptation provides robust solution and speeds up the annealing procedure reducing the convergence time.

Computational Algorithm

The algorithm to determine EELD of thermal system using adaptive SA method is described in the following computational steps.

1. Input the system data consisting of fuel cost values and emission coefficients of thermal generators, operating limits to thermal generators, penalty price factor for all the thermal units, transmission loss coefficients.
2. Randomly select a solution w such that $W \in W$, the randomly selected generating capacity is used to initialize the process.
3. Find the neighbourhood solution. If the cost of this solution J(W) is equal to the current cost accept this solution as the current solution to generate yet another candidate solution.
4. If the cost of neighbourhood solution is not less than the cost of the current solution, generate a uniform random number P_{unif} between (0, 1).
5. Calculate P_{acc} probability function as follows.
6. $P_{acc} = 1/(1+\exp(Jw^{new}) - J(w^{old})/T_{accp}\,(K_a)$
7. Check the random number P_{unif} which is less than P_{acc}. Then accept this solution as the better solution.
8. After every N_{acc} acceptance points, rescaling of the annealing time is accomplished, essentially reannealing every hundred or so acceptance-events. Rescaling is calculated in terms of the sensitivity 'S_i' which is calculated at the most current minimum value of the cost function. The sensitivity factor depends on the number of parameters considered and the current minimum of the cost function. It is a multidimensional vector, having the ith element

and the rest of the elements are all zeros. This provides the best tuning to the annealing schedule. S_i usually kept to maximum S_{max}. So scaling of each parameter generating temperature T_{gen} is scaled by a factor S_{max}/S_i and the annealing time k_i^1 is reset as:

$$k_i^1 = \frac{-1}{c}\left(\ln\frac{T_{gen}(K_a)}{T_{gen}(o)}\right)^n$$

9. Similarly $T_{acc(o)}$ is rest to the last accepted cost function, $T_{acc}(K_a)$ is rest to $J(W_{best})$ and the annealing time K_a is rescaled.

$$K_a = \frac{-1}{c}\left(\ln\frac{T_{acc}(K_a)}{T_{acc}(o)}\right)^n$$

10. After every N_{gen} generated points, annealing takes place and the annealing–time schedule is rescaled, otherwise go to step (3).
11. The searching process is terminated if the parameters have remained unchanged for a few successive reannealings or a preset maximum number of cost function evaluation has been reached. Otherwise go to step (8) and continue the process of searching.
12. Stop.

Simulation Results

The proposed method has been applied to a two test systems with three thermal generators and six thermal generators. The characteristic of thermal plants with NO_x emission equations, loss coefficients and the operating limits of the generators are considered (Hota *et al.*, 1999; Srikrishna and Palanichamy, 1991).

Case 1

In this case the computations were carried out on IEEE 14 bus system, for the load of 80 per cent, 100 per cent and 120 per cent. The performance characteristic obtained by the proposed method is compared with those obtained from most economic dispatch and the minimum emission solutions in Tables 29.1–29.3.

Table 29.1: Performance Indices Obtained from Most Economic Dispatch, Minimum Emission and Combined Economic Emission Solution

Method	*Percentage Load (MW)*	*Total Fuel Cost (Rs/hr)*			*Total Emission Level (Kg/hr)*		
		Most Economic Dispatch	*Minimum Emission*	*Combined Economic Emission*	*Most Economic Dispatch*	*Minimum Emission*	*Combined Economic Emission*
SA	80	757.8555	804.2160	798.7237	509.3563	439.5737	439.8840
	100	931.2223	985.5939	978.2851	544.6873	453.0647	453.4465
	120	1132.1194	1173.8227	1186.4550	582.0169	481.1549	484.9249
ASA	80	757.8224	804.9097	798.8604	508.7098	439.5708	439.8699
	100	931.2572	985.9275	977.8690	541.5695	453.0688	453.4036
	120	1114.2080	1175.0447	1166.3870	577.5972	481.1510	481.6305

Table 29.2: Comparison of Total Losses of SA and ASA Method

Method	*Percentage Load (MW)*	*Most Economic Dispatch Loss (MW)*	*Minimum Emission Loss (MW)*	*Combined Economic Emission Loss (MW)*
SA	80	5.5900	4.9552	5.5754
	100	8.2572	7.8559	8.1189
	120	12.1385	12.001	12.0478
ASA	80	5.5754	4.8883	4.9805
	100	8.1943	7.8308	8.1300
	120	11.6520	1.1523	1.4742

Table 29.3: Comparison of SA and ASA Results with the Reference Method for Base Load 210 MW

Method	*Total Fuel Cost (Rs/hr)*			*Total Emission Level (kg/hr)*			*Total Loss (MW)*		
	Ref (Hota et al., 1999)	*SA*	*ASA*	*Ref (Hota et al., 1999)*	*SA*	*ASA*	*Ref (Hota et al., 1999)*	*SA*	*ASA*
Most economic dispatch	962.592	931.223	931.2572	500.722	544.6873	541.5695	–	8.2572	8.1943
Minimum emission	994.424	985.9939	985.5275	453.968	453.064	453.0688	–	7.8559	7.8308
Combined economic emission	–	978.2851	977.869	–	453.4465	453.4036	–	8.1189	8.1300

Case 2

For this test case with a system load of 500 MW, 700 MW, 900 MW, 1100 MW, the most economic dispatch is conducted and the cost is calculated by the proposed algorithm and also compared with the general SA. For the same load values the NO_x emission dispatch is performed with proposed and the other methods are compared in the following Tables 29.4–29.6.

Table 29.4: Performance Indices Obtained from Most Economic Dispatch, Minimum Emission Combined Economic Emission Solution

Method	*Percentage Load (MW)*	*Total Fuel Cost (Rs/hr)*			*Total Emission Level (Kg/hr)*		
		Most Economic Dispatch	*Minimum Emission*	*Combined Economic Emission*	*Most Economic Dispatch*	*Minimum Emission*	*Combined Economic Emission*
SA	500	27562.8913	27862.4173	27592.2411	274.1810	269.6536	269.0923
	700	37144.2521	38244.0685	37268.0315	489.9646	435.667	458.1015
	900	47406.7014	49652.8316	41691.8562	806.5149	684.0336	736.4407
	1100	58384.3420	60736.1112	58840.2673	1227.8410	1026.6793	1109.5208
ASA	500	27570.9657	27832.8901	27585.2750	273.2984	262.6711	268.5792
	700	37167.8451	38244.0685	37259.5914	481.3155	435.3991	457.1015
	900	47432.2300	49738.2371	47664.9777	795.0966	684.1098	733.6878
	1100	58402.0435	60746.5101	58813.2299	1199.9424	1026.3821	1106.5024

Table 29.5: Comparison of Total Losses of SA and ASA Method

Method	Percentage Load (MW)	Most Economic Dispatch Loss (MW)	Minimum Emission Loss (MW)	Combined Economic Emission Loss (MW)
SA	500	11.7891	9.6012	11.0436
	100	22.9890	17.7411	20.4069
	900	37.7834	28.4352	32.7467
	1100	55.5406	43.4004	48.6245
ASA	500	11.5478	9.6883	11.1822
	700	22.3335	17.2709	20.4374
	900	36.0958	28.3149	32.9007
	1100	53.7508	43.1879	48.8156

Table 29.6: Comparison of SA and ASA Results with the Reference Method for Base Load 700 MW

Method	Total Fuel Cost (Rs/hr)			Total Emission Level (kg/hr)			Total Loss (MW)		
	Ref (Hota et al., 1999)	SA	ASA	Ref (Hota et al., 1999)	SA	ASA	Ref (Hota et al., 1999)	SA	ASA
Most economic dispatch	31288.10	37161.8451	37144.2521	495.348	489.39646	481.3155	26.587	22.9890	22.3335
Minimum emission	38364.50	38244.0685	38076.9141	437.966	435.6670	435.3991	20.270	17.7417	17.2709
Combined economic emission	37781.90	37268.0315	37259.5914	442.555	442.7015	441.8356	21.170	20.4069	20.4374

The calculating processes are carried out on Pentium 3, 933MHZ, coded in C language for 20 trials, 90000 iteration, for 100 accepted options considered. The CPU time of all the three methods are compared in Table 29.7. In both test cases, the calculated NO_x emission level by implementing the proposed algorithm is reduced in most economic dispatch, minimum emission and combined cost emission approaches. The values compared with the reference method shows that the cost as well as emission levels are less. Moreover, the computation time been considerably reduced than the other methods. In this process the conflicting objectives (*i.e.* cost and emission) are commensurable. The classical methods need the compromising factor such as price penalty factor (Srikrishna and Palanichamy, 1991). In the proposed algorithm there is no such compromising penalty factor is considered, and the results demonstrates best compromising solutions between emission and economic dispatch.

Table 29.7: Time

Time (Sec)	Most Economic Dispatch		Emission Dispatch		Combined Economic and Emission Dispatch	
	Case 1	Case 2	Case 1	Case 2	Case 1	Case 2
Simulated annealing	232.42	224.01	20.52	6.26	44.34	19.91
Adaptive simulated annealing	36.37	11.54	4.72	4.89	2.68	1.59

Conclusion

In this article, an adaptive simulated annealing method is proposed to solve economic emission load dispatch of thermal power plants, considering the multi objectives. The exponential annealing schedules permit resources to be spent adaptively on reannealing and on pacing the convergence in all dimensions, ensuring ample global searching in the first phase of search and ample quick convergence in the final phases.

The numerical results have shown the performance and applicability of ASA tool. Hence it may be used as an effective technique to deal conflicting objectives. The only disadvantage is that if the sensitivity factor is not properly assigned then the convergence time enormously increases because of the adaptive nature of the annealing schedule. The execution time been reduced than the conventional and general SA method by suitably selecting rescaling factor. It is proven that the proposed method gives robust results reducing the cost as well as the emission level in less convergence time.

References

Baskar, S., Subbaraj, P. and Rao, M.V.C., 2002. Hybrid genetic algorithm solution to emission economic dispatch problems. *IE (I) J.*, 82: 243–302.

Basu, M., 2005. A simulated annealing based goal-attainment method for economic emission load dispatch of fixed head hydrothermal power systems. *Electrical Power and Energy System*, 27: 147–153.

Bernow, S. and Biewald, B., 1991. Full-cost dispatch: Incorporating environmental externalities in electric system operation. *The Energy Journal.*

Eto, J. and Hilcke, G., 1990. Internalizing environmental costs: A survey of progress in estimating the external environmental costs of electricity production and of market based policies to internalize them, EVR Report for the European Commission.

Hota, P.H., Chakrabarti, R. and Chattopadhyay, P.K., 1999. Fuzzy-set based optimization technique for economic emission load dispatch. *IE (I) J.*, 80.

Ingber, L., 1995. Simulated annealing applied to combinational optimisation. *Polish J. Control and Cybernetics.*

Koomey, J., 1990. *Comparative Analysis of Monetary Estimates of External Environmental Costs Associated with Combustion of Fossils Fluids.* Lawerence Berkeley Laboratories.

Srikrishna, K. and Palanichamy, C., 1991. Economic thermal power dispatch with emission constraint. *IE (I) J.*, 72.

Venkatesh, P., Gnanadas, R. and Narayana Prasad Padhy, 2003. Comparison and application of evolutionary programming technique to combined economic dispatch with line constraints. *IEEE*, 18(2).

Chapter 30

Antibacterial and Antifungal Activity of Marine Bacteria from Gulf of Mannar Resources

N.K. Asha Devi[1], R. Gopal[2], K. Balakrishnan[3] and S. Padmavathy[1]

[1]Department of Zoology and Microbiology, Thiagarajar College, Madurai – 9, Tamil Nadu, India

[2]Reader and Head, Department of Zoology, Yadava College, Madurai – 14, Tamil Nadu, India

[3]Reader, Department of Biotechnology, Bharathidasan University, Tiruchirapalli – 24, Tamil Nadu, India

ABSTRACT

Marine microorganisms of bacterial species were isolated from coastal water of South East Zone of Rameshwaram, Dhanushkodi and Thondi, Ramanathapuram District, Tamil Nadu, India. Out of 23 bacterial isolates, 6 were identified and selected for antimicrobial activity against selected human bacterial pathogens of *Staphylococcus aureus, Pseudomonas aeruginosa, Salmonella typhi, Vibrio cholerae, Klebsiella pneumoniae* and *Aspergillus niger.* Of the 6 marine bacterial species, *Planococcus* sp. and *Bacillus* sp. showed strong antimicrobial activity. But all the marine bacterial samples showed antifungal activity against *Aspergillus niger.* The results of the present investigation reveal that the marine bacteria from coastal environment are a potent source of novel antibiotics.

Keywords: *Human pathogens, Marine bacteria, Antibacterial and Antifungal activity*

Introduction

Marine ecosystems represent 95 per cent of the biosphere and coastal regions are particularly promising, because of the highly adapted species found in these harsh environments. Marine microorganisms have become important in the search of novel microbial products. Many microorganisms such bacteria and fungi had been reported to produce antibacterial, antifungal, antiviral and antitumor substances (Bernan *et al.*, 1997; Fenical, 1997; Molinski, 1993; Okami and Okazaki, 1972; Rosenfeld and Zobell, 1947). Marine bacteria have recently been identified as a source of new bioactive metabolites (Bernan *et al.*, 1997; Jaruchoktaweechai *et al.*, 2000). There are few reports on antimicrobial activity from marine microorganisms from East Coast of India. Hence a detailed study on the antimicrobial potential microorganisms of bacteria from coastal water of South East zone of Rameshwaram, Dhanushkodi and Thondi, Ramanathapuram District, Tamil Nadu, India.

Materials and Methods

The water samples were collected in clean, sanitized and autoclaved bottles and transported to the laboratory for further analysis. Marine agar (Hi Media) was used to culturing the bacterial colonies, which were incubated at 37°C for 24–48 hours. The isolated bacteria were identified based on the colony morphology, gram staining and biochemical tests (Cappuccino, and Sherman, 1999; Holt *et al.*, 1994). Marine bacterial samples isolated were cultured in 300 ml in Marine broth for the production of secondary metabolites in 500 ml Erlenmeyer flasks. Flasks were incubated on a rotary shaker at 220 rev/min at 25°C. After 7 days of cultivation, the culture broth was centrifuged at 10000 rpm for 20 minutes to remove the cell and filtered through Millipore filter (0.45 µm) to get cell free extract. Antimicrobial activities were assayed in duplicate, using 50 µm of marine bacterial samples in well diffusion assay (Schillinger and Lucke, 1989) and the test organisms were *Staphylococcus aureus, Pseudomonas aeruginosa, Salmonella typhi, Vibrio cholerae, Klebsiella pneumoniae* and *Aspergillus niger.*

Results

Among 23 marine bacteria isolates, 6 were selected for further studies on the basis of antimicrobial activity and growth characteristics. Based on morphological, gram staining and other biochemical characteristics, they were identified as *Marinonascens* sp., *Mesophilobacter* sp., *Marinoncoccus* sp., *Planococcus* sp., *Bacillus* sp. and *Alteromonas* sp. (Table 30.1). Selected 6 bacterial filtrated samples were tested against selected human pathogens. *Planococcus sp.* showed significant antimicrobial activity and it did not show zone of inhibition in *Staphylococcus aureus* and *Klebsiella pneumoniae.* Except the gram negative bacteria of *Klebsiella pneumoniae,* other three gram negative bacteria were sensitive to *Planococcus* sp. *Bacillus* sp. also showed strong antibacterial activity against gram negative bacteria but not against *Salmonella typhi.* Of the 6 marine bacterial samples, only 2 species of *Mesophilobacter* sp. and *Marinococcus* sp. showed antibacterial activity against gram positive bacteria of *Staphylococcus aureus.* All 6 marine bacterial samples, including *Alteromonas* sp. and *Marinonascens* sp. showed significant antifungal activity against *Aspergillus niger* (Table 30.2).

Discussion

The first antibiotic from a marine bacterium was identified and characterized in 1966 (Burkholder *et al.*, 1996). The results of the growth inhibition assays against five species of pathogenic fungi provide very promising results. In this present work, *Planococcus* sp. and *Bacillus* sp. showed stronger antimicrobial activity compared to other marine bacterial samples. In recent studies wide arrays of

Table 30.1: Morphological and Biochemical Characteristics of Selected Marine Bacterial Species

Isolated Organisms	*Morphology*	*Gram Stain*	*Shape*	*Motility*	*Oxidase*	*Catalase*	*Indole Production*	*Methyl Red*	*Voges Proskaur*	*Citrate Utilisation*	*Nitrate Reduction*
Marinonascens sp.	Smooth, circular, raised yellow in colour	+	Cocci or short rods	Nonmotile	+	+	–	–	–	–	+
Mesophilobacter sp.	Irregular, smooth, pale yellow in colour	–	Cocci	Nonmotile	+	+	–	+	–	–	+
Marinococcus sp.	Circular, smooth, convex colonies, yellow orange in colour	+	Rod	Motile	+	+	–	–	–	+	+
Planococcus sp.	Wrinkled, yellow orange colonies	+	Cocci	Motile	–	+	–	+	–	+	–
Bacillus sp.	Irregular Wrinkled, orange colonies	+	Rod	Motile	+	+	–	+	–	+	–
Alteromonas sp.	Irregular, colourless wrinkled colonies	–	Cocci	Motile	+	–	–	+	–	–	–

+: Indicates positive; –: Indicates negative.

Table 30.2: Antimicrobial Activity of Marine Bacteria Against Human Pathogens

Tested Organisms	*Marine Bacteria (Zone of Inhibition in mm)*					
	Marinonascens sp.	*Mesophilobacter sp.*	*Marinococcus sp.*	*Planococcus sp.*	*Bacillus sp.*	*Alteromonas sp.*
Staphylococcus aureus	–	11	11	–	–	–
Pseudomonas aeruginosa	–	–	–	27	31	–
Salmonella typhi	–	–	–	17	–	–
Vibrio cholerae	–	–	–	27	17	–
Klebsiella pneumoniae	–	–	–	–	15	–
Aspergillus niger	17	11	13	13	11	15

–: Indicates absence of antimicrobial activity.

marine bacteria were tested for bacterial antagonistic effects and it was demonstrated that this trait appears to be a widespread feature in marine habitats and present in bacterial groups (Burgess *et al.*, 1999; Grossart *et al.*, 2004; Long and Azam, 2001).

In the present study, 33 per cent of the isolates were found to inhibit *Salmonella typhi*. This is a very encouraging and fruitful finding in the context of resistance of this pathogen to most of the widely used antibiotics. Normally, *Pseudomonas* sp. are even capable of growth in some antibiotics, such as quaternary ammonium compounds. Their resistance to more antibiotics has also been a source of medical concern (Tortora *et al.*, 2000). It is interesting to note that *Bacillus sp.* and *Planococcus sp.* response represents antibiotic potential competing microorganisms against *Pseudomonas aeruginosa* in the marine environment.

In the current study, 100 per cent of the tested marine bacterial species exhibited antifungal activity. Many marine microorganisms showed antifungal activity against *Aspergillus niger*, but not against *Candida albicans* (Kokare *et al.*, 2004). Thus the research on these marine bacteria for newer products is a timely one as the ocean appears to be a good alternative source of unique materials. For complete characterization of an antibiotic it should be isolated in pure form as a single component, but this is impractical in a screening programme like this work.

This present investigation shows that the marine bacteria from the coastal environment could be a great source for screening and isolating new antibiotics to combat many human pathogens particularly the most resistant *Pseudomonas aeruginosa* and *Salmonella typhi*, in a successful manner in the future.

Acknowledgement

Thanks are due to the Department of Zoology and Microbiology, Thiagarajar College, Madurai, Tamil Nadu, India for providing all the facilities.

References

Bernan, V.S., Greenstein, M. and Maiese, W.M., 1997. Marine microorganisms as a source of new natural products. *Adv. Appl. Micro. Biol.*, 43: 57–90.

Burgess, J.G., Jordan, E.M., Bregu, M., Means-Spragg, A. and Boyd, K.G., 1999. Microbial antagonism: A neglected avenue of natural products research. *J. Biotechnol.*, 70: 27–32.

Burkholder, P.R., Plflister, R.M. and Leitz, F.P., 1996. Production of a pyrrole antibiotic by marine bacterium. *Appl. Microbiol.*, 14: 649–653.

Cappuccino, J.G. and Sherman, N., 1999. *Microbiology: A Laboratory Manual*, 4th ed, pp. 179–182.

Fenical, W., 1997. New pharmaceuticals from marine microorganisms. *Trends Biotechnol.*, 15: 339–341.

Grossart, H.P., Schligoff, A., Bernhard, M., Simom, M. and Brinkhoff, T., 2004. Antagonistic activity of bacteria isolated from organic aggregates of the German Wadden Sea. FEMS: *Microbial Ecol.*, 70: 2560–2565.

Holt, J.G., Krieg, N.R., Sneath, P.H.A., Staley, J.T. and Williams, S.T., 1994. *Bergey's Manual of Determinative Bacteriology*, 9th ed. Williams and Wilkins, Baltimore, pp. 518–537.

Jaruchoktaweechai, C., Suwanborirux, K., Tanasupawatt, S., Kittakoop, P. and Menasveta, P., 2000. New macrolactins from a marine *Bacillus sp.* Sc. 026. *J. Nat. Products*, 63(7): 984–986.

Kokare, C.R., Mahadik, K.R. and Kadam, S.S., 2004. Isolation of bioactive marine actinomyces from sediments isolated from Goa and Maharashtra coast lines (West Coast of India). *Indian Journal of Marine Sciences*, 33(3): 248–256.

Long, R.A. and Azam, F., 2001. Antagonistic interactions among marine pelagic bacteria. *Appl. Environ. Microbiol.*, 67: 4975–4983.

Molinski, T.F., 1993. Developments in marine natural products receptor-specific bioactive compounds. *J. Nat. Prod.*, 58: 1–8.

Okami, Y. and Okazaki, T., 1972. Studies on marine microorganisms (1) Isolation from the sea. *J. Antibiot.*, 25: 456–460.

Rosenfeld, W.D. and Zobell, C.E., 1947. Antibiotic production by marine microorganisms. *J. Bacteria*, 154: 393–398.

Schillinger, U. and Lucke, F., 1989. Antibacterial activity of *Lactobacillus* strain isolated from meat. *Applied and Environmental Microbiology*, 55: 1901–1906.

Tortora, G.J., Funke, B.R. and Case, C.L., 2000. *Microbiology*, 6th ed. The Benjamin/Cummings Publishing Company Inc., pp. 300, 659–668.

Chapter 31

Modified Bio-Emf-Device Using MnO_2 Depolarizer

Bharat Mishra[1], K.M. Jain[2] and A. Hundet[2]

[1]Department of Physics, Mahatma Gandhi Chitrakoot Gramodaya Viswavidyalaya, Chitrakoot, Satna – 485 331, M.P., India

[2]Department of Physics, IEHE, Kolar Road, Bhopal – 462 016, M.P., India

ABSTRACT

A unique method or making an inexpensive low electric power source Bio-emf-device (BED) has been studied earlier. This study deals with the results on the enhancement of the voltaic contribution obtained when MnO_2 is used as depolarizer. The effect of mixing MnO_2 in different concentrations with graphite powder has been studied. After ascertaining the optimum concentration of MnO_2, various battery parameters and performance indices have been obtained,

Keywords: Bio-Emf-device, Voltaic contribution, Depolarizer, MnO_2.

Introduction

The use of biological material makes the bio-batteries unique and different from existing batteries (Jain *et al.*, 1987, 1988, 1989, 1990) and Hundet (1994). Investigation reveals that there are two main contributions in the energy generation from the bio-emf-devices (BED), these are voltaic, as well as the bio-contributions (Hundet *et al.*, 1989). The biological contributions can be increased by the proper selection of leaf, while voltaic contribution may be increased by the use of depolarizers.

Useful amount of electrical energy is obtained from the BEDs only when reasonably a large current is drawn. But the cell potential reduces to a value from its thermodynamic equilibrium potential because of irreversible losses. These losses result in a cell voltage that is less than its reversible potential. The losses which are called polarization/over potential or over voltage (Glasstone, 1956; Kordesh, 1974; Taketsugu *et al.*, 1980; Anath, 1992). The effect of polarization can be reduced by using depolarizer. A well-known depolarizer Manganese dioxide MnO_2 has therefore been used (Chemelli *et al.*, 1980; Visvanathan, 1987; Ananth, 1998; Bodoard *et al.*, 1994; Ardizzone, 1997).

MnO_2 has good conductivity, which results in minimized internal resistance, further due to its small pore-size its distribution leads to better electrolyte-electrode contact. Thus MnO_2 acts as a depolarizer as well as a higher holding capacity of electrolyte. The resistance at MnO_2 mix cathode is very complex. The main purpose however is to receive the electron continuously. The MnO_2 gets reduced to a lower form or oxide during the process (Perrault, 1974 and 1970; Udhayan *et al.*, 1992).

In the present investigation various aspects are studied for the enhancement of the voltaic contribution of bio-emf-device using MnO_2 as a depolarizer.

Materials and Methods

The graphite electrode is used for the cathodic electrode in which a cavity of the dimensions of 1.5 cms diameter and 0.3 cms depth was made, while magnesium electrode was used for the anode. MnO_2 in different concentration with graphite powder was put in it under moderate pressure. For each experiments electrode were freshly prepared. Qualigens Fine Chemicals, Bombay, supplied MnO_2 employed. The same graphite electrode was crushed and made fine to mix with MnO_2.

All the measurement has been carried out at room temperature 25±2°C using Agronic digital parameters having range 0±19.99 (mA and mV) DC. Initially, the optimum ratio of MnO_2 was determined than the BED prepared using optimum concentration of MnO_2. For the fabrication of BED the same procedure was followed as discussed by earlier researcher, except using the above mentioned cathodic electrode. *Agave-margineta* leaf was taken to employ in fabrication of BED.

Results and Discussion

Standardization of the Cathodic Mix

Effect of mixing different amount of MnO_2 with graphite powder has been obtained. The results for *Agave-margineta* leaf are presented in Table 31.1 and plotted in Figure 31.1.

Table 31.1: Effect of Varying Concentration of MnO_2 on BEDs (Leaf: *Agave-margineta*, size 1.76 sq.cm.)

Percentage of MnO_2 (by weight)	*0*	*10*	*15*	*20*	*25*	*30*	*35*	*40*	*45*	*50*
OCV±0.10 (Volt)	1.84	2.14	2.15	2.18	2.18	2.17	2.20	2.21	2.23	2.17
SCC±0.50 (mA)	2.80	2.85	2.91	3.20	3.60	3.91	3.65	3.33	2.80	2.30

The results show that the variation in open circuit voltage (OCV) of BEDs is almost constant for the entire ratio studied, while the short circuit current (SCC) value varied significantly. With increase in MnO_2 concentration the short circuit current density of BED is increased, attaining a maximum value at 25–35 per cent percentages amount of MnO_2. This indicates that at this composition the internal resistance of BED attains minimum value. The prepared using 30 per cent amount of MnO_2 shows the OCV value of 2170±100 mV and SCC values of 3.70 to 4.00 mA.

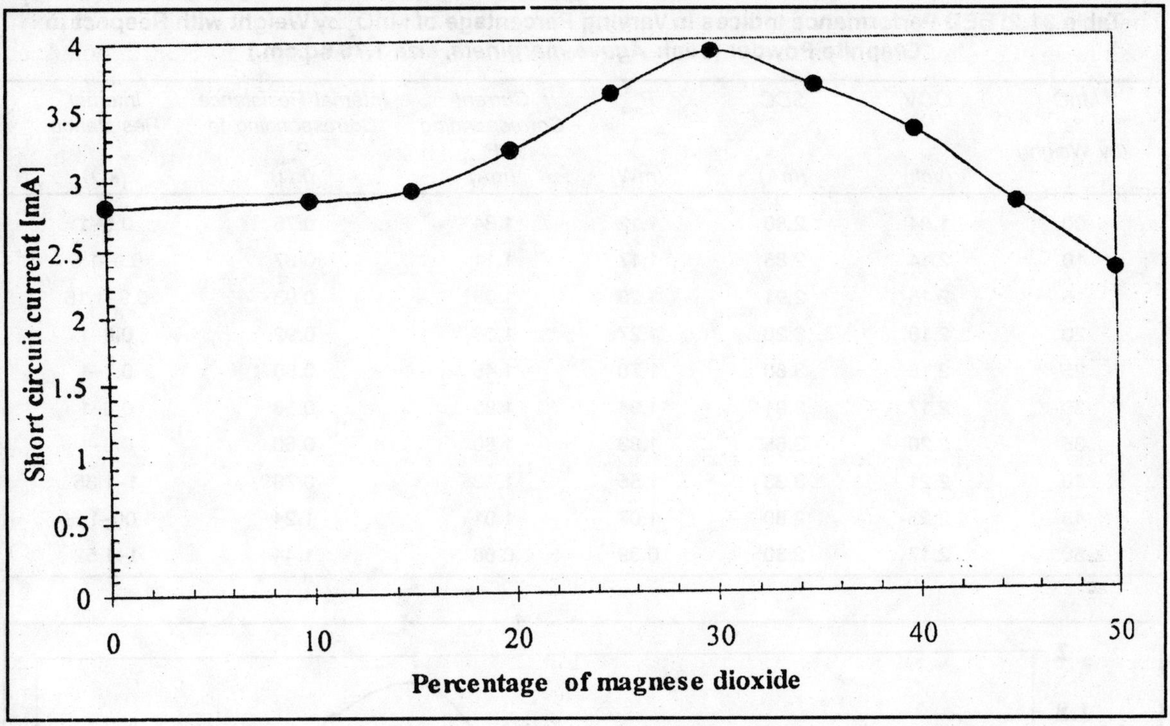

Figure 31.1: Effect of Increasing Percentage of MnO_2 with Graphite Powder on SCC of BED (Size: 1.76 sq.cm., Leaf: *Agave margineta*)

The different concentration of MnO_2 and power of BED have also been studied and presented in Table 31.2 and Figure 31.2. Result indicates that by increasing the MnO_2 concentration the power of BED is increases, attaining a maximum value at 25–35 per cent percentages. The resistance of BED corresponding to P_{max} is progressively lowered with increase in concentration of MnO_2 level and reach to an optimum for 25–35 per cent. After this optimum amount of MnO_2, internal resistance further increases and hence 30 per cent has been taken as the optimum concentration of MnO_2 by weight for the fabrication of BED.

Characteristics of Modified BED

The BEDs prepared using 30 per cent amount at MnO_2 by weight with respect to graphite powder was used for further studies. The current-voltage (I–V) characteristics of Modified-BED using leaf *Agave-margineta,* is depicted in Figure 31.3. A very interesting feature of the curves has been observed regarding the nature of slope. Result shows almost a linear fall in the cell voltage with increase in current. Slope angle of the curve is somewhat lower than of former BED. This result indicates that the cell electrode is capable of being discharged at high current drains with low polarization. Polarization effects are more pronounced below from the BED. It may be noted that cell can operate at 1 Volt at 1.3 mA current drains.

Table 31.2: BED Performance Indices in Varying Percentage of MnO_2 by Weight with Respect to Graphite Powder (Leaf: *Agave-margineta*, size 1.76 sq.cm.)

MnO_2 % (by Weight)	*OCV (Volt)*	*SCC (mA)*	*P_{max} (mW)*	*Current Corresponding to P_{max} (mA)*	*Internal Resistance Corresponding to P_{max} (kΩ)*	*Internal Resistance (kΩ)*
00	1.84	2.80	1.12	1.34	0.75	0.6–1
10	2.14	2.85	1.17	1.11	0.87	0.9–1.6
15	2.15	2.91	1.20	1.05	0.93	0.9–1.16
20	2.18	3.20	1.27	1.08	0.92	0.8–1
25	2.18	3.60	1.70	1.45	0.80	0.7–1
30	2.17	3.91	1.94	1.89	0.58	0.6–1
35	2.20	3.65	1.83	1.80	0.60	0.6–1
40	2.21	3.33	1.55	1.43	0.78	1–1.35
45	2.23	2.80	1.07	1.01	1.24	1.00–1.50
50	2.17	2.30	0.89	0.88	1.44	1–1.52

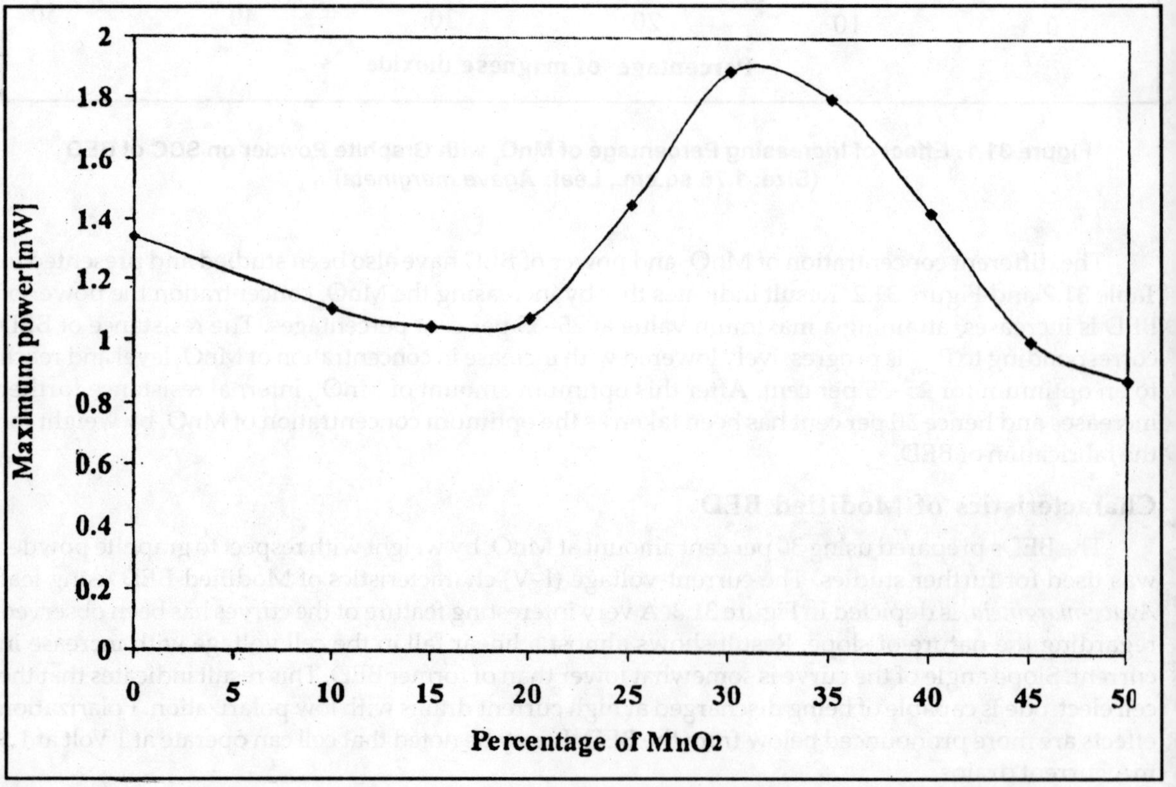

Figure 31.2: Maximum Power of BED vs. Percentage of MnO_2 w.r. to Graphite Powder (Size: 1.76 sq.cm., Leaf: *Agave margineta*)

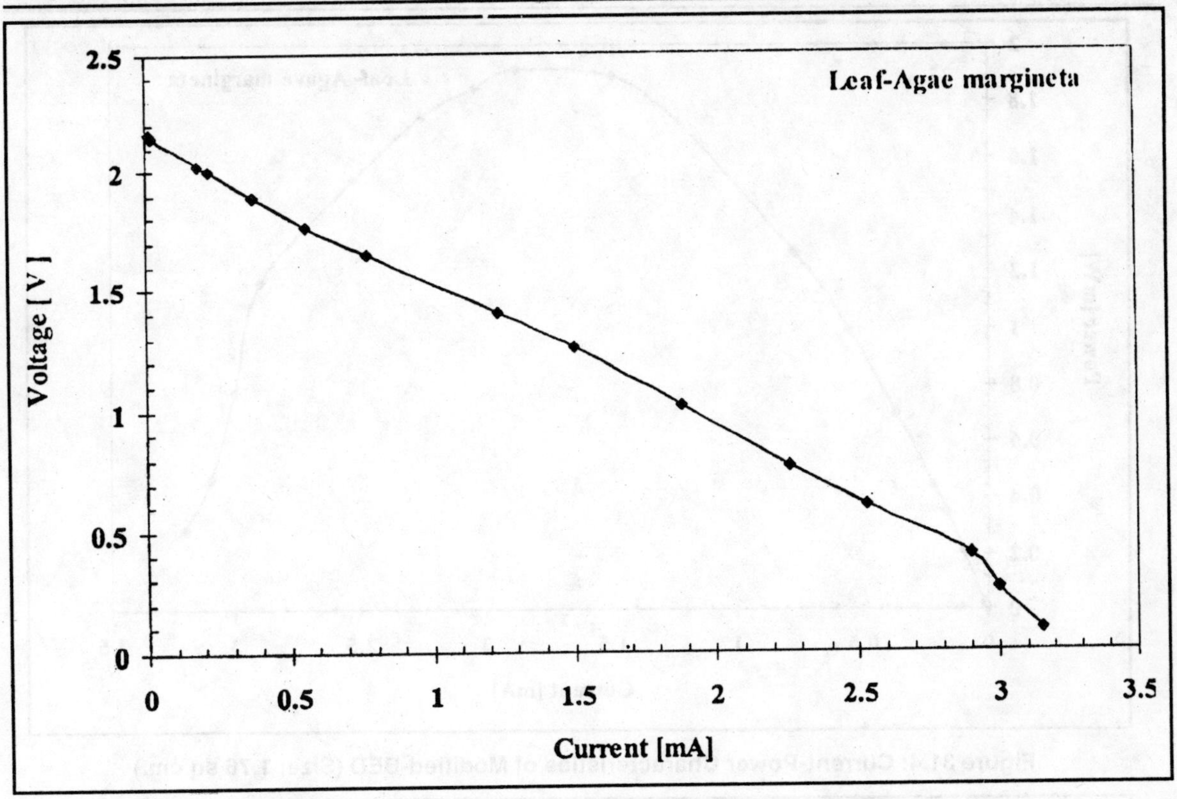

Figure 31.3: Current-Voltage Characteristics of Modified-BED (Size: 1.76 sq.cm.)

The current-power (I–P) characteristic of modified BED is shown in Figure 31.4, and it reveals that Modified-BED gave expanded maxima. The maximum power of this battery is 1.91mW, which shows that the BED may be used in different loads, at constant power. In general it was observed that, the power of Modified-BED is increased by two to three folds.

A discharge characteristic of Modified-BED across load 100 kΩ is shown in Figure 31.5. The voltage drop rate is negligible when a higher load was used. The energy density is 20 W.hr/mt². It can notice that the modified BED can work continuously for 90 hrs of discharge within a voltage range of 1.5 volt. Thus Modified-BED give higher cell voltage and longer duration of discharge at fairly flat potential-plateau.

The Measurement of Voltaic Contribution Aspect

To estimate the voltaic contribution of the Modified-BED, *Agave-margineta* leaf was dipped in molar solution of deactivating copper sulphate ($CuSO_4$) for 10 minutes period. Deactivation of living cells is achieved at this period (Hundet *et al.*, 1989).

The result under same conditions for a fresh and dipped leaf of *Agave-margineta* in a molar $CuSO_4$ solution for 10 minutes is plotted in Figure 31.6. When the physiological activities are ceased by copper sulphate, the MnO_2-BED becomes a simple voltaic cell.

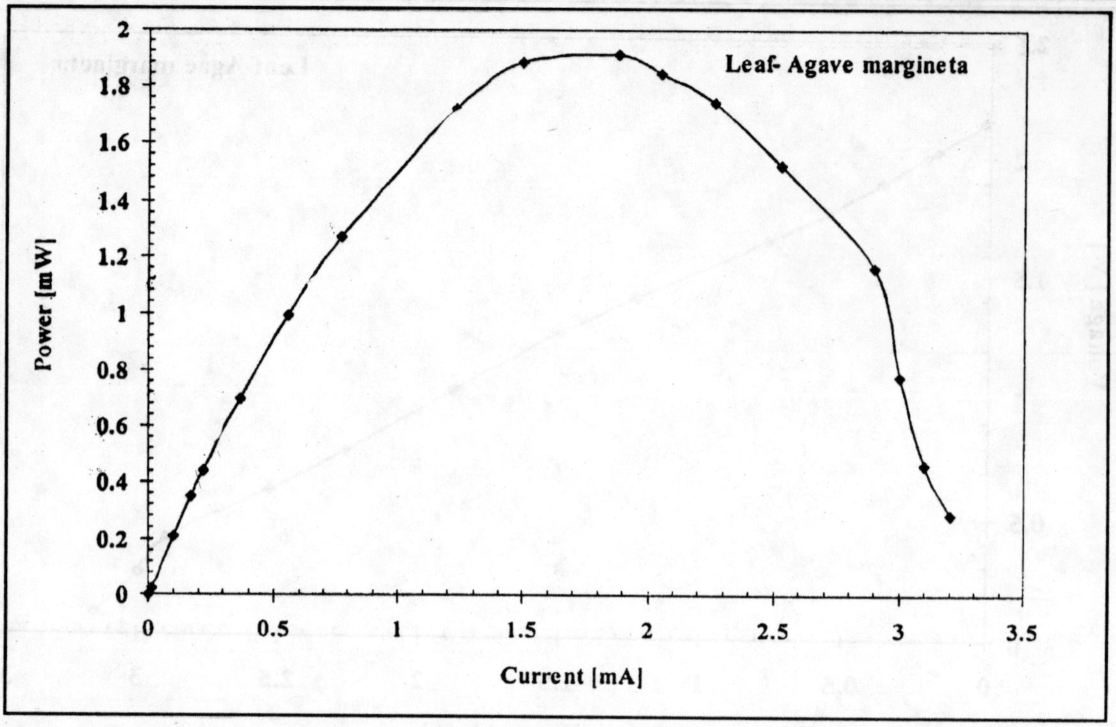

Figure 31.4: Current-Power Characteristics of Modified-BED (Size: 1.76 sq.cm.)

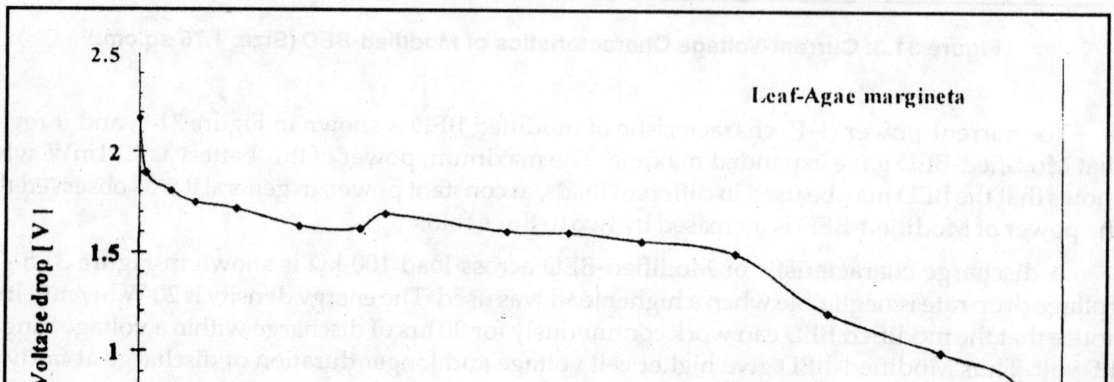

Figure 31.5: Discharge Characteristics of Modified-BED (Size: 1.76 sq.cm)

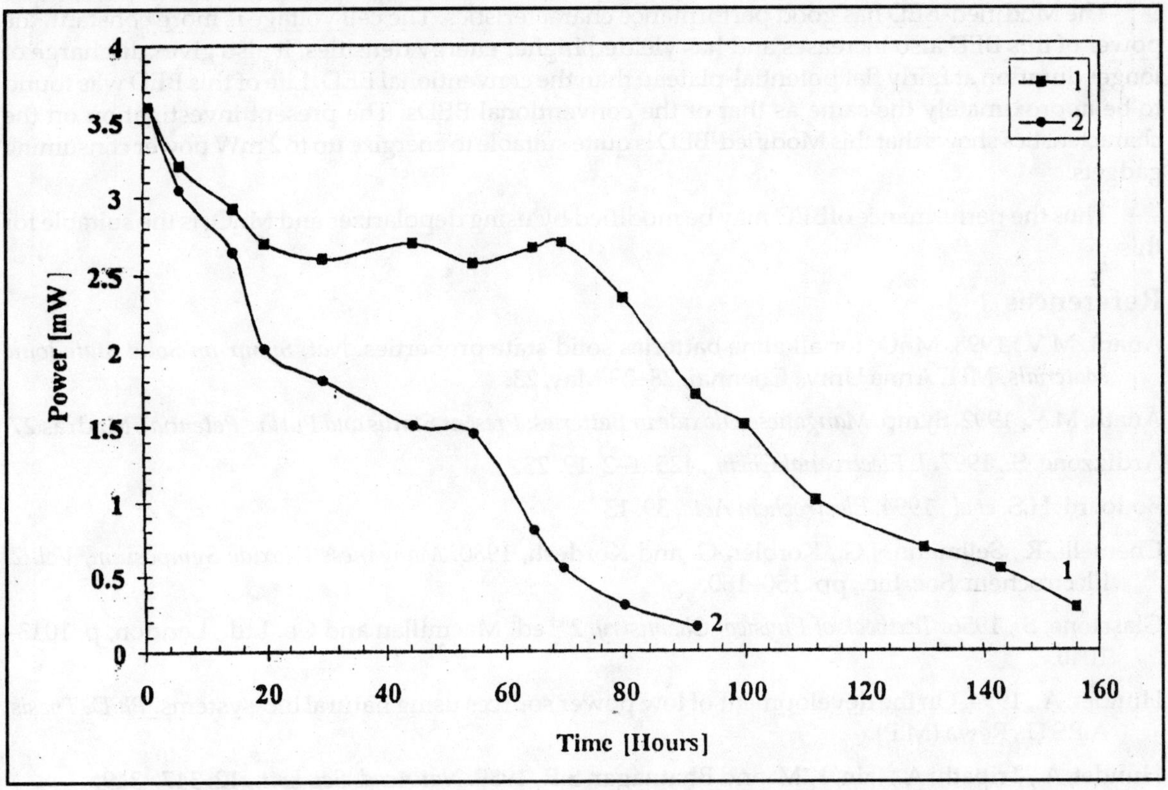

Figure 31.6: Discharge Characteristics of Modified-BED Across a Load Resistance of 100kΩ (1) Fresh Leaf (2) Leaf Dipped in IM $CuSO_4$ Solutions

Deactivated leaf BED gives total energy 1.5 mW.hr, while fresh leaf battery gives 3.25 mW.hr, this indicates that the total energy produced in case of the fresh leaf-BED is much higher than that of deactivated leaf-BED, which shows the involvement of bio-activities present in leaf tissue and non-toxicity of cathodic mix. It is clear that using MnO_2 as a depolarizer, the voltaic contribution may be enhanced while reducing ohmic polarization. Obtained results indicate that approximate 40 per cent energy is enhanced due to the use of MnO_2 as depolarizer.

These results further prove that the energy generation from BEDs has voltaic as well as bio-origin and non-toxicity of the cathodic mix.

Conclusion

The performance of a Bio-emf-device is related to proper selection of leaf, secondly, the performance can also be improved by increasing its voltaic contribution by using depolarizer. The efficiency of BED is increased by using MnO_2 as depolarizer. The presence of 30 per cent MnO_2 with respect to graphite powder is found to be the best composition of the cathodic mix. In this composition the power of BED are improved 2 to 3 times than the conventional BED.

The non-toxicity of the cathodic mix could be another attractive reason for the acceptability of MnO_2 in making Modified-BED. Living cells are not deactivated by MnO_2 due to which the contribution of its continued in the performance of BED.

The Modified-BED has good performance characteristics. The cell voltage is more constant, the power of this BED also increases and has yielded higher energy densities. It also gives discharge of longer duration at fairly flat potential-plateau than the conventional BED. Life of this BED was found to be approximately the same as that of the conventional BEDs. The present investigation on the characteristics shows that this Modified-BED is quite suitable to energize up to 2 mW power consuming gadgets.

Thus the performance ofBED may be modified by using depolarizer and MnO_2 is the suitable for this.

References

Anant, M.V., 1998. MnO_2 for alkaline batteries solid state properties. *Nat. Symp. on Solid State Ionic Materials*, MIT, Anna Univ., Chennai, 28–29 May, 23.

Anath, M.V., 1992. Symp. *Manganese Dioxide in Batteries: Present Status and Future Potential*, Madras 27.

Ardizzone, S., 1997. *J. Electroanal Chem.*, 425: 1–2, 19, 23.

Bodoard, H.S. *et al.*, 1994. *Electrochem Acta*, 39: 13.

Chemelli, R., Sellmann, J.G., Korbler, G. and Kordesh, 1980. *Manganese Dioxide Symposium, Vol. 2.* Electrochem. Soc. Inc., pp. 150–160.

Glasstone, S., 1956. *Textbook of Physical Chemistry*, 2nd ed. Macmillan and Co. Ltd., London, p. 1013–1040.

Hundet, A., 1994. On the development of low power sources using natural bio-systems. *Ph.D. Thesis*, A.P.S.U., Rewa (M.P).

Hundet, A., Tripathi A., Jain, K.M. and Bhatnagar, S.P., 1989. *Nat Acad. Sci. Lett.*, 12: 347–349.

Jain, K.M. and Abraham, S., 1988. *Trans. SAEST*, 23: 379.

Jain, K.M. and Hundet, A., 1988. *Pak. J. Sci. and Indus. Res.*, 31: 803–806.

Jain, K.M. and Hundet, A., 1988. *Pak. J. Sci. and Indus. Res.*, 31: 690–692.

Jain, K.M., 1989. *Pak. J. Sci. and Indus. Res.*, 32(1): 11.

Jain, K.M., 1990. *Bull. Electrochem.*, 6(3): 363.

Jain, K.M., Abraham, S. and Hundet, A., 1988. *Jpn. J. Appl. Phys.*, 27: 867–868.

Jain, K.M., Hundet, A., Abraham, S. and Nigam, H.L., 1987. *Bull. Electrochem.*, 3: 359.

Kordesh, K.V. (Ed.), 1974. *Batteries, Vol 1: Manganese Dioxide.* Marcel Dekker, Inc., New York.

Perrault, G.G., 1974. *J. Electroanal Chem.*, 51: 107.

Perrault, G.G., 1970. *J. Electroanal Chem.*, 27: 47.

Taketsugu Hirai, Tsao Tari and Tsatomu, 1980. *Bull. Chem. Soc., Jpn.*, 53: 1477.

Udhaan, R., Bhatt, D.P., Mathur, P.P., 1992. *J. Apl. Electrochemistry*, 22: 285–292.

Visvanathan, S., 1987. *EMD and Activated Manganese Dioxide.* 1st Nat. get together on battery industry testing and evaluation, 29th July, CECRI, Madras, 27.

Chapter 32

A Survey of Readymade Chilli Powders for Aflatoxin B_1 Contamination

K. Aruna Lakshmi*, M.V.V. Chandana Lakshmi and V. Sridevi

Department of Biotechnology, College of Engineering, Gandhi Institute of Technology and Management, Rushikonda, Visakhapatnam – 530 045, A.P.

ABSTRACT

Mycoflora associated with eleven popular brands of chilli powders were enumerated. The mycoflora present were *Aspergillus flavus, Aspergillus niger, Aspergillus candida, Fusarium, Curvularia, Gliocladium, Rhizopus* and *Mucor.* Indirect competitive ELISA revealed that 45 per cent of the samples were contaminated with aflatoxin Bland the toxin ranged from 0–173.3 µg/Kg.

Keywords: Aflatoxin B_1, Chilli powders, Aspergillus flavus.

Introduction

Substantial quantities of food grains are being affected by mycotoxins (Bosch *et al.*, 1991; Busby *et al.*, 1979; Jelinke *et al.*, 1989; Pinstrup-Anderson *et al.*, 2000; Reddy *et al.*, 2001). Health hazards due to ingestion of aflatoxin contaminated food are much greater in developing countries, as they lie in

* Corresponding Author: Residential address: 8-1-71, Chinmaya Marg, Peda Waltair, Visakhapatnam – 530 017 Phone: 0891-2840246 (off), 0891-2532194 (R); E-mail: vellurusridevi@yahoo.co.in

tropics, where temperature and relative humidities are conducive for growth of aflatoxin producing fungi–*Aspergillus flavus* and *Aspergillus parasiticus*. Due to established carcinogenic effect of aflatoxins on various animals (Katiyar *et al.*, 2000), developed countries which import food and feed, have set aflatoxin contamination limits in food stuffs ranging from 0–10 μg/kg. This has resulted in import restrictions on many foods and feeds from India.

Foodstuffs are regularly surveyed for mycotoxin contamination in the first world countries. To discourage the marketing of high toxin contaminated foods it is essential to monitor various foods periodically. Occurrence of aflatoxins in stored chillies has been reported by Reddy *et al.* (2001), Ravi Kiran *et al.* (2005). No information about contamination levels of chilli powders is available. In the present study we examined eleven brands of commercially available chilli powders to determine the incidence of mold and aflatoxin B_1 content.

Materials and Methods

Eleven popular brands of chilli powders (500 g each) were collected from shops, super markets and households from various parts of Visakhapatnam, Andhra Pradesh. Aflatoxin B I in the powders was estimated using indirect competitive ELISA method of ICRISAT (Anjaiah *et al.*, 1989; Reddy *et al.*, 2002).

Methanolic extracts of the chilli powders (20 g of each sample) diluted 1 : 10 in PBS Tween were used for analysis. ELISA plates coated with AFB_1–BSA conjugate were washed and blocked with 0.2 per cent BSA. After thorough washing, 100 μl of diluted aflatoxin B_1 (100 ng to 10 pg) and 100 μl of methanolic extract of chilli powder intended for analysis were added to each well of the ELISA plate. Antiserum (50 μl) was added to each well. About 150 μl goat anti-rabbit IgG (1 : 1000) labelled with ALP was added to each well. The substrate used was P-nitrophenylphosphate. Absorbance was measured at 405 nm in an automatic ELISA reader.

Quantification of aflatoxin was done as follows:

$$AFB_1(\mu g/kg) = \frac{ADE}{G}$$

A = AFB_1 concentration in diluted sample extract (ng/ml); D = Times dilution with buffer; E = Extraction solvent volume used (ml); G = Sample weight (g).

Results and Discussion

Different fungal species isolated from chilli powders are shown in Table 32.1. All the samples invariably harboured mycoflora. The total viable count ranged between 2×10^4 to 9×10^5/ml. Mold incidence was particularly high in some local brands (brands 3, 4, 7, 10). The predominant fungal species were *Aspergillus flavus* and *Aspergillus niger*. The incidence of these species was 100 per cent in samples 6, 8, 9 (Table 32.1). *Penicillium fusarium*, *Mucor Rhizopus*, and *Alternaria* were found in low percentage. *Fusarium* and *Alternaria* are known to be associated with stored chillies elaborating mycotoxins (Patel *et al.*, 1996; Hasmi and Thrane, 1997). *Cladosporium*, an aeroallergic fungi, and soil fungi *Trichoderma* and *Gliocladium* were found in high concentrations in chilli powders purchased from local mills suggesting cross contamination from other food grains like rice, wheat, and millet.

Table 32.1: Mycoflora and Aflatoxin B_1 Associated with Various Brands of Chilli Powders

Brand	Total Viable Count	Name of Fungi	Aflatoxin B_1 µg/kg
1	3×10^4	*A. flavus*	7.3
		Cladosporium	
		Mucor	
2	2×10^4	*A. flavus*	77
		A. niger	
		Gliocladium	
		Penicillium	
3	9.1×10^5	*A. flavus*	36
		A. niger	
		Rhizopus	
4	8×10^5	*A. candida*	5.1
		A. flavus	
		Fusarium	
		Gliocladium	
5	2×10^4	*A. niger*	0
		A. flavus	
		Mucor	
		Trichoderma	
6	2.9×10^5	*A. flavus*	38.0
		A. niger	
7	3×10^6	*A. flavus*	60.0
		A. niger	
		Gliocladium	
		Mucor	
		Rhizopus	
8	5×10^4	*A. flavus*	0
		A. niger	
9	2×10^5	*A. flavus*	173.3
10	3×10^6	*A. flavus*	4
		Cladosporium	
11	3.1×10^5	*A. flavus*	8.1
		A. niger	
		Alternaria	
		Fusarium	
		Gliocladium	
		Mucor	

Our survey found widespread and considerable contamination of chilli powders stored in houses, supermarkets and marketplaces (Raghu Kishore *et al.*, 2006. Chillies stored in cold storage were often found to be contaminated with aflatoxigenic strains of *A. flavus* and *A. niger* (Ravi Kiran *et al.*, 2005). Chilli powders prepared from contaminated chillies would be containing aflatoxins (Ravi Kiran *et al.*, 2005; Raghu Kishore *et al.*, 2006). In India, the maximum permissible level of aflatoxin in foods is 30 µg/kg. The present study revealed that 45 per cent of the chilli powders are unfit for consumption as they had greater then 30 µg/kg of aflatoxin B_1. Only 55 per cent of the samples complied with the 30 µg/kg maximum level as set by Indian standards. Reddy *et al.* (2001) surveyed the occurrence of aflatoxins B_1 in 139 samples of chillies and showed that 65 per cent of chilli are contaminated by aflatoxin B_1 and 20 per cent had non permissible levels (>30 μg/kg). According to Klieber (2000) the risk of aflatoxin contamination seems to increase for chilli powders compared to whole fruits probably due to the increased L surface areas and associated increased moisture content. Aflatoxin B_1 as high as 969 μg/kg was found in some samples of grade three chilli powders (Reddy, 2001). In the present study 173 μg/kg aflatoxin B_1 has been found in one local brand. Being cheap in price it is mostly consumed by the weaker sections in Andhra Pradesh. All batches of powders should be tested to comply with relevant legal limits (30 μg/kg). Monitoring of foodstuffs periodically to discourage the marketing of high toxin contaminated foods is therefore inevitable in India.

References

Anjaiah, V., Mehan, V.K., Jayanthi, S., Reddy, D.V.R. and Donald, D.Mc., 1989. Enzyme linked immunosorbant assay (ELISA) for Aflatoxin B_1 Estimation. In: *Aflatoxin Contamination of Groundnuts*. ICRISAT, India, pp. 183–190.

Bosch, F.X. and Press, F., 1991. Aflatoxins: Data on human carcinogenic risk and relevance to human cancer of N–Nitroso compounds. In: *Tobacco and Mycotoxin*, (Eds.) O'Neill, J.K., Chen, J. and Barstch, H. International Agency for Research on Cancer, Lyon, France, pp. 48–53.

Busby, W.F. Jr. and Wogan, G.N., 1979. Food borne mycotoxins and alimentary mycotoxicoses. In: *Food-borne Infections and Intoxications*, (Eds.) Riemann, H. and Bryan, F.B. Academic Press, New York, pp. 519–610.

Hashmi, M.H. and Thrane, U., 1990. Mycotoxins and other secondary metabolites in species of Fusarium isolated from seeds of capsicum, coriander and fenugreek. *Pakistan J. Bot.*, 22: 106–116.

Jelinke, C.F., Pohland, A.E. and Wood, G., 1989. World wide occurrence of mycotoxins in foods and feeds. *AOAC*, pp. 223–230.

Katiyar, S., Thakur, V., Gupta, R.C., Sarin, S.K. and Das, B.C., 2000. P53 tumour suppressor gene mutations in hepatocellular carcinoma patients in India, pp. 1565–1573.

Klieber, A., 2000. *Chilli Spice Production in Australia*. RIDC Publication No. 20/33.

Patel, S., Hazel, C.M., Winterton, A.G.M. and Morbey, E., 1996. Survey of ethnic foods for contamination by mycotoxins. *Food. Add. Contam.*, 13: 833–841.

Pinstrup-Andersen, P., Pandya-Lorch, R.C. and Rosegrant, M.W., 2000. *Food Security: Problems, Prospects and Policies 2020 Vision*. Discussion Paper, Washington DC, International Food Policy Research Institute.

Raghu Kishore, G., Kanthi, P., Sujatha, S., Kalyan Chakravarthy, N. and Aruna Lakshmi, K. *AJMBES* (In Press).

Ravi Kiran, D., Narayana, K.J.P. and Vijayalakshini, M., 2005. Aflatoxin B_1 production in chillies (*Capsicum annum* L.) kept in cold stores. *African Journal of Biotechnology*, pp. 791–795.

Reddy, D.V.R., Thirumala Devi, B., Reddy, S.V., Waliyar, F., Mayo, M.A., Raffia Devi, K., Ortiz, R. and Lenne, J.M., 2002. Estimation of aflatoxin levels in selected foods and feeds in India. In: *Food Safety Management in Developing Countries*, (Eds.) Hanak, E., Boutrif, E., Fabre, P. and Pinerio, M. CIRAD–FAO, Montpellies, France, pp. 11–13.

Reddy, S.V., Kiran Mayi, D., Uma Reddy, M., Thirumala Devi, K. and Reddy, D.V.R., 2001. Aflatoxins B_1 in different grades of chillies (*Capsicum annum* L.) in India as determined by indirect competitive–ELISA. *Food Additives and Contaminants*, pp. 553–558.

Chapter 33

Clinical Manifestation of Experimentally Induced Fish-Borne Trematode Infection in Experimental Animals

S.I. Shalaby[1], Olfat Anter[2], Nabila Hassan[2], M. El-Mahdy[2], Iman Mohamed[2] and Neelima Gupta[3]*

[1]Department of Complementary Medicine, National Research Center, Dokki, Cairo, Egypt

[2]Department of Parasitology, Faculty of Veterinary Medicine, Cairo University, Giza, Egypt

[3]Department of Animal Science, M.J.P. Rohilkhand University, Bareilly, India

ABSTRACT

Rats and dogs experimentally fed with viable encysted metacercariae from *Mugil cephalus* and *Tilapia nilotica* revealed the presence of seven species of trematodes. The pre-patent period was 3–10 days post-infection according to the type of trematode species. *Heterophyes hterophyes, H. heterophyes nocens, H. katsuradai, H. dispar* and *H. aequalis* were recovered from rats and dogs while *Haplorchis pumilio* and *Prohemistomum vivax* were only obtained from dogs. The number of recovered worms in rats was less as compared to dogs indicating that the former is a less favored host for the investigated trematodes. The experimentally infested animals showed signs of diarrhoea. The structural and functional mechanisms of diarrhoea induced by these parasites are discussed.

***Keywords**: Metacercaria, Trematode, Mugil, Tilapia, Diarrhoea.*

* Corresponding Author: E-mail: guptagrawal@rediffmail.com

Introduction

Fish is considered to be one of the important sources of parasitic infestations to man and fish eating mammals particularly after Increased pollution of rivers and lakes. Ooi *et al.* (1999) reported mass infection of heterophyid metacercariae in cultured Japanese eels, *Anguilla japonica* in Taiwan. However, not all final host mammals might be favorable for parasitism with trematodes that are transmitted through eating fish infested with encysted metacercariae of those parasites. Boulos (1979) reported *Pygidiopsis genata* from dogs experimentally fed with *Tilapia* sp. from lake Edku, Maryut and Manzala. Ahmed (1980) found *Heterophyes heterophyes. Haplorchis yokogawai* and *Prohemistomum vivax* in stray dogs from Mansoura, Egypt. Riffat *et al.* (1980) observed *H. heterophyes.* in puppies fed on *Mugil* sp. and *Tilapia* sp. El-Mokaddem (1982) reported *H. heterophyes* from cats in Sallum and Fayoum. Egypt; Shalaby (1982) described 11 species of trematodes from experimentally infected dogs with encysted metacercariae in *Bagrus bayad, Syndontis schall* and *Tilapia nilotica.* Fahmy *et al.* (1984) examined 45 stray dogs, 29 stray cats and 673 different rodents and recorded haplorchid, prohemistomatid and cynodiplomatid trematodes from them. Noshy *et al.* (1987) collected *Pygidiopsis genata* and *Haplorchis pumilio* from experimentally infected hamsters, mice, rats, cats and puppies with *Tilapia* fishes having encysted metacercariae. El-Reid (1994) infested puppies, albino rats, chickens, ducks and pigeons by feeding them fish muscles and recorded *Heterophyes, Haplorchis, Prohemistomum* and *Mesostephanus* spp. Shalaby and Easa (1994) observed *Phagicola vrnamentata* and *Pharyngostomum flapi* from dogs in Cairo. Moreover, Mahdy *et al.* (1995) fed kitten and puppies with infected tilapia fish in Manzala lake and obtained *Heterophyes, Ascocotyle, Haplorchis* and *Pygidiopsis* spp. Shalaby *et al.* (2003a) infected puppies and albino rats with heterophyid encysted metacercariae and could find extra-intestinal lesions due to the parasites. Also, Shalaby *et al.* (2003b) could trace the effect of parasitization by certain heterophyid species in dogs and rats.

Byong *et al.* (1980) reported that clinically. Echinostomes caused intermittent abdominal pain and nausea in a 60 years old woman. Byong *et al.* (1981) stated that *Heterophyes* spp caused eosinophilia in 7 per cent of human cases passing eggs in their stools. Line (1981) reported pain, gurgling and loose stools in a 40 years old farmer who ate improperly cooked fish. Byong *et al.* (1984) reported that the clinical picture in man as a result of heterophyid infection was represented by non-specific gastrointestinal symptoms such as epigastric pain, easy fatigability, weakness, indigestion and cardiac arrhythmia. Byong *et al.* (1985) reported that the major subjective symptoms in human volunteers and in experimental rats were diarrhoea and abdominal pain. Ana *et al.* (1991) recorded abdominal liver function tests, eosinophilia and obstructive jaundice complicated by sepsis and transhepatic cholangeogram demonstrated diffuse irregular dilatation of biliary tree with abrupt ending of terminal branches. Several small fusiform defects were noticed within small ducts. *Opisthorchis* eggs were found during aspiration of bile of this patient. Metaplasia of goblet cells and adenomatous hyperplasia of bile ducts epithelium with time followed by periductal fibrosis and thickening of ductal wall with irregular structure of its lumen were also reported. Jong *et al.* (1993) observed that *Metagonimus* spp. caused diarrhoea in infested persons. In Egypt, Shalaby (2002) stated that heterophyiasis constitutes a public health problem where people eat raw, inadequately salted or otherwise improperly prepared infested fish. This may result in chronic intermittent mucous diarrhoea which may occasionally be bloody, alternating with constipation. Moreover, Shalaby *et al.* (2002) reported that patients mostly had indigestion, abdominal pain or heart problems in the form of arrhythmia and.ventricular premature beats. Sometimes, obstructive jaundice, complicated by sepsis and disseminating intravascular coagulopathy-like syndrome was evident.

This investigation reports the suitability of the final host (mammal) for trematodes transmitted hrough fish, the effect of the numbers of parasites on the final host as well as the clinical picture in the nal hosts.

laterials and Methods

Twenty, albino rats (110–150 gm B.W.) and 12 puppies (4 weeks old) were used for experimental ıfection. Prophylactic dose of Praziquantel (50 mg/10 kg B.W.) was given once a week before xperimental infection. Each animal was given 150 encysted metacercariae collected from *Mugil ephalus* and *Tilapia* spp. Faecal examination was done daily to determine the pre-patent period. The linical picture was repol1eli and samples were taken. Finally animals were sacrificed and their ntestines were examined.

esults

Experimental feeding of rats and dogs with viable encysted metacercariae revealed the presence f seven species of trematodes. Eggs of trematodes appeared after 3–10 days post infection according the type of trematode species. Five species of digenetic trematodes were collected from experimentally ıfested rats and dogs, namely: *Heterophyes heterophyes, H. heterophyes nocens, H. katsuradai, H. dispar* nd *H. aequalis*. However, *Haplorchis pumilio* and *Prohemistomum vivax* were obtained only from dogs. he characteristic features of the trematodes were studied.

It was noticed that the number of recovered worms in albino rats were less as compared to those btained from dogs.

It was obvious from the clinical observation of the infested animals that they suffered from iarrhoea, passed abnormal fluidy faeces which was usually accompanied by increased volume and requency of defecation. The animals were nervous and their blood films showed eosinophilia.

iscussion

Day to day, new records are coming up proving that fish is a source for serious parasitic infectious iseases for man, animals and birds. Now, it is believed that some trematodes other than *Heterophyes eterophyes* can be transmitted to man and animals through fishes.

In the present investigation, five species of heterophyid trematodes were recovered when rats and logs were experimentally infected namely, *Heterophyes heterophyes, H. heterophyes nocens, H. katsuradai, I. aequalis* and *H. heterphyes dispar.* Moreover, *Haplorchis pumilio* and *Prohemistomum vivax* were obtained rom the small intestine of dogs only.

The pre-patent period ranged between 3 to 10 days post-infection according to the type of trematode pecies. Nearly similar results were obtained by Dennis and Penner (1971) (4–6 days), Galaktionov 1980) (7–8 days), Pike (1980) (3 days) and Shalaby (1985), (4–5 days) post infection. The morphological haracters of the trematodes in this study were identical with those of the original description given by iebold (1853), Sonsino (1892), Looss(1902), Onji and Nishio (1915), Ozaki and Asada (1926), Witenberg 1929), Chen (1936), Fayek *el al.* (1986) and El-Reid (1994).

Heterophyes heterophyes was found more commonly in human beings in Egypt and in the middle ast (Taraschewski, 1984) and later it was found in the Far East such as Japan and Korea (Yokogawa *t al.*, 1965; Chai *et al.*, 1985). *H. heterophyes* was detected from an Egyptian child (Fayek *et al.*, 1986). This species was reported repeatedly from man, birds and carnivorous mammals such as dogs, cats and foxes (Harford and Arlen, 1994).

Heterophyes heterophyes nocens is an intestinal trematode of dogs, cats and human beings of eastern countries like Japan and Korea (Chai and Lee, 1990), the parasite was considered as a rare species in Egypt as it was detected only by Abdel-Maksoud (1992) and El Reid (1994). However, this parasite may become endemic in Egypt, *Heterophyes dispar* was first found in dogs and cats from Egypt (Looss, 1902; Mahdy *et al.*, 1995). Also, it was reported from a variety of carnivorous mammals including foxes and wolves (Witenberg, 1929; Taraschewski, 1984). The parasite was found in 7 men from Korea (Eom *et al.*, 1985).

Heterophyes aequalis was recovered from the small intestine of house rats, domestic dogs, domestic and wild cats as well as jackal (Wells and Randell, 1956). Harford and Arlen (1994) mentioned that *H. heterophyes, H. dispar* and *H. aequalis* occur in man in the Mediterranean, Far East and Egypt, infestation with *H. pumilio* in dogs was recorded by Azim (1938), El Mokaddem (1982) and Shalaby (1982). This parasite has been shown to be of public health importance (Shalaby, 1982; Mahmolld, 1983).

Prohemistomum vivax infestation in dogs was reported by Azim (1938) and Ahmed (1980). In man, it was reported by Nasr (1941).

H. pumili and *P. vivax* were recovered from dogs only during the present study. This finding is in conformity with the results of Fahmy *et al.* (1976), Shalaby (1982) and Mahmoud (1983).

It was further noticed that the number of observed worms in albino rats were less as compared with those obtained from dogs. This may be attributed to the severe tissue reaction against the parasite in rats rendering them as less favorable hosts for these flukes and could be explained due to the difference in physiological status and acidity of the stomach of both hosts as well as other anatomical and physiological factors. These results are in conformity with those reported by Jong (1979), Shalaby *et al.* (1989) and Sung *et al.* (1990).

These flukes are responsible for causing gastrointestinal disturbances such as abdominal pain, gastric discomfort and/or diarrhoea in human hosts (Seo, 1978; Jong *et al.* (1986) and are thus of medical importance. However, certain kinds of heterophyids were reported to produce erratic parasitism in many other organs. In the present investigation, it was obvious from the clinical observation of infested animals that they suffered from diarrhoea with abnormal faeces, fluid in nature which was usually accompanied by Increased volume and increased frequency of defecation. This is a common sign occurring in many enteric malfunctions. The available data were insufficient bases for speculation on the structure and function of the intestine. Therefore, the diarrhoea produced by these helminths may be due to the mechanical and chemical stimulation of the epithelial layers of villi and/or crypts (secretory diarrhoea). A second factor in the pathogenesis of diarrhoea may be the decreased intestinal assimilation of nutrients (malabsorption diarrhoea). Malabsorption occurred mostly as a result of the reduced villous and microvillous surface (villous atrophy). The third mechanism for production of diarrhoea is the increased permeability of the mucosa which may contribute to diarrhoea by permitting increased retrograde movement of solute and fluid from the lateral intercellular space to the lumen or by facilitating transdaction of tissue fluid (filtration secretion). Effusion occurred in inflamed lamina propria due to increased vascular permeability and proprial oedema. Increased exfoliation of epithelium and erosions provide further potential sites of interstitial fluid effusion (effusive diarrhoea). These results were in agreement with Cheville (1983), Stuart and Bernard (1989) and Robert (1993).

References

Abdel-Maksoud, S., 1992. Zoonotic agents in marine fish marketed in Dumyat. *M.Sc. (Zoonoses)*, Fac. Vet. Med., Zagazig Univ.

Ahmed, F., 1980. Intestinal helminthes among stray dogs in Mansoura City, Egypt. *J. Egypt. Soc. Parasitol.*, 10: 289–294.

Ana, H., Sydney, F. and Benton, R., 1991. A case of Opisthorchiasis diagnosed by cholangiography and bile examination. *Amer. Surg. J.*, 57: 206–209.

Azim, M., 1938. Intestinal helminthes of dogs in Egypt. *J. Egypt. Med Assoc.*, 21: 105.

Boulos, L., 1979. Parasitological studies on the heterophyids in *Tilapia* fish from Edku and Maryut lakes and their transmission to laboratory animals. *Ph.D. Thesis (Parasitology)*, Fac. Med., Alex. University.

Byong, S., Sung, Y. and Jong, Y., 1980. Studies on intestinal trematodes in Korea: A human case of *Echinostoma cinetorchis* infection with an epidemiological investigation. *Seoul J. Med.*, 21: 21–29.

Byong, S., Sung, Y. and Jong, V., 1981. Studies on intestinal trematodes in Korea. III. Natural human infections of *Pygidiopsis summa* and *H. h. nocens. Seoul J. Med.*, 22: 228–235.

Byong, S., Soon, H., Sung, Y. and Jong, Y., 1984. Studies on intestinal trematodes in Korea. III. Two cases of natural human infections by *H. continua* and the status of metacercarial infection in brackish water fishes. *Korean J. Parasitol.*, 22: 51–60.

Byong, S., Kwang, S., Soon, H., Sung, Y. and Jong, Y., 1985. Studies on intestinal trematodes in Korea. XVII. Development and egg laying capacity of *Echinstoma hortense* in albino rats and human experimental infection. *Korean J. Parasitol.*, 23: 24–32.

Chai, J., Hong, S., Sohn, W., Lee, S. and Seo, B., 1985. Further case of human *Heterophyes heterophyes nocens* infection in Korea. *Seoul J. Med.*, 26: 197–200.

Chai, J. and Lee, S., 1990. Intestinal trematodes of human in Korea: *Metagoninus. Heterophyes* and echinostomes. *Korean J. Parasitol.*, 28: 103–122.

Chen, H., 1936. A study of *Haplorchidae* (Loos, 1899). Poche (1926) (Trematoda : Heterophyidae). *Parasitology*, 28: 40–55.

Cheville, N., 1983. *Cell Pathology*, 2nd edn. Ames, Iwa, USA, 223 pp.

Dennis, E. and Penner, L., 1971. *Mesostephanus yedeae* sp.n, its life history and descriptions of the developmental stages. Occasional Papers, Uhiv. Connecticut, 2: 5–15.

El-Mokaddem, A., 1982. The prevalence of heterophyids in man and animals and various fish hosts in Edku area. *M.Sc. Thesis (Zoonoses)*, Public Health Institute, Alex University.

El Reid, A., 1994. Role of marine fish in transmission of some parasites to animals and birds. *Ph.D. Thesis (Parasitology)*, Fac. Vet. Med., Zagazig Univ.

Eom, K., Yoo, H., Chung, M., Lee, J. and Rim, H., 1985. Heterophyid trematodes human infection imported from Sudan to Korea. *Korean J. Parasitol.*, 23: 360–431.

Fahmy, M., Mandour, A. and El-Naffar, M., 1976. Successful infection of dogs and cats by *P. vivax* and *H. yokogawai. J. Egypt. Soc. Parasitol.*, 6: 77–82.

Fahmy, M., Arafa, M., Abdel Rahman, A. and Mounib, M., 1984. Studies on helminth parasites in small mammals in Assiut. 1-Trematode parasites. *Assiut Vet. Med. J.*, 11: 43–50.

Fayek, S., Nada, M. and Ahmed, B., 1986. Heterophyid parasites recovered from dogs with special reference to *Phagicola longus. Vet. Med. J.*, 34: 201–210.

Galaktionov, K., 1980. The life cycle of *M. appendiculatus* (Ciurea, 1916). Vestnik Leningradskego-Univ., Biologiya, 21: 27–34.

Harford, W. and Arlen, J., 1994. *Parasitic Worms of Fish*, 1st edn. Taylor and France, London, New York and Philadelphia.

Jong, Y., 1979. Study of *Metagonimus yokogawaj* in Korea. V. Intestinal pathology in experimentally infected albino rats. *Seoul J. Med.*, 20: 104–112.

Jong, Y., Byog, S. and Soon, H., 1986. Human infection by *H. heterophyes, H. dijpar* imported from Saudi Arabia. *Korean J. Parasitol.*, 24: 82–88.

Jong, Y., Sun, H., Jae, R., Jina, K. and Soon, H., 1993. An epidemiological study of *Metagonimus* along the upper reaches of Namhan river. *Korean J. Parasitol.*, 31: 99–108.

Line, O., 1981. *Echinochasmus perfoliatus* in man. A case report. *Natural Med. J. China*, 61: 249–252.

Looss, A., 1902. Notizen Zur Helminthologie Egyptens. V. Eine Revesion der Fasciolden-gattung Heterophye. Cobb. Centralab. Bakt. Orig., 32: 886–891.

Mahdy, O., Manal, A., Essa, A. and Ecisa, M.El.S., 1995. Parasitological and pathological studies on heterophyid infections in *Tilapia* spp. from Manzala lake, Egypt. *Egypt. J. Comp. Pathol. and Clin. Pathol.*, 8: 131–145.

Mahmoud, N., 1983. Parasitic infestations of some native species of fishes in Cairo markets with special reference to parasites transmissible to man and animals. *M.V. Sc. (Hygiene and Food Control)*, Fac. Vet. Med., Cairo University, Cairo, Egypt.

Nasf, M., 1941. The occurrence of *Prohemistomum vivax* infection in man with description of the parasite. *Lab. Med. Prog.*, 2: 135–149.

Noshy, S., Magda, M., Helen, N., Nibal, A. and Laila, M., 1987. *Heterophyes metacercariae* in Maryut and Manzala lakes in Egypt. *J. Egypt. Soc. Parasitol.*, 17: 481–493.

Onji, Y. and Nishio, T., 1915. Studies on the metacercaria in *Mugil cephalus. Tokyo Tji Shinshi*, p. 2390–2395.

Ooi, H., Wang, W., Tu, C., Chang, H. and Chen, C., 1999. Natural mass infection by heterophyid metacercariae in aquacultured Japanese eel in Taiwan. *Dis. Aquat. Org.*, 35: 31–36.

Ozaki, Y. and Asada, J., 1926 A new human trematode *H. katsuradai. J. Parasitol.*, 12: 216–218.

Pike, A., 1980. Observations on the life cycle of *M. milvi* Yamaguti, 1939. *J. Nat. Hist.*, 14: 389–399.

Riffat, M., Salem, S., El-Kholy, S., Hegazi, M. and Youssef, M., 1980. Studies on the incidence of *Heterophyes heterophyes* in Dakahlia. *J. Egypt. Soc. Parasitol.*, 10: 369–373.

Robert, M., 1993. Diarrhoea in helminthic infections. *Clin. Inf. Dis.*, 16: 2120–2122.

Seo, B., 1978. *Clinical Parasitology*. Revised Ed. II Cho-kah Seoul, pp. 111–113.

Shalaby, S.I., 1982. Studies on the role played by some Nile fishes in transmitting trematodes. *M.Sc. Thesis (Parasitology)*, Cairo University, Cairo, Egypt.

Shalaby, S.I., 1985. Further studies on the role-played by catfish in transmitting some trematodes to fish eating mammals, with special reference to the morphobiology of *M. appendiculatous. Ph.D. Thesis (Parasitology)*, Cairo University, Cairo, Egypt.

Shalaby, S.I., 2002. Clinical picture of diseases transmitted to man through fish. 1st Annual International Congress of the Medical Division, N.R.C., Egypt, pp. 68.

Shalaby, S.I., Ibrahim, M., Mahmoud, N. and El Assaly, T., 1989. Parasitological and pathological studies on encysted metacercariae in the musculature and different organs of *T. nilotica. J. Comp. Pathol. and Clin. Pathol.*, 2: 186–212.

Shalaby, S.I. and Easa, M.El.S., 1994. Trematode parasites transmitted to man and fish-eating mammals through *Tilapia nilotica*. II. New trematode species. In: *Proc. 18th World Bur Cong.*, Bologna, Italy.

Shalaby, S.I., Mohamed Imam and Gupta Neelima, 2002. Clinical and parasitological studies on human fish borne trematodes with emphasis on intestinal histoenzymatic changes in Egypt: Comparative records from India. *Indian J. Environ. and Ecoplan.*, 6: 409–414.

Shalaby, S.I., Nabiha Hassan, El-Mahdy, M., Olfat Anter, Eman Mohamed and Gupta, Neelima, 2003a. Parasitoiogical and pathological studies on extra-intestinal heterophyiasis. *J. Par. Dis.*, 27: 39–46.

Shalaby, S.I., Nabiha Hassan, Raafat, A. Olfat Anter, El-Mahdy, M. Iman Mohamed and Gupta, Neelima, 2003b. Effect of parasitization by certain heterophyid species in rats and dogs in Egypt. *Biol. Memoirs*, 29: 21–28.

Siebold, C., 1853. Uber Leucochoridium paradoxum. *Zeitschr. Wissensch. Zool.*, 5: 425–436.

Sonsino, P., 1892. Studi sui parasiti di molluschi di acqua dolle nei dinotorni di Cairo in Egitto. Festochr.Z. Oebutstag R. Leue Karts Leipsing, pp. 134–146.

Stuart, K. and Bernard, F., 1989. Pathological effects of *Echinostoma caproni* in domestic chicks. *J. Helminth.*, 63: 227–230.

Sung, T., Sun, H., Woon, O. and Soon, H., 1990. Changes in histopathological and serological findings of the liver after treatment in rabbit clonorchiasis. *Seoul J. Med.*, 31: 117–127.

Taraschewski, H., 1984. Die trematoden der gattung Heterophyes taxonomie,biologie, epidemiologie. *Ph.D. Dissertation*, Universitat Hohenheim.

Wells, W. and Randell, B., 1956. New hosts of trematodes of the genus *Heterophyes* in Egypt. *J. Parasitol.*, 42: 286–292

Witenberg, O., 1929. Studies on trematodes: Family Heterophyidae. *Annals Top. Med. and Parasitol.*, pp. 231–239.

Yokogawa, M., Sano, M., Itabashi, T. and Kachi, S., 1965. Studies on the intestinal flukes. II. Epidemiological studies on heterophyid trematodes of man in Chiba Prefecture. *Japan J. Parasitol.*, 14: 577–585.

Chapter 34

Acoustic Studies on Conduction Polymers in Solutions

R. Palanivel, K. Balamurugan and M.E. Raja Saravanan
Annamalai University, Annamalainagar – 608 002, Tamil Nadu, India

ABSTRACT

The conduction polymer Polyaniline (PANi) base is soluble in selected organic solvents. The PANi becomes conductive upon partial oxidation or reduction by a process commonly known as doping. The PANi base was doped by chemical method using perchloric acid ($HClO_4$) as a dopant. The doped conducting polymer was dissolved in three selective organic solvents namely p-Cresol (PCL), Dimethyl sulphoxide (DMSO) and 2-Chlorophenol (2CPL). The variation in the ultrasonic velocity values measured using fixed frequency (3MHz) continuous wave ultrasonic interferometer technique with temperature and their related thermodynamic parameters have been discussed with a view to understand the molecular interactions of the conduction polymers in solutions. Hence the acoustic properties of PANi base and its doped polymer in solutions will pave the way to obtain possibly a new product ideas and benefits over the traditional polymers.

***Keywords**: Polyaniline (PANi), Conduction polymer, Molality (M), Ultrasonic velocity, Thermodynamic parameters.*

Introduction

In recent years the measurement of ultrasonic velocity has been adequately employed in understanding the nature of the molecular interactions of polymers in solutions. It is also noted that the ultrasonic investigations find extensive applications in identifying the characteristic aspects of molecular interactions existing in binary systems (Anwar Ali *et al.*, 2004; Moore and Uddin, 1969).

Further the ultrasonic velocity is also useful for evaluating the some important acoustic parameters Adiabatic compressibility (β), Intermolecular freelength (L_f), Relaxation time (τ) and specific acoustic impedance (Z). The physical nature and strength of the intermolecular interactions in solutions are also being obtained qualitatively from the ultrasonic velocity values and their variations with temperature (Palandhi and Singh, 1990).

In the recent past a large number of research and review articles have been published on conductivity of polymers (Pethric, 1973; Desai and Prasania, 1996). The unique stability of polymers have become so important on the development of conducting polymers as a class of synthetic metals with promising applications.

Compared to solid polymers, solutions of synthetic polymers find widespread technological applications. In solutions, it is possible to gain the details of information about the dimensions of the solute polymer molecules. Such information cannot be ascertained with so ease from the solid polymers (Kamani and Parsania, 1996).

The present article discusses the synthesis of polyaniline in base form and its behaviour in solutions of different organic solvents. Since the solubilization of PANi is important for the development of many technologies involving PANi as a class of new electronic material. The PANi is available in various forms as leuco emeraldine, emeraldine base and pemigraniline.

Materials and Methods

Polyaniline is synthesized through chemical polymerization method from its monomer aniline following the procedure reported in the literature (Bell and Pethrick, 1982; Kalyanasundaram, 1995)

The purity of the solvents selected was maintained by repeated distillation. The measured values of ultrasonic velocity (U), density (ρ) and viscosity (η) for the solvents (PCL, 2CPL and DMSO) are found as U_{2CPL} = 1377.6 ms^{-1}, ρ^2_{CPL} = 1.241 $_{Kg/m^3}$ and η_{2CPL} = 3.054 $_{Nsm^{-2}}$; U_{PCL} =1441.1 ms^{-1}, ρ_{PCL} =1.016 $_{Kg/m^3}$ and η_{PCL} =5.537 $_{Nsm^{-2}}$ and U_{DMSO} = 1487 ms^{-1}, ρ_{DMSO} =1.0955 $_{Kg/m^3}$ and η_{DMSP} = 1.989 $_{Nsm^{-2}}$ and are in agreement with the reported value (Manwar *et al.*, 2004). The accuracy of the measurement of the ultrasonic velocity, density and viscosity was within ±0.02, ±0.2 and ±0.2 per cent respectively.

The molecular weight of the doped PANi was determined using MALDI-TOF-MS technique at Indian Institute of Science, Bangalore, India with laser radiation of wavelength 300 nm of accuracy 0.001 nm as ionizing source. The molecular weight value so determined was 3479.342. The ultrasonic velocity values were measured for the conduction polymer Polyaniline, using ultrasonic interferometer. The sample Polyaniline is doped with percholoric acid (0.01 N) as the dopant. The Mittel ultrasonic interferometer was used for the polymer in solutions with PCL, 2CPL and DMSO as the solvents to measure the ultrasonic velocity values at three different fixed temperatures along with the density and velocity values respectively. The other related acoustic parameters were also calculated by applying the following relations 1–4.

$$\beta = \frac{1}{c^2 \rho} \quad \text{.......(1)}$$

$$L_f = K_T \beta^{1/2} \quad \text{.......(2)}$$

$$\tau = \frac{4}{3} \eta\beta \quad \text{.......(3)}$$

$$Z = \rho c \quad \text{.......(4)}$$

In equation (2), K_T is the temperature dependent constant = [(93.875+0.345T) × 10^{-8}] where T stands for the absolute temperature (Kannappan and Hemalatha, 2005)

Results and Discussion

The values of ultrasonic velocity, density and viscosity of all the three systems increase with increasing of concentration and decrease with increasing of temperatures. In the present investigation also the adiabatic compressibility of polymer solutions is less than that of the three solvents employed in accordance with the data found in the literatures (Moore and Uddin, 1969; Paladhi and Singh 1990)

In the process of solvation, highly polar forces result in a localized field and an incompressible region was formed by the solvating molecules which result in lesser value of adiabatic compressibility for the polymer in solutions than solvents (Kalyanasundaram *et al.*, 2000). Increase of ultrasonic velocity with concentration and decrease of adiabatic compressibility with concentration indicate that the molecules are forming a more tightly bound system (Ali and Nain, 1997)

The increase in velocity with concentration suggests that increase in cohesive forces due to powerful polymer-solvent interactions, *i.e.*, intermolecular interactions. The decrease of density and viscosity with temperature supports decrease in cohesive forces and hence decrease of intermolecular free length. Increasing thermal energy of the system causes volume expansion and hence decreases in density and viscosity while an increasing in L_f is observed.

Relaxation time (τ) for doped polyaniline has been reported in solutions at different temperatures by varying the concentration form 0.001M to 0.005M. It is found the relaxation time in all the three solvents decrease non-linearly with temperature and increase with concentration as exhibited by viscosity values as a result of the change in the thermodynamic parameters. The non-linear increase the relaxation time over a range of increase in polymer concentration suggests the existence of interaction between the polymer and solvent molecules.

Table 34.1: Acoustic Parameters of PANi in Solutions

Temp. K	*Molality M*	*Velocity m/s (v)*	*Density Kg/m³ (ρ)*	*Viscosity 10^{-3} Nsm⁻² (η)*	*Adiabatic Compressibility 10^{-10} m²N⁻¹ (β)*	*Inter Molecular Free Length 10^{-11} m (L_f)*	*Relaxation Time 10^{12} Sec (τ)*	*Specific Acoustic Impedance 10^6 Kgm⁻¹S⁻¹ (Z)*
				Polyaniline ($HClO_4$ 0.01N) + P-cresol				
313	0.001	1441.2	1007.25	4.4071	4.7798	4.3879	2.8087	1.4516
	0.002	1442.9	1008.56	4.5049	4.7624	4.3798	2.8605	1.4552
	0.003	1444.7	1009.31	4.5980	4.7470	4.3727	2.9103	1.4581
	0.004	1446.1	1010.62	4.6522	4.7317	4.3656	2.9351	1.4614
	0.005	1447.8	1011.37	4.7446	4.7171	4.3589	2.9841	1.4643
318	0.001	1427.1	1004.05	3.3380	4.8903	4.5085	2.1765	1.4329
	0.002	1428.5	1005.18	3.4262	4.8752	4.5020	2.2272	1.4359
	0.003	1430.2	1005.93	3.4804	4.8600	4.4945	2.2553	1.4387
	0.004	1431.7	1007.24	3.5456	4.8435	4.4873	2.2898	1.4421
	0.005	1433.6	1008.74	3.6540	4.8235	4.4780	2.3500	1.4461

Contd...

Table 34.1–Contd...

Temp. K	Molality M	Velocity m/s (v)	Density Kg/m³ (ρ)	Viscosity 10^{-3} Nsm^{-2} (η)	Adiabatic Compres-sibility 10^{-10} m^{2}N^{-1} (β)	Inter Molecular Free Length 10^{-11} m (L_f)	Relaxation Time 10^{12} Sec (τ)	Specific Acoustic Impedance 10^6 Kgm^{-1}S^{-1} (Z)
323	0.001	1414.2	1001.87	2.7129	4.9907	4.6203	1.8053	1.3854
	0.002	1415.4	1002.63	2.7953	4.9785	4.6138	1.8556	1.3312
	0.003	1417.4	1003.76	2.8594	4.9589	4.6046	1.8906	1.2941
	0.004	1418.8	1005.08	2.9141	4.9426	4.5970	1.9205	1.2643
	0.005	1420.3	1006.21	2.9979	4.9266	4.5896	1.9690	1.2180
Polyaniline ($HClO_4$ 0.01N) + Dimethyl Sulfoxide								
300	0.001	1498.6	1099.31	2.4224	4.1010	4.0421	1.3082	1.6474
	0.002	1499.9	1100.62	2.4788	4.0889	4.0361	1.3348	1.6508
	0.003	1501.4	1101.74	2.5462	4.0765	4.0300	1.3670	1.6542
	0.004	1503.6	1103.05	2.5813	4.0597	4.0217	1.3801	1.6585
	0.005	1505.2	1104.74	2.6534	4.0448	4.0143	1.4135	1.6628
310	0.001	1467.3	1096.25	1.9226	4.2817	4.2009	1.0861	1.6085
	0.002	1468.5	1097.38	1.9664	4.2703	4.1953	1.1079	1.6115
	0.003	1470.1	1098.69	2.0247	4.2558	4.1882	1.1369	1.6152
	0.004	1471.7	1100.19	2.0634	4.2407	4.1807	1.1545	1.6191
	0.005	1473.2	1101.69	2.1500	4.2262	4.1736	1.1989	1.6230
320	0.001	1436.5	1091.81	1.6201	4.4700	4.3553	0.9588	1.5684
	0.002	1438.2	1093.13	1.6777	4.4556	4.3483	0.9893	1.5721
	0.003	1439.9	1094.82	1.7369	4.4405	4.3409	1.0202	1.5764
	0.004	1441.3	1096.52	1.7735	4.4272	4.3344	1.0381	1.5804
	0.005	1442.8	1098.03	1.8445	4.4127	4.3273	1.0760	1.5842
Polyaniline ($HClO_4$ 0.01N) + 2-Chlorophenol								
303	0.001	1393.5	1234.04	4.9505	4.1731	4.1207	2.7545	1.7196
	0.002	1395.4	1235.16	5.0249	4.1579	4.1132	2.7858	1.7235
	0.003	1397.1	1236.66	5.1276	4.1428	4.1056	2.8323	1.7277
	0.004	1398.7	1237.97	5.1606	4.1290	4.0987	2.8411	1.7315
	0.005	1400.3	1239.28	5.2718	4.1152	4.0918	2.8926	1.7354
313	0.001	1357.7	1231.26	3.5807	4.4060	4.3075	2.1035	1.6717
	0.002	1359.2	1232.39	3.6605	4.3922	4.3008	2.1437	1.6750
	0.003	1360.2	1233.51	3.7072	4.3818	4.2956	2.1659	1.6778
	0.004	1361.8	1234.64	3.8089	4.3675	4.2885	2.2181	1.6813
	0.005	1363.4	1235.96	3.9099	4.3526	4.2812	2.2691	1.6851
323	0.001	1324.5	1228.30	3.0867	4.6408	4.4836	1.9099	1.6269
	0.002	1325.6	1229.44	3.1596	4.6288	4.4781	1.9501	1.6297
	0.003	1327.4	1230.57	3.2066	4.6120	4.4696	1.9719	1.6335
	0.004	1329.1	1231.89	3.2936	4.5953	4.4610	2.0180	1.6373
	0.005	1330.8	1233.58	3.3522	4.5773	4.4526	2.0459	1.6416

The acoustic impedance of the polymer is also increases with increase of concentration in all the three system and decreases with increase of temperature. The nonlinear variation of impedance both with concentration and temperature is the indication that there is a strong interaction between solute and solvent molecules. The specific acoustic impedance is a significant factor affecting reflection and transmission of sound waves in the medium and solvent molecules.

Acknowledgements

The authors are thankful sincerely to Dr. A.N. Kannappan, Professor and Head of the Department of Physics, Annamalai University for his consistence encouragement to execute the indented studies so gracefully in the physics laboratory of the department.

References

Ali, A. and A.K. Nain, 1997. Ultrasonic studies of formamide + 1,2-ethanediol and sodium iodide + formamide + 1,2-ethanediol mixtures at 298.15 K. *Indian J. Pure and Appl. Phys.*, 35: 729–733.

Anwar Ali, Anil Kumar Nair and Bhajan Lal, 2004. Ultrasonic study of intermolecular interactions in non-aqueous binary liquid mixtures at 30°C. *J. Pure Appl. Ultrasonic*, 26: 40–48.

Bell, W. and R.A. Pethrick, 1982. Ultrasonic investigation of linear and star shaped polybutadiene polymers in solutions of cyclohexane, hexane and ethylbenzene, polymer. *J. Polymer*, 23: 369.

Desai, J.A. and P.H. Prasania, 1996. Acoustic studies of cardo polysulfonates of 1,1′-bis (4-hydroxy phenyl) cycloxane with 1,3-benzene disulonyl chloride in different solvents at 30°C. *J. Polym. Mater.*, 13: 79.

Kalyanasundaram, S., Sundaresan, B. and J. Hemalatha, 2000. Solvation study of polymers by ultrasonic method. *J. Polm. Mate.*, 17: 155–163.

Kalyanasundaram, V., Stephen, M. and A. Gopalan, 1995. Molecular studies on poly (metylmethacrylate) in choloroform. *J. Polym. Meter.*, 12: 323.

Kamani, M.M. and P.H. Parsania, 1996. Ultrasonic velocity studies of chloroform and dichloro Methane solutions of cardo polysulfone of 1,1′-bis (3-methyl-4-hydroxy phenyl) cyclohexane and 4,4′-difluorodiphenyl sulfone. *J. Pure Appl. Ultrasonic*, 18: 46.

Kannappan, V. and G. Hemalatha, 2005. Ultrasonic studies on the molecular interaction of 1-chlorobenzotriazole with aromatic compounds in solution. *Indian J. Pure and Appl. Phys.*, 43: 849–853.

Manwar, B.G., Kavitha, S.H. and P.H. Parsania, 2004. Ultrasonic velocity study of poly (R,R′,4,4′-cyclohexane). *J. Pure and Appl. Ultrasonic*, 26 :49–57.

Moore, W.R. and M.A. Uddin, 1969. Dilute solution properties of bisphenol A polycarbonate-I Osmotic, light scattering and viscosity measurements. *Eur. Polym. J.*, 5: 185–193.

Paladhi, R. and R.P. Singh, 1990. The effect of molecular weight on adiabatic compressibility and solvation number of polymer solutions. *Eur. Polym. J.*, 26: 441–444.

Pethrick, R.A., 1973. Acoustic studies of polymer solution. *J. Macromol. Chem.*, 91.

Chapter 35

On the Occurrence of *Nostoc* spp. from Jaisamand and Surrounding Area Udaipur, Rajasthan (India)

B.L. Chaudhary and Lakhpat Meena[1]

Department of Botany, University College of Science, Mohanlal Sukhadia University, Udaipur – 313 001, Rajasthan, India

ABSTRACT

Present article discusses about systematic as well as ecology two species of *Nostoc* namely *N. punctiforme* (Kütz.) Hariot, *N. muscorum* Ag. ex Born. et Flah from Jaisamand and its neighbourhood during 2004–06, Udaipur district of Rajasthan (India).

***Keywords**: Cyanobacteria, Jaisamand, Taxonomy, Water.*

Introduction

Work on algal flora in South Rajasthan has been neglected aspect since long because of many reasons as well as the fact that work in the algal taxonomy are very few. But with reference to Cyanobacterial flora of Rajasthan very little has been done (Gupta, 1962; Gupta and Kumar, 1968; Kapoor and Arora, 1998, 1999). The present study deals with two species of *Nostoc* spp. taxonomy as well their relationship with environment factors.

* Corresponding Author: E-mail lmeenak@yahoo.com; Mobile +910414553603

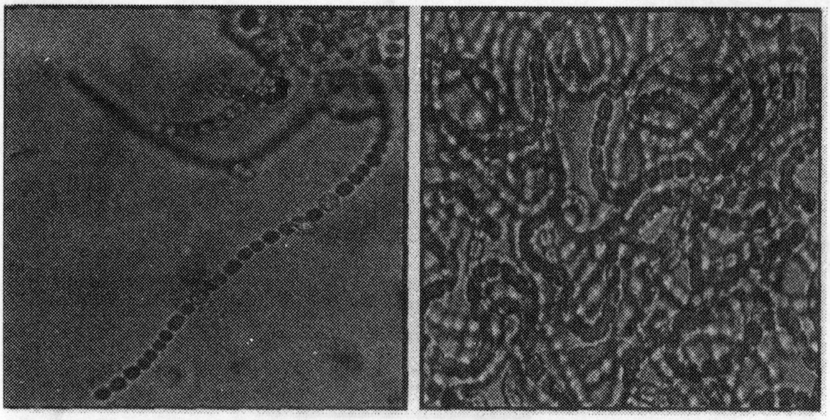

Figure 35.1–35.2: *Nostoc muscorum* Ag. ex Born. et Flah

Figure 35.3: *Nostoc muscorum* Ag. ex Born. et Flah

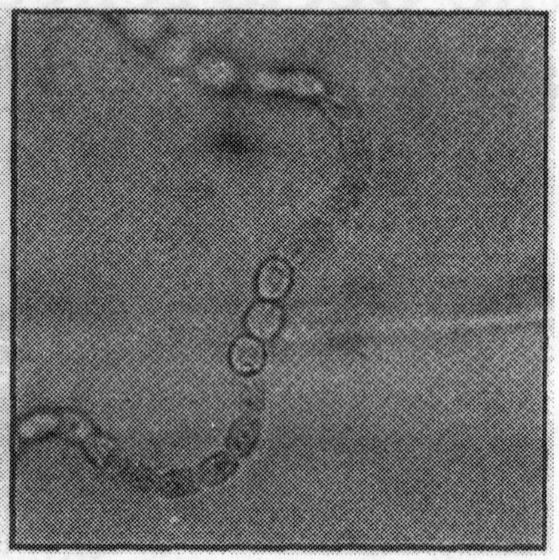

Figure 35.4: *Nostoc punctiforme* (Kutz.) Hariot

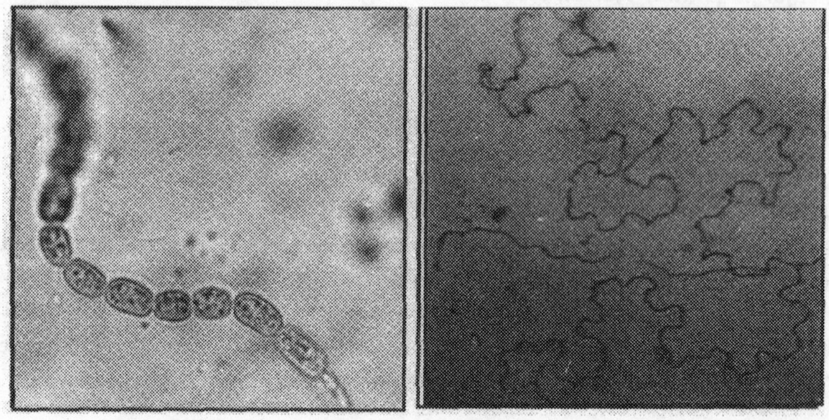

Figure 35.5–35.6: *Nostoc punctiforme* (Kutz.) Hariot

Materials and Methods

The water samples of Jaisamand lakes and its surrounding were collected at regular time intervals from different sites. Samples were brought to laboratory, then preserved in 6 per cent formalin or cultured in Allen and Arnon (1955). After that specimens were identified with the help of standard method (Desikachary, 1959).

Results and Discussions

The morpho-taxonomic results of identified species are as below.

Nostoc punctiforme (Kütz.) Hariot (Figures 35.1–35.3)

Locality: Temporary running water eastern to Jaisamand

Collection no. and Date-ALC. 23

Thallus dark-brown attached on submerged hydrophytes, filaments densely entangled, sheath delicate, hyaline, constricted at crosswall, width 4.09–5.46 μ, vegetative cells shape varies from spherical to elongated, 5.46–1.2 μ long, blue-green in colour, heterocyst both intercalary and terminal, 6.8–8.19 μ long, 5.5–6.9 μ broad. Spores formed in chain, subspherical with granulated cytoplasm, 5.46–8.19 broad and 8.19–10.9 μ long.

Ecology: Attached on submerged dead objects like twing of plants, January 2005.

Nostoc muscorum Ag. ex Born. et Flah (Figures 35.4–35.6)

Locality: Streams midway to Jaisamand

Collection no. and Date-ALC. 21

Thallus macroscopic, blue-green not in definite shape mucilaginous, yellowish on lower side, attached, 3–4 cm dia, filaments densely entangled, sheath not distinct, constricted at cross wall, trichome 2.7–4.0 μ broad, cells subspherical, End cell half conical, 4.0–4.2 μ long. Heterocyst both intercalary as well as terminal, terminal heterocyst half-conical (like vegetative cell) with single polar nodules intercalary heterocyst spherical upto 5.5 μ broad, akinetic ellipsoidal, rarely seen, upto 5.4 μ broad and 8.0 μ long.

Ecology: Attached on moist wall near running water, January 2005.

Acknowledgement

The authors are thankful to Professor N.C. Aery, Head, Department of Botany, University College of Science, M.L.S.U. Udaipur, Rajasthan, for providing research facilities and guidance and one of us (L. Meena) thanks the UGC, New Delhi for awarding the fellowship during the tenure of study.

References

Desikachary, T.V., 1959. *Cyanophyta: Monographs on Algae*. ICAR, New Delhi, India, 686 pp.

Gupta, R.S. and Kumar, H.D., 1968. Blue green algal flora of Udaipur and its neighborhood. *Revue Algol*, 9: 91–103.

Palmer, C.M., 1980. *Algae and Water Pollution*. Castle House Publication Ltd., England, 123 pp.

Rao, C.B., 1937. The myxophyceae of the United Provinces India-III. Proc. *Indian Acad. Sci.*, B–6(6): 339–375.

Tewari, G.L., 1979. A study of the blue-green algae from paddy field of soils of India. Part IV. Taxonomic consideration of Nostocales and Stigonematales, Nova. *Hedwigia*, 63: 133–159.

Chapter 36

Ultrasonic Studies on Hydroxyl Terminated Polybutadiene in Methyl Methacrylate

P. Siva Gurunathan[1], F. Liakath Ali Khan[2]
[1]Department of Physics, Annamalai University, Annamalai Nagar – 608 002, India
[2]Deparment of Physics, Islamiah College, Vaniyambadi – 635 752, India

ABSTRACT

The Ultrasonic velocity, adiabatic compressibility. relaxation amplitude and Rao's number are reported for Hydroxyl Terminated Polybutadiene (HTBP) in Methyl Methacrylate (MMA) at different concentrations and temperatures. The results show a increasing trend of ultrasonic velocity, viscosity, Rao's number, relaxation time, relaxation amplitude and decreasing trend of adiabatic compressibility with increasing HTPB concentration in the MMA indicating the interaction between MMA and HTPB.

Keywords: *Ultrasonic parameters, HTPB, MMA.*

Introduction

In recent years, extensive use of polymeric materials has necessitated the study of molecular interaction of polymers with MMA Because of their low cost, high strength to weight ratio, non-corrosiveness, polymers are finding enormous applications in our daily life. The usage of polymers are increasing many fold in aerospace, automobile, medicine, electronics, electrical, domestic, building construction and other fields as substitution materials (North and Pethrick, 1980).

The ultrasonic studies on polymer solutions provide ample evidence for inter/intra molecular forces operating between polymer solute and solvent (Kalyanasundaram *et al.*, 1996, Venkatanaidu *et al.*, 1990, Hassun *et al.*, 1985 and Vijayalakshmi *et al.*, 1996). This work is an attempt to study the molecular interactions of solutions of HTPB in MMA at several concentrations and temperatures.

Experimental Details

Methyl Methacrylate is a analyzed-grade material obtained from the Aldrich Chemical Co. It contained 6.5×10^{-5} mass fraction of hydroquinone monomethyl ether and it is used as such without any further treatment. The HTPB is obtained from Vikram Sarabhai Space Centre, Thiruvananthapuram is used as such and dissolved in MMA for solute concentrations from 0 to 9 per cent. Precautions were taken to avoid any possible degradation (Bill Meyer, 1984). Ultrasonic velocity measurements in HTPB solution have been made using a variable path interferometer (Mittal Enterprises, New Delhi) at a fixed frequency of 1 MHz, with an accuracy of 0.05 per cent. The density was measured using a specific gravity bottle. An Ubbelholde viscometer was used for viscometric studies (0.015). All the above measurements were carried out at three different temperatures 303K, 313K and 323K with an accuracy of ±0.1°C. Constant temperature of the sample was maintained by circulating water from a thermostatically controlled water bath. The molecular weight of HTPB is determined by viscosity method (Gowraiker *et al.*, 1988).

The adiabatic compressibility (β) is calculated using the formula

$$\beta = (C^2 \rho)^{-1}$$

where,

C is the ultrasonic velocity in the solution and p is the density of the solution. The Rao's number R is calculated using the formula

$$R = C^{1/3} V$$

where,

V is the molar volume of the solution. The weight average molecular weight is used to calculate the molar volume of the solution. The relaxation amplitude α/f^2 is calculated using the relation (Hassun, 1987).

$$\frac{\alpha}{f^2} = \frac{8\pi 2\eta}{3 \rho C^3}$$

and the relaxation time τ is calculated using the relation (Hassun, 1987).

$$\tau = \frac{4\eta_s}{3 \rho C^2}$$

where,

η_s is the shear viscosity of the solution.

Results and Discussion

The measured values of density (ρ), ultrasonic velocity (U), shear viscosity (η) and the calculated parameters namely, adiabatic compressibility (β), Rao's number (R), relaxation time (τ) and relaxation amplitude (α/f^2) for solute concentrations from 0 to 9 per cent at temperatures 303K, 313K and 323K are given in Table 36.1.

Table 36.1: Ultrasonic Velocity and Derived Parameters for the Solution of HTPB in MMA

Concentration (wt%)	*Ultrasonic Velocity C m/s*	*Density of Solution ρ kg m⁻³*	*Adiabatic Compressibility $\beta_s \times 10^{-10}$ m²/N*	*Rao's Number $R \times 10^3$*	*Shear Viscosity $\eta_s \times 10^{-3}$ Ns/m²*	*Relaxation Time $\tau \times 10^{-13}$ s*	*Relaxation Amplitude $\alpha/f^2 \times 10^{-14}$ s²/m*
			at 303K				
0.000	1167	932.3	7.8759	1.1277	0.5290	5.5551	0.9386
0.250	1168	931.9	7.8672	1.1284	5.6041	5.8713	0.9941
0.500	1169	931.7	7.8559	1.1292	5.9277	6.1835	1.0455
1.000	1171	931.1	7.8323	1.1309	0.6543	6.8329	1.1506
3.000	1175	930.2	7.7866	1.1689	0.8454	8.7770	1.4730
5.000	1179	929.2	7.7421	1.1904	0.9447	9.7520	1.6311
7.000	1182	928.6	7.7079	1.2195	1.2614	12.9637	2.1627
9.000	1189	927.4	7.6272	1.2487	1.5311	15.5708	2.5824
			at 313K				
0.000	1129	925.7	8.4750	1.1233	0.4531	5.1189	0.8948
0.250	1130	925.4	8.4758	1.1238	4.8201	5.4638	0.9544
0.500	1131	925.0	8.4702	1.1244	5.1437	5.8021	1.0136
1.000	1133	924.4	8.4719	1.1257	0.5752	6.4974	1.1338
3.000	1138	922.5	8.3704	1.1662	0.7402	8.2610	1.4315
5.000	1144	920.6	8.2999	1.1896	0.8235	9.1133	1.5709
7.000	1147	919.8	8.2638	1.2189	1.1373	12.5312	2.1544
9.000	1150	918.2	8.2350	1.2472	1.3092	14.3751	2.4649
			at 323K				
0.000	1106	920.3	8.8830	1.1222	0.3902	4.6227	0.8242
0.250	1107	920.0	8.8741	1.1229	0.4101	4.8569	0.8660
0.500	1108	919.7	8.8628	1.1237	0.4299	5.0692	0.9057
1.000	1110	919.2	8.8296	1.1253	0.4718	5.5551	0.9868
3.000	1115	918.6	8.7563	1.1632	0.6393	7.4639	1.3200
5.000	1120	917.2	8.6916	1.1856	0.7361	8.5305	1.5019
7.000	1125	916.8	8.6182	1.2150	1.0056	11.5554	2.0254
9.000	1131	915.2	8.5419	1.2444	1.1531	13.1333	2.2898

The ultrasonic velocity increases with increasing the HTPB concentration at all temperatures and decreases with increase in temperature. The gradual increase in ultrasonic velocity with solute concentrations are observed at all temperatures may be attributed due to molecular interaction between polymer and MMA (Hassun *et al.*, 1989, Ali-Hussain *et al.*, 1990 and Sundaresan *et al.*, 1994). The weak dipolar forces, dispersive forces in addition to the Van der Waal's forces may be responsible for interaction between HTPB and MMA. The decrease in ultrasonic velocity with increase in temperatures are observed in the present study, which may indicate the weakening of intermolecular forces due to thermal effects (Vijayataksbmi *et al.*, 1996). The adiabatic compressibility (β) decreases with increase in solute concentrations may confirm the interaction between HTPB and MMA. The viscosity, relaxation time and relaxation amplitude increases with increase of solute concentrations at all temperatures. These observations are in accordance with those reported by Kalyanasundaram and Manuel Stephan (1997). The increase in viscosity and relaxation time with increasing HTBP concentration are observed in the present study reveals the molecular association between HTBP and MMA (Hassun *et al.*, 1985) The temperature dependence of Rao's number as observed in the present study also signifies the probable interaction between the solute and MMA (Hassun *et al.*, 1990). From the results one may conclude that the molecular interactions exist between HTPB and MMA system and the degree of molecular interaction decreases with temperature. Similar observations were arrived by Vijayalakshmi *et al.* (1996).

References

Bill Meyer (Jr), F.W., 1984. *Textbook of Polymer Science*, 3rd edn. John Wiley, New York, pp. 404.

Gowraiker, V.R., Viswabathan, M.V. and Sreedhar Jyder, 1988. *Polymer Science.* Wiley Eastern, New Delhi, pp. 263.

Hassun, S.K., Al-Madfai, S.H.F. and Al-Jariah, M.M.F., 1985. *Britain Polymer Journal*, 17: 330.

Hassun, S.K., 1987. *Acoustic Letters*, 10: 110.

Hassun, S.K. and Ali-Hussain, F.M., 1989. *Acoustic Letters*, 12: 16.

Hassun, S.K., Ali-Hussain, F.M., and Al-Mehdi, HA.H., 1990. *Acoustic Letters*, 14: 17.

Kalyanasundaram, S.A., Manuel Stephan, A. and Goplan, A., 1996. *Indian Journal of Pure and Applied Physics*, 34: 353

Kalyanasundaram, S. and Manuel Stephan, A., 1997. *Acoustica*, 83: 74.

North, A.M. and Pethrick, R.A., 1980. *Development in Polymer Characterization Applied Science*, London.

Sundaresan, B. and Srinivasa Rao, A., 1994. *Polymer Journal*, 26: 1286.

Vijayalakshmi, R., Sanyal, T.K., Nambi Narayanan, Srinivasa Manja, K. and Srinivasa Rao, A., 1996. *Indian Journal of Physics*, 70A(2): 265.

Venkatanaidu, S. Varadarajulu, A. and Cbowdoji Raok, K., 1990. *Journal of Polymer Matter*, 7: 149.

Chapter 37

Study of Biochemical and Haematological Changes in the Rats Fed Different Dietary Levels of *Syzygium cuminiseeds*: A Safety Assessment

Madhuri A. Agnihotri[1] and Aqueel Khan[2]

[1]Rungta College of Dental Sciences and Research, Kohka, Kurud Road, Bhilai, C.G.

[2]Postgraduate Department of Biochemistry, RTM Nagpur University, LIT Premises, Amravati Road, Nagpur – 440 033, M.S.

ABSTRACT

The experiments were conducted to study biochemical and haematological changes in rats to assess deleterious or toxic effects, if any, of feeding of 5, 10, 15, 20 and 25 per cent *Syzygium cumini* seed diets for 21 days. The results of this study indicated that the feeding of 5, 10, 15, 20 and 25 per cent *Syzygium cumini* seed diets for 21 days did not exhibit hepatotoxic and nephrotoxic effect and did not cause gastric or intestinal haemorrhage. Blood glucose, serum cholesterol and serum triglyceride levels were found to be significantly reduced in the rats fed 5, 10, 15, 20 and 25 per cent *Syzygium cumini* seed diets. Feeding of 5, 10 and 15 per cent *Syzygium cumini* seed diets caused significant gain in body weight of rats. This body weight gain was slightly higher in the rats fed 15 per cent *Syzygium cumini* seed diet. Feeding of 20 and 25 per cent *Syzygium cumini* seed diets caused the arrest in the growth of rats.

Keywords*: Syzygium cumini seeds, Traditional antidiabetic herbal drug, Safety assessment.*

Introduction

Syzygium cumini seeds (Myrtaceae) is one of the twenty most widely used traditional antidiabetic plant, Aiman (1961), Brahmachari and Augusti (1961), Dhar *et al.* (1968), Marles and Farsworth (1988), Mukherji (1957), Nadkarni (1954), Shukla *et al.* (2000), Razzack HMA (1993), Zafar R. (1994). Antidiabetic effect of *Syzygium cumini* seeds has been reported earlier, Achrekar *et al.* (1991), Bansal and Ahmed (1981), Bhaskaran Nair and Santha Kumari (1986), Chatterjee (2000), Singh *et al.* (1986), Stanley M.P.P. *et al.* (1998), Kohli *et al.* (1993), Lal and Choudhary (1968).

Although a long history of traditional use of *Syzygium cumini* seeds as antidiabetic herbal drug may suggest its safety, the biochemical and haematological studies indicating the safety of oral administration of *Syzygium cumini* seed diets are lacking. Therefore this study was undertaken to investigate the biochemical and haematological changes in rats to assess deleterious or toxic effects, if any, feeding of different dietary levels of *Syzygium cumini* seeds.

Materials and Methods

Chemicals

Glucose oxidase, peroxidase, o-dianisidine were purchased from Sigma Chemicals Co., St. Louis, U.S.A. Casein (Vitamin free) was the product of E. Merck Co., Germany. All the other reagents and chemicals used were of analytical grade from Qualigen's Fine Co., Bombay.

Collection of Seed Sample

Dried seeds of *Syzygium cumini* seed Skeels. (Myrtaceae) were purchased from the local market and the identity of this plant material was established taking the help of a botanist. The dried seeds were washed thoroughly with distilled water 3 to 4 times. The water washing which did not show any' hypoglycaemic activity was discarded. The dried seeds were powdered in an *Syzygium cumini* seeds were prepared and used throughout the study.

Animals and Diets

The male albino rats of Wistar Strain weighing between 150–170 g were divided into six groups of six animals each designated as Group I, Group II, Group III, Group IV, Group V and Group VI. Group I served as control and the animals of this group were fed control diet. The animals of Group II, III, IV, V and VI were fed 5, 10, 15, 20 and 25 per cent *Syzygium cumini* seed diets respectively. The composition of control and experimental diets are given in Table 37.1. Animals of all the groups were fed 20 g of respective diets daily for 21 days and were observed through out the feeding period for changes in locomotion, posture, excretion and general behaviour if any. The rectal temperature of all the animals was noted periodically. The initial and final body weight of the animals of all the groups were recorded. After 21 days blood was collected by retroorbital plexus in heparinized and nonheparinized tubes and animals were sacrificed by stunning and decapitation. Immediately after excision, the internal organs were removed and blotted free of blood and weighed.

For the separation of plasma, heparinized blood was centrifuged for 10 minutes at 3000 r.p.m. Plasma was taken out by Pasteur pipette and kept in clean glass testube. Plasma protein and albumin were estimated by the modified Biuret method (Varley, 1969). Blood glucose was estimated by glucose oxidase method (Hugget, 1957). Estimation of haemoglobin was carried out by haematin method (Chanarin, 1989).

Table 37.1: Composition of Control and Experimental Diets (g/100g)

Ingredients	*Control Diet*	*Experimental 5%*	*Syzygium 10%*	*Cumini 15%*	*Seed 20%*	*Diets 25%*
Casein	20.0	20.0	20.0	20.0	20.0	20.0
Syzygium cumini seeds	–	5.0	10.0	15.0	20.0	25.0
Hydrogenated oil	10.0	10.0	10.0	10.0	10.0	10.0
Starch	65.0	60.0	55.0	50.0	45.0	40.0
Hawk's-oser salt mixture	4.0	4.0	4.0	4.0	4.0	4.0
Vitamin mixture*	1.0	1.0	1.0	1.0	1.0	1.0

Vitamin mixture* (1g/100g)

Thiamine Hcl: 10 mg, Riboflavin: 20 mg, Pyridoxine: 10 mg, Calcium pantothenate: 40 mg, Folic acid: 12.5 mg, Choline chloride: 20 mg, Biotin: 10 mg, Menadione: 5 mg, L-Tochopherol: 50 mg, Cyanocobalamine: 10µg, Vitamin A: 2000 IU, Vitamin D: 20 IU.

Serum samples were obtained by centrifuging the clotted blood and serum cholesterol, serum triglyceride were estimated by the method of Zlatkis *et al.* (1953) and Zilversmit and VanHandel respectively (1957).

Activity of SGOT and SGPT was assayed by King's method (1959) and that of serum alkaline phosphatase by King and King's method (1954). Serum bilirubin and serum creatinine were estimated by Malloy and Evelyn (1937) and Jaffe's method (Folin, 1922) respectively. Special test was done for the detection of occult blood in faecal matter by Benzidine test (Choudhary and Dubey, 1981). Total RBC, WBC count were determined by haemocytometer and differential leucocyte count (DLC) was done after the blood slides were stained with Leishman stain.

The results were statistically analyzed using students "t" test.

Results and Discussion

The rats fed 5, 10 and 15 per cent *Syzygium cumini* seed diets showed significant increase in the body weight and no change in normal posture, locomotion, excretion and the rectal temperature was observed in these rats.

Our results clearly indicate that as far as growth is concerned, the most favourable effect was observed in the rats fed 15 per cent *Syzygium cumini* seed diet as much significant increase in the body weight was observed in these rats.

Absence and loss of body weight gain in rats fed 20 and 25 per cent *Syzygium cumini* seed diets respectively suggest that the growth rate was arrested in these rats (Table 37.2). Thus feeding of 15 per cent *Syzygium cumini* seed diets showed most beneficial effect on the physiological system without any toxic effect on growth rate of rats.

The occult blood test of faecal matter of rats of experimental groups was not positive thereby indicating that *Syzygium cumini* seed diets did not cause any gastrointestinal bleeding. The feeding of 5, 10 and 15 per cent *Syzygium cumini* seed diets did not produced any significant change in haematological parameters (Table 37.3).

Table 37.2: Changes in the Body Weight of Rats after Feeding of Different Dietary Levels of *Syzygium cumini* Seeds for 21 Days

Groups	*Initial Body Weight (g)*	*Final Body Weight (g)*
Group I Control diet fed	155±5.37	180±8.01
Group II 5 per cent *Syzygium cumini* seed diets fed	158±6.00	198*±8.68
Group III 10 per cent *Syzygium cumini* seed diets fed	160±6.75	200**±9.91
Group IV 15 per cent *Syzygium cumini* seed diets fed	161±6.05	214**±8.40
Group V 20 per cent *Syzygium cumini* seed diets fed	168±6.19	168^{NS}±6.35
Group VI 25 per cent *Syzygium cumini* seed diets fed	170±5.95	129**±4.51

*: P < 0.05; **: P < 0.01; NS: Not Significant. Final weight compared with the initial body weight of rats.

Values indicate mean body weight±SD of 6 rats in each group.

Table 37.3: Effect of Feeding of Different Dietary Levels of *Syzygium cumini* Seeds for 21 Days on Haemogram of Rats

Treatment	*Hb (per cent)*	*Total WBC × 10^3/cu.m.m.*	*Total RBC 10^6/cu.m.m*	*Differential Leucocyte Count (DLC per cent)*			
				N	*E*	*M*	*L*
Group I (6) Control diet fed	14.70 ± 3.55	7.69 ± 0.61	5.30 ± 0.55	64.46 ± 5.621	2.90 ± 0.038	5.40 ± 0.705	77.24 2.250
Group II (6) 5 per cent *Syzygium cumini* seed diets fed	14.63 ± 3.51	7.70 ± 0.63	5.27 ± 0.51	63.66 ± 5.551	2.89 ± 0.037	5.50 ± 0.718	27.95 2.24
Group III (6) 10 per cent *Syzygium cumini* seed diets fed	14.60 ± 3.42	7.65 ± 0.59	5.24 ± 0.41	65.00 ± 5.668	2.85 ± 0.037	5.30 ± 0.692	26.85 2.111
Group IV (6) 15 per cent *Syzygium cumini* seed diets fed	14.51 ± 3.39	7.50 ± 0.54	5.10 ± 0.42	63.50 ± 5.537	2.83 ± 0.037	5.00 ± 0.653	28.67 2.354
Group V (6) 20 per cent *Syzygium cumini* seed diets fed	11.85** ± 1.73	7.70 ± 0.62	4.21 ± 0.48	64.24 ± 5.601	2.87 ± 0.037	5.40 ± 0.705	27.49 2.353
Group VI (6) 25 per cent *Syzygium cumini* seed diets	11.30** ± 1.51	7.68 ± 0.59	3.45* ± 0.48	64.30 ± 5.607	2.80 ± 0.036	5.51 ± 0.719	27.39 2.291

The results presented in table are means±SD of 6 rats each. Number of animals used are given in parenthesis. The data were analysed statistically by students 't' test. The haemoglobin, total WBC count, total RBC count and differential leucocyte count of normal treated rats compared with their. normal control rats for statistical analysis.

*: P < 0.05; **: P < 0.01; N: Neutrophil; E: Eosinophil; M: Monocyte; L: Lymphocyte.

Table 37.4: Effect of Feeding of Different Directory Levels of *Syzygium cumini* Seeds for 21 Days on Biochemical Parameters of Rats

Treatment	Blood Glucose (mg/100 ml)	Serum Cholesterol (mg/100 ml)	Serum Triglyceride (mg/100 ml)	Serum Creatinine (mg/100 ml)	Serum Bilirubin (mg/100 ml)	Total Plasma Protein (g/100 ml)	Plasma Albumin (g/100 ml)	SGOT (U/L)	SGPT (U/L)	Alkaline Phosphatase (U/L)
Group I Control diet fed	97.80 ± 8.60	180.00 ± 11.25	120.35 ± 9.00	0.90 ± 0.11	0.75 ± 0.09	7.40 ± 0.85	3.53 ± 0.40	127.44 ± 43.94	55.36 ± 20.02	140.12 ± 49.70
Group II 5 per cent *Syzygium cumini* seed	79.0* ± 7.40	155.00** ± 9.60	105.41 ± 8.99	0.81 ± 0.09	0.63 ± 0.08	7.44 ± 0.49	3.45 ± 0.33	126.53 ± 41.71	59.55 ± 20.00	140.33 ± 49.00
Group III 10 per cent *Syzygium cumini* seed diets fed	74** ± 7.10	44.23** ± 8.15	97.63** ± 7.87	0.78 ± 0.08	0.61 ± 0.07	7.25 ± 0.77	3.40 ± 0.30	125.44 ± 40.89	60.30 ± 21.27	139.89 ± 43.23
Grpup IV 15 per cent *Syzygium cumini* seed diets fed	68.00*** ± 5.9	132.65*** ± 6.32	90.00*** ± 7.16	0.65 ± 0.07	0.60 ± 0.07	6.99 ± 0.68	2.98 ± 0.27	124.69 ± 40.13	61.98 ± 20.99	141.47 ± 47.65
Group V 20 per cent *Syzygium cumini* seed diets fed	69.90* ± 6.2	140.19 ± 6.99	68.72* ± 5.80	0.60 ± 0.07	0.52 ± 0.03	6.90 ± 0.60	2.68 ± 0.20	123.51 ± 35.24	63.88 ± 22.13	141.66 ± 48.21
Group VI 25 per cent *Syzygium cumini* seed diets fed	82.14* ± 9.2	160.54* ± 10.61	79.55* ± 5.47	0.50 ± 0.03	0.39 ± 0.01	6.86 ± 0.55	2.50 ± 0.17	124.30 ± 40.73	65.62 ± 25.81	142.08 ± 49.98

*: $P < 0.05$; **: $P < 0.01$; ***: $P < 0.0001$. The blood glucose, serum cholesterol and serum triglycerides levels of treated rats compared with control rats.

Values are mean per cent SD of 6 rats in each group.

The Hb concentration of these treated rats did not indicate any symptoms of anaemia. Feeding of 20 and 25 per cent *Syzygium cumini* seed diets showed significant decrease in total RBC count and Hb concentration which indicate an adverse effect on erythropoiesis in these animals. Blood glucose, Serum cholesterol and Serum triglyceride levels were found to be significantly reduced in the rats fed 5, 10, 15, 20 and 25 per cent *Syzygium cumini* seed diets (Table 37.4). The hypoglycaemic effects observed in our study are confirmed by earlier reports. Aiman (1961), Brahmachari and Augusti (1961), Bhaskaran Nair and Santha Kumari (1986), Shrotri *et al.* (1963). But there is no report available on the hypocholesterolaemic effect of *Syzygium Cumini* seeds.

These result suggest that the *Syzygium cumini* seeds at all dose levels incorporated in the diets had hypoglycaemic and hypolipidaemic effects.

No alternation in serum bilirubin concentration indicate that no liver or biliary disorders occurred due to the feeding of different levels of *Syzygium cumini* seeds in the diets. No change observed in the activity of the enzymes SGOT, SGPT and serum alkaline phosphatase in all the treated rats indicate that there was no hepatocellular damage(Table 37.4).

The total plasma protein and albumin levels remained within the normal range in the rats fed different *Syzygium cumini* seed diets showing that there was no indication of any kind of serious ailments of liver in rats. No significant change in serum creatinine level in treated rats as compared to control rats indicated that renal excretion was not impaired. These observations are suggestive of no nephrotoxic effect in these rats (Table 37.4).

Table 37.5: Effect of Feeding of Different Dietary Levels of *Syzygium cumini* Seeds on Weight of Vital Organs/100 g of Body Weight of Rats After 21 Days of Treatment.

Groups	*Liver*	*Spleen*	*Kidney*	*Heart*	*Lungs*	*Adrenals*
Group I N (6) Control diet fed	3.343 ± 0.236	0.270 ± 0.030	0.741 ± 0.07	0.394 ± 0.04	0.988 ± 0.079	0.029 ± 0.006
Group II N (6) 5 per cent *Syzygium cumini* seed diets fed	3.509* ± 0.249	0.265 ± 0.028	0.722 ± 0.05	0.389 ± 0.03	0.982 ± 0.075	0.027 ± 0.004
Group III N (6) 10 per cent *Syzygium cumini* seed diets fed	3.821* ± 0.252	0.269 ± 0.255	0.701 ± 0.035	0.385 ± 0.01	0.975 ± 0.071	0.025 ± 0.003
Group IV N (6) 15 per cent *Syzygium cumini* seed diets fed	4.612* ± 0.222	0.255 ± 0.021	0.698 ± 0.04	0.387 ± 0.01	0.940 ± 0.063	0.023 ± 0.015
Group V N (6) 20 per cent *Syzygium cumini* seed diets fed	3.340 ± 0.100	0.268 ± 0.018	0.743 ± 0.04	0.392 ± 0.012	0.980 ± 0.060	0.029 ± 0.020
Group VI N (6) 25 per cent *Syzygium cumini* seed diets fed	3.346 ± 0.109	0.274 ± 0.020	0.740 ± 0.037	0.390 ± 0.021	0.981 ± 0.059	0.028 ± 0.008

Weight of vital organ, liver of normal treated rats compared with the weight of liver of normal control rats for statistical analysis.

*: $P < 0.05$

Values indicates mean weight of vital organs±SD of 6 rats in each groups.

Animals fed 5, 10 and 15 per cent *Syzygium cumini* seed diets showed increase in liver weight but no appreciable change was observed in the weights of vital organs such as spleen, heart, lungs, kidney and adrenals (Table 37.5). Increased liver weights may be due to an increase liver protein biosynthesis.

The results of this study suggest that feeding of *Syzygium cumini* seeds at the dose levels of 5,10 and 15 per cent in the diet was found to be quite safe as evident by the haematological and biochemical investigations and growth rate of rats. No gross behavioural changes were detected in these animals.

The 5, 10, 15, 20 and 25 per cent *Syzygium cumini* seeds showed hypoglycaemic and hypolipidaemic effects. These effects were more pronounced in the rats fed 15 per cent *Syzygium cumini* seed diet.

References

Achrekar, S., Kallij, G.S., Pote, M.S. and Kelkar, S.M., 1991. Hypoglycaemic activity of *Eugenia jambolana* and *Ficus bengalensis*. *In vivo*, 2: 193.

Aiman, R., 1961. Indigenous antidiabetic substances of plant origin. *Ind. J. Pharmacol.*, 23(4): 115.

Bansal, R., Ahmad, N. and Kidwai, R., 1981. Effects of Oral administration of *Eugenia jambolana* seeds and chlorpropamide on blood glucose level and pancreatic cathepsin B in rat. *Ind. J. Biochem. Biophysics*, 18: 377.

Bhaskaran Nair, R. and Santha Kumari, G., 1986. Antidiabetic activity of the seed kernel of *Syzygium cumini* Linn. *Ancient Sciences of Life*, 2: 80.

Brahmachari, H.D. and Augusti, K.T., 1961. Hypoglycaemic agent from onions. *J. Pharm. Pharmacol.*, 13: 381.

Chanarin, I., 1989. *Laboratory Haematology*. Churchill Livingstone, Longman Group, U.K. Ltd., p. 3.

Chatterjee, K., 2000. *Herbal Options*. Books and Allied (P) Ltd., Calcutta, 3: 57.

Choudhary, K. and Dubey, K.M., 1981. *Biochemical Techniques.* Medical Publishers, Jaypee Brothers, New Delhi.

Dhar, M.L., Dhar, M.M., Dhawan, B.N., Mehrotra, B.N. and Ray, C., 1968. Screening of Indian plants for biological activity. *Ind. J. Exp. Biol.*, 6: 232.

Folin, J., 1922. *J. Biol. Chem.*, 51: 377.

Hugget, A.G. and Nixon, D.A., 1957. *Lancet*, 2: 368.

King, J., 1959. *J. Med. Lab. Tech.*, 17: 1.

King, P.R.N. and King, E.J., 1954. *J. Clin. Path.*, 7: 322.

Kohli, K.R., Naik, S., Tripathi, R., Zaraptkar, M. and Kenkare, D., 1993. *The J. Gen. Med.*, 6(1): 47.

Lal, B.N. and Choudhary, K.D., 1968. Observations on *Momordica charantia* (Karvellaka) and Jambu as antidiabetic remedies. *J. Fes. Ind. Med.*, 2(2): 161.

Malloy, H.T. and Evelyn, K.A., 1937. *J. Biol. Chem.*, 119: 481.

Marles, J.R. and Farsworth, R.N., 1988. Plants as sources of antidiabetic agents. *Economic and Medicinal Plant Research*, 6: 149.

Mukherji, B., 1957. Indigenous Indian drugs used in the treatment of diabetes. *J. Sci. Industr. Res.*, 16 A Supplement 1: 1.

Nadkarni, K.M., 1954. *Indian Materia Medica*. Popular Book Depot, Bombay, pp. 544.

Razzak, H.M.A., 1993. Therapeutic value of Indian medicinal plants. In: *Biotechnological Application for Food Security in Developing Countries*, (Ed.) H.C. Shrivastava. Oxford and IBH Publishing Co. Pvt. Ltd., New Delhi, pp. 335.

Shrotri, D.S., Kelkar, M., Deshmukh, V.K. and Aiman, R., 1963. Investigations of the hypoglycaemic properties of *Vinca rosea, Cassia auriculata* and *Eugenia jambolana. Ind. J. Med. Res.*, 51(3): 464.

Shukla, R., Sharma, S.B., Puri, B., Prabhu, K.M. and Murthy, P.S., 2000. Medicinal plants for treatment of diabetes mellitus. *Indian J. Clin. Biochem.*, 15 (Suppl.): 169.

Singh, N., Tyagi, S.D., Agarwal, S.C., 1991. Antidiabetic effect of ethanolic extract of *Syzygium cumini* seed powder. *Biosphere*, 3(1): 29.

Stanley, M.P.P., Menon, V.P. and Pari, L., 1998. Hypoglycaemic activity of *S. cumini* seeds. Effect on lipid peroxidation in alloxan diabetic rats. *J. Ethanopharmacol.*, 61: 1.

Van Handel, E. and Zilversmith, D.B., 1957. *J. Lab. Clin. Med.*, 50: 152.

Varley, H., 1969. *Practical Clinical Biochemistry*. ELBs and HMC Ltd. White Bridge Press, London, pp. 236.

Zafar, R., 1994. *Medicinal Plants of India*. CBS Publishers and Distributors, New Delhi, pp. 105.

Zlatkis, A., Zak, B. and Boyle, J., 1953. A new method for direct determination of serum cholesterol. *J. Lab. Clin. Med.*, 41: 481.

Chapter 38

Variability and Heritability in Castor (*Ricinus communis* L.)

S. Rajamani*
Scientist (Plant Breeding), Agricultural Research Station, Lam – 523 247, Andhra Pradesh

ABSTRACT

Seven genotypes were evaluated for genetic variation studies for six characters *viz.*, plant height, days to 50 per cent flowering, nodes up to primary spike, spike length, number of nodes up to primary spike, number of spikes and yield. The variance analysis revealed that the high genetic variability present in the quantitative characters studied. Yield, days to 50 per cent flowering and plant height has high GCV, PCV, heritability and genetic advance reveals prevalence of additive gene action and simple selection is effective for improvement of these characters.

Keywords: *Castor, Genetic variance, Heritability, Genetic advance.*

Introduction

The two basic requirements of plant breeding are presence of genetic variation and exploitation of variation through selection. Presence of genetic variation in a character is a must for any crop improvement in that character. The nature and magnitude of genetic variability in the population is of most valuable for executing efficient breeding programme to improve the yield potential of the genotypes. Genotypic coefficient of variation in addition with other heritable estimates are useful for improvement

* Present Address: Scientist (Br.), Cotton Section, RARS, Lam, Guntur – 522 034

in genotypic values over the population. Knowledge on heritability with genetic advance will be of much value for selection. Hence, considering these aspects genetic variability studies were carried with diverse castor genotypes.

Materials and Methods

Field experiment was conducted with seven castor genotypes were sown in randomized block deign with three replications at Agricultural Research Station, Darsi. The data was recorded on characters *viz.*, plant height, days to 50 per cent flowering, nodes up to primary spike, spike length (cm), number of spikes and yield (Kg/ha). Data was subjected to statistical analysis and various genetic parameters such as GCV, PCV, heritability and genetic advance were estimated as per Johnson *et al.*, 1995 and Hanson 1963.

Results and Discussion

Analysis of variance revealed that highly significant differences among the genotypes for seed yield and other characters shows considerable genetic variation in the genotypes (Table 38.1). Study of genetic parameters (Table 38.2) revealed that plant height, yield, days to 50 per cent flowering and spike length (cm) had high GCV and PCV values indicating that these characters are under the influence of genetic control. Moderate GCV and PCV values were observed for spike length and number of nodes up to primary spike.

Table 38.1: Analysis of Variance for Seed Yield and Yield Attributing Characters in Castor

Sl.No.	Character	Mean Sum of Squares	
		Replications	Treatments
1.	Plant height	7.64	82.18*
2.	No. of nodes up to primary spike	0.20	2.12*
3.	Spike length (cm)	4.08	28.22*
4.	Days to 50 per cent flowering	1.71	112.63*
5.	No. of spikes	0.038	4.513*
6.	Yield (Kg/ha)	1225	181753*

*: Significant at 5 per cent level.

Table 38.2: Mean, Range, GCV, PCV, Heritability, Genetic advance and Genetic Advance as per cent Over Mean for Seed Yield and Yield Attributes in Castor Genotypes

Sl.Nc.	Character	Mean	Range	GCV	PCV	Heritability	Genetic Advance	GA as % Over Mean
1.	Plant height	50.50	40.80–60.40	26.08	29.18	0.89	27.02	53.51
2.	No. of nodes up to primary spike	14.08	12.20–15.20	10.75	11.57	0.92	3.08	21.94
3.	Spike length (cm)	24.73	24.20–33.40	17.58	21.35	0.81	9.01	36.43
4.	Days to 50% flowering	50.42	42–60	37.00	37.66	0.98	38.33	76.03
5.	No. of spikes	5.23	3.30–7.80	13.85	15.29	0.90	1.48	28.35
6.	Yield (Kg/ha)	1564.28	1050–1920	1900.7	2008.75	0.94	60846.52	3890.44

The phenotypic coefficient of variation was higher than GCV indicating that presence of environmental influence in the expression of these characters studied. Since variability estimates are influenced by environment the use of genetic material and their utility is limited unless we study the genetic advance in association with heritability.

High heritability and low genetic advance was observed in number of spikes, number of primary spikes and spike length indicating that there is prevalence of non additive gene action. The high heritability is being exhibited due to favourable influence of environment rather than genotype. Selection for such traits may not be rewarding. Moderate heritability and high genetic advance was observed in plant height indicating that the character is governed by additive gene action. The low heritability is being exhibited due to high environmental efforts. Selection may be effective in such cases. The characters yield and days to 50 per cent flowering shows high heritability and high genetic advance indicates that most likely that the heritability is due to additive gene effects and selection may be effective.

From these discussions, it can be concluded that high GCV, FCV, heritability and genetic advance as per cent over mean were observed for yield, days to 50 per cent flowering and plant height indicating that prevalence of additive gene action in the control of these characters and simple selection may be effective to improve these characters. These results are in accordance with Reddy *et al.* (1999).

References

Hanson, W.O., 1963. *Heritability in Statistical Genetics and Plant Breeding.*

Hanson, W.O. and Robinson, H.F (Ed), 1982. National Academy of Science and National Research Council, Washington, D.C.

Johnson, H.W., Robinson, H.F. and Comstock, R.E., 1955. Estimates of genetic and Environmental variability in soybean. *Agronomy Journal*, 47: 314–318.

Reddy, F.R., Sankar, G.R.M., Vanaja, M., Rao, C.M., Venkateswarulu, S. and Eastin, J.D., 1999. Heritability and genetic advance in castor (*Ricinus communis*) over varied soils and dates of sowing. *Indian J. Agri. Sci.*, 69(2): 371–375.

Chapter 39

Effect of Potash and Zinc Sulphate on Greengram

S.V. Patil, A.V. Solanke, T.K. Tiwari and B.J. Shete
Department of Agronomy, Mahatma Phule Krishi Vidyapeeth, Rahuri – 413 722, India

ABSTRACT

A field experiment was conducted during 2003–2004 at Rahuri, to study the effect of potash and zinc on greengram. Both potash and zinc affected the seed yield and yield attributes significantly. Potash @ 25 kg ha^{-1} and zinc 15 kg ha^{-1} were found optimum.

***Keywords**: Greengram, Zinc, Potash, Yield.*

Introduction

Among short duration pulses, greengram has great potential for crop intensification and diversification both under rainfed and irrigated conditions. It is recognized as excellent catch crop, cover crop and intercrop. Being photo-insensitive it is cultivated throughout the country in different seasons, however its productivity continues to be low due to various biotic and abiotic stresses. Among, abiotic stresses no or inadequate use of plant nutrients, particularly potash and zinc is prominent. A significant response of applied K has been reported by Duke *et al.* (1980) and that of zinc sulphate by Singh *et al.* (2001). However, the study on combined effect of potash and zinc simultaneously has not largely been studied particularly in Maharashtra. The present experiment was therefore planned to study the effect of varying levels of potash and zinc on yield and agro-physiological parameters of green gram in clayey soils of western Maharashtra.

Materials and Methods

The investigation was carried out at experimental research farm of Mahatma Phule Krishi Vidyapeeth, Rahuri (M.S.), for rainy season of 2003–2004 in clayey soil. It had 178 kg ha^{-1} available nitrogen, 16.8 kg ha^{-1} available phosphorus, 448 kg ha^{-1} available potassium and 0.42 mg kg^{-1} zinc. The study was carried out in factorial randomized block design. The treatments consisted of 5 levels of potash (0, 12.5, 25.0, 37.5 and 50.0 kg ha^{-1}) and 2 levels of zinc (0 and 15 kg ha^{-1}), thus forming 10 treatment combinations replicated 3 times. Muriate of potash and zinc sulphate were the source of potash and zinc, respectively. Full amount of potash and zinc according to treatments was applied at planting below the seed. Nitrogen @ 25 kg ha^{-1} and phosphorus @ 50 kg ha^{-1} were also applied basal at planting below the seed. FYM @ 5 t ha^{-1} was applied to all the plots during land preparation. Cultivar 'Vaibhav' of greengram was planted on 24th July whereas final picking of pods was done on 12th October. The crop received total precipitation of 193 mm (14 rainy days) during the crop season. Other agronomical practices were performed as per the recommendations of crop treating the field homogeneous.

Results and Discussion

Effect of Potash on Yield and Yield Attributes

Potash influenced the number of branches per plant significantly and it registered significant response over control. Doses of 25, 37.5 and 50 kg ha^{-1} however did not differ to each other statistically. Number of pods per plant was also influenced significantly by potash. Doses of 25, 37.5 and 50 kg ha^{-1} being at par to each other registered significantly higher number of pods per plant over the control. Seed yield per plant was influenced significantly and again both the doses, *i.e.* 25, 37.5 and 50 kg ha^{-1} being at par registered a significant response over the control. The 1000 seed weight and harvest index were influenced by potash significantly. Potash @ 25 kg ha^{-1} proved as good as 37.5 and 50 kg K_2O ha^{-1} for yield attributing characters (Table 39.1). The significant response of potash on seed yield of greengram was restricted only up to 37.5 kg ha^{-1} but application of 25 kg K_2O ha^{-1} yielded statistically identical seed yield of greengram. The restricted response of potash in the present study on yield and yield attributes may be due to higher potassium status of soil. Response of potash on greengram upto 20 kg ha^{-1} has also been reported by Saxena *et al.* (1996).

Table 39.1: Effect of Potash and Zinc Sulphate on Yield and Yield Components of Green Gram

Treatment	Branches per Plant	Pods per Plant	Seeds Yield per Plant (g)	1000 Seed Weight (g)	Harvest Index (%)	Seed Yield (q ha^{-1})
Potash levels (kg ha^{-1})						
K_0 : 0	2.81	9.80	2.62	44.33	39.20	7.49
K_1 : 12.5	2.95	10.90	2.81	44.55	39.97	8.07
K_2 : 25.0	3.28	13.40	3.46	47.97	39.85	8.98
K_3 : 37.5	3.48	13.83	3.61	49.62	39.33	9.23
K_4 : 50.0	3.38	13.60	3.52	48.43	39.85	9.10
S.Em. ±	0.10	0.29	0.10	1.10	0.48	0.17
CD at 5 per cent	0.30	0.86	0.30	3.29	N.S.	0.51

Contd...

Table 39.1–Contd...

Treatment	Branches per Plant	Pods per Plant	Seeds Yield per Plant (g)	1000 Seed Weight (g)	Harvest Index (%)	Seed Yield (q ha^{-1})
			Zinc sulphate (kg ha^{-1})			
Z_0 : 0	3.04	11.33	2.95	45.91	39.50	8.12
Z_1 : 15	3.32	13.28	3.45	48.04	39.78	9.03
S.Em. ±	0.06	0.18	0.06	0.70	0.30	0.11
CD at 5 per cent	0.19	0.54	0.19	2.08	N.S.	0.32

Effect of Zinc on Yield and Yield Attributes

Significant effect of zinc was noticed on the number of branches per plant, number of pods per plant, seed yield per plant and 1000 seed weight. The effect on harvest index was not significant. Application of 15 kg Zn ha^{-1} recorded significantly the highest number of branches per plant, pods per plant, seed yield per plant and 1000 seed weight (Table 39.1). The response of greengram crop to the application of 15 kg Zn ha^{-1} was positive and the seed yield obtained in these plots was significantly more than the control without zinc application. The beneficial effects of zinc may be attributed to its role in metabolic activities mainly in protein synthesis in plants as well as in N fixation. The results confirm the findings of Singh *et al.* (2001).

References

Duke, S.H., Collings, M. and Solenlaske, R.M., 1980. Effect of K-fertilization on N-fixation, node enzyme of nitrogen metabolism in alfalfa. *Crop Sci.*, 20: 213–219.

Saxena, K.K., Varma, H.R. and Saxena, H.K., 1996. Effect of phosphorus and potassium on greengram (*Phaseolus radiatus*). *Indian J. Agron.*, 41: 81–87.

Singh, R.C., Singh, Mehar, Kumar, Rakesh, Singh, M. and Kumar, R., 2001. Response of mungbean [*Vigna radiata* (L.) Wilczek] to phosphorus application with and without FYM and $ZnSO_4$. *Haryana Agric. Univ. J. Res.*, 31: 89–92.

Chapter 40

Efficacy of *Hyptis suaveolens* (Lamiaceae) and *Melochia chorcorifolia* (Sterculiaceae) Fractions Against *Pericalia ricini* (Lepidoptera : Noctuidae)

K. Elumalai, C. Prabhahar and S. Dhanasekaran
Department of Zoology, Annamalai University, Annamalai Nagar – 608 002, Tamil Nadu, India

ABSTRACT

Effective crude extracts of *Hyptis suaveolens* and *Melochia chorcorifolia* were further fractionated and tested against *Helicoverpa armigera*. Fraction I of diethylether extract of *H. suaveolens* was showed maximum antifeedant activity. Fraction III of ethyl acetate extract of *H. suaveolens* was found to have significant ovipositional deterrent and ovicidal activity. Fraction II of methanol extract of *H. suaveolens* showed increased larvicidal activity than *M. chorcorifolia*.

Keywords: *Plant extracts, Hyptis suaveolens and Melochia chorcorifolia, Pericalia ricini.*

Introduction

Castor defoliator *Pericalia ricini* is a serious pest of castor in the Indian subcontinent and south east Asia (Reed and Pawar, 1982; King, 1994). Widespread appearance of resistance to chemical insecticides, including the widely used pyrethroids, in the late 1980s caused an increase in losses to this pest and has made control by chemicals increasingly unreliable and expensive (Armes *et al.*, 1992). In recent years the chemical resistance problem has worsened, resulting in estimated crop losses due to this pest in India in 1997–1998 reaching 158 million US$ (Russell, 1999) and causing widespread hardship especially amongst poorer farmers. This has stimulated efforts to develop alternative control methodologies including the use of chemicals derived from plants (botanicals) (Abdul Kareem, 1999). Botanicals have the advantage of being biodegradable without causing pollution and also safer to the environment (Cunat *et al.*, 1990). Use of the pesticides that are least harmful to beneficial organisms can potentially ameliorate this problem (Jepson, 1989; Croft, 1990; Abdul Kareem 1999). The objective of the present study was to evaluate antifeedant, insecticidal, oviposition deterrent and ovicidal activity of different fractions of *H. suaveolens* (Lamiaceae) and *M. chorcorifolia* (Sterculiaceae) against *P. ricini.*

Materials and Methods

Collection and solvent extraction of selected plants: *Hyptis suaveolens* and *Melochia chorcorifolia* leaves were collected and shade dried in 15 days under 30 ± 2°C in the laboratory, department of zoology, after 15 days the plant materials were powdered using electric blender and sieved through kitchen strainer. 100 g of the plant powder was loaded in glass column (60 cm height and 3.3 cm diameter) and extracted with increasing polarity of the solvents such as hexane (HE), diethylether (DEE), dichloromethane (DCM), ethylacetate (EA), methanol (ME) and water. The effective crude extracts were identified and fractionated using TLC system (Tables 40.1 and 40.2). Further more sufficient quantity of the fractions were obtained from the corresponding crude extracts, condensed *in vacuo* at 50° C and the solvent free fractions were tested for their efficacy against *P. ricini* procured from our stock culture.

Table 40.1: Crude Extracts of *Hyptis suaveolens* (*Hs*) and *Melochia chorcorifolia* (*Mc*) Tested Against *Pericalia ricini*

Solvents Used for Crude Extraction	*Antifedancy*		*Ovipositional*		*Ovicidal*		*Larvicidal*	
	Hs	*Mc*	*Hs*	*Mc*	*Hs*	*Mc*	*Hs*	*Mc*
Hexane								
Dietheyl ether	+	+		+				+
Dichloromethane								
Ethyl acetate			+		+	+		
Methanol							+	
Water								

+: Significant activity was recorded from the plant crude extracts.

Antifeedant Activity

Antifeedant activities of the fractions were studied using leaf disc no choice method (Isman *et al.*, 1990). Fresh castor leaf discs of 3 cm diameter were punched using cork borer and dipped in 250, 500, 1000 and 2000 ppm concentration of plant extracts. Respective solvent and without treatment were considered as negative control. In each plastic petridish (1.5 cm × 9 cm) wet filter paper was placed to avoid early drying of the leaf disc and single fourth instar larva of *P. ricini* was introduced. Progress consumption of leaf area by the larva after 24 hrs was recorded in control and treated discs using leaf area meter. Leaf area consumed in plant extract treatment was corrected from the solvent control leaf. Five replicates were maintained for each concentration and percentages of antifeedant index was calculated (Blanney *et al.*, 1984).

Table 40.2: Number of Fractions Identified and Isolated from the Selected Plants Using TLC and Column Chromatography

Solvents	*H. suaveolens*		*M. chorcorifolia*	
	NF	*Solvents Ratio*	*NF*	*Solvents Ratio*
Diethyl ether	4	(Benzene : Chloroform–1 : 1)	3	(Benzene : Chloroform– 1 : 1)
Ethyl acetate	4	(Chloroform : Methanol–9.5 : 0.5)	5	(Chloroform : Methanol–9.5 : 1)
Methanol	3	(Benzene : Chloroform–1 : 1)	4	(Benzene : Chloroform : Methanol–1 : 1 : 1)

NF: Number of fractions obtained.

Oviposition Deterrent Activity

Oviposition deterrent activities of the fractions were studied at 250, 500, 1000 and 2000 concentration. They were individually sprayed on fresh cotton plants (cultured in pots) along with selected controls as mentioned in antifeedant activity and placed inside the cage (60 cm × 45 cm × 45cm) covered with mosquito net. Ten pairs of *P. ricini* moths were introduced in a cage and 10 per cent (w/v) sucrose solution with multivitamin drops was provided for adult feeding. After 48 hrs the number of eggs laid on treated and control leaves were recorded and the percentage of oviposition deterrence was calculated (Williams, 1986).

Ovicidal Activity

P. ricini eggs were collected from the stock culture and 100 eggs were separated in 5 lots and dipped in selected concentrations of the fractions as mentioned earlier. Number of hatched eggs in control and treatments were recorded and the percentage of ovicidal activity was calculated using corrected Abbott's formula (Abbott, 1925).

Insecticidal Activity

Fresh cotton leaves were treated with 250, 500, 1000 and 2000 concentration of fractions and the controls were provided as that of antifeedant studies. Fourth instar larvae of *P. ricini* were introduced individually in petridish (size as mentioned in antifeedant studies). Five replicates were maintained for all the concentration and the number of dead larvae was recorded after 48 hrs. Percentage of larval mortality was calculated and corrected by Abbott's formula (Abbott, 1925).

Results and Discussion

Antifeedant activity of diethylether extract fractions of *H. suaveolens* and *M. chorcorifolia* against *P. ricini* are presented in Table 40.3. Among the four fractions obtained from *H. suaveolens,* fraction III was found to inhibit the feeding ratio of the *P. ricini* and it is apparent from the Table. Moreover, in all concentrations (250–2000 ppm) the same fraction showed increased antifeedant activity over the other fractions. While in *M. chorcorifolia* only three fractions have been obtained, among them fraction II was found to induced more feeding deterrent activity in all the concentrations with the maximum antifeedant activity at 2000 ppm concentration. Besides, the antifeedant activity of fractions 1 and 2 are less significant in 500–2000 ppm as evidenced from the table (LSD; P = 0.005). In the recent past environmental degradation due to consecutive application of chemicals to control the various insect pest, threatened the scientific community worldwide. The use of botanicals are one of the components of Integrated Pest Management (IPM) proved to be the promising tool in controlling the storage as well as the field pests. In the present study indicates that the active principles present in the plants dissolved in this solvent may inhibit the larval feeding behavior or makes the food unpalatable. Earlier Hermawan *et al.* (1994) reported that the plant extract of *Andrographis paniculata* effectively reduced the feeding of fourth stadium diamond black moth larvae (*Plutella xylostella*). Further antifeedant agent from the neem seed kernel extract (NSKE) reduced feeding rate of *H. armigera* resulted in lowest pod and grain damage (Sarode *et al.*, 1995; Shanower *et al.*, 1997). Substances present in the various fractions of both the leaf extracts showed significant antifeedant activity against the fourth instar larvae of *P. ricini.* This finding is substantiated not only by the low consumption of the treated leaf disks, but also by the expression of some morphological deformities such as pupa with irregular shape, size and adult moths with deformed wings. The active antifeedant substances present in *H. suaveolens* and *M. chorcorifolia* could play an important role for managing *P. ricini.* In addition they do not pose risks to humans and other mammals since the leaves of *H. suaveolens* are used as an anticancer (Mabberley, 1990) and antifertility agent (in females) (Oliver-Bever, 1986), for the treatment of colics, stomach ache, and fever (Dalziel, 1937) and also possess antimicrobial activity (Asekun, 1999). Where as *M. chorcorifolia* is used as fodder to cattle.

Table 40.3: Antifeedant Activities of Different Fractions of Selected Plants Against Fourth Instar Larvae of *Pericalia ricini*

Fractions Tested	*Hyptis suaveolens (DEE)*				*Melochia chorcorifolia (DEE)*			
	Concentrations (ppm)							
	250	*500*	*1000*	*2000*	*250*	*500*	*1000*	*2000*
Fraction I	20.64c (26.99)	34.35bc (35.85)	54.69bc (47.64)	76.77a (61.14)	16.5bc (23.97)	26.25c (30.79)	37.15c (37.52)	52.51[b] (46.43)
Fraction II	18.85cd (25.7)	30.7cd (33.65)	47.35d (43.45)	65.64d (54.09)	27.2b (31.44)	38.5b (38.35)	51.6b (49.52)	70.85 (57.29)
Fraction III	27.63b (31.69)	42.65b (40.74)	63.74b (52.95)	83.43c (65.96)	19.35b (26.06)	28.6c (32.33)	40.2c (39.35)	55.56 (48.16)
Fraction IV	24.25b (29.47)	39.45b (38.88)	57.63b (49.37)	61.8d (51.83)	– –	– –	– –	–
Azadirachtin 1% (2ml/l)	74.64a (59.74)	74.64a (59.74)	74.64a (59.74)	74.64a (59.74)	74.64a (59.74)	74.64a (59.74)	74.64a (59.74)	74.64 (59.74)

Values represent means of five replications and parenthesis hold angular transformed values. Within the column different alphabets are statistically significant as per LSD test (P = 0.005). –: Indicates fractions were not obtained.

The ovipositional deterrent activity of *H. suaveolens* and *M. chorcorifolia* against *P. ricini* are presented in Table 40.4. Ethylacetate extract of *H. suaveolens* yielded four fractions and diethylether extract of *M. chorcorifolia* yielded five fractions. In *H. suaveolens* fraction III showed increased ovipositional deterrent activity against the gravid moths of *P. ricini* at 2000 ppm concentration, when compared to the control the activity was less but in comparison with other fraction it showed significant increase as evidenced from the Table 40.4. The ovipositional deterrent activity of different fractions of *M. chorcorifolia* revealed that, fraction II was found to have more deterrent activity against the gravid moths of *P. ricini* and their significance are apparent in 1000 and 2000 ppm concentrations. In the same way the oviposition of *P. ricini* was significantly deterred by Fraction III of ethyl acetate extract of *H. suaveolens* and fraction II of diethyl ether extract of *M. chorcorifolia*. It may due to the consequence volatile present in the solvent extract which makes malfunctioning of the ovariole in female moths. Earlier, Hemawan *et al.* (1998) reported that the leaf extract of *Andrographis paniculata* significantly reduced the egg laying performance of the diamond back moth *Plutella xylostella*. Our findings are also corroborating with the reports of Jeyabalan and Murugan (1997), Patel *et al.* (1994) and Mehta *et al.* (1994).

Table 40.4: Ovipositional Deterrent Activity of Different Fractions of Selected Plants Against *Pericalia ricini*

Fractions Tested	*Hyptis suaveolens (EA)*				*Melochia chorcorifolia (DEE)*			
	Concentrations (ppm)							
	250	*500*	*1000*	*2000*	*250*	*500*	*1000*	*2000*
Fraction I	24.66b (29.73)	30.75b (33.65)	36.44c (37.11)	42.66d (40.74)	9.5bc (17.95)	17.55c (24.73)	29.3c (32.77)	42.55 (40.69)
Fraction II	19.5b (26.21)	33.65b (35.43)	37.6c (37.82)	47.75d (43.68)	13.65b (21.64)	27.65b (31.69)	42.7b (40.51)	55.61[b] (48.22)
Fraction III	15.75bc (23.34)	26.8c (31.18)	49.5b (44.71)	64.73d (53.55)	8.75c (17.15)	14.66c (24.46)	27.7cd (31.76)	38.85 (38.23)
Fraction IV	20.7b (27.06)	27.6bc (31.69)	32.2c (34.57)	40.61de (39.58)	17.5b (24.73)	23.56b (29.00)	32.47c (34.7)	47.25 (43.39)
Fraction V	– –	– –	– –	– –	6.75c (15.00)	12.8cd (20.96)	25.95cd (30.59)	35.77 (36.69)
Azadırachtin 1% (2ml/l)	65.88a (54.21)	65.88a (54.21)	65.88a (54.21)	65.88a (54.20)	63.75a (52.95)	63.75a (52.95)	63.75a (52.95)	63.75 (52.95)

Values represent means of five replications and parenthesis hold angular transformed values. Within the column different alphabets are statistically significant as per LSD test (P = 0.005). –: Indicates fractions were not obtained.

Ovicidal activity against the eggs of *P. ricini* by the four fractions obtained from the ethylacetate extract of *H. suaveolens*. As per the data fraction III showed statistically significant ovicidal activity. It is also interesting to note that the same fraction was exerted strong ovipositional deterrent activity. Whereas, ethylacetate extract of *M. chorcorifolia* yielded five fractions of which, fraction III showed increased ovicidal activity over the other fractions at higher concentration (Table 40.5). It is interesting to note that fraction III of *H. suaveolens* and *M. chororifolia* are promising in their ovicidal activity against *P. ricini*. This might be due to interference with the egg contour and cytoplasmic disintegration. The disturbance of these extracts with egg morphology may plug the micropyles of the chorion thereby preventing the airflow in and out vice versa. The disturbances with egg cytoplasm was reflected in the form of dead eggs with black spot stage and it seems to be arresting of further development of embryo inside the egg. Bhatnagar and Sharma (1994) noticed similar anatomical and physiological disturbances of plant extracts on maize stem borer, *Chilo partellus*.

Methanol extract of *H. suaveolens* yielded three fractions and their larvicidal activity against the fourth instar larvae of *P. ricini* is presented in Table 40.6. Larvicidal activity obtained from the fraction II was significant than the other two fractions of the same extract. The larvicidal activity exhibited by the five fractions of diethyl ether extract of *A. chorcorifolia* is shown in Table 40.6. It is clear that fraction

III at 1000 and 2000-ppm concentration was found to have increased activity than the other fractions. Methanol extract of *H. suaveolens* (fraction II) was found to have significant larvicidal activity against the fourth instar larvae of *P. ricini.* Whereas fraction III of diethylether extracts of *M. chorcorifolia* showed increased mortality against *H. armigera.* Larvicidal activity was mainly depended on the presence of toxic materials in the plant extracts. Both the plants selected in the present study is non toxic and the plant materials are used for cattle feed in many Indian villages. Further synergistic studies may enhance the. insecticidal activity of the plant extracts and reduced the concentration of pesticides. Manoharan and Uthamasamy (1993) reported that addition of *A. indica* oil to endosulfan and phosalone increased the mortality of larvae compared with the insecticides alone. Rajasekhar *et al.* (1996) reported that *Parthenium* leaf extract in combination with endosulfan and cypermethrin used for the control of *H. armigera.* In addition Durairaj and Venugopal (1995) were also reported that neem oil 2 per cent + Diflubenzuron 0.03 per cent was found to have significant antifeedant effect and early larval mortality in *H. armigera.* Jaglan *et al.* (1997) reported that the neem seed and leaf extracts of chloroform : methanol (9 : 1) extract was the most promising in causing larval mortality with adverse morphogenic effects on various biological parameters of *H. armigera.* In our studies we used only one solvent at time for extraction. Further combination of more than solvents in certain ratio may enhance the larvicidal activity of the selected plants. Present study methanol extract showed some insecticidal activity it may due to reduction in the total protein content which is a major component for the metamorphosis of the larval instars, this was clear that the dead larvae showed the symptom of improper metamorphosis from one instar to another instar. This result is also coinciding with the findings of Krishnayya and Rao (1995) who had been reported that the application of plumbagin greatly reduced the protein concentration of *H. armigera.* In the present study *H. suaveolens* and *M. chorcorifolia* exhibited one or more statistically significant activity against *P. ricini.* It may include IPM program for controlling this important lepidopteran pest. The present findings further suggest the value of exploring other Indian plants in search for new, environmentally acceptable pest control agents for *P. ricini.*

Table 40.5: Ovicidal Activity of Different Fractions of Selected Plants Against the Eggs of *Pericalia ricini*

Fractions Tested	*Hyptis suaveolens (EA)*				*Melochia chorcorifolia (EA)*			
	Concentrations (ppm)							
	250	*500*	*1000*	*2000*	*250*	*500*	*1000*	*2000*
Fraction I	6.25c (14.42)	10.6c (19)	15.45c (23.11)	20.83cd (27.13)	8.65bc (17.05)	14.3b (22.22)	20.5cd (26.92)	24.45 (29.6)
Fraction II	4.9c (12.79)	12.75c (20.88)	18.35c (25.33)	23.71c (29.13)	10.15b (18.53)	15.45b (23.11)	28.88b (32.46)	30.44 (33.46)
Fraction III	16.85b (24.2)	20.35b (26.78)	23.64b (29.06)	42.53b (40.69)	8.9b (17.36)	13.85b (21.81)	24.44c (29.6)	37.55 (37.76)
Fraction IV	5.1c (13.05)	9.5c (17.95)	16.77c (24.12)	25.44c (30.26)	12.5b (20.7)	15.5b (23.18)	21.25c (27.42)	27.9c (31.88)
Fraction V	– –	– –	– –	– –	7.5b (15.89)	12.55b (20.7)	18.6cd (25.55)	23.88 (29.2)
Azadirachtin 1% (2ml/l)	74.88a (59.87)	74.88a (59.87)	74.88a (59.87)	74.88a (59.87)	73.82a (59.21)	73.82a (59.21)	73.82a (59.21)	73.82 (59.21)

Values represent means of five replications and parenthesis hold angular transformed values. Within the column different alphabets are statistically significant as per LSD test (P = 0.005). –: Indicates fractions were not obtained.

Table 40.6: Larvicidal Activities of Different Fractions of Selected Plants Against Fourth Instar Larvae of *Pericalia ricini*

Fractions Tested	*Hyptis suaveolens (ME)*				*Melochia chorcorifolia (DEE)*			
	Concentrations (ppm)							
	250	*500*	*1000*	*2000*	*250*	*500*	*1000*	*2000*
Fraction I	20.13b (26.64)	26.88bc (31.18)	30.19bc (33.27)	63.82b (53.01)	10.03b (18.43)	20.06b (26.57)	30.09b (33.21)	53.48[b] (46.95)
Fraction II	16.81b (24.2)	36.94b (37.41)	40.26b (39.35)	77.2a (61.48)	6.72b (15.00)	13.34b (21.39)	33.41b (35.3)	60.2[b] (50.89)
Fraction III	13.48b (21.47)	47.01b (43.28)	50.33b (45.17)	53.75b (47.12)	0bc (0.00)	16.75b (24.12)	36.82b (37.35)	63.51 (52.83)
Fraction IV	– –	– –	– –	– –	3.31b (10.47)	10.03bc (18.43)	30.1b (33.27)	53.47[b] (46.95)
Fraction V	– –	– –	– –	– –	5.33b (13.31)	8.47bc (16.85)	15.89c (23.4)	33.41 (35.3)
Azadirachtin 1% (2ml/l)	79.62a (63.15)	79.62a (63.15)	79.62a (63.15)	79.62a (63.15)	78.94a (62.65)	78.94a (62.65)	78.94a (62.65)	78.94 (62.65)

Values represent means of five replications and parenthesis hold angular transformed values. Within the column different alphabets are statistically significant as per LSD test (p = 0.005). –: Indicates fractions were not obtained.

References

Abbott, W.S., 1925. A method of computing the effectiveness of an insecticide. *J. Econ. Entomol.*, 18: 265–266.

Abdul Kareem, 1999. Biopesticides and insect pest management. In: *Biopesticides in Insect Pest Management*, (Eds.) Ignacimuthu, S., Sen, A. Phonix Publishing House Pvt. Ltd., New Delhi, p. 1–6.

Annes, N.J., Jadhav, D.R., Bond, G.S., and A.B.S. King, 1992. Insecticide resistance, in *Helicoverpa armigera* in south India. *Pesticide Sci.*, 34: 35–34.

Asekun, O.T., 1999. Antimicrobial activity of the essential oil of *Hyptis suaveolens* leaves. *Fitoterapia*, 70: 440–442.

Bhatnagar, A. and V.K. Sharma, 1994. Insecticidal and ovicidal activities of plant extracts against maize stem borer, *Chilo partellus* (Swinhoe). *Plant Protect. Bull.*, 46: 12–16.

Blaney, W.M., M.S.J. Simmonds, S.V. Evans and L.E. Fellows, 1984. The role of the secondary plant compound 2,5-dihydroxymethyl 3,4-dihydroxypyrrolidine as a feeding inhibitor for insects. *Entomol. Exp. Appl.*, 36: 200–216.

Cunat, P., Primo, E., Sanz, I., Garcera, M.D., March, M.C., Bowers, W.S. and R. Martinez Pardo, 1990. Biocidal activity of some Spanish Mediterranean plants. *J. Agric. Food Chem.*, 38: 497–500.

Dalziel, J.M., 1937. *The Useful Plants of West Tropical Africa.* The Crown Agents for the Colonies, London, pp. 461.

Durairaj, C. and M.S. Venugopal, 1995. Combined effect of margosa oil and diflubenzuron on tobacco caterpillar (*Spodoptera litura*) gram pod borer (*Helicoverpa armigera*). *Ind. J. Agric. Sci.*, 65: 301–304.

Hermawan, W., Kajiyama, R. and S. Tsukuda, 1994. Antifeedant and anti oviposition activities of the fractions of extract from a Tropical plant, *Andrographis paniculata* (Acanthaceae), against the diamondback moth, *Plutella xylostella* (Lepidoptera : Yponomeutidae). *Appl. Entomol. Zool.*, 29: 533–538.

Hermawan, W., Ritsuko, T., Shuhei, N., Kenji, F. and N. Fusao, 1998. Oviposition, deterrent activity of *Andrographolide* against the diamondback moth (DBM), *Plutella xylostella* (Lepidoptera : Yponomeutidae). *Appl. Entomol. Zool.*, 33: 239–241.

Isman, M.B., Koul, O., Luczynski, A. and J. Kaminski, 1990. Insecticidal and antifeedant bioactives of neem oil and their relationship to azadirachtin content. *J. Agric. Food Chem.*, 38: 1406–1411.

Jaglan, M.S., Khokhar, K.S., Malik, M.S. and R. Singh, 1997. Evaluation of neem (*Azadirachta indica* A. Juss) extracts against American bollworm, *Helicoverpa armigera* (Hubner). *J. Agric. Food Chem.*, 56: 114–129.

Jepson, P.C., 1989. *Pesticides and Non-Target Invertebrates.* Intercept, Wimborne, Dorset, UK.

Jeyabalan, D. and K. Murugan, 1997. Effect of neem limonoids on feeding reproduction of *Helicoverpa armigera* (Hubner) (Lepidoptera : Noctuidae). *Entomon.*, 22: 15–20.

King, A.B.S., 1994. *Heliothis/Helicoverpa* (Lepidoptera : Noctuidae). In: *Insect Pests of Cotton,* (Eds.) Matthews, C.A., Tunstall, J.P. CAB International, Cambridge, p. 39–106.

Krishnayya, P.V. and P.J. Rao, 1995. Plumbagin effects on *Helicoverpa armigera* Hubner (Lepidoptera : Noctuidae) IV. Final instar haemolymph trehalose, cations nucleic acids. *Entomon.*, 20: 197–202.

Mabberley, D.J., 1990. *The Plant Book.* Cambridge University Press, London, pp. 209.

Manoharan, T. and S. Uthamasamy, 1993. Evaluation of plant products for synergistic action with insecticides for control of gram-pod borer (*Helicoverpa armigera*). *Ind. J. Agric. Sci.*, 63: 759–761.

Mehta, D.M., Patel, J.R. and N.C. Patel, 1994. Ovicidal ovipositional deterrent effect of botanicals individually in combination with endosulfan on *Helicoverpa armigera. Indian J. Plant Prot.*, 22: 215–216.

Oliver-Bever, B., 1986. *Medicinal Plants in Tropical West Africa.* Cambridge University Press, London, pp. 225.

Patel, M.S., Yadav, D.N. and A.B. Rao, 1994. Neemark (Azadirachtin) as its ovipositional deterrent against cotton pests. *Pestology*, 18: 17–19.

Rajasekhar, P., Rao, N.V. and M. Venkataiah, 1996. Utility of insecticide additives for managing *Helicoverpa armigera* on cotton. *Pesticide Res. J.*, 8: 119–123.

Reed, W. and C.S. Pawar, 1982. *Heliothis:* A global problem. In: *Proceedings of an International Workshop on Heliothis Management*, Patancheru, India, 15–20 November 1981, ICRISAT, pp. 9–14.

Russel, D., 1999. Insecticide resistance and its management in *Helicoverpa armigera* (Lepidoptera : Noctuidae) in India. In: *Proceedings of the 1998 Meeting of the Plant Protection Society*, Hyderabad, India, 22nd December.

Sarode, S.V., Jumde, Y. S., Deotale, R.O. and H.S. Thakare, 1995. Evaluation of neem seed kernel extract (NSKE) at different concentrations for the management of *Helicoverpa armigera* (Hub.) on pigeon pea. *Ind. J. Entomol.*, 57: 385–388.

Shanower, T.G., Yoshida, M. and Peter, A.J., 1997. Survival, growth, fecundity, behavior of *Helicoverpa armigera* (Lepidoptera : Noctuidae) on pigeonpea two wild *Cajanus* species. *J. Econ. Entomol.*, 90: 837–841.

Williams, A., Mitchell, L.E.R., Heath, R.R. and C.S. Barfield, 1986. Oviposition deterrents for fall army worm (Lepidoptera : Noctuidae) from larval frass, II, corn leaves, and artificial diet. *Environ. Entomol.*, 15: 327–330.

Chapter 41

Identification of Rice Root-Knot Nematode Resistant Cultivars for Eastern India

Lotan K. Bose, L. Behera, H.N. Subudhi and Sarat C. Sahu
Central Rice Research Institute, Cuttack, Orissa, India

ABSTRACT

One hundred twenty five rice cultivars/land races were evaluated for root-knot nematode in the net house. Only one land race, Kalarata and two elite cultivar Ramakrishna and Rasi showed highly resistant to root-knot nematode. Cultivars, such as Patnai-6, Daya, Udaya, Akasi, Annada and Kiran reported earlier as highly resistant/resistant to root-knot nematode are becoming susceptible in the present evaluation.

***Keywords:** Screening, Root-knot nematode, Rice varieties.*

Introduction

The root-knot nematode is one of the major constraints of upland rice production in India. Of the six species of the root-knot nematode affecting rice, *Meloidogyne graminicola* Golden and Birchfield (1968) is extensively present in the rice growing states of eastern India. Control of the nematode has been attempted through use of organic amendments, crop rotation or pesticides/nematicides but without much success. Deployment of host resistance in endemic areas seems to be a better alternative to control the pest. Although a number of varieties are reported resistance to the root-knot nematode, they do not show promising results in the field tests either due to lack of proper characterization or understanding about the nature of the resistance gene. So, it is high time to find out resistant varieties/ donors in order to develop nematode resistant varieties and popularization of varieties in the endemic areas. No work has yet been carried out to study the nature and mechanism of inheritance of the resistance gene. It is also highly essential to study the genetics of root-knot nematode resistance in rice in order to map the gene via linkage to DNA markers for use in marker-assisted breeding programs.

Materials and Methods

One hundred twenty five varieties including sixty one indica land races, fifty-nine elite cultivars and five javanicas were screened in the net house for their reaction to the rice root-knot nematode (Table 41.1). Seedlings were raised singly in plastic cups containing 250 cc autoclaved mixture of sand: clay loam soil (50 : 50). Ten days after germination, each seedling was inoculated with 5000 freshly hatched second stage larvae. The experiment was replicated thrice and the variety Annapurna was always taken as the susceptible check. The plants were uprooted twenty five days after inoculation, roots carefully washed and gall index was calculated based on the report of the international Meloidogyne Project, 1985; 0: immunue, 1: highly resistant, 2: resistant, 3: moderately resistant, 4: susceptible and 5: highly susceptible. The gall index score and reaction of the cultivar as per Sasser and Carter (1985). Varieties/land races found highly resistant were transferred to 9" diameter earthen pots and grown till maturity. Selfing was done by bagging the inflorescences before flowering to avoid cross pollination. The self seeds were grown and again screened for resistance.

Table 41.1: Rice Varieties Screened for Root-knot Nematode

Varieties/Land Races	*Total Nos. Screened*	*Immune*	*Highly Resistant*	*Resistant*	*Moderately Resistant*	*Susceptible*
Assam rice collection	40	0 (0)	0 (I)	2 (2)	12 (11)	26 (26)
Koraput rice collection	21	0 (0)	0 (0)	4 (3)	11 (11)	6 (7)
Elite cultivars	59	0 (0)	2 (6)	13 (15)	25 (20)	19 (18)
Javanicas	5	0 (0)	0 (0)	1 (1)	4 (4)	– –

Figures in parenthesis denote results of 2nd screening.

Results and Discussion

Only one land race, Kalarata and two elite cultivars Ramakrishna and Rasi showed consistently high resistance. Cultivars, such as Patnai-6, Daya, Udaya, Akasi, Annada and Kiran reported earlier as highly resistant/resistant to root-knot nematode, were all turned out to be susceptible. This may be due to increase in sand content of the experimental soil that facilitated movement of the infective larvae in the rhizosphere. The donors will be utilized to develop mapping populations to map the resistance gene via linkage to DNA markers for use in marker-assisted breeding programs. The varieties Ramakrishna and Rasi should be cultivated in the root-knot nematode endemic areas and also utilised in the varietal development programme.

References

Golden, A.M. and Birchfield, W., 1968. Rice root-knot nematode (*Meloidogyne graminicola*) as a new pest of rice. *Pl. Dis. Reptr.*, 52: 423.

Sassar, J.N. and Carter, C.C., 1985. *An Advanced Treatise on Meloidogyne, Vol. 1.* North Caroline State University Graphics, 422 pp.

Chapter 42

Effect of Age and Sex on Behavioural Responses of Female Lesser Bandicoot Rat *Bandicota bengalensis* (Gray and Hardwicke) Towards Male Urine

Gurpreet Kaur* and V.R. Parshad
Department of Zoology, Punjab Agricultural University, Ludhiana – 141 004, India

ABSTRACT

The detailed study on the lesser bandicoot rat *Bandicota bengalensis* was started in this department for the first time. The study shows that reproductive rates relates with the age of the rat. It is observed that the maximum reproductive rate is in the age group of less than one year. In the present study lesser bandicoot rat urine of male was taken as the parameter. It is found that with the age, the attraction power in urine also decreases. It may be due to the hormonal changes.

Keywords: *Age, Bandicota bengalensis, Urine, Rat, Pheromone.*

Introduction

For animals that rely on chemical communication for the regulation of social and sexual interactions, there is some indication that an animal's age can be determined by its scent. Ferkin (1999)

* Corresponding Author.

showed that both male arid female meadow voles are attracted to the odours of older conspecifics. That odour changes with age is suggested by the research of Robinson and colleagues who showed that the composition of mouse urine varies with age (Robinson *et al.*, 1976 and Miyashita and Robinson, 1980). Short lived animals such as rats, house and voles reach sexual maturity rapidly, mate soon after reaching puberty and show a dramatic decline in reproduction as they become aged, the post reproductive period occurs early in the life cycle (Spruijt *et al.*, 1989 and Vega Matusozyk *et al.*, 1994). The age of an individual will also affect the way conspecifics respond to it and how the individuals responds to conspecifics (Forslund and Part, 1995).

For many rodents odour communication is an essential element of their social biology. Conspecifics odours affect the reproductive physiology of individuals and animals use odours to indicate their presence in an area and to signal their individual identity, willingness to mate, social status and condition or quality to conspecifics (Halpin, 1986, Drickamer, 1992 and Hurst, 1993). Urine the principal medium by which metabolic waste is eliminated from an organism, functions centrally in the lives of many species of rodents as a source of chemical signals. The lesser bandicoot rat *Bandicota bengalensis* which is the most predominant agricultural pest in South Asia (Parshad, 1999) breeds throughout the year with large litters and maintains its large populations (Parshad *et al.*, 1989). The rats exhibited strong discrimination between two odours and this discrimination was affected by age of the animal (Peterson, 1986 and Solomon and Rumbaugh, 1997).Preliminary studies on the effect of age on urine attraction were available on *Mus musculus* (Drickamer, 1989, Drickamer and Brown, 1998 and Osada *et al.*, 2003) and *Microtus pennsylvanicus* (Ferkin, 1999). However no information is available on effect of age and sex on urine attraction in *Bandicota bengalensis.* This paper describes the effect of age and sex on behavioural responses of female bandicoot rat *Bandicota bengalensis* towards male urine.

Materials and Methods

Rat Collection

Adult *Bandicota bengalensis* caught from the rice fields were paired (n = 6) in laboratory cages for breeding. Mixture of cracked wheat, powdered milk, powdered sugar and groundnut oil (WMSG, 85 : 5 : 5 : 5) food and water were provided *ad libitum* to each pair. Youngones were weaned from the mother at 22 days of age but kept together until 30 days of age before housing singly in laboratory cages. They were provided WMSG, food and water *ad libitum.*

Grouping of Rats

In this experiment, 21 laboratory bred rats of known age were used. Of these, 9 rats (9 female) were used as recipients and 12 (12 males) as donors. All were individually housed in cages. The recipients were further divided in 3 groups of 3 rats with 3 females in each group. Every group each was tested for behavioural responses of individual rats to urinary odours of their opposite sex donor rats.

Out of 12 male donors 3 rats each were used for collection of urine at 4 months, 12 months, 18 months, and 24 months of age. Grouping of recipient and donors at the start of the experiment was done in a manner so that in each group individuals were unrelated to each other and had comparable body weights(g).

Collection of Urine

For the collection of urine samples, sheet of plastic was placed under each cage of the donors (Gao, 1991). Rats urinated on the plastic sheet and then the samples were collected with a clean glass pipette. Fresh urine samples were used for the experiment daily by pooling the urine of 3 rats of each group.

Experimental Design

All recipient rats were individually housed from day 30 of their age onwards in, experimental cages. Each cage consisted of a set of 2 laboratory cages joined through their one of the passage so that the rat had free access to each cage A and B. The door connecting the cages had a perforated shutter. Food and water were provided *ad libitum* in cage A. In cage B 2 small bowls, one (empty food bowl) for test samples of urine and other for water were placed at 6 cm distance from each other.

Odour testing experiments were carried out in experimental cages which were home cage of the subject and bowls for urine and water samples were used also were similar to food and water bowls used before the start of the experiment so as to avoid the affects of neophobia or neophilia on odour discrimination behaviour of each rat. During the odour testing no other cage/animal were placed in the experimental room except for the observer for whom the rats had developed familiarity during daily care and feeding. Odour testing experiment was carried out between 0930 to 1030 hrs under normal temperature and day light conditions for 3 days on each group of rats at their different age. The perforated barrier between the cages was closed 10 minutes before the experiment with rat separated in cage A. Three non-kin female rats each at 6, 12, 24 months of age were presented urine sample (0.05 ml) of donors of different age (as given in Table 42.1) in choice to same quantity of distilled water for one day in order to determine the discrimination ability of recipients between water and urine. Urine samples of 2 donors (n = 3) of different age (Table 42.2) were presented to opposite sex recipients in order to determine the discrimination ability for urine samples of donors of different age.

Table 42.1: Details of Age and Sex of Recipient and Donor Rats Used in Study of Behavioural Responses of *Bandicota bengalensis* to Urinary Signals in Choice to Distilled Water

Recipients		Donor Rats			
Sex	Age (Months)	Sex	Age (Months) of Rats and Urine Samples Presented to Recipients in Choice to Distilled Water (DW)		
			I	II	III
F	6	M	DW v/s 12	DW v/s 18	DW v/s 24
F	12	M	DW v/s 12	DW v/s 18	DW v/s 24
F	24	M	DW v/s 12	DW v/s 18	DW v/s 24

Table 42.2: Details of Age and Sex of Recipient and Donor Rats Used in Study of Behavioural Responses of *B. bengalensis* to Urinary Signals of Different Age

Recipients		Donor Rats			
Sex	Age (Months)	Sex	Age (months) and Urine Samples Presented to Recipients		
			I	II	III
F	6	M	4 v/s 12	4 v/s 18	4 v/s 24
F	12	M	4 v/s 12	4 v/s 18	4 v/s 24
F	24	M	4 v/s 12	4 v/s 18	4 v/s 24

The freshly collected urine sample (0.05 ml) was smeared inside the empty bowls. The 3 minute urine odour test began 10 minutes after the bowls were placed in cage B and then the perforated barrier between the cages was removed. Criteria for investigation of urine odour were that the rat licked or sniffed a stimulus urine and/or its nose was directed towards the urine odour source. The distance between the two bowls was about 6cm. The location of the bowls was rotated daily. The experimenter wore disposable latex gloves to minimize human scent transfer while handling bowls as also adopted earlier by Ferkin (1999).

Observations

When the bowls containing urine were placed in the cage, the observations were taken for 3 minutes period. The frequency of behavioural acts was noted with the help of a stop watch. The following observations were recorded:

First response: First response was given to urine to which the rat attracted first.

Time spent: The total time the recipient rat spent at each bowl.

Number of visits: Number of times recipient rat visited urine source for 3 minutes.

Statistical Analysis

The time spent per visit and number of visits were analyzed by non parametric test namely Wilcoxon paired-sample test (Zar, 1974).

The data between the age group was analyzed by Friedman's analysis of variance by ranks.

Results and Discussion

Effect of Age and Sex: Comparison Between Water and Urine

The lesser bandicoot rat exhibited strong discrimination between urine and water and this discrimination was affected by both age and sex of the individual. During the odour test, after perceiving the odour from a distance of about 30 cm, the rat frequently directed its movement accompanied by sniffing towards the source of odour and accordingly showed more stronger first response towards the urine odour of opposite sex than to water followed by significantly more number of visits during the test period (Table 42.3).

Table 42.3: Differential Behavioural Responses of Female *Bandicota bengalensis* of Different Age Toward Water and Urine of Opposite Sex

Age of Recipients (Months)	Sex	First Response by Rats (n = 9)		Number of Visits/ 3 Minutes		Time Spent/Visit (Seconds)	
		Distilled Water	Urine of Donor of Opposite Sex	Distilled Water	Urine of Donor of Opposite Sex	Distilled Water	Urine of Donor of Opposite Sex
6	F(3)	3	8	11.11±0.11	8.66±1.60*	2.66±1.08	6.14±0.48*
12	F(3)	2	7	2.44±0.52	3.44±0.47*	5.25±1.99	10.24±2.68*
24	F(3)	3	6	1.77±0.32	3.55±0.60*	3.55±0.60	6.86±1.33*

Asterisks indicate statistical significance by Wilcoxon signed-ranks test ($P<0.05$).

In general, irrespective of age and sex, the number of visits and time spent/visit were significantly more toward the urine of the opposite sex donors. Relationship of age and reproduction is well known in rodents and the active period of reproduction in *B. bengalensis* may be in between 6–12 months old whereas breeding activity may decrease in 24 month old rats. During the present study number of visits of 6 and 12 months old females to the urine of opposite sex donors were more than by 24 months old rats. The difference in the number of visits between 24 month old females and 12 month old females revealed the effect of age which relates to their decreased reproductive activity at this age. Similarly, in voles sensitivity of 24 months old males to female urine decreased (Corwin *et al.*, 1995).

The more number of responses towards urine as compared to distilled water clearly showed that urine plays an important role in olfactory communication in *B. bengalensis.* Its attraction towards urine from a distance of about 30 cm clearly showed that urine of *B. bengalensis* may have certain volatile compounds which were earlier identified like in Swiss male mice 3-cyclohexane-l-methanol, 3 amino-5-triazole, 4-ethyl phenol, 3 ethyl-2,7 dimethyl octane and 1-iodoundecane(Chiraman and Archunan, 2002).

Relationship Between Age of Donors and Recipients

Exploratory data analysis showed age related ability of bandicoot rats to discriminate between urinary odours of rat of opposite sex and of different age. They determined the age from the urinary odours without actually exploring the animal. This ability depended upon the age- related sensitivity of the recipients as shown in case of the 6 months old female bandicoot rats which were attracted more towards the urine of older male bandicoots. The time spent and number of visits were significantly more ($P < 0.05$) towards the older male urine as compared to 4 months male urine ($P < 0.05$) except in 18 months old male urine (Table 42.4). Present studies clearly showed that the female bandicoots were able to discriminate the age of male urine from about 30 cm distance as the first response and number of visits/3 minutes were more towards the older male urine as compared to that of younger males. It means that the urine of the older male bandicoots contains information about their longevity that is absent in the urine of younger males. The differences in time spent/visit ($P < 0.50$) and number of visits towards male urine of 12 months, 18 months and 24 months were significantly different ($P < 0.75$). It means female bandicoots discriminated the urine by frequent visits (Kaur, 2005).

The responses of 12 months old females were more towards urine of 4 months old male urine as compared to 12 months, 18 months and 24 months, the mean time spent was significantly more ($P < 0.05$) towards 12 months, 18 months and 24 months male urine. It is evident that the female bandicoots which are at their peak reproductive age were attracted more towards older males as compared to younger males. It may be due to the greater reproductive performance of older males than younger males or 1 year old females may choose older males as a indirect mate as also revealed by Kokko and Lindstrom (1996).

The 24 months old female bandicoots were attracted more towards urine of 12 months males as compared to that of 18 months and 24 months old males but the number of visits were significantly more ($P < 0.05$) toward the urine of 18 months old males. It may be either due to decreased reproductive activity of female bandicoots or decrease in male urinary pheromone with age as previous studies on mice revealed age-related decrease in the volatile compounds like 2-phenylacetamide and methyl butyric acid in urine (Osada *et al.*, 2003). The 24 months female bandicoot's reproductive activity may decline due to age related pulsatile release of luteinizing hormone (Mobbs *et al.*, 1984), loss of ovarian cyclicity (Aschheim, 1965) small oestrogen-induced luteinizing hormone surge (Depaolo and Chappel, 1986).

Table 42.4: Behavioural Responses of Female *Bandicota bengalensis* Towards Male Urine Donors of Different Age

Age of Recipient Females (Months)	*Age of Donor Males*			*First Response by Rats (n = 9)*		*Number of Visits/ 3 Minutes*		*Time Spent/ Visit (Seconds)*	
	A	*v/s*	*B*	*A Donor*	*B Donor*	*A Donor*	*B Donor*	*A Donor*	*B Donor*
6	4	v/s	12	3	6	2.66±0.62	5.66±1.59*	5.32±1.36	5.48±0.75*
	4	v/s	18	4	5	3.55±0.68	4.88±1.29*	6.48±1.20*	5.02±0.68
	4	v/s	24	2	7	4.66±1.16	6.00±1.4*	5.78±1.07	6.24±1.13
12	4	v/s	12	8	1	3.33±0.86	5.00±1.17*	5.66±1.64	5.75±1.09
	4	v/s	18	1	8	3.33±0.44*	2.88±0.45	4.24±0.8	4.73±0.71*
	4	v/s	24	9	0	3.22±0.46*	2.22±0.27	4.74±0.6	6.48±1.66*
24	4	v/s	12	7	2	4.44±1.04	4.77±1.22*	5.55±1.07	6.55±1.59*
	4	v/s	18	2	7	2.44±0.47	6.88±0.69	5.10±1.59	5.39±0.47*
	4	v/s	24	5	4	4.88±0.82*	1.22±0.32	5.90±1.01*	5.77±1.65*
	4	v/s	12			(χ^2=0.33 0.50 < P < 0.75)		($\chi^2$0.59 0.25 < P < 0.50)	
	4	v/s	18			(χ^2=2.99 0.05 < P < 0.10)		(χ^2=0.33 0.50 < P < 0.75)	
	4	v/s	24			(χ^2=0.92 0.25 < P < 0.50)		(χ^2=0.59 0.25 < P < 0.50)	

*: Significant differences (P < 0.05) between responses of recipients to urine of two donors as determined by Wilcoxon signed rank test.

χ^2 based on Friedman's analysis of variance between age groups of recipients towards the urine of same age male donors.

The differences in the attraction of recipients of different age groups towards urine of 4 v/s 24 months old male donors were related more than between other age groups of donors. It may be due to reduced catecholamine levels that decline with age in both male and female mice (Finch, 1978).

It is concluded that 6 months and 12 months females were attracted more towards the older male urine as compared to that of younger males. It means that the urine of the older male bandicoots contain information about their longevity that is absent in the urine of younger males but 24 months old female bandicoots were attracted more towards urine of 12 months males as compared to that of 18 months and 24 months old males. It may be either due to decreased reproductive activity of female bandicoots or decrease in male urinary pheromone with age as previous studies on mice revealed age related decrease in the volatile compounds in urine(Osada *et al.*, 2003).Further study is necessary to know the chemical changes taking place in urine with the age to control rat population.

Acknowledgement

Authors are thankful to Dr. M.H. Ferkin, Department of Ecology and Organismal Biology, University of Memphis, Memphis, TN 38152, USA, Dr. L.C. Drickamer, Department of Biological Sciences, Northern Arizona University, Flagstaff, Arizona 86011, Dr. G.K. Beauchamp, Monell Chemical Senses Center, Philadelphia, PA 19104-3308, USA, Nancy G. Solomon, Department of Zoology, Miami University, Oxford, Ohio, 45056 and Anthony R. Isles, Laboratory of Cognitive and Developmental Neuroscience, Neurobiology Programme, The Babraham Institute, Babraham, Cambridge, CB2 4A T, United Kingdom for providing literature.

References

Aschheim, P., 1965. Resultats foumis par la greffe heterochrone des ovarie dan 'st' etude de la regulation hypothalamus-hypophysoovarienne de la ratte senile. *Gerontologia*, 10: 65–75.

Chiraman, S.A. and Archunan, G.A., 2002. Characterization of urinary volatiles in Swiss male mice (*Mus musculus*), bioassay of identified compounds. *J. Bio. Sci.*, 27: 679–686.

Corwin, J., Loury, M. and Gilbert, A.N., 1995. Workplace, age and sex as mediators of olfactory function: Data from a national geographic smell survey. *J. Gerontol.*, 50: 179–186.

Depaolo, L.V. and Chappel, S.C., 1986. Alterations the secretion and production of follicle stimulating hormone precede age related lengthening of estrous cycles in rats. *Endocrinology*, 118: 1127–1133.

Drickamer, L.C., 1989. Odor preferences of wild stock female house mice (*Mus musculus*) tested at three ages using urine and other cues from conspecific males and females. *J. Chem. Ecol.*, 15: 1971–1987.

Drickamer, L.C. 1992. Behavioural selection of odor cues by young female mice affects age of puberty. *Dev. Psychobiol.*, 25: 461–470.

Drickamer, L.C. and Brown, P.L., 1998. Age-related changes in odour preferences by house mice living in seminatural enclosures. *J. Chem. Ecol.*, 24: 1745–1756.

Ferkin, M.H., 1999. Attractiveness of opposite-sex odour and responses to it vary with age and sex in meadow voles (*Microtus pennsylvanicus*). *J. Chem. Ecol.*, 24: 1745–1756.

Finch, C.E., 1978. Reproductive senescence in rodents: Factors in the decline of fertility and loss of estrous cycles. In: *The Aging Reproductive System*, (Eds.) Schneider, E. Raven Press, New York, pp. 193–212.

Forslund, P. and Part, T., 1995. Age and reproduction in birds-hypothesis and tests. *Tree*, 10: 374–378.

Gao, Y., 1991. Behavioural responses of rats to the smell of urine from conspecifics. *Anim. Behav.*, 42: 506–508.

Halpin, Z.T., 1986. Individual odors among mammals: Origins and functions. *Adv. Study Behav.*, 16: 39–70.

Burst, J.L., 1993. The priming effects of urine substrate marks on interactions between male mice, *Mus musculus domesticus* Schwatz and Schwarz. *Anim. Behav.*, 45: 55–81.

Kaur, G., 2005. Urinary signals in the lesser bandicoot rat *Bandicota bengalensis* (Gray), and their application for management of its population. *Ph.D. Dissertation*, Punjab Agricultural University, Ludhiana, India.

Kokko, H. and Lindstrom, J., 1996. Evolution of female preferences for old mates. *Proc. R. Soc.*, London, 263(B): 1533–1538.

Miyashita, K. and Robinson, A.B., 1980. Identification of compounds in mouse urine vapor by gas chromatography and mass spectrometry. *Mech. Ageing Dev.*, 13: 177–184.

Mobbs, C.V., Gee, D.M. and Finch, C.E., 1984. Reproductive senescence in female C 57 BL/6J mice: Ovarian impairments and neuroendocrine impairments that are partially and reversible and delayable by ovariectomy. *Endocrinology*, 115: 1653–1662.

Osada, K., Yamazaki, K., Curran, M., Bard, J., Smith, B.P.C. and Beauchamp, G.K., 2003. The scent of age. *Proc. R. Soc.*, London, 270: 929–933.

Parshad, V.R., 1999. Rodent control in India. *Pest. Mgmt. Rev.*, 4: 97–126.

Parshad, V.R., Kaur, P. and Guraya, S.S., 1989. Reproductive cycles of mammals. In: *Reproductive Cycles of Indian Vertebrates*, (Eds.) Saidapur, S.K. Allied Publishers Limited, New Delhi, pp. 347–408.

Peterson, S.L., 1986. Age related changes in plasma oestrogen concentration, behavioural responsiveness to oestrogen and reproductive success in female gray-tailed voles, *Microtus canicaudus*. *J. Reprod. Fertil.*, 78: 57–64.

Robinson, A.B., Dirren, H., Sheets, A., Miquel, J. and Lundgreen, P.R., 1976. Quantitative aging pattern in mouse urine vapor as measured by gas-liquid chromatography. *Exp. Gerontol.*, 11: 11–16.

Solomon, N.G. and Rumbaugh, T., 1997. Odour preferences of weanling and mature male and female pine voles. *J. Chem. Ecol.*, 23: 2133–2143.

Sproijt, B.M., Mayerson, B.J. and Hoglund, U., 1989. Aging and sociosexual behaviour in male rat. *Behav. Brain Res.*, 32: 51–61.

Vega, M.J., Appa, R.S. and Larson, K., 1994. Age-dependent variations in the sexual preference of male rats. *Physiol. Behav.*, 55: 827–830.

Zar, J.H., 1974. *Biostatistical Analysis*. Prentice Hall, Inc., Englewood Cliffs, N.J.

Chapter 43

Effect of *Lantana camara* as Growth Inhibitor

Padma Chatterjee*, Nisikta Banerjee and Urbi Bhattacharya
Plant Physiology and Biochemistry Laboratory, Department of Botany, University of Kalyani, Kalyani – 741 235, West Bengal, India

ABSTRACT

Lantana camara is an aggressively growing herb dominating as a large population covering almost the entire field area of a specific region. The successful growth of this plant may be due to its allelopathic effect. This property of *Lantana camara* as a phytotoxic growth inhibitor towards other plant species was evaluated in this article. 10 per cent aqueous leaf extract of *Lantana camara* was evaluated against three common pulses, *Lens esculanta, Phaseolus aureus, Phaseolus mungo.* As demonstrated by common petridish germination method, it was revealed that the leaves of *Lantana camara* cause strong growth inhibition in early seedling growth of all the three pulse seeds (compared to corresponding controls).

Keywords: *Lantana camara, Lens esculanta, Phaseolus aureus, Phaseolus mungo, Phytotoxicity.*

Introduction

Weeds are unwanted plants, which interfere with the utilization of land and water resources. In agricultural field weeds compete with other beneficial vegetation reducing the yield and quality of production (Kohli *et al.,* 1998; Quasem and Foy, 2001). This interference has been seen especially in

* E-mail: schatterjeecal2003@yahoo.co.in

weed communities (Muller, 1969). The chemicals which are responsible for bringing such effects are released through leachation, volatilization, root exudation or decomposition (Rovira, 1969). The leachation is the most common method. *Lantana camara* is a very well known weed among plant kingdom and it is very common in different parts of India.

Present study involve the effect of 10 per cent aqueous leaf extract of *Lantana camara*, on early seedling growth of three common pulses, *Lens esculanta, Phaseolus aureus, Phaseolus mungo*.

Materials and Methods

The leaves of *Lantana camara* was collected from Kalyani of West Bengal during March–April. The seeds of pulses (*Lens esculanta, Phaseolus aureus, Phaseolus mungo*) were collected from B.C.K.V., Mohanpur, West Bengal in March, 2006.

To study the phytotoxicity, leachates of sun-dried powdered leaves of *Lantana camara* was extracted in distilled water. For this, 10 grams of sun-dried leaves were soaked in 100 ml of distilled water for 24 hours and then filtered. The filtrate was taken to study the early seedling growth of *Lens esculanta, Phaseolus aureus, Phaseolus mungo*. For this experiment, 5 seeds each of three pulses were pre-soaked for 5 hours in the filtrate and simultaneously the control sets were maintained using distilled water. Then the seeds were placed in petridishes lined with filter paper moistened with 3 ml of filtrate and separately also with 3 ml of distilled water to serve as control.

The results were determined by measuring the length of primary root and main shoot on 7th day of experiment and data were statistically analyzed.

The ratio of root and shoot elongation were calculated as suggested by Rho and Kil (1986).

$$\text{Relative Elongation Ratio of Root} = \frac{\text{Mean Root Length of Tested Plant}}{\text{Mean Root Length of Control}} \times 100$$

$$\text{Relative Elongation Ratio of Shoot} = \frac{\text{Mean Shoot Length of Tested Plant}}{\text{Mean Shoot Length of Control}} \times 100$$

Results and Discussion

It is clear from the results (Table 43.1 and 43.2) that the leaf extract of *Lantana camara* cause inhibition in both root and shoot elongation of *Lens esculanta, Phaseolus aureus, Phaseolus mungo* seeds.

Table 43.1: Root Elongation (cm) of *Lens esculanta, Phaseolus aiueus, Phaseolus mungo* to Control and the Leaf Extract of *Lantana camara*

Treatment	Agricultural Crops		
	Lens esculanta	*Phaseolus aurueus*	*Phaseolus mungo*
Control	7.13±0.71	7.32±0.59	838±0.80
Leaf extract of *Lantana camara*	2.43±0.14	1.03±0.25	2.35±0.17

Previous reports indicated that *Lantana camara* has allelopathic effect on *Ipomea tricolor* (Casado, 1995). Oudhia, Kolhe and Tripathi (1998) also showed the negative allelopathic effect *Lantana camara* on linseed.

Table 43.2: Shoot Elongation (cm) of *Lens esculanta, Phaseolus aureus, Phaseolus nungo* to Control and the Leaf Extract of *Lantana camara*

Treatment	Agricultural Crops		
	Lens esculanta	*Phaseolus aiueus*	*Phaseolus mungo*
Control	11.33±1.53	24.63±1.73	20.97±0.60
Leaf extract of *Lantana camara*	3.38±0.10	2.73±0.28	3.55±0.33

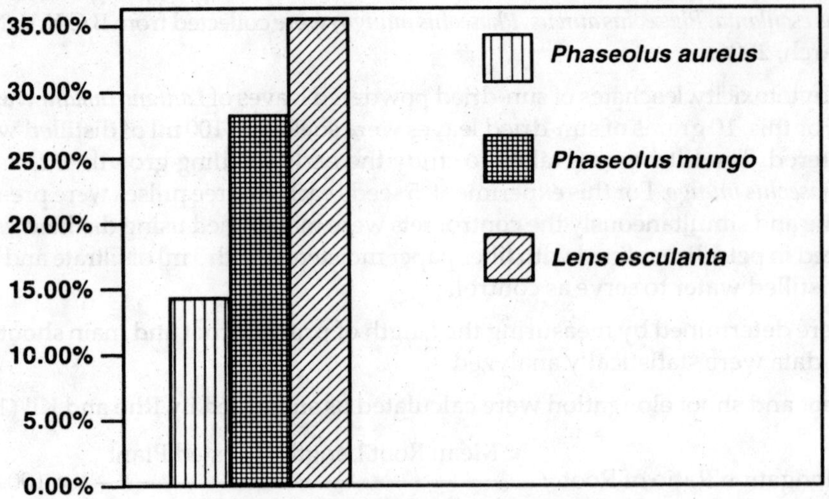

Figures 43.1: Relative Elongation Ratio of Root

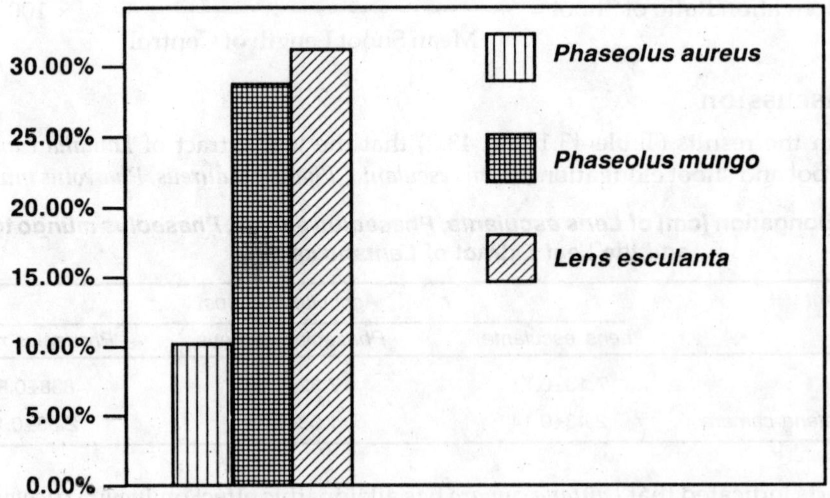

Figures 43.1: Relative Elongation Ratio of Shoot

The present study revealed that, the maximum inhibition in both root and shoot elongation was found in case of *Phaseolus aureus* seeds. The relative root and shoot elongation in *Phaseolus aureus* were 14.07 per cent and 11.08 per cent respectively. Such findings may be cited as to be one of the reasons for the dominant growth of *Lantana camara* over other plants of the neighboring areas.

References

Casado, C.M., 1995. Allelopathic effects of *Lantana camara* (Verbenaceae) on morning glory (*Ipomea tricolor*). *Rhodora*, 97: 264–274.

Kohli, R.K., Batish, D. and Singh, H.P., 1998. Allelopathy and its implications in agro ecosystems. *J. Crop. Prod.*, 1: 169–202.

Muller, C.H., 1969. Allelopathy as a factor in ecological process. *Vegetatio.*, 18: 348–357.

Oudhia, P., Kolhe, S.S. and Tripathi, R.S., (1998c). Negative (stimulatory) allelopathy of *Lantana camara* L. on Linseed var. Kiran In: (Abstract) *International Symposium on Microbial Biotechnology for Sustainable Development and Productivity*, Rani Durgavati University, Jabalpur (India) 14–16 November, p. 64.

Qasem, J.R and Coy, C.L., 2001. Weed allelopathy: Its ecological impacts and future prospects: A review. *J. Crop Prod.*, 4: 43–119.

Rovira, A.D., 1969. Plant root exudates. *Bot. Rev.*, 35: 35–59.

Chapter 44

Water Quality and Zooplankton Diversity of Ramgarh Lake, Jaipur

P.V. Paulose and Kamlesh Maheshwari
Environmental Biology Laboratory, Department of Zoology, University of Rajasthan, Jaipur – 302 004, Rajasthan, India

ABSTRACT

The present article focuses on the limnological characteristics of Ramgarh lake, Jaipur. This lake is an important source of drinking water to the inhabitants of Jaipur city. The lake was studied for the physico-chemical parameters and zooplankton composition. The samples were collected from five different sites considering the intensity of human activities. Water was found alkaline and hard. All chemical parameters studied were found within permissible limit for drinking water. The highest zooplankton density was 40.21×10^6/L in the month of October. The zooplankton number was decreased gradually till 4.30×10^6/L in the month of February and again increased up to 11.52×10^6 L in the month of March. Copepods and Cladoceras constituted majority of Zooplanktons. The rest was constituted mainly by Protozoans, Rotifers, Ostracods and Insect larvae. Distribution and abundance of zooplankton population is discussed with respect to changes in physico-chemical factors.

Keywords: *Zooplankton, Diversity, Physico-chemical factors, Ramgarh lake.*

Introduction

Water is a precious commodity especially in arid and semi arid regions of Rajasthan. Jamawa Ramgarh Lake is an important source of drinking water supply to the people of Pink City, Jaipur, besides being a source of water for irrigation purpose. Limnological studies of some reservoirs of Rajasthan have been reported by some researchers (Jakher *et al.*, 1989; Khatri, 1992; Bahura, 1997). But Ramgarh Lake has not been studied for physico-chemical or biological aspects at least for the last three decades. In the present study, it has been decided to investigate the physico-chemical aspects and zooplankton diversity of this important aquatic ecosystem.

Ramgarh Lake is situated about 30 km away from Jaipur City in the northeast direction. It lies between 27′6 N latitude and 72′4 E longitude. It is a medium size impoundment with a mean depth of 3.43 m and the total reservoir area is about 1260 ha. Annual rainfall ranges between 20–22″ which is maximum in the month of August.

Materials and Methods

During the study period (October'05–March'06) samples of water and zooplankton were collected monthly from five different locations including areas of maximum and minimum human activities.

Planktons were collected by a plankton net made up of bolting silk (No. 25; mesh size 55μ). The samples were then transferred to narrow mouthed plastic bottles, preserved in 5 per cent formalin and brought to laboratory. Zooplankton samples were identified and counted under a microscope using a zooplankton counting chamber. Zooplanktons were identified with the help of Edmondson (1965), Needham and Needham (1978), Tonapi (1980) and APHA (1995). The results of zooplankton count were expressed as No. X 10^6/L.

The physico-chemical factors were studied according to APHA (1995). Some of the parameters such as temperature and water transparency were analysed on the spot itself.

Results and Discussion

Physico-chemical Factors

Various physico-chemical parameters studied are summarized in the Table 44.1. In the present study, water temperature ranged between 29 to 16°C. The surface water temperature was slightly lower than the atmospheric temperature. Transparency during the sampling period ranged between 0.18 M and 0.37 M.

pH remained in the alkaline range (7–8.1) throughout the study period. This may be the due to the high buffering capacity of the system (Dhakad and Choudhary, 2005). In the present study a direct relationship was observed between pH and temperature. As the temperature decreased the pH was also found decreased. Such a relationship was also reported by Bandela *et al.* (1998) in Barul Dam water and Shaikh and Yeragi (2003) in Tansa River of Thane District. The electrical conductivity (EC) remained between 243–260 μmhos cm^{-1}.

The alkalinity ranged between 120–155 mg/L. The alkalinity is contributed by both carbonate and bicarbonate ions and the increased ions are correlating with the higher pH also Barret (1953) have suggested that total alkalinity value of 60 mg/L or more indicate the hardness of water. The lake water was hard (110–118 mg/L) due to the higher concentration of calcium (30.0–34.4 mg/L). Similar findings were noted by Bagde and Verrna (1985) in the JND lake at New Delhi.

Fluctuation in dissolved oxygen was noticed with highest value of 8.7 mg/L during the month of March. The suitability of water resources for irrigation use in agriculture depends on its salt concentration, especially chloride content. Salinity gradually increased from 3.12 mg/L in the month of October to 14.2 mg/L in the month of March. This may be due to high evaporation of Lake Water.

In the present study at Ramgarh lake nitrate was not reported and phosphate was in the range of 0.04–0.08 mg/L. This indicates the oligotrophic nature of the reservoir. The concentration of sulphate ions ranged between 3.5 mg/L and 8.7 mg/L. According to Ganapati (1960), non-polluted waters are generally deficient in nitrate and phosphate contents.

Table 44.1: Average of Physico-chemical Factors

Parameters	*October*	*November*	*December*	*January*	*February*	*March*
Temperature (°C)	29°C	23°C	20°C	16°C	22°C	24°C
Transparency (meter)	0.18	020	0.30	0.37	0.22	0.33
pH	7.5	7.4	7.2	7.0	7.8	8.1
Conductivity (± μhos cm^{-1})	243	245	245	249	255	260
Dissolved O_2 (mg/L)	3.8	2.9	6.5	2.9	1.21	8.7
Free CO_2 (mg/L)	2.2	8.3	5.5	8.8	0.0	0.0
Alkalinity (mg/L)	120	132	138	150	150	155
Acidity (mg/L)	7.5	12.5	8.5	10.0	15.0	10.0
Total hardness (mg/L)	110	113	115	118	124	115
Calcium hardness (mg/L)	30.0	31.2	32.8	34.4	33.6	33.6
Magnesium hardness (mg/L)	7.0	7.1	7.3	7.8	8.28	6.8
Salinity (mg/L)	3.12	11.5	12.7	11.3	11.3	14.2
BOD (mg/L)	–	–	1.5	1.9	2.0	2.2
COD (mg/L)	–	–	4.0	6.0	5.2	6.5
TSS (mg/L)	–	–	0.1	0.1	0.2	0.2
TDS (mg/L)	–	–	1.4	2.0	2.0	3.4
TS (mg/L)	–	–	1.5	2.1	2.2	3.6
Nitrate (mg/L)	–	–	0.0	0.0	0.0	0.0
Phosphate (mg/L)	–	–	0.05	0.06	0.04	0.08
Sulphate (mg/L)	–	–	8.79	5.8	4.76	3.54

The siltation rate in a reservoir depends on the presence of total solids in it. In the present study the total solids content during monitoring was in the range of 1.5 mg/L and 3.6 mg/L and it shows minimum siltation in reservoir. Bhosale *et al.* (1994) and Khabade *et al.* (2002) have found higher values of total solids in freshwater body.

Zooplankton Diversity

A total of 37 species belonging to Copepoda (12), Cladocera (7), Rotifera (8), Protozoa (5), Ostracoda (3) and Insecta (2) were observed during the study period as presented in Table 44.2. Protozoa were represented by five species and their total density ranged from 1.18–3.0 × 10^6/L. The contribution of protozoan was highest (38.3 per cent) in the month of February, when the total zooplankton population was at the lowest level.

Crustacean zooplankton belonged to three sub-classes, Copepoda, Cladocera and Ostracoda of which Copepoda (45.6 per cent) and Cladocera (39.9 per cent) were the dominant groups followed by Ostracoda (1.6 per cent). *Mesocyclops* sp. and *Diaphnosoma* sp. constituted majority of Copepods. *Daphnia* sp. and *Polyphemus* sp. showed poor abundance throughout the study period. This study indicates that Crustaceans dominated the zooplankton population in the Ramgarh lake. In many other Indian waters also Crustaceans (particularly, copepods) generally dominate (Mitra and Patra, 1990; Shyam, 1991; Varghese and Naik, 1992).

Table 44.2: List of Zooplankton Group and Genera

Group	Genera
Rotifera	1. *Asplanchna pridonta*
	2. *Filinia longiseta*
	3. *Brachionus forficula*
	4. *B.rubens*
	5. *B. calyciflorus*
	6. *Testudinella* sp.
	7. *Horaella brahmi*
	8. *Lecane luna*
	9. *Cypris* sp.
Ostracoda	10. *Heterocypris* sp.
	11. *Stenocypris* sp.
	12. *Ceriodaphnia reticulata*
Cladocera	13. *Daphnia carinata*
	14. *Diaphanosoma excisum*
	15. *Moina* sp
	16. *Simocaphalus vetulus*
	17. *Sida* sp.
	18. *Polyphemus* sp.
	19. *Mesocyclops hyalinus*
Copepoda	20. *M. eukart*
	21. *Pyllodiaptomus annae*
	22. *Heliodiaptomus viduus*
	23. *Rhinediaptomus indicus*
	24. *Neodiaptomus schmackeri*
	25. *N. handeli*
	26. *Eubranchiopus* sp.
	27. *Mysis* sp.
	28. *Canthocaptomus* sp.
	29. *Zoa larvae*
	30. *Nauplius larvae*
Protozoa	31. *Arcella discoidae*
	32. *A. vulgaris*
	33. *Paramecium Aurelia*
	34. *Vorticella* sp.
	35. *Stentor* sp.
Insecta	36. *Larvae of Ephemeroptera*
	37. *Larvae of Diptera*

Insecta were represented only by the larvae of Ephemeroptera and Diptera. Average monthly fluctuation in planktonic population in Ramgarh lake is shown in Figure 44.1.

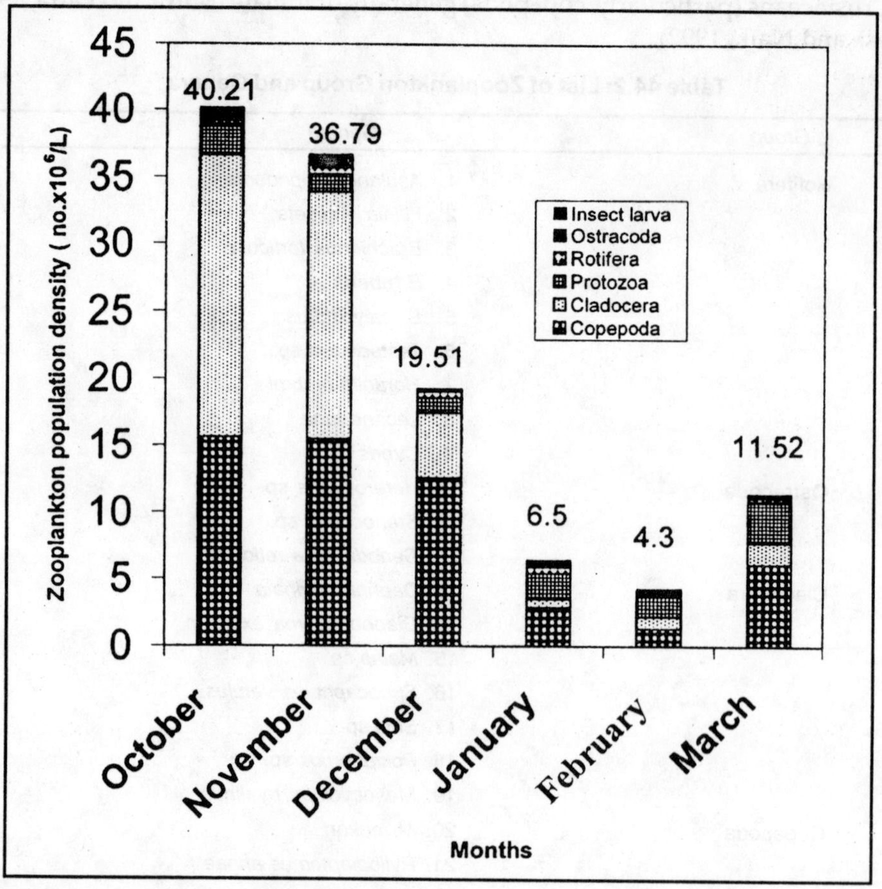

Figure 44.1: Average Zooplankton Density in Various Months

Moitra and Bhowmik (1968) observed members of three main zooplankton groups such Rotifera, Cladocera and Copepoda, which dominate, in freshwater fishpond in Kalyani, West Bengal. In Ramaua reservoir four genera of Rotifera and Cladocera and two genera under Copepoda were observed by Agrawal (1980). Pathak and Mudgal (2002) reported 19 species of Protozoa, Copepoda, Cladocera, Ostracoda and Rotifera of Virla reservoir of Khargone (M.P.). In the present study 37 common species were recorded. So the species diversity appears to be higher in Ramgarh lake as compared to the above studies. The possible cause may be due to good water quality, which has had a positive impact on the diversity and abundance of zooplankton population.

Maximum population of Zooplankton was observed in the month of October constituting 40.21×10^6/L. The Zooplankton number was decreased gradually to 4.30×10^6/L in the month of February and followed by an increase in March. Isaiarasu *et al.* (2001) reported a zooplankton density of 127 no./L. in a tropical pond near Sivakasi, Tamil Nadu. Bohra (1997) has reported a zooplankton density of 16 to 185 no./L in Gajner Lake, Bikaner, Rajasthan. So, from the present study, it is clear that

Ramgarh Lake is highly productive and this water body can also be utilized for fish rearing purpose in addition to a source of drinking water.

Acknowledgement

Authors are thankful to CAS and Head of Department of Zoology, University of Rajasthan, Jaipur for providing necessary facilities.

References

Agrawal, S.S., 1980. Some aspects of limnology of Ramaua Dam with special reference to phytoplankton and zooplankton. *Ph.D. Thesis*, Jiwaji University, Gwalior.

APHA, 1995. *Standard Methods for the Examination of the Water and Wastewater*, 19th edition. New York, U.S.A.

Bagde, V.S. and Verma, A.K., 1985. Physico-chemical characteristics of waters of JND lake at New Delhi. *Indian J. Ecol.*, 12(1): 151–156.

Bahura, C.K., 1997. Limnological studies of Gajner lake, Bikaner. *Ph.D. Thesis*, Rajasthan University, Jaipur (Rajasthan).

Bandela, N.N., Vaidya, D.P. and Lomate, V.S., 1998. Seasonal temperature changes and their influence on the level of carbon dioxide and pH in Barul Dam water. *J. Aqua. Biol.*, 13(1): 43–46.

Barret, P.H., 1953. Relation between alkalinity and adsorption and segmentation of added phosphorus in fertilized trout lakes. *Trans Ajmer Soc.*, 82: 78–90.

Bhosale, L.J., Sabale, A.B. and Mulik, N.G., 1994. Survey and status report on some wetlands of Maharashtra. Final report of project submitted to Shivaji University, Kolhapur, India, pp. 60.

Dhakad, N.K. and Choudhary, P., 2005. Hydrobiological study of Natnagra pond in Dhar district (M.P.) with special reference to water quality impact on potability, irrigation and aquaculture. *Nature Environment and Pollution Technology*, 4(2): 269–272.

Edmondson, W.T. (Ed) 1965. *Freshwater Biology*. John Wiley and Sons, Inc., New York.

Ganapati, S.V., 1960. Ecology of tropical waters. In: *Proceedings of Symposium on Algology*, (Ed.) D. Raghavan and P. Kachrro. ICAR, New Delhi, pp. 204-218.

Isaiarasu, L., Murugavel, P. and Murugan, N., 2001. Biodiversity of zooplankton observed in a seasonal, tropical pond near Sivakasi, Tamil Nadu, India. *Ecol. Env. and Cons.*, 7(3): 257–261.

Jakher, G.R. *et al.*, 1989. Comparative limnology of Sambhar and Didwana lakes (Rajasthan, NW India). *Hydrobiologia*, 197: 245–256.

Khatri, T.C., 1992. Seasonal distribution of zooplankton in Lakhotia lake. *Env. Ecol.*, 10(2): 317–322.

Khabade, S.A., Mule, M.B. and Sathe, S.S., 2002. Studies on physico-chemical parameters of Lodhe reservoir from Tasgaon Tahsil (Maharashtra). *Indian J. Environ. and Ecoplan.*, 6(2): 301–304.

Mitra, A. and Patra, K.C., 1990. Ecology of planktonic copepods in the Mandarmani Creek of West Bengal, India. *J. Mar. Sci.*, 19(4): 278–281.

Moitra, S.K. and Bhowanik, M.L., 1968. Seasonal cycle of rotifers in freshwater fish pond in Kalyani, West Bengal. *Proc. Symp. Recent Adv. Trop. Ecol.*, pp. 359–367.

Needham, J.G. and Needham, P.R., 1978. *A Guide to the Study of Freshwater Biology*, 5th edn. Holden Day Inc. Publ. San Francisco, pp. 12–30.

Pathak, S.K. and Mudgal, L.K., 2002. A preliminary survey of zooplankton of Virla reservoir of Khargone (M.P.), India. *Indian J. Environ. and Ecoplan.*, 6(2): 297–300.

Shaikh, N. and Yeragi, S.G., 2003. Seasonal temperature changes and their influence on free carbon dioxide, dissolved oxygen and pH in Tansa River of Thane District, Maharashtra. *J. Aqua. Biol.*, 18(1): 73–75.

Shyam, R., 1991. Comparative dynamics of zooplankton in a tropical freshwater swamp and in newly constructed pond in swampy area. *J. Appl. Zoo. Res.*, 2(2): 80–84.

Tonapi, G.T., 1980. *Freshwater Animals of India: An Ecological Approach.* Oxford and IBH Publishing Co., New Delhi.

Varghese, M. and Naik, L.P., 1992. Hydrobiological studies of a domestically polluted tropical pond. *Poll. Res.*, 11(2): 101–102.

Chapter 45

Prevalence of House Dust Mites in the Homes of Asthmatic Patients

Navpreet Kaur Gill and Parminder Singh Dehar

Department of Zoology, Punjabi University, Patiala – 147 002, Punjab

ABSTRACT

House dust mites have been shown to be strongly associated with allergic asthma in the world over. These associations were mostly made with reference to the mites of *Dermatophagoides* Spp. But large number of storage mites were also present in dwelling of asthmatic patients and are responsible for aggravating asthma. During present studies 24 sps. belonging to the 7 genera from 4 orders *i.e.* Astigmata Prostigmata, Mesostigmata and Cryptostigmata have been identified from the homes of 100 Asthmatic patients.

Keywords: *Allergic asthma and Dermatophagoides spp.*

Introduction

Many scientists have focused attention on the significance of house dust and harbouring acarofauna in the development of asthma and other allergic diseases in human beings (Kern, 1921; Cooke, 1922; Dekkar, 1928, 1930; Voorhorst *et al.*, 1969; Plattis *et al.*, 1989, 1992; Mark, 1998; Nelson *et al.*, 1999; Peat *et al.*, 1999). These allergic diseases, such as asthma, rhinitis, eczema and food allergies, are reaching epidemic proportions in both the developed and developing world (Peat *et al.*, 1999). India has an estimated 15–20 million asthmatics, rough estimates indicate a prevalence of between 10 per cent and 15 per cent in 5–11 years old children (WHO, 2000).

The development of allergies (allergic asthma, rhinitis, eczema) have been related with the exposure of human to both house dust and house dust mites (Maunsell *et al.*, 1968; Spieksma and Voorhorst, 1969; Cuthbert *et al.*, 1979; Custovic *et al.*, 1996; Nelson *et al.*, 1999). Mites that commonly subsist in human dwellings and provide a major source of multiple potent allergens are referred to as house dust mites. The most important of these, because of their cosmopolitan occurrence and abundance in homes, are *Dermatophagoides farinae, Dermatophagoides pteronyssinus* along with certain spp. of mites. Highest concentration of these species are usually found in high-use areas, such as beds, overstuffed furniture and carpeted floors were shed skin scales collect and serve as a food source (Arlian, 1989). Choovoravech (1974), Spieksma and Spieksma-Boezeman (1967) have also suggested that house dust contained mites of the genus *Dermatophagoides* which are the major source of allergens in house dust. Sears *et al.* (1993) and Peat *et al.* (1996) found a direct relationship between severity of asthma and their exposure to dust mites.

Materials and Methods

A total 100 dust samples along with data collected from mattresses, sofas, carpets and by dusting clothes of asthmatic patients. Data regarding occurrence of disease, type of house, floor and income of family was collected by using a questionnaire, which was prepared on the pattern of the questionnaire used by American Thoracic Society-Division of Lung Diseases Respiratory Symptoms and the questionnaire used by Health and Welfare Canada for a community based study of Children's Health (Ferris, 1978; Stern *et al.*, 1989).

The mites were extracted from dust with the help of modified Berlese funnel (Haarlov, 1947 and Mac Fadyen, 1953). It was found that on average it took 24–36 hrs to isolate the maximum number from a sample. The specimen were preserved in a mixture of 70 per cent alcohol and glycerin in 10 : 1 ratio in small vials for microscopic studies the mites were cleared in 60 per cent lactic acid 10-15 specimens were transferred to a cavity block containing 60 per cent lactic acid and were kept 3–5 days at 37°C in an oven. Gentle warming will accelerate the rate of clearing. The cleared mites were mounted in Hoyer's medium (Krantz, 1978) for preparing permanent whole mounts.

Results and Discussion

Analysis of dust from the homes of asthmatic patients, have shown that the mites belong to *Dermatophagoides* spp. were found in 81.57 per cent of the samples. Other pyroglyphid and non pyroglyphid mites spp. present in the dust samples were *Acarus* spp., *Glycyphagus* spp., *Cheyletus* spp., *Tyrolichus* spp., *Tyrophagus* spp., *Thyreophyus* spp. *Rhizoglyphus* spp., *Aleuroglyphus* spp., *Calpoglyphus* spp., *Suidasia* spp., *Demanyssus* spp. and *Oribatida* sp. were found in significantly less number (Graph 45.1, Figures 45.1–45.8). These findings were in consistent to the findings of Arlian *et al.*, 1984; 1984; Cuthbert *et al.*, 1980; Green *et al.*, 1978; Ingram *et al.*, 1979; Mumcuoglu *et al.*, 1977; Warren *et al.*, 1974; Woodcock *et al.*, 1980; Wraith *et al.*, 1979. According to these workers the mites belonging to *Aleuroglyphus, Carpoglyphus, Suidasia, Dermanyssus, oribatida,* Genera were less prevalent in house dust therefore they are less important in house dust mite allergy.

The percentage of mite positive homes and mite density within the homes are generally chiefly determined by relative humidity and availability of adequate food. The homes with lose tile floor and without proper ventilation had more mites than the homes with marble floor and proper ventilation. These findings are supported by the finding of Arlian *et al.* (1978, 1984). The percentage of asthmatic patients belonging to low socio-economic status were quite high than the people belonging to high socio-economic status. The reason for this can be associated with physical factor in the homes. Hart

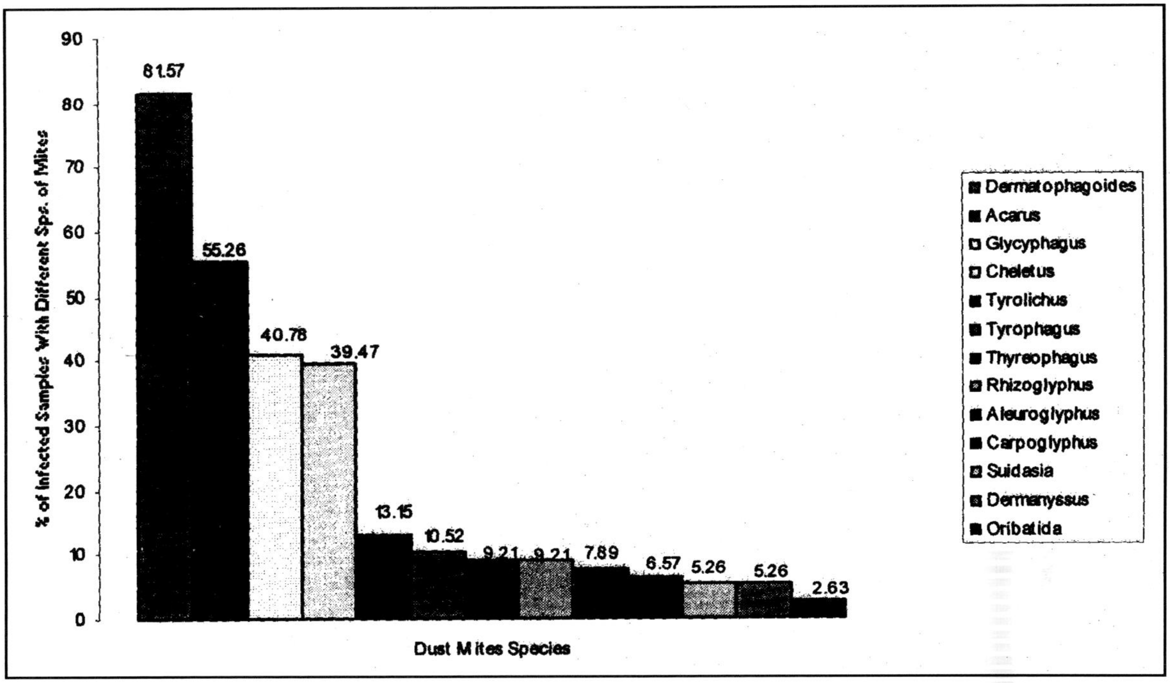

Graph 45.1: Showing the Percentage of Different Species of Mites Present in Dust Samples

(1998) has given the same observations. According to him the people belonging to low socio-economic status are more prone to allergic diseases. Exposure to dust mite is one of the most important factor for aggravating asthma (Platts-Mill, 1992; Peat *et al.*, 1996; Custovic *et al.*, 1996) and severity of asthma varies with exposure to dust mite (Mark, 1998 and Nelson *et al.*, 1999).

Following species have been identified from the house dust of asthmatic patients during present studies.

Order: Astigmata

Family: Acaridae

Genus: *Acarus siro* L. 1758 (Figure 45.1).
Acarus gracilus Hughes, 1957. *Acarus* sp.
Acarus sp.

Genus: *Tyrophagus putrescentiae* (Schrank, 1781) (Figure 45.2).
Tyrophagus tropicus Robertson, 1959.
Tyrophagus sp.

Genus: *Tyrolichus casei* Oudemans, 1910 (Figure 45.3).

Genus: *Aleuroglyphus ovatus* (Trofepeau, 1878).

Genus: *Rhizoglyphus robini* (Lapareds, 1869).

Genus: *Thyreophagus entomophagus* (Laboulbene, 1852) (Figure 45.4).

Genus: *Suidasia* sp.

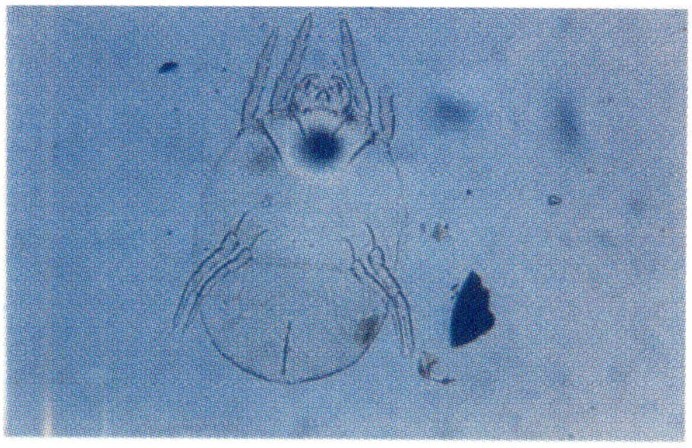

Figure 45.1: *Acarus siro*

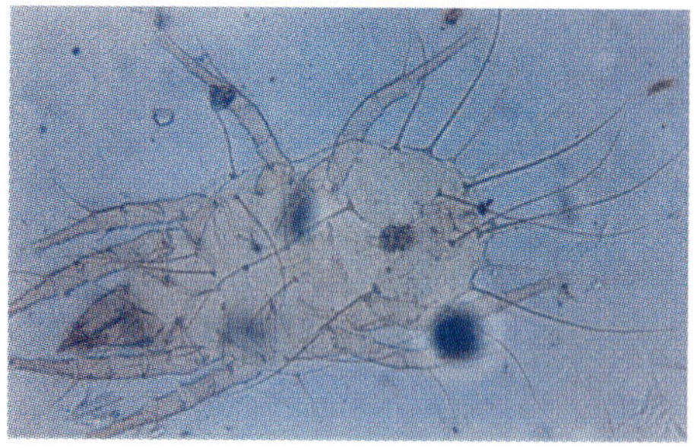

Figure 45.2: *Tyrophagus putrescentiae*

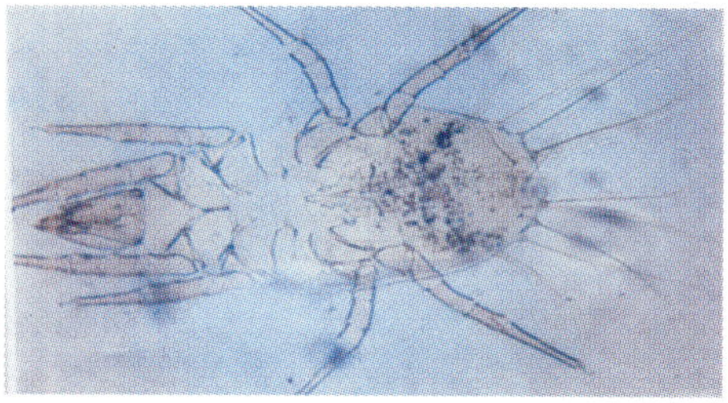

Figure 45.3: *Tyrolichus casei*

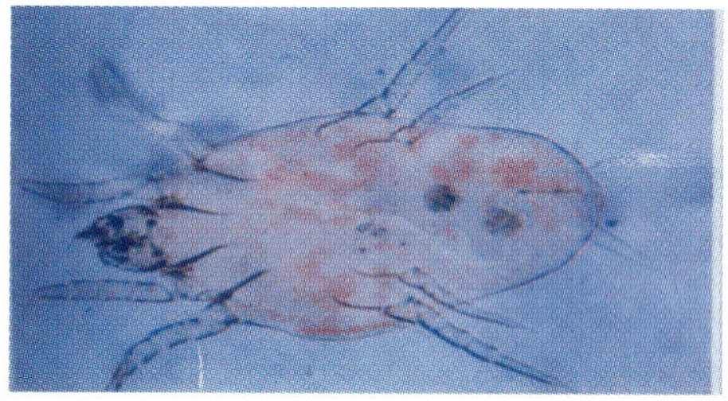

Figure 45.4: *Thyreophagus entomophagus*

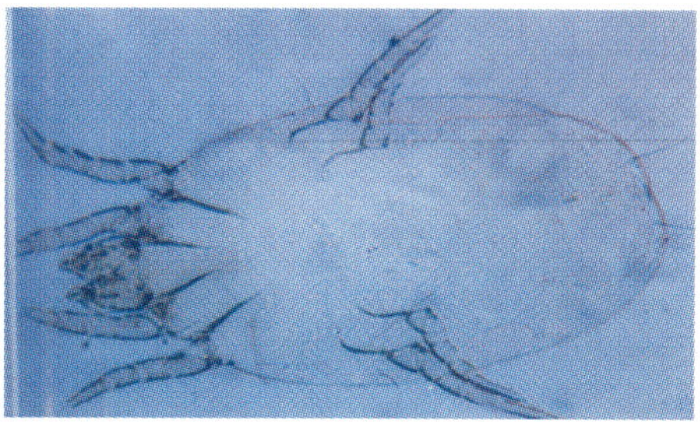

Figure 45.5: *Glycyphagus ornatus*

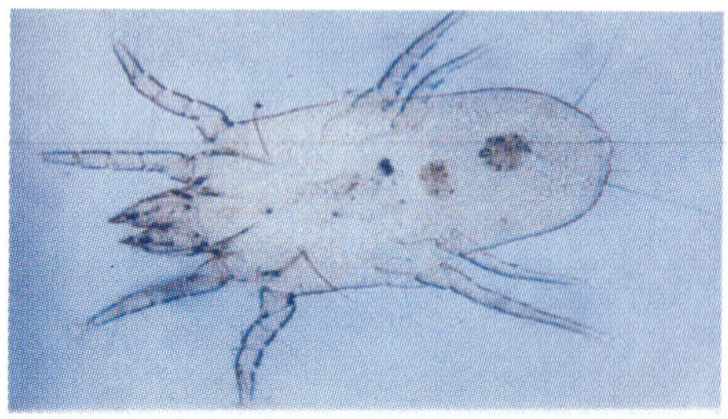

Figure 45.6: *Dermatophagoides farinae*

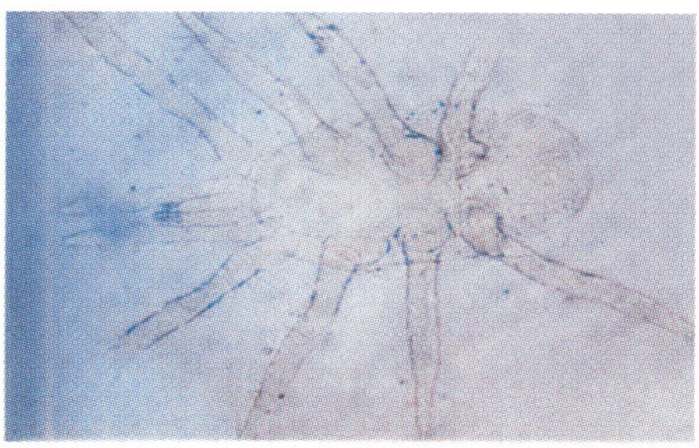

Figure 45.7: *Dermanyssus gallinae*

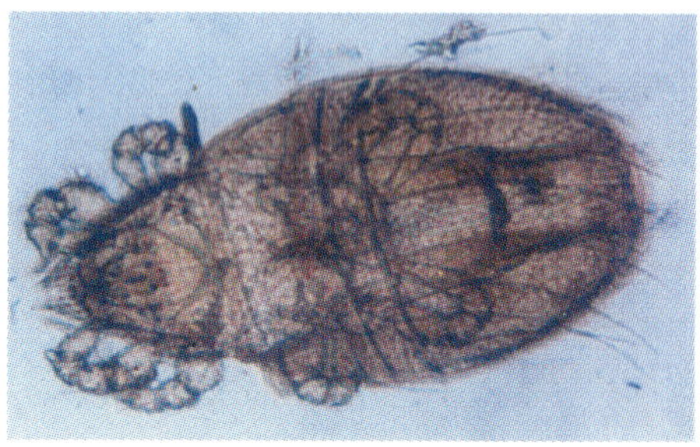

Figure 45.8: *Oribatida* sp.

Family: Glycyphagidae

Genus: *Glycyphagus ornatus* Kramer, 1881 (Figure 45.5).
Glycyphagus destructor (Schrank, 1781).
Glycyphagus sp.

Family: Carpoglyphidae

Genus: *Carpoglyphus lactis* (L., 1758).

Family: Pyroglyphidae

Genus: *Dermatophagoides farinae* Hughes, 1961.
Dermatophagoides pteronyssinus (Trouessart, 1897) (Figure 45.6).

Order: Prostigmata

Family: Cheyletidae

Genus: *Cheyletus aversor* Rohdendorf, 1940.
Cheyletus eruditus (Schrank, 1781).
Cheyletus trouessarti Oudemans, 1903.
Cheyletus sp.

Order: Mesostigmata

Family: Dermanyssidae

Genus: *Demanyssus gallinae* (De Geer, 1778) (Figure 45.7).

Order: Cryptostigmata

Super Family: Galumnoidae

Genus: *Oribatida* sp. (Figure 45.8).

References

American Thoracic Society, 1996. Health effects of outdoor air population. *Am. J. Respir. Crit. Care Med.*, 153: 3–50.

Arlian, L.G., Brandt R.L. and Bernstein, R., 1978. Occurrence of house dust mite, *Dermatophagoides* spp. (Acari : Pyroglyphidae) during the heating season. *J. Med. Entomol.*, 15: 35.

Arlian, L.G., Geis D.P., Vyszenski-Moher, D.L., 1984. Antigenic and allergenic properties of the storage mite *Tyroohagus putrescentiage*. *J. Allergy Clin, Immunol.*, 74 : 166.

Choovoravch, 1974. Effect of immunotherapy in bronchial asthma: treatment with extracts of house dust and mite. *J. Med. Assoc. Thailand*, September: 445–450.

Cook, R.A., 1922. Studies in specific hypersentiveness, JV. New etiological factors in bronchial asthma. *J. Inmunol.*, 7: 147.

Custovic, A., Simon, C.D., Taggart, H.C., Martin, D., Champan and Ashely, W., 1996. Exposure to house dust mite allergens and the clinical activity of asthma. *J. Allergy Clin. Immun.*, 98(1): 64–72.

Cuthbert, O.D.J., Brostoff, D.G., Wraith and Brightens W.D., 1979. 'Brain allergy', asthma and rhinitis due to storage mites. *Clin. Allergy*, 9: 229–336.

Dekker, H., 1928. Asthma and Milber. *Munch. Med. Wschr.*, 75: 615.

Dekker, H., 1930. Allergisches Asthma. In Prktikum der Allergischen Krankheiten. Montana-Verlag, Leipzig.

Ferris, B.G., 1978. Epidemiology standardization project. *Am. Rev. Respir. Dis.*, 118: 36–47.

Green, W.F. and Woolcock, A.J., 1978. *Tyrophagus putrescentiae*: An allergenically important mite. *Clin. Allergy*, 8: 135.

Hart, B.J., 1998. Life-cycle and reproduction of house-dust mites: Environmental factors influencing mite populations. *Allergy*, 53 (Suppl. 48): 13–17.

Ingram, C.G., Jeffrey, I.G. and Symington, I.S., 1979. Bronchial provocation studies in farmers allergic to storage mites. *Lancet*, 22/29: 1330.

Kern, A., 1921. Dust sensitigation in bronchial asthma. *Med. Clin. N. Amer.*, 5: 751.

Krantz, G.W., 1978. *A Manual of Acrology*, 2nd edn. Oregon State University Books Store, Inc.

Marks, G.B., 1998. House-dust mite exposure as a risk factor for asthma: Benefits for avoidance. *Allergy* (*Copenhagen*), 53 (Suppl. 48): 108–114.

Maunsell, K., D.G. Wraith and Cunnington, A.M., 1968. Mites and house dust allergy in bronchial asthma. *Lancet*, 1: 1267– 1270.

Peat, J.K. and Li, J., 1999. Reversing the trend: reducing the prevalence of asthma. *J. Allergy Clin. Immunol.*, 103: 1–10.

Peat, J.K., Tovey, E., Toelle, B.G., Haby, M.M., Gray, E.J. and Mahmic, A., 1996. House dust mite allergens: A major risk factor for childhood asthma in Australia. *Am. J. Respir. Crit. Care Med.*, 153: 141–146.

Plattis-Mills, T.A.E., 1992. Dust mite allergens and asthma: Report of a second international workshop. *J. Allergy Clin. Immunol.*, 89: 1046–1060.

Plattis-Mills, T.A.E. and De Weck, A., 1989. Dust mite allergens and asthma a worldwide problem, *J. Allergy Clin. Immunol.*, 83: 416–427.

Sears, M.R., Burrows, B., Herbison, G.P., Holdaway, M.D. and Flannery, E.M., 1993c. Atopy in childhood III: Relationship with pulmonary function and airway responsiveness. *Clin. Exp. Allergy*, 23: 957–963.

Spieksma, F.T.M. and Spieksma-Boezman, M.I.A., 1967. The mite fauna of house dust with particular reference to *Dermatophagoides pteronyssinus* (Trouessart, 1897) (Psoroptidae : Sarcoptiformes). *Acarologia*, 9: 226–241.

Spieksma, F.Th.M. and Voorhorst, R., 1969. Comarisons of skin reactions to extracts of house dust, mites and human skin scales. *Acta Allergol.*, 24: 124–144.

Voorhorst, R., F.Th.M. Spieksma and VareKamp, H., 1969. House dust atopy and the house dust mite *Dermatophagoides pteronyssius* (Trouessart 1897). Staffleu's Scientific Publishing Co., Ledien.

Warren, P., Cherniack, R.M. and Tse, K.S., 1974. Hypersensitivity reactions to grain dust. *J. Allergy Clin. Immunol.*, 33: 139.

Woodcock, A.A. and Cunnington, A.M., 1980. The allergenic importance of house dust and storage mites in asthmatic in Brunei, S.E. Asia. *Clin. Allergy*, pp. 10–609.

World Heaith Organisation, 2000. *The World Health Report 2000*. Bronchial asthma. WHO, Geneva.

Wraith, D.G., Cunnington, A.M. and Seymour, W.M., 1979. The role and allergenic importance of storage mites in house dust and other environments. *Clin. Allergy*, 9: 545.

Chapter 46

Effect of Mulches on Soil Properties and Yield of Maize in Mizoram

U.K. Sahoo and Nigel J.S. Kharmalki
Department of Forestry, School of Forestry and Earth Sciences, Mizoram University, Post Box 190, Aizawl – 796 001, Mizoram, India

ABSTRACT

The effect of four mulches *viz.*, rice straw, weeds (*Eupatorium*), *Leucaena* leaves and polythene with and without alder (*Alnus nepalensis* D. Don) were studied on soil properties and yield of maize in Mizoram during 2004. The N, P and K levels in the soil were increased due to addition of *Leucaena* and *Eupatorium* leaf mulches whereas the soil pH decreased by these mulches. The maize productivity was significantly ($P < 0.05$) higher under *Leucaena* leaf mulched plots than the other treatments. The study depicted an important role played by alder in improving soil properties and enhancing maize yield.

Keywords: *Mulches, Soil properties, Maize yield, Alder, Mizoram.*

Introduction

Maize (*Zea mays* L.) is one of the most important cereal crops and is grown either as a sole crop or with other species over 8260 ha of land in Mizoram. The rapid use of chemical fertilizers was reported to decrease the yield of maize but growing of maize with different nitrogen fixing tree species in recent years have resulted in enhanced yield (Mathew *et al.*, 1992, Khola *et al.*, 1997, Sahoo *et al.*, 2005). Shifting cultivation (locally known as jhum) which is traditionally practiced widely in the hilly terrain causes numerous problems like heavy soil and water erosion and consequent low maize productivity. The productivity of maize can be increased by adopting improved package of practices, particularly *in situ* moisture conservation by mulching. The soil properties may also be affected by use of different mulches in the system (Sharma and Pawar, 1998; Mc Intyre *et al.*, 2000). The present investigation therefore seeks to determine the effect of three vegetative mulches (rice straw, weed leaves, *Leuceana leucocephala*) and one artificial mulch (polythene) on soil moisture retention, pH, nitrogen, phosphorus, potassium and organic carbon and productivity of maize during 2004 in Mizoram.

Materials and Methods

Study Site

Study was carried out at Tanhril which is located on the south-western part of Aizawl city, the capital of Mizoram and is about 22 km from Aizawl city and lies between 92°38′ to 92°42′ E longitude and 23°42′ to 23°46′ N latitude. The site is at an altitude of 350 m with an average rainfall of 250 cm. The site is moderately slope and covers an area of 167.5 sq. m. The experiment plot consists of one year old established Alder (*Alnus nepalensis* D. Don) on recently cleared forest stand. Maize (*Zea mays* L.) was selected as an agriculture crop introduced between the inter-spaces of the tree species.

Design and Layout of Experiment

Field experiment was carried out following the Randomized Block Design (RBD). The experimental plot covers a total area of 167.5 sq.m. This plot was divided into two equal parts (plot 1 and plot 2). In plot 1 maize was raised along with alder trees, while in plot 2 maize was planted without the Alder trees. In both plot 1 and plot 2, four types of surface mulch were applied to maize. A control (without mulch) was also maintained in each of the plots, thus there were a total of ten (10) treatments and each treatment was replicated thrice. The treatments included: T_1: Maize + *Alnus* + Rice Straw, T_2: Maize + *Alnus* + Weed (*Eupatorium*), T_3: Maize + *Alnus* + *Leuceana leucocephala* leaves, T_4: Maize + *Alnus* +Artificial Mulch (polythene), T_5: Maize + *Alnus*, T_6: Maize + Rice Straw, T_7: Maize + Weed (Eupatorium), T_8: Maize + *Leuceana leucocephala* leaves, T_9: Maize + Artificial mulch (polythene) and T_{10}: Only maize crop. The alder trees were maintained at a uniform spacing of 1.5 × 1.5 m. Maize was planted at a uniform spacing of 60 × 40 cm. Thus there were a total of sixty alder trees and six hundred and seventy maize plants at the experimental site. Each category of mulch was incorporated at the time of maize planting at the rate of 10 tons per hectare. During the period of the experiment, three weeding were carried out. The first weeding was done in mid-April, the second in mid-June and the last in the first week of August, prior to harvesting of maize. The crop was raised purely under rain fed condition.

Soil moisture was determined on a monthly interval from the samples collected at a depth of 30 cm. Soil moisture loss on drying to a constant weight was determined for 100 g fresh soil. The other soil properties such as soil pH, nitrogen, phosphorus, potassium, and organic carbon were determined at monthly intervals as per Aliens *et al.* (1974). Crop productivity was assessed at the time of crop harvest. The data were subjected to analysis of Variance (ANOVA) to see the effect of mulches on soil moisture retention, soil properties and maize yield.

Results and Discussion

Soil Moisture Retention

Soil moisture retention was lowest in month of May and highest in July in all the treatments. However, straw mulch was found to conserve more moisture as compared to other mulches although there was ample variation in the soil moisture content from the various treatments (Table 46.1). The higher moisture retention by straw mulch could be attributed to its slower rate of decomposition than the other mulch materials. Least moisture conservation was recorded for the control plots without mulch. This was found to be in consonance with the findings of Mohan *et al.* (2004). The vegetable mulch materials may act like some cover crops on hill slopes (Khera and Singh, 1995, 1998) and help not only in reducing the run off and nutrient losses (Hadda and Sur, 1989) but also in improving soil water regime (Unger, 1978).

Table 46.1: Effect of Mulches on Soil Moisture Retention (%) in the Site During the Study Period

Treatments	Months				
	April	May	June	July	August
T_1	9.70	8.87	9.05	9.27	9.13
T_2	8.97	7.97	8.72	9.07	9.05
T_3	8.92	8.00	8.79	9.12	9.06
T_4	9.25	8.23	8.96	9.12	9.09
T_5	8.10	7.81	8.30	9.00	8.89
T_6	9.73	8.80	9.04	9.23	9.11
T_7	8.92	7.98	8.68	9.08	9.06
T_8	8.94	7.98	8.97	9.06	9.06
T_9	8.87	8.09	8.91	9.10	9.10
T_{10}	8.11	7.84	8.31	9.29	8.92
S.E. m ± n = 24	0.10	0.07	0.05	0.032	0.02
C.D. (P = 0.05)	0.35	0.25	0.20	NS	0.11

Soil Properties

The mean monthly records of soil pH of the various treatments (Table 46.2) showed highest pH in the month of April and lowest in August. The treatments with plastic mulch and control raised without Alder trees showed the highest pH. However, the lowest pH values were recorded for treatments with *Leuceana* and weed (*Eupatorium*) mulch. The soil pH gradually declined with decomposition time of mulch materials and this was true in all the cases, possibly due to release of more bases to the soil. This is clearly reflected from the fact that the pH did not change in control (Table 46.2). The highest soil nitrogen (per cent) levels were recorded in May and lowest in August although it did not vary significantly between the months. Soil nitrogen was higher in *Leuceana* and *Eupatorium* (weed) mulch applications compared to other mulch materials (Table 46.3). Incorporation of plant residues as mulch increased the concentrations of nitrogen in the soil. The foliage of *Leucaena* and *Eupatorium* containing considerable nitrogen when mixed with the soil improved the soil nitrogen status. Although there was no significant variation in soil phosphorus between treatments, a decreasing trend was observed from May to August (Table 46.4). Highest P (33.78 kg/ha) was found in the plot treatment with *Leuceana* mulch and lowest (30.47 kg/ha) in control. During May to August, the microbial population must have multiplied frenetically consuming sizeable amount of phosphorus resulting in its lower level in all treatments. The levels of potassium gradually decreased form the initiation to the end of the study period. Highest K content (146.99 kg/ha) was recorded for treatment with *Leuceana* mulch in May and the lowest (142.64 kg/ha) was for the control in August (Table 46.5). Soil organic carbon (per cent) increased significantly ($P < 0.05$) compared to the control towards the end of study period which could have been due to faster decomposition of most of the mulch materials (Table 46.6). This explains the role played by decomposing mulches in bringing some changes in organic carbon (per cent) in all treated plots especially at the end of study period. The organic carbon levels were highest in the month of April and lowest in the months of July and August. The highest organic carbon (7.08 per cent) in the month of April was for treatment with weed (*Eupatorium*) mulch and the lowest (5.35 per cent) was in control during the month of August.

Table 46.2: Effect of Mulches on Soil pH in the Site During the Study Period

Treatments	*Months*				
	April	*May*	*June*	*July*	*August*
T_1	5.77	5.50	5.20	5.06	4.98
T_2	5.42	5.11	5.03	4.92	4.76
T_3	5.41	5.25	5.03	4.87	4.81
T_4	6.07	5.71	5.53	5.01	4.98
T_5	6.07	5.72	5.56	5.08	5.00
T_6	6.13	5.59	5.30	5.09	4.94
T_7	6.04	5.23	4.96	4.93	4.87
T_8	5.65	5.27	4.93	5.03	4.85
T_9	6.25	5.77	5.15	5.06	5.00
T_{10}	6.25	6.05	6.14	6.04	6.08
S.E. m ± n = 24	0.06	0.05	0.05	0.02	0.02
C.D. (P = 0.05)	0.10	0.20	0.28	0.32	0.36

Table 46.3: Effect of Mulches on Soil Nitrogen (%) in the Site During the Study Period

Treatments	*Months*				
	April	*May*	*June*	*July*	*August*
T_1	0.20	0.19	0.19	0.18	0.18
T_2	0.21	0.20	0.19	0.18	0.19
T_3	0.20	0.19	0.19	0.19	0.19
T_4	0.20	0.19	0.18	0.16	0.15
T_5	0.20	0.19	0.17	0.16	0.15
T_6	0.20	0.19	0.18	0.18	0.19
T_7	0.21	0.19	0.18	0.18	0.18
T_8	0.20	0.20	0.19	0.19	0.19
T_9	0.20	0.19	0.17	0.16	0.16
T_{10}	0.20	0.20	0.17	0.16	0.16
S.E. m. ± n = 24	0.001	0.0013	0.0022	0.0024	0.0028
C.D. (P = 0.05)	NS	NS	NS	0.014	0.012

Table 46.4: Effect of Mulches on Soil Phosphorus (kg/ha) in the Site During the Study Period

Treatments	*Months*				
	April	*May*	*June*	*July*	*August*
T_1	33.47	32.41	32.25	31.83	31.53
T_2	33.38	33.17	32.84	32.26	32.04
T_3	33.78	33.13	32.80	32.04	31.59
T_4	33.33	32.19	32.15	31.53	30.89
T_5	33.32	32.14	32.09	31.40	30.60
T_6	33.43	32.49	32.56	31.57	31.15
T_7	33.37	33.16	32.37	31.49	32.09
T_8	33.47	32.72	32.36	32.04	31.98
T_9	33.21	32.36	31.8	30.96	30.50
T_{10}	33.19	31.95	31.95	31.00	30.47
S.E. m ± n = 24	0.05	0.12	0.12	0.10	0.12
C.D. (P = 0.05)	NS	NS	NS	NS	0.54

Table 46.5: Effect of Mulches on Soil Potassium (kg/ha) in the Site During the Study Period

Treatments	*Months*				
	April	*May*	*June*	*July*	*August*
T_1	146.78	146.19	145.47	144.65	144.60
T_2	146.82	146.31	145.92	145.27	145.03
T_3	146.96	146.42	145.59	145.16	144.97
T_4	146.34	146.07	145.06	144.04	142.89
T_5	146.26	146.09	145.17	143.44	143.13
T_6	146.72	145.90	145.45	144.57	144.52
T_7	146.76	146.00	145.83	145.00	144.76
T_8	146.99	146.87	145.57	144.92	145.02
T_9	146.29	146.18	145.56	143.57	143.09,
T_{10}	146.14	146.04	145.54	143.48	142.64
S.E. m ± n = 24	0.08	0.10	0.11	0.15	0.19
C.D. (P = 0.05)	NS	NS	NS	0.98	0.87

Crop Productivity

Maximum number of cobs was obtained under treatment with *Leuceana* mulch that was grown along with Alder trees (*Alnus nepalensis*) and minimum number of cobs was from maize grown under control without Alder trees (Table 46.7). It was observed that the mean maximum weight of grain per cob was 78.07 g under *Leuceana* mulch (with alder trees) and the lowest mean weight of grain per cob was from maize in the control plots (70.61 g). The mean maximum number of grains was 405.13 under

Leuceana mulch treatment along with Alder trees. The mean least number of grains per cob (367.73) was recorded in the control sub-plot without alder trees.

Table 46.6: Effect of Mulches on Soil Organic Carbon (%) in the Site During the Study Period

Treatments	*Months*				
	April	*May*	*June*	*July*	*August*
T_1	7.00	6.88	6.45	6.35	6.41
T_2	7.08	6.96	6.51	6.42	6.50
T_3	6.98	6.89	6.59	6.51	6.46
T_4	6.99	6.85	6.26	5.68	5.38
T_5	6.96	6.0	5.89	5.67	5.35
T_6	7.00	6.84	6.40	6.40	6.45
T_7	7.03	6.88	6.41	6.35	6.40
T_8	6.96	6.96	6.50	6.45	6.45
T_9	6.94	6.82	6.06	5.59	5.50
T_{10}	6.96	6.69	5.78	5.30	5.00
S.E. m ± n = 24	0.01	0.01	0.02	0.03	0.03
C.D. (P = 0.05)	NS	NS	0.15	0.16	0.15

Table 46.7: Effect of Mulches on Crop Productivity of Maize in the Site

Treatments	*Parameters for Crop Productivity*		
	Average No. of Cobs per Plant	*Average Weight of Grains per Cob (gm)*	*Average No. of Grains per Cob*
T_1	5.67	74.14	394
T_2	6.73	77.03	402
T_3	6.81	78.07	405
T_4	4.87	71.28	377
T_5	4.7	70.61	375
T_6	5.53	73.6	389
T_7	6.6	75.93	400
T_8	6.7	76.92	401
T_9	4.6	70.82	370
T_{10}	4.43	70.63	368
S.E. m ± n = 24	0.18	0.53	2.81
C.D. (P = 0.05)	0.59	1.38	13.99

The role of organic waste as compost and mulch in increasing various crop yield such as rice (Bawm, 1990), corn yield (Gill *et al.*, 1992, Ortega *et al.*, 2000) pigeon pea (Gajera *et al.*, 1998), potato (Acharya and Kapur, 2001) and ginger (Agarwal *et al.*, 2003) have been reported. The N availability in the soil due to stubble and vegetable waste materials may help in fertility management (Bawm, 1990,

Singh, 1995). Therefore it is suggested to practice tillage and mulching combining for improved crop productivity (Hadda and Sur, 1989, Moitre *et al.*, 1994).

The overall yield of maize was found to be significantly ($P < 0.05$) higher in the treatments with *Leuceana, Eupatorium* (weed) and straw mulch raised along with *Alnus nepalensis*. This is in accordance with the findings of Mannering and Mayor (1963) and Singh *et al.* (2002) in which growth of maize was highest under vegetative mulch treatments. The present study depicts better growth and higher grain yield of maize by application of mulch as compared to the un-mulched crops. Further, mulching has conserving effect on soil hydrothermal regime; hence it is a very important management practice for wide spaced maize in sloppy lands under rainfed conditions.

Acknowledgement

We wish to acknowledge the grant received from ICAR, New Delhi (Sanction No.5(29)197-SW&DF) for carrying out this work.

References

Acharya C.L. and Kapur, O.C., 2001. Using organic wastes as compost and mulch for potato (*Solanum tuberosum*) in low water retaining hill soils of North-West India. *Ind. J. Agric. Sci.*, 71: 306–309.

Agarwal, S., Korla, B.N. and Raina, J.N., 2003. Effects of mulch on soil hydrothermal regimes, weed incidence, yield and quality of ginger. *J. Ind. Soc. Soil Sci.*, 51: 65–67.

Allen, S.E., Grime S., Parkinsons, I.A. and Quarmby, G., 1974. *Chemical Analysis of Ecological Materials*. Blackwell Scientific Publication, Oxford.

Bawn, P.E., 1990. Effects of stubble and N fertilization management on N availability and up take under successive rice (*Oyza sativa* L.) crops.*Plant and Soil*, 121: 11–19.

Gajera, M.S., Ahlawat, R.P.S. and Ardashna, P.B., 1998. Effect of irrigation schedule, tillage depth and mulch on growth and yield of winter pigeon pea (*Cajanus cajan*). *Ind. J. Agron.*, 43: 689–693.

Gill, K.S., Arshad, M.A., Chivunda, B.K. and Gumbo, M., 1992. Influence of residue mulch, tillage and cultural practices on weed and corn yield from three field experiments. *Soil and Tillage Res.*, 24: 211–223.

Hadda, M.S. and Sur, H.S., 1989. Effect of mulch on rate of runoff, sediment and nutrient losses from a sandy soil under untilled and tilled conditions. *J. Res. P.A.U., Ludhiana*, 26: 37.

Khera, K.L. and Singh, G., 1995. Effect of paddy straw mulch and rainfall intensity on runoff and soil loss under simulated rainfall. *Ind. J. Soil Conserv.*, 23: 20–23.

Khera, K.L. and Singh, G., 1998. Effect of crop cover and field slope on soil on soil erosion in northern plain hot sub-humid Punjab. *Ind. J. Soil Conserv.*, 26: 19–21.

Khola, O.P.S., Dube, R.K. and Sharma, N.K. 1997. Biological and economic feasibility and intercropping of legumes with maize (*Zea mays* L.) on slopping valley lands. *Ind. J. Soil Cons.* 25: 141.

Mannering, J.V. and Mayer, L.D., 1963. The effect of various rates of surface mulch on infiltration and erosion: empirical studies on soil conservation technique and design procedure. *Soil Conserv. RPC, Morgan*, 19.

Mathew, R.B., Lungu, S., Volk, J., Holden, S.T. and Solberg, K., 1992. The potential of alley cropping in improvement of cultivation systems in high rainfall areas of Zambia. II. Maize production. *Agroforestry System*, 17: 241.

Mc Intyre, B.D., Speijer, P.R., Riha, S.J. and Kizito, F., 2000. Effects of mulching on biomass, nutrients and soil water in banana inoculated with nematodes. *Agron. J.*, 91: 1081–1085.

Mohan, R., Lalsanglura, C. and Sahoo, U.K., 2004. Effect of different mulches on soil, moisture and crop productivity of maize in an agroforestry system of Mizoram. *Sci. Vis.*, 4: 14–17.

Moitra, R., Ghosh D.C., Majumdar S.K. and Sarkar, S., 1994. Influence of tillage and mulching on growth, yield and economics of rainfed yellow (*Brassica rapa* var. *glauca*) in laterite acid belt of West Bengal. *Ind. J. Agric. Sci.*, 64: 156–159.

Ortega, A.L., Sayre, K.D. and Francis, C.A., 2000. Wheat and maize yields in response to straw and nitrogen under a bed planting system. *Agron. J.*, 92: 295–302.

Sahoo, U.K., Mohan, R., Vanlalhluna, P.C. and Sangma, M.M., 2005. Effect of *Leucaena leucocephala* (Lam.) de Wit on the growth and yield of *Zea mays* L. in the humid subtropics of Mizoram, India. *J. Nat. Conserv.*, 17: 177–183.

Sharma, P.P. and Parmar, D.K., 1998. Effect of phosphorus and mulching on the uptake of secondary nutrients and productivity of wheat at different growth stages on an alfisol from Western Himalayas. *J. Ind. Soc. Soil Sci.*, 46: 67–70.

Singh, H., 1995. Nitrogen mineralization, microbial biomass and crop yield as affected by wheat residue placement and fertilizer in a semi-arid tropical soil with minimum tillage. *J. Appl. Ecol.*, 35: 588–595.

Unger, P.W., 1978. Straw mulch rate effect on soil water storage and sorghum yield. *Soil Sc. Soc. Am. J.*, 42: 486–491.

Chapter 47

Parametric Studies of Pollutants from Copper Coated Spark Ignition Engine with Catalytic Converter with Gasoline Blended with Methanol

M.V.S. Murali Krishna, K. Kishor, P.R.K. Prasad, A. Sahithy and G.V.V. Swathy

Department of Mechanical Engineering, Chiatanya Bharathi Institute of Technology, Gandipel, Hyderabad – 500 075, Andhra Pradesh

ABSTRACT

The major pollutants emitted from spark ignition engine run with gasoline fuel are carbon monoxide (CO) and unburnt hydrocarbons (UBHC), which are hazardous to human beings and environment. However, if the engine is run with alcohol, aldehydes are also to be checked which are carcinogenic in nature and cause eye irritation, severe headache and vomiting sensation and hence control of these pollutants call for immediate attention. Experiments are carried out for controlling the pollutants from a variable-compression ratio, copper-coated spark ignition engine run with methanol-blended gasoline (20 per cent V/V) fitted with catalytic converter containing sponge iron catalyst. The influence of the engine parameters such us speed, compression ratio and contiguration of the engine with and without catalytic converter on the pollutants is studied. The speed of the engine has marginal effect, while compression ratio has strong influence on reduction of pollutants. Air Injection into the catalytic converter has further reduced the pollutants. Copper-coated spark ignition engine with methanol-blended gasoline found to decrease the pollutants considerably when compared to conventional engine with pure gasoline operation.

Keywords: *Spark ignition engine, Pollutants, Catalytic converter, Air injection, Copper coating.*

Introduction

CO is major pollutant from spark ignition engine, breathing of which causes many healh disorders, like reduction of haemoglobin content in the blood, increases dizziness, breathing and respiratory problems, eye irritation, loss of appetite etc. (Fulekar, 1999). It also cause detrimental effects on other animal and plant life besides, environmental disorders (Sharma, 1996). This pollutant is considerably high during idling and peak load operation of the engine. Hence globally, stringent regulations are made for permissible CO levels in the exhaust of 2 and 4-stroke spark ignition engines. The formation of unburnt hydrocarbons is due to incomplete combustion. The two important reasons for incomplete combustion of the fuel are cool metal surfaces of the combustion chamber and imperfect mixture ratio. Aldehydes are intermediate compounds of partial combustion. They are carcinogenic in nature. They too cause health problems. Of many methods available for reduction of these emissions, the one employing a catalytic converter is more effective (Vara Prasad *et al.*, 1997). The use of platinum group metals as catalysis is quite expensive and hence efforts are on for search of cheaper catalysts (Luo and Zheng, 1999; Murali Krishna *et al.*, 2000). Further modification of engine design (Nedunchezhian and Dhandapani, 2000) and fuel composition (Pulldir and Abraham, 1985; Sharma *et al.*, 1985) is also found to be advantageous in controlling the pollutants in the exhaust of the engine. The use of catalysts to promote combustion is an old concept. More recently copper is coated over piston crown and inside of cylinder head wall (Dhandapani, 1991) and it is reported that the copper coating improves the fuel economy and increased combustion stabilization. In the context of depletion of fossil fuels due to increase of fuel consumption, the search for alternate and renewable fuels has also become pertinent. The properties of methanol are very close to those of gasoline. In addition, no major modification in the engine is required if low quantities of methanol are blended with gasoline ill spark ignition engine. In the present study, the effect of various engine parameters on the control of these emissions is reported with different versions of the engine such as conventional engine and copper coated engine with catalytic converter with sponge iron as catalyst and methanol blended gasoline (methanol 20 per cent V/V) as fuel.

Experimental Programme

The experimental set-up employed in the present study is shown in Figure 47.1. A four-stroke, single-cylinder, water-cooled, spark ignition engine of brake power 2.2 kW at rated speed of 3000 rpm is used. The engine is coupled to an eddy current dynamometer for measuring its brake power. The compression ratio of the engine is varied from 3 to 9 with the change of the clearance volume by adjustment of cylinder head, threaded to the cylinder of the engine. The engine speeds are varied from 2200 to 3000 rpm. In the present investigations, in order to improve the effective combustion of fuel and to reduce the emission of pollutants, the piston crown and inside surface of the cylinder head is coated (Nedunchezhian and Dhandapani, 2000) with copper by plasma spraying. A bond coating of NiCoCr alloy is applied for a thickness of about 100 microns using an 80 kW METCO plasma spray gun. Over the bond coating, copper (89.5 per cent), aluminium (9.5 per cent) and iron (1.0 per cent) is coated for 300 microns thickness. The coating has very high bond strength and does not wear off even after 50 hrs of operation. A catalytic converter shown in Figure 47.2 is fitted to the exhaust pipe of the engine.

Provision is made to inject a definite quantity of air into the catalytic converter. The converter is filled with sponge iron catalyst with void ratio of 0.7 : 1 (void ratio is the ratio between the volume occupied by the catalyst to the volume of the catalytic chamber), where the pollutants are found (Murali Krishna *et al.*, 2000) to be minimum. The CO and UBHC emissions in the exhaust of the engine

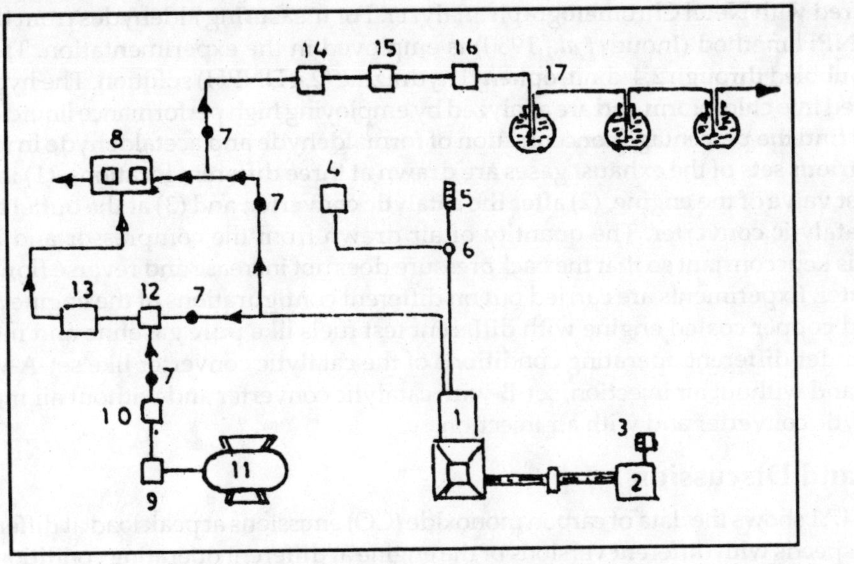

Figure 47.1: Experimental Set-up

(1) Engine, (2.) Eddy current dynamometer, (3) Loading arrangement, (4) Fuel tank, (5) Burette, (6) Three-way valve, (7) Directional valve, (8) CO and UBHC analyzer, (9) Rolometer, (10) Heater, (11) Air-compressor, (12) Air chamber, (13) Catalyst chamber, (14) Rotometer, (15) Heater, (16) Round bottom flask containing 2-DNPH solution

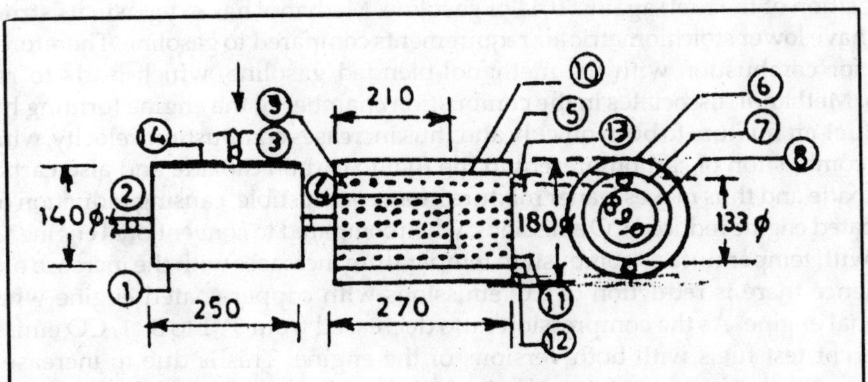

All dimensions are in mm

Figure 47.2: Details of Catalytic Converter

(1) Air chamber, (2) Inlet for air chamber from the engine, (3) Inlet for air chamber from the compressor, (4) Outlet for air chamber, (5) Catalytic chamber, (6) Outer cylinder, (7) Intermediale-cylinder, (8) Inner-cylinder, (9) Inner sheet, (10) Intermediate sheet, (11) Outer sheet, (12) Outlet for exhaust gases, (13) Provision to deposit the catalyst and (14) Insulation

are measured with Netel Chromatograph analyzer. For measuring aldehydes from the exhaust of the engine, DNPH method (Inoue *et al.*, 1980) is employed in the experimentation. The exhaust of the engine is bubbled through 2,4-dinitrophenyl hydrazine (2,4 DNPH) solution. The hydrazones formed are extracted into chloroform and are analyzed by employing high performance liquid chromatography (HPLC) to find the percentage concentration of formaldehyde and acetaldehyde in the exhaust of the engine. Various sets of the exhaust gases are drawn at three different locations, (1) immediately after the exhaust valve of the engine, (2) after the catalytic converter, and (3) at the outlet after air injection into the catalytic converter. The quantity of air drawn from the compressor and injected into the converter is kept constant so that the backpressure does not increase and reverse flow is not created in the converter. Experiments are carried out on different configurations of the engine *i.e.*, conventional engine and copper coated engine with different test fuels like pure gasoline and methanol blended gasoline under different operating conditions of the catalytic converter like set-A-without catalytic converter and without air injection, set-B-with catalytic converter and without air injection and set-C with catalytic converter and with air injection.

Results and Discussion

Table 47.1 shows the data of carbon monoxide (CO) emissions at peak load at different compression ratios and speeds with different versions of the engine at different operating conditions of the catalytic converter. A reduction in CO emissions is found to increase with increase of the speed of the engine for all configurations with different test fuels. Improved combustion with the increase of turbulence decreased CO emissions. It is observed that at each speed, the CO content in the exhaust decreased considerably with the use of catalyst, which is more pronounced with the air injection into the converter. Methanol blended gasoline decreased CO emissions considerably when compared to pure gasoline. The combustion of alcohol produces more water vapor than free carbon atoms as methanol has lower C/H ratio of 0.25 (where C and H represent the number of carbon and hydrogen atoms respectively in the composition of the fuel) against 0.44 of gasoline. Methanol has oxygen in its structure and hence its blends have lower stoichiometric air requirements compared to gasoline. Therefore more oxygen is available for combustion with the methanol-blended gasoline, which leads to reduction of CO emissions. Methanol dissociates in the combustion chamber of the engine forming hydrogen, which helps the fuel-air mixture to burn quickly and thus increases combustion velocity, which brings about complete combustion of carbon present in the fuel to carbon dioxide and also carbon monoxide to carbon dioxide and thus makes leaner mixture more combustible, causing reduction of CO emissions. Copper-coated engine reduced CO emissions; when compared to conventional engine. Catalytic activity increases with temperature as combustion temperature increases with the increase of the speed of the engine. Hence there is reduction of CO emissions with copper-coated engine when compared to conventional engine. As the compression ratio decreased from 9 : 1 to 8 : 1, CO emissions decreased with different test fuels with both versions of the engine. This is due to increase of exhaust gas temperature with the decrease of compression ratio leading to oxidation of CO in the exhaust manifold with different versions of the engine.

The data of un-burnt hydro carbon (UBHC) emissions are shown in Table 47.2 while Table 47.3 and Table 47.4 show the data of formaldehyde and acetaldehyde emissions at peak load for different test fuels at different compression ratios, speeds with different versions of the engine respectively. The trend exhibited by UBHC emissions is the same as that of CO emissions. As speed decreased from 3000 rpm to 2600 rpm, the turbulence of combustion decreased and hence the speed of the flame decreases. The gas layer is entrapped between the piston and combustion chamber walls leading to increase in quench area. The presence of quench area inhibits the spreading of the flame, thereby increasing the hydrocarbon emissions. Formaldehyde concentration is observed to be more in comparison with acetaldehyde concentration in both versions of the engine when they are run with methanol-blended gasoline, as it is evident from the Table 47.3 and Table 47.4. Copper coated engine decreased aldehydes emissions considerably when compared with conventional engine. Catalytic converter with air injection drastically decreased aldehyde emissions in both versions of the engine.

Table 47.1: Data of Carbon Monoxide (CO) Emissions (%) at Peak Load for Different Test Fuels at Different Compression Ratios, Speeds with Different Versions of the Engine

Fuels →		*Gasoline*						*Methanol Blended Gasoline (20 per cent V/V)*					
Engine Version →		*CE*			*CCE*			*CE*			*CCE*		
Speed (rpm) →		*2600*	*2800*	*3000*	*2600*	*2800*	*3000*	*2600*	*2800*	*3000*	*2600*	*2800*	*3000*
Set A	8 : 1	3.55	3.35	3.25	2.72	2.62	2.5	2.32	2.22	2.11	1.53	1.43	1.33
	9 : 1	3.95	3.85	3.75	3.23	3.13	3.0	2.81	2.71	2.61	2.04	1.94	1.83
Set B	8 : 1	2.05	1.95	1.85	1.66	1.56	1.4	1.33	1.23	1.12	0.72	0.61	0.51
	9 : 1	2.55	2.35	2.25	2.02	1.92	1.8	1.74	1.64	1.52	1.12	1.01	0.91
Set C	8 : 1	1.52	1.42	1.3	1.24	1.14	1.0	0.83	0.72	0.61	0.52	0.45	0.34
	9 : 1	1.71	1.61	1.5	1.43	1.33	1.2	1.02	0.92	0.81	0.71	0.65	0.54

Set A: Without catalytic converter and without air injection; Set B: With catalytic converter and without air injection; Set C: With catalytic converter and with air injection; CE: Conventional engine; CCE: Copper coated engine.

Table 47.2: Data of Unburnt Hydro Carbon (UBHC) Emissions (ppm) at Peak Load for Different Test Fuels at Different Compression Ratios, Speeds with Different Versions of the Engine

Fuels →		*Gasoline*						*Methanol Blended Gasoline (20 per cent V/V)*					
Engine Version →		*CE*			*CCE*			*CE*			*CCE*		
Speed (rpm) →		*2600*	*2800*	*3000*	*2600*	*2800*	*3000*	*2600*	*2800*	*3000*	*2600*	*2800*	*3000*
Set A	8 : 1	550	510	450	425	375	325	320	295	270	195	175	155
	9 : 1	610	560	500	445	420	375	370	345	320	240	225	205
Set B	8 : 1	340	300	250	240	215	155	165	140	95	102	80	65
	9 : 1	390	350	300	310	265	206	205	180	135	142	120	105
Set C	8 : 1	240	200	150	310	165	80	125	100	70	81	60	45
	9 : 1	290	250	200	195	152	105	145	120	90	103	80	65

Set A: Without catalytic converter and without air injection; Set B: With catalytic converter and without air injection; Set C: With catalytic converter and with air injection; CE: Conventional engine; CCE: Copper coated engine.

Conclusions

The pollutants of CO, UBHC and aldehyde emissions at peak load decreased by 20–45 per cent relatively with the change of the engine configuration from conventional version to catalytic coated engine with different test fuels. The pollutants decreased by 25–45 per cent relatively with the change of fuel from gasoline to methanol-blended gasoline in both versions of the engine under different sets. They increased by 10–13 per cent relatively with the change of compression ratio from 8 : 1 to 9 : 1 while they decreased by 3–12 per cent relatively with the change of speed from 2600 rpm to 3000 rpm with different test fuels in both configurations of the engine under different operating conditions of the catalytic converter. Air injection decreased the emissions by 60 per cent with different test fuels with different configurations of the engine.

Table 47.3: Data of Formaldehyde Emissions (%) at Peak Load for Different Test Fuels at Different Compression Ratios, Speeds with Different Versions of the Engine

Fuels →		*Gasoline*						*Methanol Blended Gasoline (20 per cent V/V)*					
Engine Version →		*CE*			*CCE*			*CE*			*CCE*		
Speed (rpm) →		*2600*	*2800*	*3000*	*2600*	*2800*	*3000*	*2600*	*2800*	*3000*	*2600*	*2800*	*3000*
Set A	8 : 1	8.5	7.5	6.5	6.5	5.5	4.5	16.2	17.5	16.5	11.4	10.5	9.5
	9 : 1	9.2	8.2	7.5	7.1	6.2	5.5	19.1	18.2	17.4	12.2	11.2	10.4
Set B	8 : 1	6.6	5.5	4.5	4.4	3.5	2.5	9.3	8.5	7.5	7.5	6.5	5.5
	9 : 1	7.5	6.5	5.5	5.4	4.5	3.5	10.4	9.5	8.5	8.4	7.5	6.5
Set C	8 : 1	4.5	3.5	2.5	2.4	1.5	0.75	7.3	6.5	5.5	5.4	4.5	3.5
	9 : 1	5.0	4.0	3.0	3.1	2.0	1.25	8.1	7.0	6.0	6.1	5.0	4.0

Set A: Without catalytic converter and without air injection; Set B: With catalytic converter and without air injection; Set C: With catalytic converter and with air injection; CE: Conventional engine; CCE: Copper coated engine.

Table 47.4: Data of Acetaldehyde Emissions (%) at Peak Load for Different Test Fuels at Different Compression Ratios, Speeds with Different Versions of the Engine

Fuels →		*Gasoline*						*Methanol Blended Gasoline (20 per cent V/V)*					
Engine Version →		*CE*			*CCE*			*CE*			*CCE*		
Speed (rpm) →		*2600*	*2800*	*3000*	*2600*	*2800*	*3000*	*2600*	*2800*	*3000*	*2600*	*2800*	*3000*
Set A	8 : 1	7.5	6.5	5.5	5.5	4.5	3.5	10.5	9.5	8.5	8.4	7.5	6.5
	9 : 1	7.9	7.2	6.5	5.9	5.2	4.5	10.9	10.2	9.5	8.9	8.2	7.5
Set B	8 : 1	5.2	4.5	3.5	3.5	2.5	1.5	6.5	5.5	4.5	4.5	3.5	2.5
	9 : 1	6.2	5.5	4.5	6.5	4.5	2.5	7.0	6.0	5.0	5.0	4.0	3.0
Set C	8 : 1	3.2	2.5	1.5	0.75	0.5	0.25	4.4	3.5	2.5	1.75	1.5	1.25
	9 : 1	3.7	3.0	2.0	1.5	1.0	0.5	5.1	4.0	3.0	2.25	2.0	1.75

Set A: Without catalytic converter and without air injection; Set B: With catalytic converter and without air injection; Set C: With catalytic converter and with air injection; CE: Conventional engine; CCE: Copper coated engine.

Acknowledgements

The authors are thankful to the authorities of Chaitanya Bharathi Institute of Technology, Hyderabad, for the facilities provided. The financial assistance provided by Andhra Pradesh Council of Science and Technology (APCOST) for this project is greatly acknowledged.

References

Dhandapani, S., 1991. Theoretical and experimental investigation of catalytically activated lean burn combustion. *Ph.D. Thesis*, Indian Institute of Technology, Chennai.

Fulekar, M.H., 1999. Chemical pollution: A threat to human life. *Indian J of Env. Prot.*, 1(3): 353–359.

Inoue, T., Oishi, K. and Tanaka, T., 1980. Determination or aldehydes in the automobile exhaust by HPLC. *Toyota Gijistu*, 21(4): 500–506.

Luo, M.F. and Zheng, X.M., 1999. CO oxidation activity and TPR characteristics of CeO_2: Supported manganese oxide catalyst. *Indian J. Chem.*, 38(1): 703–707.

Murali Krishna, M.V.S., Vara Prasad, C.M. and Ramana Reddy, Ch.V., 2000. Studies on control of carbon monoxide emissions in spark ignition engine using catalytic converter. *Ecol. Env. and Conser.*, 6(4): 377–380.

Nedunchezhian, N. and Dhandapani, S., 2000. Experimental investigation of cyclic variation or combustion parameters in a catalytically activated two-stroke SI engine combustion chamber. *Eng. Today*, 2(1): 11–18.

Pundir, B.P. and Abraham, M., 1985. Performance of methanol gasoline blends in Indian passenger cars. In: *Proc. 9th National Conference on I.C. Engine and Combustion*. Indian Institute of Petroleum. Dehradun.

Sharma, B.K., 1996. *Engineering Chemistry*. Pragathi Prakashan (P) Ltd., Meerut.

Sharma, J. *et al.*, 1985. Fleet study of small two stroke engine vehicles fuelled with methanol gasoline blends. In: *Proc. 9th National Conference on I.C. engines and Combustion*.

Vara Prasad, C.M., Murali Krishna, M.V.S. and Prabhakar Reddy, C., 1997. Reduction of CO in petrol engine exhaust using catalytic converter. In: *Proc. 15th National Conference on IC Engines and Combustion*, College of Engineering, Gundi, Anna University, Chennai, pp. 372–377.

Chapter 48

A New Nematode *Hysterothylacium ritai* from Freshwater Fish *Rita rita* Hamilton

I. Rajya Lakshmi and K. Sreeramulu
Department of Zoology, Andhra University, Visakhapatnam – 530 003

ABSTRACT

The present article deals with description of a new species of nematode of the family Anisakidae Railliet et Henry, 1912 from the stomach of freshwater fish, *Rita rita* Hamilton from Krishna river at Vijayawada (Andhra Pradesh). Description is based on two male and four female specimens from a large number of hosts examined. The new species is characterized by body size, location of nerve ring and excretory pore, length of ventriculus, oesophageal appendix and intestinal caecum, number and arrangement of caudal papillae, spicules length, vulva position and shape of tail.

The new species is designated as *Hysterothylacium ritai.*

Keywords: *Hysterothylacium ritai n.sp., Nematode, Anisakidae, Stomach. Freshwater fish, Vijayawada.*

Introduction

In the course of study of nematode parasites of freshwater fishes, a new species of the genus *Hysterothylacium* Ward and Magath, 1917 was obtained from the stomach of *Rita rita* Hamilton. This study gives the description and status of the new species.

Materials and Methods

The freshwater fishes were collected from Krishna river (Vijayawada). Fishes were examined thoroughly, out of 29 fishes examined, 4 were infected. Nematodes were fixed in hot 70 per cent alcohol and were preserved in 10 per cent glycerin alcohol. Two clearing media, glycerol and creosote were used in order to clear and elucidate the specific characters. Internal organs of the specimens could be seen clearly in pure glycerol. Creosote has been found suitable to clear the cuticular structures such as striations and papillae. Specimens cleared in creosote were washed in acid alcohol before returning to 70 per cent alcohol.

For enface-view study, anterior end of the female was cut with a sharp blade and brought into the desired position under the coverglass in creosote on the slide.

All measurements are given in millimetres.

Results

Family: Anisakidae Railliet et Henry, 1912

Genus: *Hysterothylacium* Ward and Magath, 1917

Hysterothylacium ritai n.sp. (Plate 48.1, Figures 1–4; Plate 48.2, Figures 1–4)

Material: Two males and four females

Host: *Rita rita* Hamilton

Location: Stomach

Locality: Vijayawada, Andhra Pradesh, India

These are medium sized worms with their bodies tapers anteriorly, being thicker near posterior end in both sexes. Striations are absent in anterior end of male and female but these are deep and small grooves in posterior end of male and small grooves in posterior end of female. Lips are clearly separated from each other and the space between the adjacent lips occupied by very prominent interlabia. Dorsal lip has two round ends which fit into the cavities of the two subventral lips. Interlabia large and covered with thick cuticle. Oesophagus long, ending in a small muscular and globular ventriculus. A large and posteriorly directed oesophageal appendix is given off with ventriculus, which ends in a circular dilation. Intestine thick, a long caecum projects anteriorly. Excretory pore situated at the level of never ring. Tail end conical and pointed in both male and female.

Male

Measures 24–26 in length and maximum width 0.360–0.384. Body widest in posterior region and narrowest just behind lips. Head diameter 0.132–0.144. Nerve ring at 0.372–0.396 from anterior end. Oesophagus 1.836–1.860 length and 0.060–0.072 width. Ventriculus 0.084–0.108 length and 0.120–0.132 width. Oesophageal appendix 0.792–0.816 and intestinal caecum 4.800–4.860.

Tail short, conical and pointed 0.120–0.132 length. Posterior end of male curved. Altogether 7 pairs of sessile papillae in caudal region, 6 pairs precloacal and pair postcloacal. Caudal alae absent. Spicules unequal, left spicule 0.876–0.912 and right spicule 0.696–0.720.

Female

Measures 28–30 in length with maximum width of 0.456–0.492. Head diameter 0.168–0.192. Nerve ring at 0.444–0.468 from anterior end. Oesophagus 1.956–2.004 length and 0.084–0.096 width. Ventriculus 0.108–0.120 length and 0.120–0.132 width. Oesophageal appendix 0.816–0.852 and intestinal caecum 4.932–4.992.

Tail conical, 0.528–0.552 length. Vulva 16–17 distance from anterior end (postequatorial). Eggs spherical, and measure 0.012–0.024 in diameter.

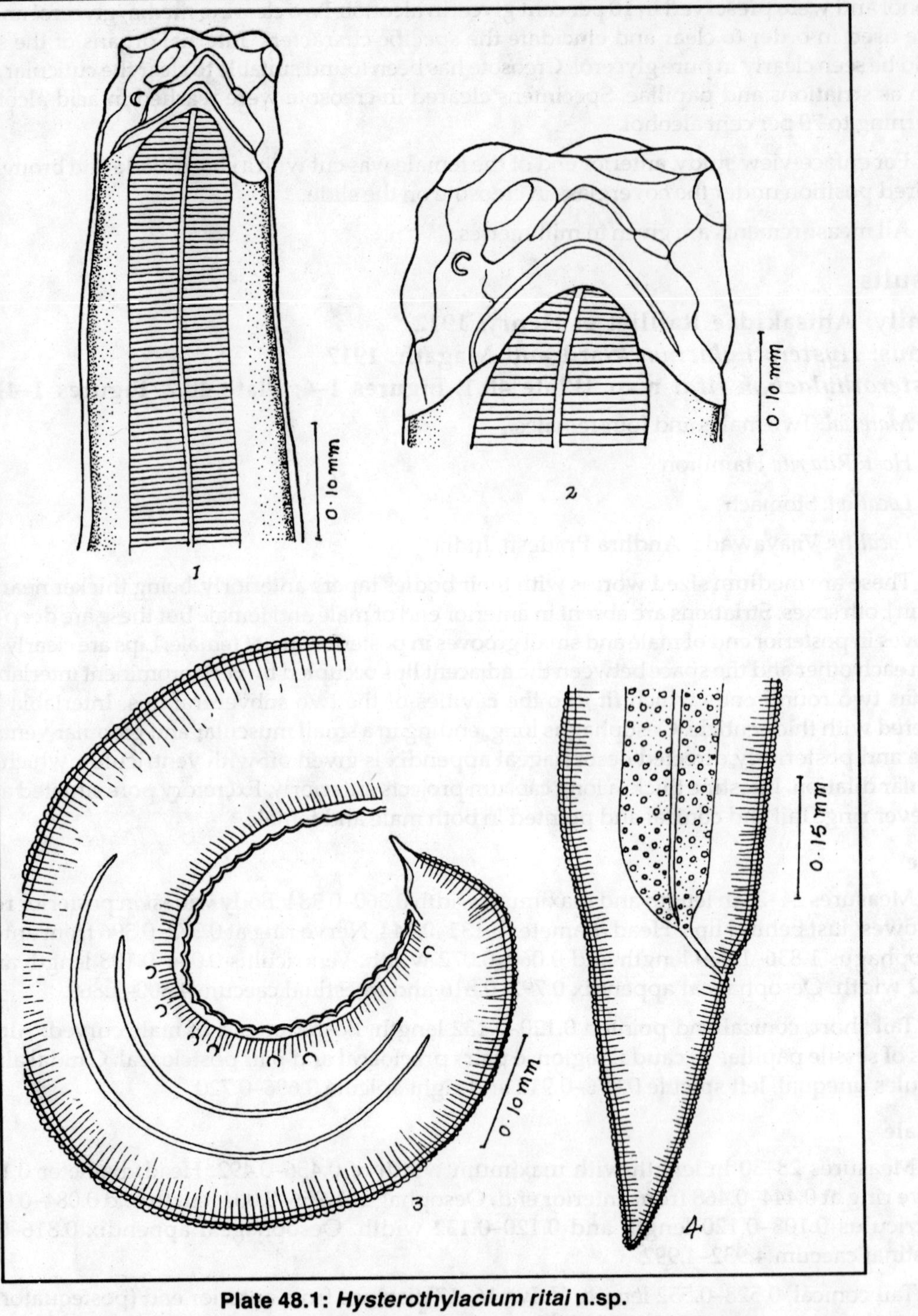

Plate 48.1: *Hysterothylacium ritai* n.sp.

Figure 1: Anterior end of male; Figure 2: Head region-enlarged; Figure 3: Posterior end of male; Figure 4: Posterior end of female.

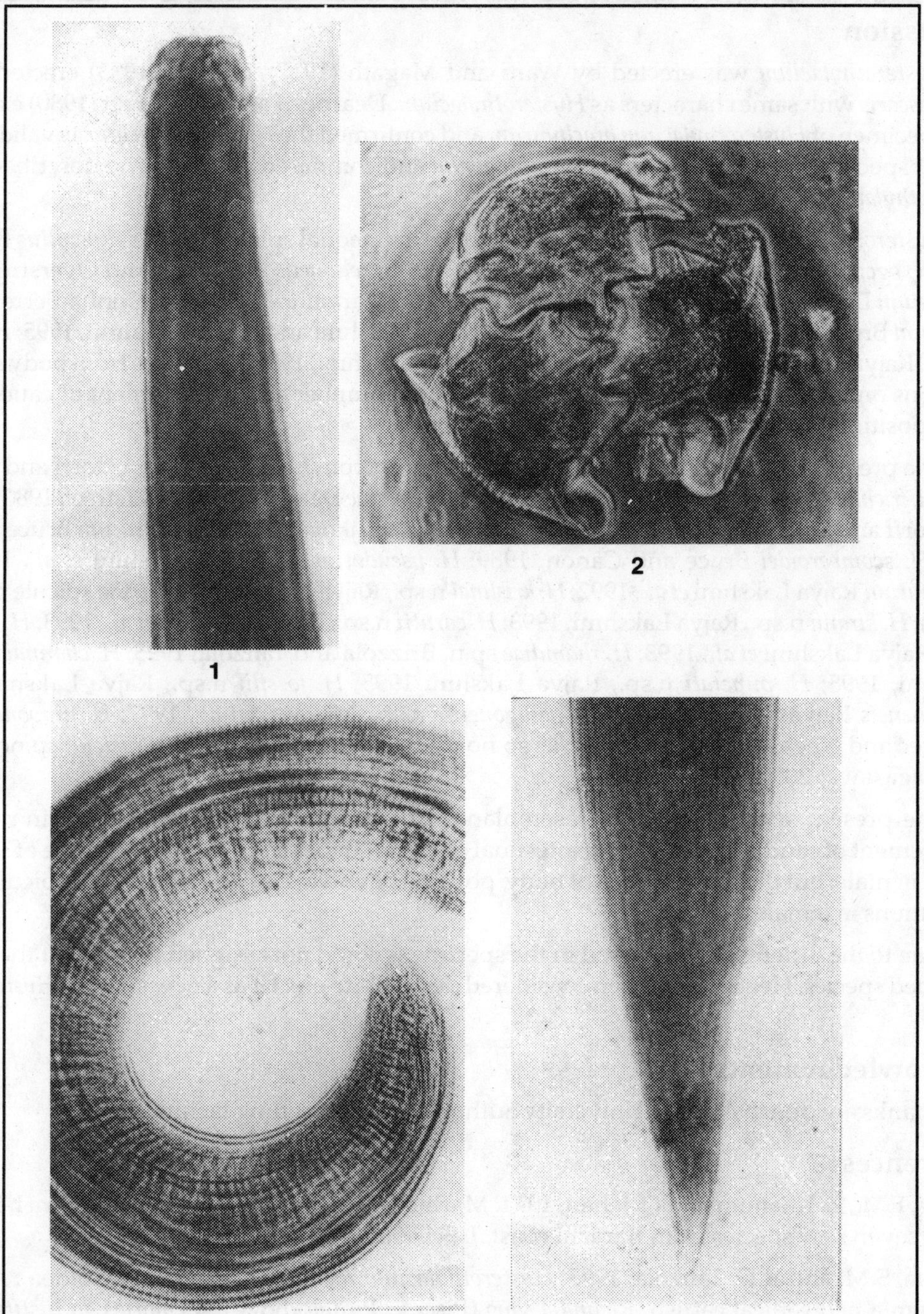

Plate 48.2: *Hysterothylacium ritai* n.sp.

Figure 1: Anterior end of male; Figure 2: Head of male-enface-view; Figure 3: Posterior end of male; Figure 4: Posterior end of female.

Discussion

Hysterothylacium was erected by Ward and Magath (1917). Dollfus (1933) erected the genus *Thynnascaris* with same characters as *Hysterothylacium.* Deardorff and Overstreet (1980) examined the type specimen of *Hysterothylacium brachyurum* and confirmed the *Hysterothylacium* as valid genus. The present specimens considered as *Hysterothylacium* and compared with hitherto described species of *Hysterothylacium.*

Hysterothylacium ritai n.sp. shows resemblance in unequal spicules to *H. synpapillus* Bilqees *et al.*, 1971; *H. ogcocephali* Deardorff and Overstreet, 1980; *H. reliquens* Deardorff and Overstreet, 1980; *H. corrugatum* Deardorff and Overstreet, 1980; *H. aduncum* Carvalho-Varela and Cunha-Ferreira, 1987; *H. pelagicum* Bruce and Cannon, 1989; *H. elurensis* Rajya Lakshmi and Vijaya Lakshmi, 1995; *H. nellorensis* sp.nov. Rajya Lakshmi, 1996 and *H. japonicum* Rajya Lakshmi, 1996 but differs from body size, type of striations on the cuticle, position and size of spicules, number and arrangement of caudal papillae, vulva position and shape of tail.

The present species differs in body measurements from *H. incurvum* Deardorff and Overstreet, 1980; *H. frichiuri* Deardorff and Overstreet, 1980; *H. fortalezae* Deardorff and Overstreet, 1980; *H. chaunaxi* Deardorff and Overstreet, 1980; *H. fabric* Petter and Radujkovic, 1986; *H. cornutum* Bruce and Canon, 1989; *H. scomberoidei* Bruce and Canon, 1989; *H. pseudotumbili* Rajya Lakshmi *et al.*, 1991 and *H. neocomutum* Rajya Lakshmi *et al.*, 1992; *H. krishnai* n.sp., Rajya Lakshmi, 1992 (one spicule project from cloaca); *H. kiranii* n.sp., Rajya Lakshmi, 1993; *H. carutti* n.sp., Rajya Lakshmi *et al.*, 1993; *H. longicaecum* n.sp., Rajya Lakshmi *et al.*, 1993; *H. rhamdiae* sp.n. Brizzola and Tanzola, 1995; *H. channai* n.sp., Rajya Lakshmi, 1995; *H. punctati* n.sp., Rajya Lakshmi 1995; *H. fossilii* n.sp., Rajya Lakshmi 1996; *H. narayanensis* Rajya Lakshmi, 1997; *H. patagonense* n.sp., Moravec *et al.*, 1997; *H. nipponense* nom.n. Moravec and Nagasawa, 1998; *H. geschei* sp.nov. Torres *et al.*, 1998 and *H. physiculi* sp.nov. Moravec and Nagasawa, 2000.

The present species has close resemblance to *Hysterothylacium carutti* n.sp. in number and arrangement of caudal papillae, long intestinal caecum, shape of tail in male and type of striations on cuticle in male, but differs from size of body, position of vulva, size and position of spicules and type of striations in female.

Due to the differences discussed in the specimens could not assigned to many of the previously described species. Hence, it has been considered desirable to erect it as a new species, *Hysterothylacium ritai.*

Acknowledgements

Thanks are due to Andhra University authorities for providing facilities.

References

Bilqees, F.M., Z. Khanum and Q. Jehan, 1971. Marine fish nematodes of West Pakistan. I. Description of seven new species from Karachi coast. *J. Sci. Karachi*, 1: 175–184.

Brizzola, S.M. and R.D. Tanzola, 1995. *Hysterothylacium rhamdiae* sp.n. (Ascaridoidea : Anisakidae) from a neotropical catfish, *Rhamdia sapo* (Pisces : Pimelodidae). *Memorias do Instituto Oswaldo Cruz.*, 90(3): 349–352.

Bruce, N.L. and L.R.G. Cannon, 1989. *Hysterothylacium, Iheringascaris* and *Maricostula* new genus, nematodes (Ascaridoidea) from Australian Pelagic marine fishes. *J. Nat. Hist.*, 23: 1397–1441.

Carvalho-Varela, M. and V. Cunha-Ferreira, 1987. Helminth parasites of the common sole, *Solea solea*, and the Senegalese sole, *Sele senegalensis* on the Portuguese continental coast. *Aquaculture*, 67: 135–138.

Deardorff, T.L. and R.M. Overstreet, 1980. *Hysterothylacium* and *Iheringascaris* (both previously = *Thynnascaris*) (Nematoda : Anisakidae) from the northern Gulf of Mexico. *Proc. Biol. Sac. Wash.*, 93: 1035–1079.

Dollfus, R.Ph., 1933. *Thynnascaris legendrei* n.gen., n.sp. de l'estomac du German, Germoalalonge (Gmel). *Bull. Soc. Zool. France*, 58: 7–12.

Moravec, F. and K. Nagasawa, 1998. *Hysterothylacium nipponense* nom.n. (Nematoda : Anisakidae) for *H. japonicum* Moravec et Nagasawa, 1998 preoccupied by *Hysterothylacium japonicum* Rajya Lakshmi, 1996. *Folia Parasitologica*, 45(4): 328.

Moravec, F. and K. Nagasawa, 2000. Some anisakid nematodes from marine fishes of Japan and the North Pacific Ocean. *J. Nat. Hist.*, 34(8): 1555–1574.

Moravec, F., S. Urawa and C.O. Coria, 1997. *Hysterothylacium patagonense* n.sp. (Nematoda : Anisakidae) from freshwater fishes in Fatagonia, Argentina, with key to the species of *Hysterothylacium* in American freshwater fishes. *Systematic Parasitology*, 36(1): 31–38.

Petter, A.J. and B.M. Radujkovic, 1986. (Nematode parasites offish in Adriatic Sea) Nematodes parasites de poissons de la mer Adriatique. *Bulletin du Museum, Nacional d'Histoire Naturelle A (Zoologie, Biologie et Ecologie Animales)*, 8: 487–499.

Rajya Lakshmi, I., 1992. Description of new species of *Hysterothylacium* Ward and Magath, 1917 (Nematoda : Anisakidae) from the intestine of the freshwater fish, *Anabas testudineus* (Bloch). *Rivista di Parassitologia*, 9(48) No.3: 299–309.

Rajya Lakshmi, I., 1993. *Hysterothylacium kiranii* n.sp. (Nematoda : Anisakidae) from the body cavity of *Channa* (Ophiocephalus) *punctatus. Indian J. Parasti.*, 16(2): 83–83.

Rajya Lakshmi, I., 1995. *Hysterothylacium punctati* sp.nov. (Nematoda : Anisakidae) from the stomach of the freshwater fish, *Channa punctatus* Bloch. *Geobios New Reports*, 14(2): 123–129.

Rajya Lakshmi, I., 1995. *Hysterothylacium channai* n.sp. (Nematoda : Anisakidae) form the intestine of *Channa punctatus* (Bloch) from Kolleru lake (Andhra Pradesh). *Rivista di Parassitologia*, 12(46)N–2: 221–228.

Rajya Lakshmi, I., 1996. Description of a new species of *Hysterothylacium* Ward and Magath, 1917 (Nematoda : Anisakidae) from the intestine of marine fish, *Nemipterus japonicus. Rivista di Parassitologia*, 13(47): 53–60.

Rajya Lakshmi, I., 1996. Description of new species of *Hysterothylacium* Ward and Magath, 1917 (Nematoda : Anisakidae) from the intestine of the freshwater catfish *Heteropneustes fossilis* (Bloch). *Rivista di Parassitologia*, 13(47)N–2: 169–176.

Rajya Lakshmi, I., 1996. *Hysterothylacium nellorensis* sp.nov. (Nematoda : Anisakidae) from freshwater fish, *Labeo bogut* (Sykes). *Geobios New Reports*, 15(2): 77–81.

Rajya Lakshmi, I., 1997. *Hysterothylacium narayanensis* n.sp. (Nematoda : Anisakidae) from the body cavity of *Barbus ticto* (GUnther). *The Uttar Pradesh Zool. Soc.*, 17(2): 151–155.

Rajya Lakshmi, I., K. Hanumantha Rao and K. Shyamasundari, 1991. *Hysterothylacium pseudotumbili* n.sp. (Nematoda : Anisakidae) from the marine fishes of Visakhapatnam (Bay of Bengal). *Bol. Chil. Parasitol.*, 46(3–4): 69–73.

Rajya Lakshmi, I., K. Hanumantha Rao and K. Shyamasundari, 1992. *Hysterothylacium neocornutum* n.sp. (Nematoda : Anisakidae) from the freshwater fish, *Anguilla bicolor bicolor* McClalland of Godavari river (Rajahmundry). *Riv. Parassit.*, 9(43)N–2: 157–169.

Rajya Lakshmi, I., K. Hanumantha Rao and K. Shyamasundari, 1993. *Hysterothylacium longicaecum* n.sp. (Nematoda: Anisakidae) from the intestine and stomach of freshwater fishes of Visakhapatnam. *J. Parasit. Appl. Animal Biol.*, 2(1): 1–5.

Rajya Lakshmi, I., K. Hanumantha Rao and K. Shyamasundari, 1993. *Hysterothylacium carutti* n.sp. (Nematoda : Anisakldae) from the marine fish *Johnius carutta* Bloch of Bay of Bengal, Visakhapatnam. *Uttar Pradesh J. Zool.*, 13(1): 1–10.

Torres, P., P. Andrade and R. Silva, 1998. On a new species of *Hysterothylacium* (Nematoda : Anisakidae) from *Cauque mauleanum* (Pisces : Atherinidae) by bright field and scanning electron microscopy. *Memorias do Instituto Oswaldo Cruz.*, 93(6): 745–752.

Ward, H.B. and Magath, 1917. Notes on some nematodes from freshwater fishes. *J. Parasitol.*, 3: 57–65.

Chapter 49

Evaluation of Rice Varieties for Stagnant Flooding

H.N. Subudhi, L.K. Bose and J. Meher
Crop Improvement Division, Central Rice Research Institute, Cuttack, India

ABSTRACT

Evaluation of 117 rice varieties for stagnant flooding was conducted under natural conditions in the farm of Central Rice Research Institute, Cuttack during 2003 wet season. The plots were completely submerged for 10 days due to heavy rainfall (from 28th August to 7th September, 2003). After receding of flood, survival score was recorded according to IRRI standard evaluation system. It was observed that the varieties *viz.*, Dubraj, Utkal prava, CSR-30 and Jaladhi 2 were tolerant having the SES score 1. FR 13A was used as check. Besides these, some other lowland varieties *viz.*, Sabita, Nalini, Neerja, Bhanja, Purnendu, Durga, Sarala, Saraswati and Ketekijoha etc were moderately tolerant having SES score 2. So these varieties should be popularised in the flood prone areas of Eastern India and used as donor in varietal development programme.

Keywords: *Evaluation, Rice varieties, Stagnant flooding, Natural condition.*

Introduction

India is having 17 million ha rainfed lowland rice of which 10 million ha spread in Eastern India (Mishra *et al.*, 1986). Rainfed low lands are very fragile due to flash flood, stagnant flooding and submergence. The water depth usually varies from 20 cm to 1 m because of heavy rainfall and over flooding of rivers. Most of the high yielding varieties are susceptible to submergence/stagnant flooding. The variety FR 13A is the only submergence tolerant variety but it is very low yielder, poor combiner and inferior grain quality. So breeding of new alternative genetic stock is highly required. Due to lack of high yielding varieties, Rainfed lowland farmers depend on tall, traditional photosensitive varieties, which have better adaptation to this environment than modern varieties (Mallik *et al.*, 1987). Yield stability is a major issue in stagnant flooding of Rainfed Lowland Ecosystem because hydrology varies from field to field (Mallik *et al.*, 1988; Mallik 1994). Besides these, submergence is the 3rd important limitation to rice production of all the abiotic stresses (Widawsky and Toole, 1990). The productivity of this ecology is about 1.5 t/ha. Due to this low production, the economy of the farmer is very poor. It is the high time to find out submergence/stagnant flooding tolerant varieties to develop the economy of rainfed lowland farmers. Efforts are on to develop submergence tolerant variety on priority.

Keeping this in view, evaluation of 117 released varieties of different states were undertaken in the farm of Central Rice Research Institute, Cuttack to find out tolerant varieties for flood prone areas of Eastern India which will be popularized in rainfed lowland areas and used as donor in varietal development programme.

Materials and Methods

The experiment was conducted during 2003 wet season in Central Rice Research Institute, farm Cuttack. Transplanting was done on 8th August 2003. Due to severe rainfall the farm was completely submerged for 10 days from 28th August to 7th September 2003. After that, water receded slowly. One hundred seventeen varieties were transplanted in RBD design with two replications and spacing was 15 × 20 cm. The plot size was 10 sq m^2. Data on survival was recorded after 10 days of submergence following IRRI SES Score (Anonymous, 1996).

Table 49.1: Reaction of Rice Varieties for Stagnant Flooding

SES Score	*Reaction (%)*	*Promising Varieties*	*Remark*	*No. of Varieties*
1	100	FRI3A, CSR30,Utkalprava, Jaladhi-2, Dubraj	T	5
3	95–99	Sabita, Hhanja, Nalini, Neerja, Purnendu, Hhagirathi, Suresh, Saraswati, Kanchan, Ketakijoha, Durga, Sarala	MT	12
5	75–94	Amulya, CSR13, Pokkali, Jogen, Hanseswari, Triveni, Sarnridhi, Golak, Surendra, Pankaj, Bhudev, Vijeta, ASD 16, Vytilla-2, Matangini, IR64, BLB-2, Ambika, IR72, Gouri, Jaya, CSR27, Kranthi, Ranjeet, Phalguna, Bharathi, Samalei, Tulasi, Gayatri, CR1014, Panidhan	MS	31
7	50–74	Taroari, Basumati, Jitendriya, Safri 17, Kakatiya, Jalaprava, Biraj, GR4, Narendra-I, Ghanteswari, Satabdi, Himalaya, Neela, Savitri, Dharitri	S	14
9	0–49	Gour 3, Mahendra, Sashyree, IR8, Rohini, VLK8, Ajay, Panvel 3, Jogen, Kasturi, Bindli, Jolia, Basmati 370, Pusa Sugandha 2, IR 36, Pushkala, Pusasugandh 3, Biraj, Jagannath, Komal-9, Samanta, Dhusura, Annapurna, Sugandha, Meher, Panke, PY3, Sashi, Swarna, Pathara, Pantdhan-11, Canning 7, V Ldhan 16, Krishnabeni, Parijat, Samanta, Sashi, TN 1, CSR 4, PR 104, Ramakrishna, Pusa Basumati I, Khitish, Gouri, Poorva, Pusa 44, Sambhamasuri, IR 50, Daya, ADT 39, VLD 61, Karuna, Konark, Gajapati, Pusa 33	HS	55

N.B.: T: Tolerant, MT: Moderate tolerant, MS: Moderately susceptible, S: Susceptible, HS: Highly susceptible.

Results and Discussion

It was observed that most of the upland varieties were completely damaged. The recovery score was very good in some lowland varieties. The details are given in Table 49.1. The varieties viz FR13A,

CSR30, Utkalprava, Jaladhi2, Dubraj are having 100 per cent recovery, of these CSR30 is a saline tolerant scented variety. Dubraj, the most popular scented variety is tolerant to stagnant flooding. Similarly, the varieties *viz.*, Sabita, Bhanja, Nalini, Neerja, Purnendu, Bhagirathi, Suresh, Saraswati, Kanchan, Ketekijoha, Durga and Sarala are moderately tolerant and are having 95–99 per cent recovery score. The low land varieties *viz.* Arnulya, CR 1014, Pokkali, Tulsi, Golak, Gayatri, Hanseswari, Ambika and popular irrigated varieties *viz.*, IR64, IR72, Jaya, Gouri, Kranthi showed moderate susceptible and the recovery score was 75–94 per cent. It is interesting to note that the most popular varieties *viz.* IR 64, Jaya, Lalat are having 77.0, 85.7 and 74.3 per cent recovery respectively. But Swarn the most popular and wide spread variety of Eastern India is having 0 per cent recovery. So it is not suitable for flood prone areas. The popular varieties *viz.*, IR 36, Pathara, Parijat, Sashi, PR 104, Khitish, Pusa 44, Sambhamahsuri, ADT 39, Daya, Konark, Gajpati, and Sankar etc are highly susceptible to submergence. Tolerant/moderately tolerant varieties should be popularised in flood prone areas of Eastern India and these varieties should be used in varietal development programme as donors.

References

Anonymous, 1996. Standard evaluation system for rice. IRRI, Manila, Philippines.

Mallik, S., 1992. Consortia for strategic and applied research-rainfed lowland rice, paper presented at the ICAR collaborative research and training programme 6–7 July, New Delhi.

Mallik, S., Biswas, S. and De, R.N., 1987. Rainfed lowland rice: a research priority in Eastern India. *Phytobredon*, 3(2): 69–82.

Mallik, S., Mitra, N.K., Mandal, B.K., Roy, S.K.B. and Biswas, S., 1998. Recently released deepwater rice varieties in West Bengal, India. In: *Proceeding of International Deepwater Rice Workshop*. International Rice Research Institute, P.O. Box 933, Manila, Philippines, pp. 121–123.

Mishra, B., Mukhopadhya, S.K. and Flinn, J.C., 1986. Production constraints of rainfed lowland rice in Eastern India. In: *Progress in Rainfed Lowland Rice*. International Rice Research Institute, P.O., Box 933, Manila. Philippines, pp. 25–52.

Mohanty, H.K. and Choudhary, R.C., 1996. Breeding for submergence tolerance in rice in India. In: *Progress in Rainfed Lowland Rice*. IRRI, Manila, Philippines, pp. 191–200.

Widawsky, D.A. and Toole, J.C., 1990. Prioritising the rice biotechnology research agenda for Eastern India. The Rockefeller Foundation, New York.

Chapter 50

Green Economics and Green Accounting: An Overview

Amar Kant Poddar
Department of Economics, A.N. College, Dumka – 814 101, Jharkhand

ABSTRACT

Green economics is the economics of the real world–the world of work, human needs, the earth's materials and how they mesh together most harmoniously. It is primarily about use value, not exchange value or money. This article analyses the various dimensions of green economy. This article outlines the importance of green accounting in environmental preservation and conservation and the possible ways of attaining sustainable development. The cost of green accounting and problems of quantifying the benefit of environmental resources are spelt out in this article. This article concludes with some policy measures to eradicate the negative effects on green economy the life support system.

Introduction

Green accounting is a system in which economic measurements take into account the effects of production and consumption on the environment. It incorporates environmental benefits and costs into economic decision making and it includes the depreciation of natural resources and the environment into estimates of net domestic product of net national product. Sometimes green accounting is said to be natural resource accounting or environmental accounting.

In general comprehensive national income calculation takes into account human as well as environmental capital. The purpose of income calculation in practical affairs is to give people an indication of the amount which they can consume without impoverishing themselves. It is significant to note that Gross National Product (GNP) and other national income aggregates are imperfect measures of welfare. Hence, there is a need to correct the deficiency in estimate of national income. At present there is no reason to arrest general economic growth to conserve natural resources, although there is good reason to provide proper economic incentives to conserve resources which currently cost their users less than true social costs.

Green accounting can be explained with the help of a model by treating air emission as cumulative pollutants. It devices measure of green net national product. The model implies that pollution emissions are properly valued at their marginal costs in the measurement of green Net National Product (NNP). It is observed that the level of genuine savings, net savings less the value of resource depletion and pollution emission is shown to provide a useful indicator of sustainable development (Hamilton, Kirk and Atkinson, Giles 1996).

Dimensions of Green Economy

The green economy is based on the primary use value, intrinsic value and quality. It is a service economy focused on end use and human and environmental needs. Matter is a means to the end of satisfying real need and can be radically conserved. Green economy follows natural flows. The economy not only flows with solar, renewable energies, but also with natural hydrological cycles with regional vegetation and food webs and with local materials. As society becomes more ecological, political and economic boundaries tend to coincide with ecosystem boundaries; it becomes bioregional. The waste recycled or treated can be useful.

Green economy deals with diversity. In a world of constant flux, health and stability seems to depend on diversity. This applies to all levels including diversity of species, ecosystems and regions and to social as well as ecological organization. Green economy depends on self-reliance, self-organization and self-design since it is a complex system. In an economy complex systems move with ecosystem processes causing tremendous scope for local response, design and adaptation must be provided.

The successful maintenance of green economy depends on people's participation. In this context ecological organization and new information communication technologies can provide the means to deeper levels of participation in the decision that count in society. Human creativity and developments are the interrelated aspects of green economy. It could be noted that displacing resources from production and turning into the spontaneous productivity of nature require tremendous creativity. It requires all round human development that entails great qualities of nurture. These are qualities of giving and real service that have been suppressed especially in men by the social and psychological conditioning of the industrial order. In green change, the personal and political, the social and ecological go hand in hand. Social aesthetic and spiritual capacities become central to attaining economic efficiency and become important goals in themselves. The strategic role of the green environment, the landscape and spatial design is an important aspect of green economy. It could be noted that the greatest efficiency gains can often be achieved by a simple spatial rearrangement of system component.

Importance of Green Accounting

Green accounting is a management tool used for a variety of purposes such as improving environmental performance, controlling costs investing in cleaner technologies, developing greener processes and products and informing decisions related to product mix, product relation and product mixing (Paul E Bailey, 1991). Green accounting has a lot of utility to mankind. Hence, there is a need for a brief discussion on it. Green accounting helps one to make sustainable use of natural resources such as water, soils, and forests and also it helps to conserve non renewable natural resources through efficient use and careful planning.

Biospheres can be protected through green accounting process. This type of accounting will reduce and make contingent progress toward eliminating the release of any substances that may cause environmental damage to the air water or the earth its inhabitants. Green accounting measures will safeguard all habitat affected by human operation and will protect open spaces and wilderness while preserving biodiversity.

Green accounting measures give a clue to reduce waste and disposal of waste as per the analyses of existing conditions. Through this measures environmental planners will reduce waste and wherever possible eliminate waste through source reduction and recycling. All waste will be handed and disposed through safe and responsible method. Energy utilization level can be identified through green accounting and it enables planners to conserve energy and improve the energy efficiency. Further it helps to use environmentally safe and sustainable energy sources.

Green accounting enables the planners to minimize the environmental, health and safety risks to industrial workers and the communities in which people operate through safe technology facilities and operating procedures. The production of safe products and services can be achieved through green accounting process. Through this process planners will reduce and if possible eliminate the use, manufacture or sale of products and services that cause environmental damage or safety hazards, so that policy makers can inform the people of the environmental impact of the products. Green accounting helps achieve environmental restriction. Through this accounting planners could promptly and responsibly correct conditions that people have caused, endangering health safety of the environment.

Green Power

Many governments and corporate institutions are beginning to explore how to use green power as an approach for reducing the climate and air impacts of their business. World Resources Institute has observed that power comes from renewable resources such as solar, wind, geothermal, biomass and landfill gas can provide many environmental benefits. It could be noted that relative to conventional fossil fuels, renewable sources, emissions of Greenhouse gases such as carbon dioxide (CO_2) reduce other significant pollutants such as sulfur dioxide (SO_2) nitrogen oxide (NO_2) and mercury (Hg) that affect the environment and human health. Hence there is a need for policy framework. For instance an emission trading system for Greenhouse gases could be an efficient and transparent mechanism for placing financial value on the possible attributes of green power, consequently reducing the cost of green power. Allocating emission allowances or awarding credits to renewable energy generators could help overcome current market price barriers. National and state policy makers are beginning to explore these and other policy frameworks that could enable renewable markets to succeed and contribute to a clear energy future.

Developing on emissions inventory is particularly important for companies that want to use green power as a means to reduce greenhouse gases emissions. Here inventories record and quantify the emission benefit or switching over to cleaner energy sources such as solar, photovoltaic systems, landfill gas, geothermal, wind-generated electricity and power from hydrogen fuel cells.

Greenhouse gases account should be based on the following guiding principles. First one has to see its relevance in terms of Greenhouse gases emissions from business and the decision making needs. Then one has to account for all emisions sources and activities within organizational and operational boundaries. Consistency is needed in greenhouse gases accounting. Here a meaningful comparison could be made with respect to emission performance overtime and clearly state the changes to the basis of reporting to enable to the valued comparison. The next issue is transparently. Here one has to address all relevant issues in a factual and coherent manner based on a clear audlit.

Green Accounting Costs

Green accounting is the process of identification, prioritisation quantification and incorporation of environmental costs into business decisions. Cost of green accounting is classified into various types. These include environmental cost. It refers to the use, release and regulation of materials in production of goods and services. It covers environmental management costs and it includes all expenditures directly associated with the environmental management function of the facility and in keeping a facility in regulatory compliance. Opportunity costs are also a part of environmental cost which is meant as lost value of process and other lost material costs associated with unoptimal use of raw materials in the production process. Contingent costs constitute a part of environmental costs that are not certain to occur in the future but depend on uncertain events and image costs are less tangible costs incurred to affect subjective perceptions of management, customers, employees, communities and regulations.

Costs Issues in Green Accounting

In green accounting a comparison could be made with respect to supplier packaging, recycling and product packaging management from one firm to another. In green accounting environmental cost accounting refers to the addition of environmental cost information on the existing cost accounting procedures and recognizing embedded environmental cost and allocating them to appropriate products or processes. Full cost analysis usually means the allocation of all direct and indirect costs of a product or product line for the purposes of convention valuation, profitability analysis and pricing decisions. Here it highlights environmental elements. The total cost assessment has come to represent the process of integrating environmental costs into a capital budgeting analysis.

Life cycle assessment is a process to identify the environmental consequence of a product, process or activity through its entire life cycle and to identify opportunities for achieving environmental improvements. It includes major stages in the life cycle of a product, process, or activity such as raw materials acquisition, manufacturing, consumer use, reuse, and maintenance and recycle/waste management.

In green accounting, life cycle accounting is a term used to describe the assignment and analysis of product, specific costs within a life cycle framework including casual, hidden liability and less tangible costs. In general life cycle cost deals with all the costs incurred for a product from its inception to abandonment, as a part of its product cost. In green accounting this includes cost of extraction, intermediate manufacturing, transportation product recycling intake back disassembling, reverse distribution, restocking used materials, disposing of waste etc.

There is a confusion with respect to the term life cycle costs. It could be noted that some people view life cycle costing as referring only to private costs, while others view it as private and societal costs. Some apply a life cycle perspective to capital budgeting while others apply life cycle concept to process and product design. In general private costs are those costs a business incur or for which a business can be held responsible. These are the costs that directly affect a firm's bottom live whereas societal costs are the costs of industrial impact on the environment and society for which the business is not financially responsible. These costs do not directly affect a firm's bottom live, societal costs may also be referred to as external costs or externalities.

Conclusion and Policy Measures

It could be seen clearly from the above discussion that green economy and green accounting provide valuab!e information to sustainable life support system. In order to increase the performance of green accounting system, the following suggestions can be considered.

1. Green accounting should consist of all environmental costs such as environmental management costs, opportunity costs. Contingent costs, direct costs, indirect costs and so on.
2. Efforts to make sustainable use of renewable natural resources such as waste, soils and forests.
3. A proper assessment of the cost of reducing pollution and sustainable pollution level. Green national accounting should be motivated by its apparent promise to provide an indicator of sustainable development.
4. A proper assessment of trade in environmental services.
5. Promote inter-disciplining and trans-disciplining researches in evaluation of the status of green economy.
6. Green accounting should be done by environmental damage valuation graphics.

References

Asheim, Geir B., 200. Green National Accounting: Why and How. *Environmental and Development Economics*, 5(1–2): 25–48.

Baile, Paul E., 1991. *Full Cost Accounting for Life Cycle Costs: A Guide for Engineers an Financial Analysis, Environmental Finance*, pp. 13–29.

Hartwick, John M., 1990. Natural resources national accounting and economic depreciation. *J. Pub. Econ.*, 43: 291–304.

Hecht, Joy E., 1999. Environmental accounting where we are now, where we are heading. *Resources*, 135(99)14–17.

Kirk, Hamiton and Atkinson, Giles, 1996. Air pollution and green accounts. *Energy Policy*, 24(7): 675–684.

Peter, Bartelmus, 1997. *The Value of Nature: Valuation and Evaluation in Environmental Accounting*. Department for Economic and Social Information and Policy Analysis, United Nations, Working Paper Service, No. 15.

Chapter 51

Studies on the Impact of Phytopesticide, Nimbecidine on Histological Changes in the Testis of Adult Male, Insect *Sphearodema rusticum* (Heteroptera : Belostomatidae)

C. Elanchezhiyan, V. Shoba and Selvisabhanayakam
Department of Zoology, Annamalai University, Annamalainagar – 608 002, Tamil Nadu, India

ABSTRACT

Insects are highly evolved groups since Ordovician period. They have wide adaptation to different habitats. Hence the insects are the chief biological organisms not only causing severe damage to crops on land but also causes severe injury to economically important game fishes such as Catla, Rohu etc. For successful aquaculture, the control of predator insects are taken into consideration, works in relation to the control of predator insects by using phytopesticide like nimbecidine are rare. Therefore, it has been programmed in the present study to find out the effect of nimbecidine on the histological changes in the testis of *Sphaerodema rusticum*. The sublethal concentration was found out about 0.13 per cent for nimbecidine. The insects were exposed to nimbecidine for 7 days, 14 days and 21 days have revealed certain histopathological changes in the testis. The testis consisting apical cells, gonial cells, primary and secondary spermatocytes and sperm bundles in the control insects. In treated insects, several histopathological changes have been observed such as atrophy of apical cells, swollen spermatogonial cells, pycnotic and necrotic testis follicle, sperm bundles were disorganized and looks like whorl than 7 days and 14 days exposure. It is suggested that the phytopesticide causing several histopathological damages in the testis of *S. rusticum* and inturn causes severe reproductive potentiality of *S. rusticum*, a predator insect in aquaculture.

Keywords: *Nimbecidine, Sphaerodema rusticum, Testis.*

Introduction

In India, neem extracts are used as insect repellent to protect the agricultural crops, household and stored grains. The Neem Seed Kernel Suspension (NSKS) at 2 per cent level function as a dependable repellent in *Spodoptera littura* (Joshi *et al.*, 1975). The neem products are used in agricultural field to reduce terrestrial pollution and thus prevent food poisoning. The neem based pesticides are safe, non-toxic, non-polluted and avoid the accumulation of residual effects on agricultural products (Hedge, 1993). Azadirachtin enriched neem extract is used in the management of human and animal pests, management of stored products pests, management of garden and horticultural pests and management of crop pests. However, interest in botanical pesticides received during recent years because of some of the draw backs of synthetic insecticides, including lack of selectivity, impact on the environment and the emergence and spread the pest resistance. At present time, isolation, identification and development of natural products are the focus of numerous research programmes around the globe. Extracts of the Indian neem–tree, *Azadirachtica indica* are gaining prominence, the foremost biotoxical insecticides of this decade (Prabhakar *et al.*, 1986 and Stone, 1992).

Several studies have been made on histological changes in the testis due to pesticide treatment in few orders such as Coleoptera (Alallah and Newson, 1965); Isoptera (Rajinder *et al.*, 1965, Hamstead, 1966); Diptera, (Kennedy, 1947; Beard 1965; Benschoter, 1967; Choudhry and Tripathi, 1975; Deepak and Choudhry, 1978) and Hemiptera (Redfren *et al.*, 1980).

In recent years, there has been a spurt in research concerning the reproduction of invertebrates, as they exhibit complexity in their reproductive physiology. The reproductive system of insect is a complicated one. Testicular development and spermatogenesis have been studied in many lepidopteran insects (Numata and Hidake, 1980 and Lai-Fook, 1982).

With this in view, several studies have been undertaken to understand the histological changes in the testicular tissues of insects due to treatment with heavy metal, pesticide and other chemicals such as endrin (Irrinmount, 1970); alpholate (Gangrade and Sand, 1970); penfulron (Redfren *et al.*, 1980); phosphamidon (Balakrishnan, 1990 and Nirmala Devi 1990); nimbeciline (Thiruvasagam, 1994); *Pongamia glabra* leaf extract (Ramanathan, 1995); heavy metal mercuric chloride (Palanisamy, 1997); monochrotophos (Ravichandran, 1996) and endosulfan (Sumathi, 2002).

However, the action of nimbecidine, which is a insecticide compound with relatively high toxicity has not been well documented with reference to testicular tissues in the adult male insect, *Sphaerodema rusticum*, a predator insect in aquaculture.

Materials and Methods

Sphaerodema rusticum is found in freshwater pond and streams, clinging to waterweeds. The insects were daily fed with mosquito larvae, aquatic plants and pieces of earthworms. The troughs were cleaned properly every alternative days include changing the water.

Histopathological Preparations

The control and treated insects were kept separately. After 48 hours they were dissected under binocular microscope by using insect Ringer solution (Ephrassi and Beadle, 1936). The Ringer was subsequently removed and the tissue was fixed in Bouin's fluid for 24 h, later the tissue was processed by adopting standard histological techniques. (Gurr, 1958).

Results and Discussion

A transverse section of the testis composed of a large number of cysts compactly held together. Each follicle has a thin wall consists of large apical cells, which were mainly meant for providing

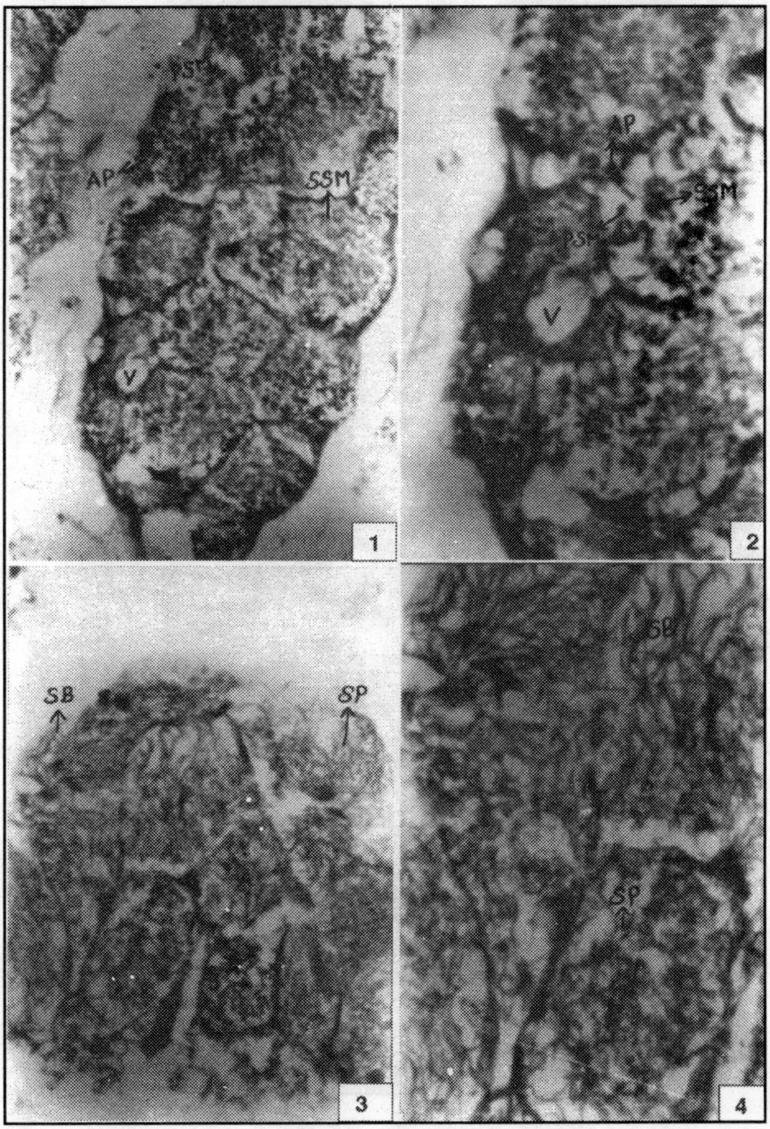

Figures 51.1–51.4

nourishment for the germ cells called trophocytes. The spermatogonia was invariably found closely associated with the germinal epithelium. The secondary spermatocytes were comparatively more in number than the primary spermatocytes where, reduction division taken place and given rise to spermatids and further, it developed into sperms. The sperms were elongated and have hair like structure arranged in the form of bundles (Figures 51.1 and 51.2).

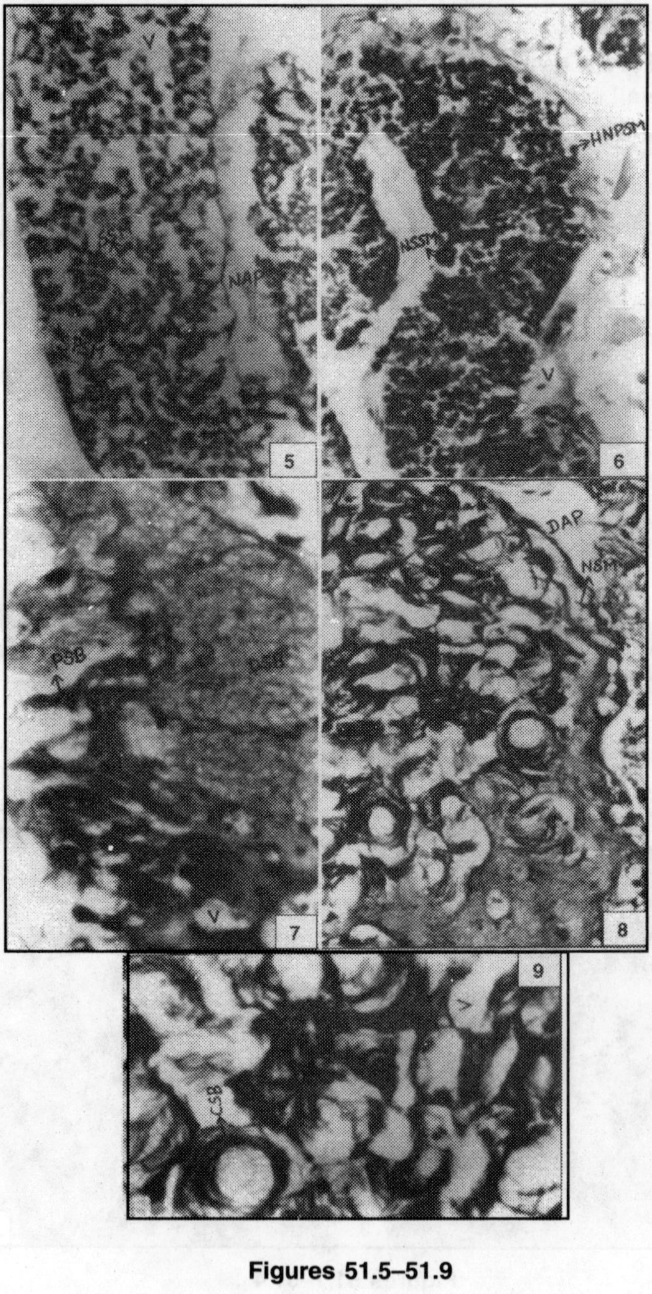

Figures 51.5–51.9

The nimbecidine treated insects exhibited remarkable histological changes in the testis such as the occurrence of pycnotic and necrotic peritoneal membrane with disintegrated nutritive cells indicate, the less amount of nutrition for the developing germ cells and spermatids. The spermatids seemed to be less in number and clumped. The primary and secondary spermatocytes appeared to be increased in their size and volume (Figures 51.3 and 51.4).

The chronic exposure of nimbecidine to the testis of *Sphaerodema rusticum* showed certain histopathological changes such as the occurrence of apical cells were highly disorganized and disintegrated. The primary and secondary spermatocytes were highly pycnotic and necrotic. The sperm bundles were found to be disintegrated, disorganized and highly pycnotic indicate that the sperms were not active which impaired the reproductive potentiality of the insect (Figure 51.5). The follicle of the treated insects contained more vacuoles. The sperm bundles were coiled, pycnotic and necrotic. The spermatids were broken into pieces. It indicates that the sperms were not considered as an effective male gametes which reduced the reproductive potential than the control (Figure 51.6.) Highly obliterated testis with necrotic and pycnotic spermatocytes. The sperms were highly nectrotic, coiled and conglomerated (Figures 51.7–51.9) indicates that sperms were not viable seems to be sterile than compared to the control.

In the present study, it has been observed that there are several changes in the morphology of the treated testis such as disintegration of primary and secondary spermatocytes, spermatids and chumping of sperms compared to the control. These findings are in parallel with the works on *Odontopus varicornis* (Jayakumar, 1988); *Pheropsophus lissoderous* (Balakrishnan, 1990); *Periplaneta americana* (Ramanathan, 1995); *Sphaerodema rusticum* (Rathinam, 1997), *Oxya nitidula* (Prasanna, 1998) and *Sphaerodema rusticum* (Rajathi, 2004), respectively.

It has been observed that the changes in the testis of treated insects for 7, 14 and 21 days were more obvious than the insects exposed for 48 h. The sperm bundles were coiled, pycnotic and necrotic sperms affect the motility and ultimately caused the sterility of insects. Thus, it is evident from the present observations that chronic exposure of the phytopesticide, nimbecidine which act as antifertility agent brought about certain reproductive impairment in this test insect, *Sphaerodema rusticum.*

Acknowledgement

The authors are very grateful to the authorities of Annamalai University and to the HOD of Zoology for providing all necessary facilities to complete this work.

References

Atallah, Y.H. and L.D. Newson, 1965. Ecological and Nutritional studies on *Coleonigilla Maculata* De Gew (coleoptera) III. The effect of DDT, toxaphere and endrin on the reproductive and survival potential. *J. Econ Ent.*, 59: 1181–1187.

Balakrishnan, S., 1990. Studies on the effect of Roger on the histology and histochemistry of the testis, fat body and accessory reproductive glands of the adult male *pherosophus lissoderus. M.Phil. Thesis*, Annamalai University.

Beard, R.L., 1965. Ovarian suppression by DDT and resistance in the Housefly, *Musca domestica. Entamon. Exp. Appl.*, 8: 193–204.

Benschoter, C.A., 1967. Effect of dietary biotin on reproduction of the Housefly. *J. Econ. Ent.*, 60.

Choudhry, H.S. and C.P.M. Tripathi, 1975. Effect of thiourea on the Nucleocytoplasmic relations of Nurse cells and developing occyte in sarcophagi ruficornis of all India insect. In. *Proceedings Chemostoriliant Research Workers Conference*. U.N.D.P. ICAR Project on Plant Production U.A.S. Bangalore, p. 78–81.

Deepak, V. and H.S. Chaudhry, 1978. Histopathological effects of thiourea on ovarion development of *Calliphora erythrocephala* (Diptera : Calliphoridae). *Appl. Ento. Zool.*, 13: 34–37.

Epherussi, B. and G.N. Beadle, 1936. A technique of transplantation of Drosophila. *Amer. Nat.*, 70: 218–225.

Gangrade, G.A. and N.C. Sant, 1970. Effect of apholate on reproductive organs of almond Moth *Cadra caultella* (Walker) (Lepidoptera : Phycitidae). *Indian J. Ent.*, 32(2): 134–139.

Gurr, E., 1958. *Methods of Analytical Histology and Histochemistry*. Leonard Hill (Brooks) Ltd., London.

Hamstead, E.O., 1966. Reduction in reproductive capacity of European Red Mite by Niagra, 9203. *J. Econ. Ent.*, 59(2): 481.

Hedge, N.G., 1993. Neem production and development constraints at grass root level. *World Neem Conference*, p. 1–7.

Irvinmount, D., 1970. Chronic effects of endrin on blunt-nose minnouws, (pipe pholasnotatus) and *Quippies octiculatus*). *Ann. Fish. Soc.*, 96: 185.

Jayakumar, C., 1988. Effect of dimethoate (Roger) on the accessory reproductive gland, testis, fat body, haemolymph neurosecretory cells and corpus allatum in *Odontopus varicornis* (Dist) (Hemiptera : Pyrrhocoridae). *Ph.D. Thesis*, Annamalai University.

Joshi, B.G., Ramaprasad, G. and Sitharamaiah, S., 1975. Effect of neem seed kernel suspension on *Telenomus remus* an egg parasite of *spodoptera littura*. *Phytoparasitica*, 10: 16–63.

Kennedy, J.S., 1947. The excitant and repellant effects on mosquitoes of sublethal contacts with DDT. *Bull. Ext. Res.*, 37: 593–607.

Lai-Fook, K., 1982. The role of male accessory glands in insects reproduction. *A. Rev. Ent.*, 21: 199–221.

Nirmala Devi, 1990. Effect of phosphamidon on the histology and Histochemistry of the testis and accessory gland of the adult male *catacanthus incarnates* (Spin) (Hemiptera : Pentatomidae). *M.Phil. Thesis*, Annamalai University.

Numata, H. and Hidaka, T., 1980. Development of male sex cells in the swallow tail *papilio xuthus* L. (Lepidoptera : Papilonidae) in relation to pupal diapause. *Appl. Ent.*, 200(15): 851–859.

Palanisamy, K., 1997. Studies on the effect of heavy metal (Mercuric chloride) on histopathological and biochemical changes in the selected tissue of the adult male Laecoterphes rubber (Linn) (Heteroptera : Nepidae). *M.Phil. Thesis*, Annamalai University.

Prabhakar, N.D.C., Condrict, A.N. Kishaba and D.E. Negar Dirk, 1986. *J. Beon. Entomol.*, 79: 39–41.

Prasanna, V., 1998. Effect of nimbeciline on the histopathological changes in the testis of the adult male paddy pest, *oxya nitidula* (walker) (Orthoptera : Acrididae). *M.Sc. Dissertation*, Annamalai University.

Rajathi, V., 2004. Impact of heavy metal mercury in the adult male reproductive physiology of *sphaeromderma rusticum* (Heteroptera : Blastomatidae). *Ph.D. Thesis*, Annamalai University.

Rajinder, S., Saini and L.K. Cutkomp, 1965. The effects of DDT and sublethal dose of bicofol on reproduction of the two spotted spider mite. *J. Econ. Ent.*, 49: 249–254.

Ramanathan, B., 1995. Studies on the effect of leaf extract of *Pongamia glabra* (vent) (Fabaceae) on histological and biochemical changes in testis and fat body of the adult male *Periplaneta americana* (Linn) (Orthoptera : Blattidae). *M.Phil. Thesis*, Annamalai University.

Rathinam, 1997. Impact of Nimbeciline on the histological changes in the testis of *Sphaerodema rusticum* (Duf.) (Hemiptera : Belastomatidae). *M.Sc. Dissertation*, Annamalai University.

Ravichandran, S., 1996. Toxicological studies on the ovary, fat body and haemolymph of the adult female *Laceotrephes rubber* (Linn) (Heteroptera : Nepidae). *Ph.D. Thesis*, Annamalai University.

Redfren, R.E., A.B. Demilo and A.B. Borkovec, 1980. Large milk weed bug: Effect of penfulron and its analogoues on reproduction. *J. Econ. Entomol.*, 73: 682–683.

Stone, C., 1992. *Science*, Washington, 255: 1070–1071.

Sumathi, 2002. Studies on the impact of endosulfan on certain selected tissues of the adult male insect *Gryllotapa africana* (Palisot de Beavrols) (Orthoptera : Gryllotalpidae) in relation to reproduction. *Ph.D. Thesis*, Annamalai University.

Thiruvasagam, M., 1994. Studies on the effect of nimbeciline on the histology, histochemistry and biochemistry of male accessory reproductive gland, fat body and haemolymph in the adult male *Aspongopus janus* (Fabr.) (Hemiptera : Pentatomidae). *M.Phil. Thesis*, Annamalai University.

Chapter 52

Biological Treatment of Azodyes

P. Dhevagi[1] and K. Sujatha[2]
[1]Department of Environmental Science, Tamil Nadu Agricultural University, Coimbatore – 3
[2]Kongu Nadu Arts and Science College, Coimbatore – 29

Introduction

Textile mill uses raw water about 61–646 1/kg of cloth with an average of 172 1/kg of cloth processed (50–81 per cent of water consumed, Bal 1999). Colored effluents are continuously released into sewage and adjoining water resources which render a catastrophic effect on the environment. The wastewater generated contains dyes at concentration of 10 to 200 mg/l and about 10 to 20 per cent goes along with other organic materials such as fats, waxes, pectin solid fragment, starch derived from the cloth during its processing and constitute major source of pollution (Zollinger, 1961; Anliker, 1979; Chudgar, 1985; Lewis, 1991 and Ashoka *et al.*, 2002). Soluble reactive dyes which are being used in increasing quantities are known to hydrolyze during application without a complete fixation results in larger proportion of these dyes being released into the environment (Carliell *et al.*, 1996). The recalcitrance of the azodyes to biological degradative processes results in severe contamination of the rivers and groundwater in those areas of the world with a high concentration of dyeing industries (Namasivayan and Yamuna, 1992).

Azo dyes are largely used in textile industries, can also be found in the food, pharmaceutical, paper and printing, leather and cosmetics industries (King-Thom Chung *et al.*, 1978 and Sudhakar *et al.*, 2002). Many of these dyes find their way into the environment via wastewater facilities. These compounds retain their colour and structural integrity under exposure to sunlight, soil, bacteria and sweat; they also exhibit a high resistance to microbial degradation in wastewater treatment systems. The development of synthetic fabrics such as nylon, Lycra, rayon and polyester has required the production of new dyes that can effectively bond to these materials. Azo dyes must be continually updated to produce colors that reflect the trends dictated by changing social ideas and styles. Brighter, longer lasting colors are often necessary to satisfy this demand. There are more than 8000 chemical products associated with the dyeing process listed in the Colour Index (1998). While over 1,00,000 commercially available dyes (7×10^5 metric tons) are produced annually (Zollinger, 1987).

A multitude of dyes were used in textile industry are usually aromatic and heterocyclic (Vyas and Motiforis, 1995) and some are toxic and carcinogenic. The different classes of dye include azo, acid, reactive, metal complex, disperse, vat, mortant, direct, basic, suphur etc. (Vijaya *et al.*, 2003).

Effective and economic treatment of a diversity of effluents containing azo group has become a problem. No single treatment system is adequate for degrading the various dye structures. Physico-chemical and biological process has been investigated extensively for treating dye-bearing effluents. Even though chemical coagulation appears to be effective in treating the dye bearing effluents (Venkata Mohan *et al.*, 1999), the toxic sludge handling problem associated and the chemical cost retarded its usage. Adsorption employing activated carbon and various other adsorbents has been quite widely studied; however, the cost of carbon and the regeneration limited its application. Chemical oxidation is sometimes coupled with UV light exposure to increase the color removal. Other techniques involve electrochemical or wet oxidation, activated carbon adsorption, reverse osmosis, or coagulation/ flocculation (Edwards, 2000). Many of these technologies are cost prohibitive, therefore are not viable options for treating large waste streams. Therefore, scientists have been forced to become very creative in the development of effective treatment technologies for the Azo dye removal.

Now the science and technology comes handy with nature's waste disposer–the microbes, in cleaning up the environmental pollutants. Bioremediation of pollution is mandatory for an ecofriendly environment. Recent stress is on microbial based biological methods which are considered to be sustainable, ecofriendly, cost effective and also easy manageable effluent treatment system.

Understanding the basic composition of azo dyes and specifically fiber-reactive azo dyes is necessary to envision how these molecules can be destroyed. By understanding the general chemical structure of these compounds, consideration can be given to the toxic potential that they pose to the environment.

Azo Dyes and its Intermediates

Azo dyes contain a least one nitrogen-nitrogen (N=N) double bond, however many different structures are possible (Cripps *et al.*, 1990 and Zollinger, 1991). Monoazo dyes have only one N=N double bond, while diazo and,triazo dyes contain two and three N=N double bonds, respectively. The azo groups are generally connected to benzene and naphthalene rings, but can also be attached to aromatic heterocyclic or enolizable aliphatic groups (Zollinger, 1991). These side groups are necessary for imparting the color of the dye, with many different shades and intensities being possible. A common example of an azo dye can be seen in Figure 52.1 (Emitiazi, 2000). When describing a dye molecule, nucleophiles are referred to as *auxochromes,* while the aromatic groups are called *chromophores.* Together, the dye molecule is often described as a *chromogen.* The absorption and reflection of visible and UV irradiation is ultimately responsible for the observed color of the dye. Synthesis of most azo dyes involves diazotization of a primary aromatic amine, followed by coupling with one or more nucleophiles. Amino- and hydroxyl-groups are commonly used coupling components (Zollinger, 1991). Because of the diversity of dye components available for synthesis, a large number of structurally different azo dyes exist and are used in industry (McCurdy, 1991).Worldwide production of organic dyes is currently estimated at nearly 4,50,000 tons, with 50,000 tons being lost in effluents during application and manufacture (Lewis, 1999).

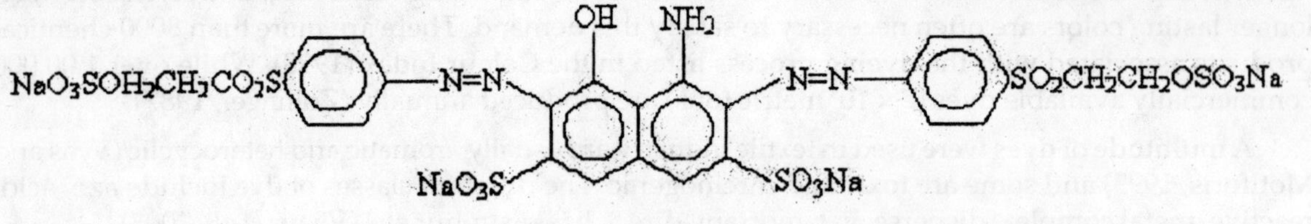

Figure 52.1: Example of an Azo Dye Structure (Remazol Black 5) (Emitiazi, 2000)

Eighty to ninety-five per cent of all reactive dyes are based on the azo chromogen (Zollinger, 1991 and Edwards, 2000). Reactive dyes are colored compounds that contain one or two functional groups capable of forming covalent bonds with the active sites in fibers. A carbon or phosphorous atom of the dye molecule will bond to hydroxyl groups in cellulose, amino, thiol and hydroxyl groups in wool, or amino groups in polyamides. Most fiber-reactive azo dyes are used for dyeing cellulosic materials, such as cotton and are a major source of dye waste in textile effluents. Fiber-reactive azo dyes exhibit a high wet-fastness, due to their ability to covalently bond to substrates. However, dyes that hydrolyze in solution prior to bonding to a substrate are often lost in the washing processes (Loyd, 1992). Schematic diagram of fiber reactive azo dyes is given in Figure 52.2.

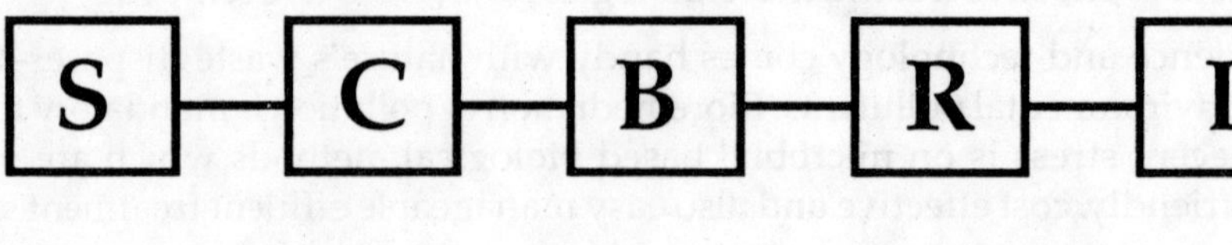

Figure 52.2: Schematic Diagram of Fiber Reactive Azo Dye

where,

S: Water solubiling group; C: Chromogen; B: Bringing group (can be part of chromogen); R: Reactive group; L: Fiber reactive group

The bridging group serves to combine the chromogen with the reactive group of the dye molecule. The bridging group must be stable, soluble in water and exhibit a certain degree of flexibility. Amino and alkylamino groups are generally used for this purpose. The reactive group serves to bond the dye molecule to a substrate via nucleophilic substitution or addition. Mono-, di- and trichlorotriazinyl are all examples of reactive functional groups. Approximately 200 different reactive groups are listed in the patent literature. Twenty-two new reactive azo dye structures were submitted for patent in early 1998, which was nearly five times as many from other classes of azo dyes (Freeman and Sokolowska, 1999). With such a disproportionate production rate, it is not surprising that a large percentage of dye pollution problems are related to fiber-reactive azo dyes.

Toxicity Considerations

The potential for toxic effects to the environment and humans, resulting from the exposure to dyes and dye metabolites, is not a new concern. As early as 1895 increased rates in bladder cancer were observed in workers involved in dye manufacturing (Rehn, 1895 and Rehn *et al.*, 1994). Since that time, many studies have been conducted showing the toxic potential of azo dyes. A complete review of these studies is beyond the scope of this study; however, a broader understanding of the problem can be found in the works of Levine (1991) as well as Brown and DeVito (1993). Both papers indicate that the problem associated with azo dyes is created by the dye metabolites.

Azo dyes have been identified as potentially carcinogenic agents (IARC, 1996). While most azo dyes themselves are non-toxic a significantly larger portion of their metabolites are toxic (Ganesh, 1992). An investigation of several hundred commercial textile samples revealed that nearly 10 per cent were mutagenic in the Ames test (McCarthy, 1997). Another study conducted on 45 combined effluents from textile finishing plants showed that 27 per cent of the wastewater samples were mutagenic in the Ames test.

Many textile effluents contain heavy metals that are complexes in the azo dyes. Little *et al.*, 1974 reported that all cationic dyes are more toxic and vat dyes are less toxic. High concentrations of salt are often used to force fiber-reactive dyes out of solution and onto substrates (Zollinger, 1991). The azo dyes reduced by cytoplasmic reductases, results in the formation and accumulation of aromatic amines (Chung and Stevens, 1993). These reduction producıs may be toxic, mutagenic and/or carcinogenic to animals and humans (Levine, 1991) and reduces photosynthesis in plants (Saranaik and Kanekar, 1995). Reduced compounds can cause high electrolyte and conductivity concentrations in the dye wastewater, leading to acute and chronic toxicity problems equal to that of other metals (Venkata Mohan *et al.*, 2002).

Aromatic amines, which are known to be human carcinogens, have been found in the urine of dyestuff workers and test animals following the administration of azo dyes. Azo dye compounds are linked to bladder cancer in humans and to hepato carcinoma and nuclear anomalies in intestinal epithelial cells in mice.

Understanding the dye structures and how they are degraded is crucial to understanding how toxic by-products are created. Brown and DeVito (1993) have proposed the biological mechanisms thought to be responsible for carcinogenic activation of azo dye compounds. This list is based on an extensive review of the literature regarding azo dye toxicity. Brown and DeVito (1993) postulate that:

1. Azo dyes may be toxic only after reduction and cleavage of the azo linkage, producing aromatic amines.
2. Azo dyes with structures containing free aromatic amine groups that can be metabolically oxidized without azo reduction may cause toxicity.
3. Azo dye toxic activation may occur following direct oxidation of the azo linkage producing highly reactive electrophilic diazonium salts.

An example of a reductive pathway that is mediated by intestinal anaerobic bacteria and leads to the formation of benzidines is shown in Figure 52.3 (Brown and DeVito, 1993).

Similar reactions and by-products would be possible in an anaerobic treatment system. Examples of toxic aromatic amines, which could be created from the degradation of azo dye compounds and its human effects (Cartwright, 1983, Levine, 1991, Brown and Devito, 1993 and Chung, 2000), were given in Table 52.1.

Treatment Options

The treatment of textile effluents is of interest due to their toxic and esthetic impacts on receiving waters. While much research has been preformed to develop effective treatment technologies for wastewater containing azo dyes, no single solution has been satisfactory for remediating the broad diversity of textile wastes. Despite being aware of the problem, many textile manufactures have failed

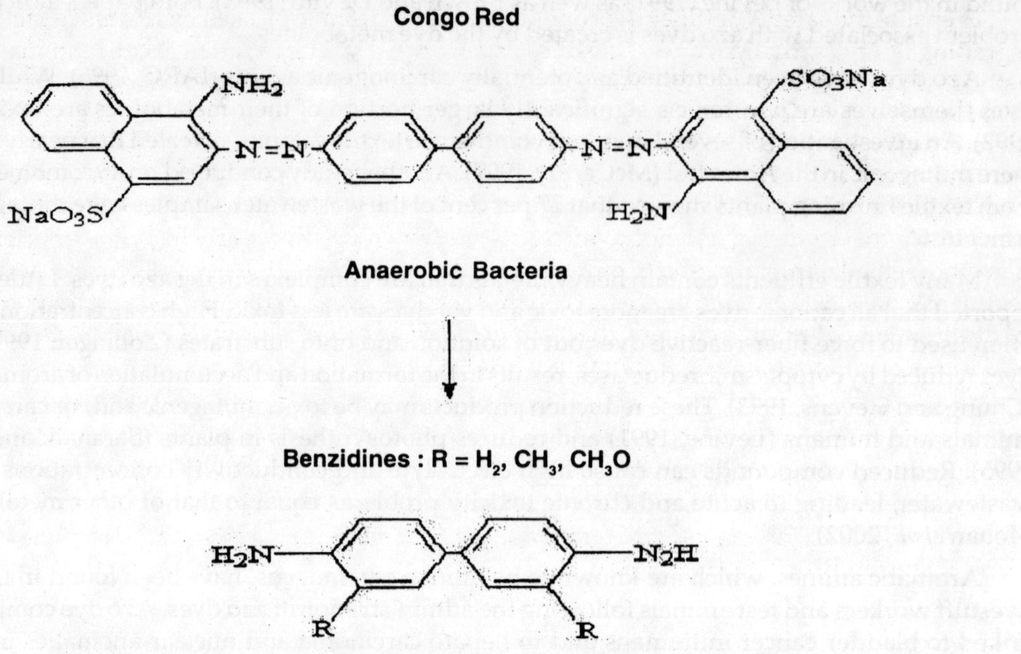

Figure 52.3: Example of Azo Dye Reduction via Anaerobic Bacterium

to adequately remove azo dye compounds from their wastewaters. Until dye and textile manufactures are able to develop efficient technologies, allowing for increased dye-fiber bonding and lower dye house losses (Lewis, 1999), the problem of treating these types of wastes will fall to the wastewater treatment facilities.

Table 52.1: Effect of Aromatic Amines and Potential Dye Metabolites

Sl.No.	*Compound*	*Human Effect*
1.	1-Napthylamine	Slight/mixed
2.	2-Napthylamine	Good
3.	2,5-Diamiotoluene	Slight
4.	3,3'-Dichlorobenzidine	Slight/mixed
5.	3,3'- Dimethoxy benzidine	Slight /mixed
6.	3,3'-Dimethylbenzidine	Slight
7.	4-Biphenylamine	Good
8.	4-Nitrobiphenyl	Slight/mixed
9.	4,4'-Methylenebis(2-chloroaniline)	Slight
10.	Auramine	Slight
11.	Benzidine	Good
12.	N,N-Bis(2-chloroetlhyl)-napthylamine	Good
13.	Magenta	Slight

To ensure the safety of effluents, proper technologies need to be used by treatment facilities when degrading fiber-reactive azo dyes. Azo dyes usually resist biodegradation in conventional treatment methods (Pagga and Brown, 1986, Shaul *et al.*, 1991 and Willmott *et al.*, 1998). Previous research efforts have focused on various biological, chemical and physical techniques for treating azo dye wastes. There is evidence that all three areas have potential for remediating dye house wastes. However, chemical treatment is often cost and application limited, while physical removal can lead to extra solid wastes and increased overhead (Yamini and Lalithakumari, 2001). Biological treatment has been effective in reducing dye house effluents and when used properly has a lower operating cost than other remediation processes. Combinations of chemical and biological or physical and biological treatment have also proven to be effective. Because this research explores the use of microbial sludges for the treatment of a textile effluent, the focus of this literature review will be on biological techniques used for degrading wastewaters containing azo dyes.

Because of their recalcitrant nature, azo dyes often pass through activated sludge facilities with little or no reduction in color (Cariell *et al.*, 1995, 1996; Pagga and Brown, 1986). Although some researchers have observed slight color reductions, their findings are largely outweighed by those who have not (Loyd, 1992, Zissi *et al.*, 1997). Reductions in the carbon content and oxygen demand of azo dye wastewaters following aerobic treatment are well cited (McCurdy, 1991, Horning 1977, Loyd, 1992, Pagga and Brown, 1986).

The anaerobic reduction of azo dyes to simpler compounds has been well researched (Chinwetkitvanich *et al.*, 2000, Razo-Flores *et al.*, 1997, Brown and Laboureur, 1983, Chung *et al.*, 1978). These and other studies, have all demonstrated the ability of anaerobic microbes and sludges to effectively reduce azo dyes to their structures, thus destroying the apparent color. Many of these intermediates are aromatic amines with constituent side groups. By reducing the dye compounds to their intermediates, the problem of aesthetic pollution is solved, but a larger and more deleterious problem may be created. Most azo dyes are non-toxic, but a higher percentage of their intermediates have been identified as carcinogens (Brown and DeVito, 1993,Wong and Yuen, 1996). Because of the toxic potential of many aromatic amines, further degradation of the dye compound is necessary if toxicity is to be eliminated or reduced (Levine, 1991 and Brown and DeVito, 1993).

Aerobic Treatment

Conventional activated sludge treatment of wastes is often an effective and highly economic system for reducing organic pollutants in wastewater. A fair amount of research has been conducted assessing the viability of using activated sludge to treat textile effluents (Pagga and Brown, 1986, Shaul *et al.*, 1991, Loyd, 1992, Brown and DeVito, 1993 and Zissi *et al.*, 1997). However, aerobic treatment of azo dye wastes has proven ineffective in most cases, but is often the typical method of treatment used today (Edwards, 2000, Yang *et al.*, 1998). Because aerobic microbes cannot reduce azo linkages, their ability to destroy dye chromogens is less than anaerobic bacterium. However, aerobic sludges have been successfully used to stabilize dye metabolites (Brown and Laboureur, 1983). Pagga and Brown (1983) conducted a study on 87 commercial dye stuffs. Some of the dye stuffs were in a technically pure form, while others were in a sales form containing organic substances such as wetting agents. The tests were performed in a reactor designed to simulate the conditions of and adapted activated sludge wastewater treatment plant. The samples were tested for color and DOC removal following 42 days of treatment. Kumar *et al.*, 2006 studied the decolorisation, biodegradation and detoxification of Direct Black-38, a benzidine based azo dye, by a mixed microbial culture isolated from an aerobic bioreactor treating textile wastewater. Pagga and Brown indicated that the primary mechanism for removal of dyes in activated sludge systems may occur by adsorption onto the cell walls (Knapp *et al.*, 1995 and Banat *et al.*, 1996). In an earlier study by Brown and Laboureur (1983a), the aerobic biodegradability of aniline, o-toludine, p-anisidine, p-phenetidine, o-dianisidine and 3,3'-dichlorobenzidine, was investigated. These compounds are all lipophilic aromatic amines and possible by-products of azo dyes. Because many aromatic structures are non biodegradable in anaerobic environments and are not hydrophilic, they can accumulate in the adipose tissues of organisms.

Brown and Laboureur (1983) indicated that azo dyes may be broken down to their intermediate structures in a reductive environment, but were not amenable to further degradation by anaerobes. Brown and Laboureur (1983a) concluded that aniline, p-anisidine, p-phenetidine and o-toludine were readily biodegradable by aerobes, while o-dianisidine and 3,3'-dichlorobenzidine were inherently biodegradable. They suggested that these compounds could be stabilized if released into the environment or directly from a dye house into a conventional wastewater treatment plant. Shaul *et al.* (1991) conducted a study on 18 dyes to determine their fate in the activated sludge process. Of these dyes, 15 were acid azo dyes and three were direct azo dyes. The dyes were spiked into pilot-scale treatment systems. Eleven of the dyes passed through the activated sludge system substantially untreated, four were significantly absorbed into the sludge and three were apparently biodegraded.

Loyd (1992) also performed activated sludge treatment tests on two wastewaters. The first was a textile dyeing and finishing process water that contained Reactive Navy 106 and the second was a municipal wastewater consisting predominately of textile effluents. Both effluents were fed to laboratory-scale activated sludge reactors. Loyd concluded that aerobic treatment of the azo dye wastewaters provided significant biodegradation with minimal decolorization and the biodegradation did not include the azo dyestuffs.

Moosvi *et al.*, 2005 developed consortium exhibited 94 per cent decolourization ability within 37 h under a wide pH range from 6.5 to 8.5 and temperature ranging from 25 to 40 °C. The bacterial consortium was able to grow and decolourize RV5 under static conditions in the presence of glucose and yeast extract and also showed an ability to decolourize in the presence of starch in place of glucose. Maximum decolourization efficiency was observed at 200 ppm (mg/l) concentration of RV5. Bacterial consortium RVM11.1 had the ability to decolourize 10 different dyes tested.

Idaka *et al.*, 1987 described the reductive fission of azo bonds by *P. cepacia* and aerobic reduction of simple azo compounds by *Aeromonas hydrophilia.* The ability of bacteria to utilize simple azo dye compounds as sole source of carbon and energy was shown by Dykes *et al.*, 1994 and Yatome *et al.*, 1993 and *Pseudomonas* K22 strain capable of utilizing carboxy orange II was well established by Kulla 1981, Zimmermann *et al.*, 1982, Kulla *et al.*, 1983 and Coughlin *et al.*, 1997 and 1999.

Finally, Zissi *et al.* (1997) investigated the biological oxidation of p-aminoazobenzene (pAAB) by *Bacillus subtilis* and results proved that *Bacillus subtilis* could cometabolize pAAB in the presence of glucose, breaking the N=N double bond and producing aniline and p-phenylenediamine. The degradation of the dye was the direct result of an oxygen insenstive azo reductase enzyme found to be present in the soluble fraction of the biomass. This enzyme was also synthesized independently of the presence of pAAB.

Bacterial cultures namely *Vibrio logei* and *Pseudomonas nitroreducens* from a wastewater treatment plant degraded a toxic azo dye (methyl red) by decolourization. Complete decolourization using a mixed-culture was achieved at pH 6, 30°C within 6 h at 5 mg/l methyl red concentration and 16 h at 20–30 mg/l (Adedayo *et al.*, 2004).

Azo Dye Degradation by Fungi

The inappropriate disposal of dyes in wastewater constitutes an environmental problem and can cause damage to the ecosystem. Alternative treatments have been reported that fungi are particularly effective in the decolorization of textile effluents.

The wood degrading whit rot basidiomycete *Phanerochaete chrysosporium* was found to degrade a number of Azo dyes (Glenn and Gold, 1983 and Paszczynski *et al.*, 1992, Spadaro *et al.*, 1992). The extracellular peroxidases produced by *P. chrysosporium*, Laccase produced by *Pyricularia orison*, Mn peroxidase produced by *Aspergillus terreus* were found to initiate azo dye degradation (Glenn *et al.*, 1986, Chivukula Cripps *et al.*, 1990 and Renganathan, 1995). The degradation pathway of copper-pthalocyanine dyes by *P. chrysosporium* was given by spectrometry and polarography (Conneely *et al.* 1999). Anthraquinone dye degradation by *Geotricum candidum* was reported by Kim *et al.*, 1995 and Kim and Shoda, 1999.

The decolorization of dyes with different molecular structures by *Cunninghamella elegans* was evaluated under several media conditions. The decolorization profile was highly dependent upon the incubation time, the molecular structure of the dye and presence or absence of co-substrates. The presence of sucrose or both sucrose and peptone significantly increased the decolorization of the solutions, however, the presence of only the nitrogen source suppressed it (Ambrosio and Campos-Takaki 2004).

Basidiomycete PV 002, a recently isolated white-rot strain from decomposed neem waste displayed high extracellular peroxidase and rapidly decolorized azo dyes. White rot strain PV 002 efficiently decolorized Ranocid Fast Blue (96 per cent) and Acid Black 210 (70 per cent) on day 5 and 9 respectively under static conditions. The degradation of azo dyes under different conditions was strongly correlated with the ligninolytic activity. The optimum growth temperature of strain PV 002 was 26°C and pH 7.0 (Pradeep Verma and Datta Madamwar 2005).

Algal Biodegradation

The commonly available algal spirogyra species was found to degrade/decolorize the aromatic basic dye-stuff containing azine group in batch scale operation (Venkata Mohan *et al.*, 2002). The decrease in aqueous phase pH below 3 during initial phase reduces the surface charge of algal cell, which sorbed the dye molecule in the aqueous environment through diffusion. This was observed in the initial stages of the study and about 10 per cent of the dye removal may be attributed to this mechanism (Biosorbtion). Once the dye molecules adsorbed onto the surface of the cell diffuses into the cell and stored in vacuole and participate in the respiration/photo conversion and subsequently releases polysaccharides and related organic compounds (bioconversion). The released metabolic intermediates possess excellent coagulation capacity and the dye remaining in the aqueous phase tend to absorb on the surface of the polymers and settles (Biocoagulation).

Actinomycete Biodegradation

A variety of actinomycetes like *Streptomyces viridosporrns* (Ramachandra *et al.*, 1988), *Thermomonospora mesophila* and *Streptomyces badius* (Godden *et al.*, 1992) and *Streptomyces chromofuscus* A11 (Pasti-Grigsby *et al.*, 1996) involving in azo dye degradation have been reported. These *Streptomycetes* sp. produce extracellular peroxidases that contribute to their ability to decolorize the dyes. Evidence does show that the aerobic biodegradation of azo dye intermediates is possible and is perhaps an effective treatment process for stabilizing these compounds after anaerobic reduction.

Anaerobic Treatment

Anaerobic reduction of azo dyes using microbial sludges can be an effective and economic treatment process for removing color from dye house effluents. Previous studies have demonstrated the ability of anaerobic bacteria to reductively cleave the azo linkages in reactive dyes (Chung *et al.*, 1978; Brown and Laboureur, 1983; Brown and Hamburger, 1987; Loyd, 1992; Ganesh, 1992; Razo-Flores *et al.*, 1997 and Chinwetkitvanich, 2000). Although this effectively alters the chromogen and destroys the observed color of the dye, many aromatic groups are not susceptible to anaerobic reduction. However, there is evidence that some azo dye metabolites may be fully stabilized in anaerobic environments (Weber and Wolfe, 1987 and Razo-Flores, 1997).

During anaerobic treatment, it is expected that the dye molecules would cleave via reduction of the azo bonds. While this is often effective for reducing the color of the dye, it does not usually lower the carbon content (Loyd, 1992; Ganesh, 1992). Brown *et al.* (1993) studied the degradability of various azo dyes in both anaerobic and aerobic systems. In the first study, Brown and Laboureur (1983) investigated the anaerobic degradability of 22 commercial dyes. Of the dyes studied, four monoazo and six diazo-dyes showed substantial biodegradation, while two polyazo dyes showed moderate to variable reductions. Later in 1987, Brown and Hamburger conducted a study on 14 azo dyes subjected to anaerobic sludge digestion followed by aerobic treatment. Brown and Hamburger's results confirmed the findings from earlier research, showing decolorization of the azo dyes. Confirming the cleavage of azo linkages, the production of metabolites was also observed, but at less than theoretical concentrations.

Brown and Hamburger (1987) Loyd (1992) and Ganesh (1992), who all reported good color losses, but low carbon removals following the anaerobic treatment of various azo dyes. It is hypothesized that the dye molecules undergo only partial degradation, limiting the carbon loss, but not the color removal. More recently, Razo-Flores *et al.* (1997) investigated the fate of Mordant Orange 1 (MO1) and Azodisalicylate (ADS) under methanogenic conditions using continuous Upflow-Anaerobic-Sludge-Blanket (UASB) reactors and demonstrated the ability of an anaerobic consortium to completely mineralize some azo dye compounds. Razo-Flores *et al.* (1997) observed the complete mineralization of ADS with and without a cosubstrate, indicating the possibility for aromatic amine destruction in methanogenic environments. The compound, 1,4-phenylenediamine was not observed to degrade in either test reactor, indicating the specificity of aromatic amine utilization by anaerobes.

In studies conducted by Loyd (1992) and Ganesh (1992), the anaerobic reduction of textile mill effluents and the azo dyes Reactive Black 5 and Navy 106 were investigated, respectively. In both cases, good decolorization with minimal nutrient removal was observed. Chinwetkitvanich *et al.* (2000) performed a study on various reactive dye bath effluents concluded that higher initial color concentrations might be deleterious to acid forming bacteria, resulting in a lower dye removal. Additionally, the authors suggest that sulfate-reducing bacteria might out-compete other anaerobic microorganisms for available organic carbon, but contribute minimally to decolorization. This could serve to limit the reduction equivalents necessary for dye degradation.

Anaerobic/Aerobic Sequential Step-Treatment

Anaerobic bacteria are often able to reduce the azo linkages, but are generally unable to further stabilize the dye metabolites; it would seem advantageous to follow anaerobic treatment processes with an aerobic treatment step. As mentioned previously, aerobic organisms can oxidize aromatic ring compounds to simpler-molecules. A substantial amount of research has been conducted on ANA/AER sequential step-treatment systems used for degrading textile wastewaters (Brown and Hamburger,

1987; Loyd, 1992; Seshadri and Bishop, 1994; Fitzgerald and Bishop, 1995 and O'Neill *et al.*, 2000). Brown and Hamburger (1987) conducted a study on the ultimate biodegradability of various dye stuffs. Fourteen azo dyes and two other dye types were studied using lab-scale anaerobic and activated sludge reactors. Metabolite production was observed following anaerobic treatment, indicating the presence of eight identifiable aromatic amines. Results from the aerobic treatment phase corresponded with the work performed by Brown and Laboureur (1983a), showing reductions in DOC for most of the dyes. Some sulphonated aromatic amines were found to be non- biodegradable during this test.

Seshadri and Bishop (1994) investigated the fate of azo dyes Acid-Orange 7, Acid-Orange 8, Acid-Orange 10 and Acid-Red 14 in an ANA/AER sequential step- treatment system in a bench-scale Fluidized-Bed Anaerobic Reactor (FBR) followed by a bench-scale activated sludge reactor as a sequenced second stage treatment step. Their results indicated that the transformation of all the dyes to intermediates was readily achieved via anaerobic reduction; and was assumed to be the result of azo bond cleavage. Complete mineralization was not observed, however, COD and color removals were greatest at HRT's of 12 and 24 hours, with one hour being the shortest HRT tested and 24 hours the longest. However, the largest cumulative removals occurred in the first two hours of treatment. This would be expected based on the higher nutrient concentrations available for biodegradation during that period. The Acid-Orange 10 appeared to inhibit dye removal at concentrations of 15 mg L^{-1}. The authors believed the production of aromatic amines was responsible for the toxicity, citing the work of Chakrabarti *et al.* (1988). All other dyes tested did not exhibit inhibition at a concentration of 15 mg L^{-1}. COD removal was variable depending on the dye, but reductions were seen in both the anaerobic and aerobic phases. Aerobic oxidation of dye intermediates was necessary to decrease COD levels to an acceptable range.

Results from Zissi and Lyberatos (1996) confirmed that the anoxic biodegradation of p-aminobenzene, a simple azo dye, is effective in removing its apparent color. The findings from this research suggest that an ANA/AER sequential step-treatment system is the most effective method for biologically treating textile wastewaters.

Fitzgerald and Bishop (1995) used three lab-scale reactor systems to study the degradation of the azo dyes Acid-Orange 10, Acid-Red 14 and Acid-Red 18. The reactor system included an anaerobic fluidized bed system in the first stage, which was followed by an aerobic swisher reactor in the second stage. Their results indicated a high degree of degradation of the Acid Red 18 and Acid Red 14 in the anaerobic stage, with decolorization greater than 90 per cent. Acid Orange 10 was only decolorized by 70 per cent, however. Furthermore, analysis of the dye intermediates suggested a high degree of removal in the anaerobic stage. This is interesting, as most of the literature reports poor metabolite reduction in anaerobic environments. Low color loss and COD removal was measured in the aerobic stage. The authors do state that further investigation is necessary to verify if the dye intermediates biodegraded in the anaerobic reactor.

Recently, O'Neill *et al.* (2000) conducted a study on degradation of reactive azo dye Procion Red H-E7B in an ANA/AER sequential step-treatment system comprised of a lab-scale UASB reactor and an activated sludge tank. Based on this research, O'Neill concluded that Procion Red H-E7B could be qualitatively shown to degrade to aromatic amine derivatives after anaerobic treatment with subsequent oxidation of these derivatives following aerobic treatment.

Factors Affecting Azo Dye Biodegradation

Due to the highly variable nature of biological treatment systems and especially textile effluents, there are a number of factors that may affect the biodegradation rate of azo dyes. Non-dye related parameters such as temperature; pH, dissolved oxygen or nitrate concentrations, type and source of reduction equivalents, bacteria consortium and cell permeability can all affect the biodegradation of azo dyes and textile effluents. Dye related parameters such as class and type of azo dye (*i.e.*, reactive-monoazo), reduction metabolites, dye concentration, dye side groups and organic dye additives could also affect the biodegradability of azo dye wastewaters.

Wuhrmann *et al.* (1980) investigated the effects of pH, temperature, type and concentration of respiration substrates and oxygen tension on the rate of biological reduction of a variety of azo dyes. A consortium of microbes was used, including *Bacillus cereus, Sphaerotilus natans* and two others isolated from sewage-activated sludge. Also, activated sludge was used in experiments with mixed biocenoses. Temperatures, which are too high or too low, can result in the exclusion of a particular group of microorganisms. Using activated sludge, Wuhrmann *et al.* (1980) determined that temperature has an increasing liner relationship with the reduction rate of Orange II and Lanasyl violet up to 28°C. In general however, most studies have been conducted at set temperatures, offering minimal data on temperature effects. The wastewater pH can affect the proper functioning of both anaerobic and aerobic organisms (Grady *et al.*, 1999). Wuhrmann *et al.* (1980) also investigated the effect of pH on dye reduction rates, but were unable to conclusively establish a relationship. However, they did state that an exponential increase in the decolorization rate was observed by decreasing the pH, but this relationship depended on the dye being tested. Loyd (1992) observed an indirect increase in the rate of decolorization of Navy-106, with decreased pH values in anaerobic batch tests. No statistical data was performed to verify this result; however Nitrate and especially oxygen may play an important role in determining the rate of dye reduction. The presence of oxygen generally inhibits the degradation of azo dye chromogens. Interestingly, Wuhrmann *et al.*, demonstrated that obligate aerobes might actually de colorize azo compounds under temporary anoxic conditions. However, high nitrite or nitrate concentrations in the mixed liquor of activated sludge plants could significantly inhibit dye removal. Zissi and Lyberatos (1996) observed *Bacillus subtilis* to degrade p-aminoazobenzene under anoxic conditions.

Without the necessary reduction equivalents to optimize bacterial respiration and growth, dye reduction may be inhibited. Often, bacterial cultures are unable to proliferate when an azo dye is the sole carbon and nitrogen source. Therefore, additional, readily biodegradable sources may be necessary. Wuhrmann *et al.* (1980) indicated that in the absence of oxygen an azo compound will act as the sole oxidant, and its reduction rate will be governed by the rate of formation of the electron donor. This may be a problem in WWTPs that receive a high loading of dye waste with a low, readily available, organic carbon content. Gingell and Walker (1971) state that the presence of oxygen may compete with the azo dye as the preferred oxidizer of reduced electron carriers in the respiration chain and thus limit the reduction of azo linkages. The type of bacteria or consortium used for dye biodegradation will undoubtedly affect the reduction rate. In general however, aerobic microbes do not have the ability to substantially decolorize azo dyes, but can oxidize the dye metabolites. Ganesh (1992) concluded that very little of the dye added to a biological reactor will be leached from the biomass when placed in a landfill. This might suggest that the dye is effectively reduced after adsorption to the cell wall or that very little dye is actually adsorbed. Cell permeability might play an important role in dye biodegradation.

Jo-Shu Chan *et al.*, 2004 reported that the addition of supernatants containing metabolites from growth and decolorization cultures triggered an enhancement of decolorization rates. Although extracellular metabolites play a crucial role for stimulating decolorization, they still cannot enable decolorization alone without involvement of biodecolorizers (*E. coli* strain NO_3).

In the study conducted by Wuhrmann *et al.* (1980), all dyes that were not reduced by whole cells were effectively degraded by cell extracts from both facultative, anaerobes and obligate aerobes. This suggests that many cells might be capable of dye biodegradation, but are limited by the permeability of their cell walls. The azo dye structure can play a significant role in the dye biodegradation rate. Depending on the number and placement of the azo linkages, some dyes will biodegrade more rapidly, than others. In general, the more azo linkages that must be broken will cause the reduction rate to be slower. Brown and Laboureur (1983) observed that two poly-azo dyes showed only moderate to variable biodegradation as compared to four monoazo and six diazo dyes. The authors indicate that poly-azo dyes are less likely to degrade than mono- or diazo dye types. Fiber-reactive azo dyes often contain solubilizing side groups, as well as a nucleophilic reactive group. Depending on the nature of these groups, biodegradation might be inhibited.

According to Ganesh (1992), sorption of dye to sludge depends on the type, number and position of the constituents in the dye molecule. Hydrophilic sulfo groups reduce the dye removal through sorption and conversely, sorption is increased by the presence of hydroxyl, nitro and azo groups in the dye molecule. The production of toxic byproducts or the presence of toxic dye additives may also inhibit biodegradation. High, salt concentrations are not uncommon in textile effluents and may result in adverse conditions for biodegradation. Dispersing and solubilizing agents may also create inhibitory conditions for dye reduction (Carliell, *et al.*, 1994).

In various studies (Wuhrmann *et al.*, 1980; Donlagic' and Levec, 1998, and O'Neill *et al.*, 2000) the production of inhibitory dye metabolites is cited as causing a decrease in biodegradation. O'Neill *et al.* concluded from respiration-inhibition testing that anaerobic degradation of simulated textile effluent generated metabolites that were toxic to some aerobic organisms, Donlagic' and Levec concluded that the distribution of dye intermediates plays an important role in determining the aerobic biodegradability of Orange II. They further state that low dye removal may be attributed to the presence of intermediates that are less susceptible to microbial degradation or that act as inhibitory agents. Wurhmann *et al.* (1980) attributes a decline in decolorization rates to the accumulation of reduction products in the test medium. Inhibition of various microorganisms to dye metabolites is frequently cited throughout the literature and is assumed to be a key factor when treating wastewaters containing azo dyes.

A final and important factor to evaluate is the initial dye concentration of the wastewater. Seshadri and Bishop (1994) performed a study investigating the effect of different influent dye concentrations on the color removal efficiency. They concluded that elevated dye concentrations may cause a drop in per cent dye removal. Furthermore, the inhibition may be directly related to the effects of increased dye formation due to higher dye concentrations. Less pronounced reductions were seen at lower concentration levels. It should be noted that tolerable influent concentrations are likely specific to individual or related groups of dyes. Cariell *et al.* (1995) also states that toxicity assays showed that C.I. Reactive Red 141 was inhibitory to anaerobic organisms at concentrations greater than 100 mg/l, however prior biomass adaptation increased their resistance to elevated dye concentrations.

The Research Need

Biodegradation of azo dyes by a single species of microorganism requires that species to possess all the necessary enzymes involved in the pathway and also genetic information. Continual exposure of microbes to azo dyes can result in the evaluation of novel metabolic processes which enables the microorganisms to wholly or partially degrade the azo dyes. There are two possible ways by which the microbes acquire new traits. They are (1) Natural mutation and natural gene transfer and (2)

Artificial gene transfer. Natural mutations occur at the rate of 1 mutation per 100 cell divisions, mostly non beneficial to the cell. Many genes that encode key enzymes in the catabolism of azo dyes are plasmid encoded. Such plasmids can be transmitted from one cell to another by the processes of conjugation and transformation. In cells possessing plasmids which only encode for a few enzymes for azo dye degradation must be complemented with chromosomal genes so that plasmid encoded pathway is linked to a metabolic pathway. Encoding of degradative genes on plasmids gives the possibility for the construction of microbes by genetic engineering with the metabolic requirements for the treatment of azo dye containing wastes.

References

Adedayo, O.S., Javadpour, C., Taylor Anderson, W.A. and Moo-Young, M., 2004. Decolourization and detoxification of methyl red by aerobic bacteria from a wastewater treatment plant. *World J. Microbiol. Biotechnol.*, 20(6): 545–550.

Ambrosio, S.T. and Campos-Takaki, G.M., 2004. Decolorization of reactive azo dyes by *Cunninghamella elegans* UCP 542 under co-metabolic conditions. *Biores. Technol.*, 91(1): 69–75.

Anliker, R., 1979. Ecotoxicology of dyestuffs: A joint effort by industry Ecotox. *Environ. Safety*, 3: 59–74.

Ashoka, C., Geetha, M.S. and Sullia, S.B., 2002. Biobleaching of compost textile dye effluent using bacterial consortia. *Asian J. Microbiol. Biotech. Env. Sci.*, 4: 65–68.

Bal, A.S., 1999. Waste water management for textile industry: An overview. *Indian J. Env. Hlth.*, 41: 264–290.

Banat, I.M., Nigan, P., Singh, D. and Marchanto, R., 1996. Microbial decolourisation of textile dye containing effluents: A review. *Biores. Technol.*, 58: 217–227.

Brown, D. and Hamburger, B., 1987. The degradation of dyestuffs, Part III: Investigations of their ultimate degradability. *Chemosphere*, 16(7): 1539–1553.

Brown, D. and Laboureur, P., 1983. The degradation of dyestuffs, Part I: Primary biodegradation under anaerobic conditions. *Chemosphere*, 12(3): 397–404.

Brown, D. and Laboureur, P., 1983a. The aerobic biodegradability of primary aromatic amines. *Chemosphere*, 12(3): 405–414.

Brown, Mark A. and Stephenm C. DeVito, 1993. Predicting azo dye toxicity. *Crit. Rev. Environ. Sci. Tech.*, 23(3): 249–324.

Brown, M.S. and Devito, S.C., 1993. Predicting azo dye city. *Crit. Rev. Environ. Sci. Tech.*, 23: 249–342.

Cariell, C.M., Barclay, S.J. and Buckley, C.A., 1996. Treatment of exhausted reactive dye bath effluent using anaerobic digestion: Laboratory and full scaletrials. *Water SA*, 22(3): 225–233.

Cariell, C.M., Barclay, S.J., Naidoo, N., Buckley, C.A., Mulholland, D.A. and Senior, E., 1994. Anaerobic decolorisation of reactive dyes in conventional sewage treatment processes. *Water SA*, 20(4): 341–344.

Canell, C.M., Barclay, S.J., Naldoo, N., Buckley, C.A., Mulholland, D.A. and Senior, E., 1995. Microbial decolourisation of a reactive azo dye under anaerobic conditions. *Water SA*, 21(1): 61–69.

Cartwright, R.A., 1983. Historical and modern epidemiological studies on populations exposed to N-substituted aryl compounds. *Env. Health Perspect.*, (49): 13.

Chakrabarti, T., Subrahmanyam, P. and Sundarsan, B., 1988. Biodegradation of recalcitrant industrial wastes. *Biotreatment Systems Vol. II.*, (Ed.) D. Wise. CRC Press, Inc., Boca Raton, FL.

Chinwetkitvanich, S., Tuntoolvest, M. and Panswad, T., 2000. Anaerobic decolorization of reactive dye bath effluents by a two-stage UASB system with tapioca as a co-substrate. *Water Research*, 34(8): 2223–2232.

Chivukula, M. and Renganathan, V., 1995. Phenolic azo dye oxidation by laccase from *Pyricularia oryzae*. *Appl. Environmental Microbiol.*, 61: 4374.

Chudgar, R.J., 1985. Azo dyes. In: *Kirk–Othmer Encyclopedia of Chemical Technology*, 4th edn, (Ed.) Kroschwitz. Wiley, New York, pp. 821–875.

Chung, King-Thom and Stevens, S.E., 1993. Degradation of azo dyes by environmental microorganisms and helminthes. *Environ. Toxicol. Chem.*, 12: 2121–2132.

Chung, King-Thom, Fulk G.E. and Egan, M., 1978. Reduction of azo dyes by intestinal anaerobes. *Appl. Environmental Microbiol.*, 35(3): 558–562.

Chung, K.T., 2000. Mutagenictiy and carcinogenicity of aromatic amine metabolites produced from azo dyes. *Enviro Carcino. and Ecotox. Rev.*, 18: 51–74.

Colour Index International: Additions and Amendments. 1998. *The Society of Dyers Colourists*, 109: 2860–2895.

Conneely, A., Smyth W.F. and McMullan, G., 1999. Metabolism of the pthalocyanine textile dye turquoise blue by *Phanerochaete chrysosporium*. *FEMS Microbiol. Lett.*, 179: 333–337.

Coughlin, M.F., Kinkle, B.K. and P.L. Bishop, 1999. Degradation of azo-dyes containing aminonaphthol by *Spingomonas* sp. strain ICX. *Indian J. Microbiol. Biotechnol.*, 23: 341–346.

Coughlin, M.F., B.K-.Kinkle, K. Tipper and P.L. Bishop, 1997. Characterisation of aerobic azo-dye degrading bacteria and their activity in biofilms. *Water Sci. Tech.*, 36: 215–220.

Cripps, C., J.A. Bumpus and S.D. Aust, 1990. Biodegradation of azoheterocyclic dyes by *Phanerochaete chrysosporium*. *Appl. Environ. Microbiol.*, 56: 1114–1118.

Donlagic', Jelka and Janez Levec, 2000. Comparison of catalyzed and noncatalyzed oxidation of azo dye and effect on biodegradability. *Environ. Sci. Technol.*, 32: 1294–1302.

Dykes, G.A., Timm, R.G. and Von Holy, A., 1994. Azo reductase activity in bacteria associated with the greening of instant chocolate pudding. *Appl. Environ. Microbiol.*, 60: 3071–3074.

Edwards, Jesscia C., 2000. Investigation of color removal by chemical oxidation for three reactive textile dyes and spent textile dye wastewater. *Masters Thesis*, Virginia Polytechnic Institute and State University, 56 pp.

Emitiazi, G., 2000. Decolorisation and biodegradation of dyes by *Aspergillus terreus* grown on wheat straw with Mn peroxidase activity. *Poll. Res.*, 19: 31–35.

Fitzgerald, Sean W. and Paul L. Bishop, 1995. Two stage aerobic/aerobic treatment of sulfonated azo dyes. *J. Environ. Sci. Health*, 30(6): 1251–1276.

Freeman, Harold S. and Sokolowska, J., 1999. Developments in dyestuff chemistry. *Review of Progress in Coloration and Related Topics*, pp. 29.

Ganesh, R., 1992. Fate of azo dyes in sludges. *Masters Thesis*, Virginia Polytechnic Institute and State University. 193 pp.

Gingell, R. and Walker, R., 1971. Mechanisms of azo reduction by *Streptococcus faecalis* II: The role of soluble flavines. *Xenobiotica*, 1: 231–239.

Glenn, J.K. and Gold, M.H., 1983. Decolorisation of several polymeric dyes by the lignin degrading basidiomycete *Phanerochaete chrysosporium*. *Appl. Environ. Microbiol.*, 45: 1741–1747.

Glenn, J.K., Akileswara, L. and Gold, M.H., 1986. Mn (II) oxidation is the principal function of the extra cellular Mn peroxidase from *Phanerochaete chrysosporium*. *Arch. Biochem. Biophys.*, 251: 688–696.

Godden, B.A., S. Ball, Helvenstein, P.A., Carthy J. Mc and Penninckx, M.J., 1992. Towards elucidation of the lignin degradation pathway in actinomycetes. *J. Gen. Microbiol.*, 138: 2441–2448.

Grady, C.P.L. Jr., Daigger, G.T., Lim, H.C., 1999. *Biological Wastewater Treatment*. Marcel Dekker, Inc., New York, NY, 1076 pp.

Horning, R.H., 1977. Characterization and treatment of textile dyeing wastewaters. *Textile Chemist and Colorist*, 9(3): 73–76.

IARC, World Health Organization International Agency for Research on Cancer. 1982. Monographs on the Evaluation of the Carcinogenic Risk of Chemicals to Humans. *Some Industrial Chemicals and Dyestuffs*, Vol. 29, Lyon, France. Jager, I., *et al.*, 1996. *Melliand Textilber*, 77: 72.

Idaka, E., Ogawa, T. and Horitsu, H., 1987. Some properties of azo reductase produced by *Pseudomonas*. *Bull. Environ. Contam. Toxicol.*, 39: 982–989.

Jo-Shu Chan, Bor-Yann Chen and Yung Sheng Lin, 2004. Stimulation of bacterial decolorization of an azo dye by extracellular metabolites from *Escherichia coli* strain NO_3. *Biores. Technol.*, 91(3): 243–248.

Kim, S.J. and M. Shoda, 1999. Purification and characterization of a novel peroxidase from *Geotrichum candidum* December 1 involved in decolorisation of dyes. *Appl. Environ. Microbiol.*, 65: 1029–1035.

Kim, S.J., Ishikawa, K., Hirai, M. and Shoda, M., 1995. Characteristics of a newly isolated fungus, *Geotrichum candidum* December 1 which decolorizes various dyes. *J. Fermentation and Bioengineering*, 79: 601–607.

King-thom, Chung, George E. Fulk and Mary Egan, 1978. Reduction of azodyes by intestinal anaerobes. *Appl. Env. Microbiol.*, pp. 558–562.

Knapp, J.S. and Newby, P.S., 1995. Decolourization of dyes by wood rotting basidiomycete fungi. *Water Research*, 29: 1807–1809.

Kumar, K., Saravana Devi, S., Krishnamurthi, K., Gampawar, S. Nilesh Mishra, Pandya G.H. and Chakrabarti, T., 2006. Decolorisation, biodegradation and detoxification of benzidine based azo dye. *Biores. Technol.*, 97(3): 407–413.

Kulla, H.G., Klausner, U. Meyer, B., Ludeke B. and Leisinger, T., 1983. Interference of aromatic sulfo groups in the microbial degradation of the azo dyes Orange I and Orange II. *Arch. Microbiol.*, 135: 1–7.

Kulla, H.G., 1981. Aerobic bacterial degradation of azo dyes in microbial degradation of xenobiotics, recalcitrant compounds, (Eds.) T. Leisinger, A.M. Cook, J. Nuesh and R. Hutter. Academic Press, London, pp. 387–399.

Levine, Walter, G., 1991. Metabolism of azo dyes: Implications for detoxification and activation. *Drug Metabolism Reviews*, 23(3–4): 253–309.

Lewis, David, M., 1999. Coloration for the next century. *Review of Progress in Coloration and Related Topics*, 29: 23–28.

Longbottorn, J.E. and Lichtenberg, J.J., 1982. *EPA Test Methods*. EPA–600/4–82–057.

Little, L.W.J., Lamd, C., Chillingworth, M.A and Durkin, W.B., 1974. Acute toxicity of selected commercial dyes to the fathead minnow and evolution of biological treatment for reduction of toxicity. In: *Proc. 29th Purdue Ind. Waste Conf.* Ann. Arbor Science, Ann Arbor, Michigan.

Loyd, Chapman K., 1992. Anaerobic/aerobic degradation of a textile dye wastewater. *Masters Thesis,* Virginia Polytechnic Institute and State University, 184 pp.

McCarthy, B.J., 1997. Biotechnology and Coloration. *Review of Progress in Coloration and Related Topics*, 27: 26–31.

McCurdy, Michael W., 1991. Chemical reduction and oxidation combined with biodegradation for the treatment of a textile dye wastewater. *Masters Thesis,* Virginia Polytechnic Institute and State University, pp. 181.

Moosvi, S., Haresh Keharia and Datta Madamwar, 2005. Decolourization of textile dye Reactive Violet 5 by a newly isolated bacterial consortium RVM 11.1. *World J. Microbiol. Biotechnol.*, 21(5): 667–672.

Namasivayam, C. and Yamuna, R.T., 1992. Removal of congored from aqueous solution by biogas waste slurry. *J. Chem. Tech. Biotechnol.*, 53: 153–157.

O'Neil, C., Lopez, A., Estevez, S., Hawkes, F.R., Hawkes, D.L. and Wilcox, S., 2000. Azo-dye degradation in an anaerobic-aerobic treatment system operating on simulated textile effluent. *Appl. Microbiol. Biotech.*, 53: 249–254.

Pagga, U. and Brown, D., 1986. The degradation of dyestuffs, Part II: Behaviour of dyestuffs in aerobic biodegradation tests. *Chemosphere*, 15(4): 479–491.

Pasti-Grigsby, M.B., Burke, N.S., Goszezynski S. and Crawford, D.L., 1996. Transformation of azo dye isomers by *Streptomyces chromofuscus*. *Appl. Environ. Microbiol.*, pp. 1814–1817.

Paszcezynski, A., Pasti-Grigsby, M., Goszceynski, S., Crawford, R. and Crawford, D.L., 1992. Mineralization of sulfonated azo dyes and sulfanilic acid by *Phanerochaete chrysosporium* and *Streptomyces chromofuscus*. *Appl. Environ. Microbiol.*, 58: 3598–3604.

Pradeep Verma and Datta Madamwas, 2005. Decolorization of azo dyes using Basidiomycete strain PV 002. *World J. Microbiol. Biotechnol.*, 21(4): 1573.

Ramachandra, M., Crawford, B.L. and Hertel, G., 1988. Characterization of extracellular lignin peroxidase of the lignocellulose actinomycetes *Streptomyces viridosporus. Appl. Environ. Microbiol.*, 54: 3057–3063.

Razo-Flores, E., Luijten, M., Donlon, B., Lettinga, G. and Field, J., 1997. Biodegradation of selected azo dyes under methanogenic conditions. *Water Sci. Technol.*, 36(6–7): 65–72.

Rehn, L., 1895. Blasengeschwulste bei Fuchsin Arbeitem. *Arch. Klin. Chir.*, 50: 588.

Saranaik, S. and Kanekar, P., 1995. Bioremediation of colour of methyl violet and phenol from a dye industry waste effluent using *Pseudomonas* sp. isolated from factory soil. *J. Appl. Bacteriol.*, 79: 459–469.

Seshadri, S., Bishop, P.L. and Agha, A.M.1994. Anaerobic/aerobic treatment of selected azo dyes in wastewater. *Waste Management*, 14(2): 127–137.

Shaul, Glenn M., Holdsworth, T.J., Dempsey, C.R. and Dostal, K.A., 1991. Fate of water soluble azo dyes in activated sludge process. *Chemosphere*, 2(1–2): 107–119.

Spadaro, J.T., Gold, M.T. and Renganathan, V., 1992. Degradation of azo dye by the lignin degrading fungus *Phanerochaete chtysosporium*. *Appl. Environ. Microbiol.*, 58: 2391–2401.

Sudhakar, P., Palaniappan, R. and Gowrishankar, R., 2002. Degradation of azo dye (Blach E) by an indigenous bacterium *Pseudomonas* sp. BSP-4. *Indian J. Microbiol. Biotech. Env. Sci.*, 4: 203–208.

Venkata Mohan, S., Charidrasakar Rao, N. and Krishna Prasad, K., 2002. Biological decolorisation of stimulated basic dye effluents by algal Spirogyra species. *Asian J. Microbiol. Biotech. Env. Sci.*, 4: 107–111.

Vijaya, P., Padmavathy, P. and Sandiya, S., 2003. Decolourisation and biodegradation of reactive azodyes by mixed culture. *Indian J. Biotech.*, 2: 259–263.

Vyas, B.R. and Molitoris, H.P., 1995. Involvement of an extracellular H_2O_2-dependent lignolytic activity of the white-rot fungus, *Pleurotus ostreatus* in the decolorization of Remazol Brilliant Blue R. *Appl. Environ. Microbiol.*, 61: 3919–3927.

Weber, Eric J. and Lee Wolfe, N., 1987. Kinetic studies of the reduction of aromatic azo compounds in anaerobic sediment/water systems. *Environmental Toxicology and Chemistry*, 6: 911–919.

Willmott, N., Guthrie, J. and Nelson, G., 1998. The biotechnology approach to colour removal from textile effluent. JSDC, 114: 38–41.

Wong, P.K. and Yuen, P.Y., 1996. Decolourization and biodegradation methyl red by *Klebsiella pneumoniae* RS-13. *Water Research*, 36: 1744–1746.

Yamini, R. and Lalithakumari, D., 2001. Aerobic treatment of textile mill effluent and decolorisation of commercial azo dyes. *Asian J. Microbiol. Biotech. Env. Sci.*, 3: 353–358.

Yang, Y., Wyatt, D.T. II and Bahorshky, M., 1998. Decolorization of dyes using UV/H_2O_2 photochemical oxidation. *Textile Chemist and Colorist*, 30: 27–35.

Yatome, C., Matsufuru, H., Taguchi, T. and Ogawa, T., 1993. Degradation of 4-dimethylamino azobenzene-2-carboxylic acid by *Pseudomonas stuzeri*. *Appl. Microbiol. Biotechnol.*, 39: 778–781.

Zimmermann, H., Kulla, H.G. and Leisinger, T., 1982. Properties of purified orange li azo reductase enzyme initiating azo dye degradation by *Pseudomonas* KF46. *Env. J. Biochem.*, 129: 197–203.

Zissi, U. and Lyberatos, G., 1996. Azo-dye biodegradation under anoxic conditions. *Wat. Sci. Tech.*, 34(5–6): 495–500.

Zissi, U., G. Lyberatos and S. Pavlous, 1997. Biodegradation of p-aminobenzene by *Bacillus subtilis* under aerobic conditions. *J. Industrial Microbiol. Biotechnol.*, 19: 49–55.

Zollinger, H., 1961. *Azo and Diazo Chemistry Aliphatic and Aromatic Compounds.* Interscience Publishers, New York, London.

Zollinger, H., 1987. *Colour Chemistry: Synthesis Properties and Applications for Organic Dyes and Pigments*, pp. 92–102. VCH Publishers, Inc., New York,

Zollinger, Heinrich, 1991. *Color Chemistry: Syntheses, Properties and Applications of Organic Dyes and Pigments.* 496 pp.

Chapter 53

Marine Actinomycetes: A Potential Source for L-asparaginase

P. Dhevagi and E. Poorani

Assistant Professor, Department of Environmental Sciences, Tamil Nadu Agricultural University, Coimbatore – 3, Tamil Nadu

Introduction

Microorganisms are miniature chemical factories as they have the capacity to convert a variety of raw materials into a series of value added products. The main products are antibiotics, polysaccharides, proteins, oils, fatty acids, enzymes, pigments etc. The success and effective exploitation of microbes to large extent rely on the diversity of microorganisms in the environment. Particularly the pharma industry recognizes that diversity is an essential and potentially important element in the development of new drugs for the disease control. Actinomycetes occur in a wide range of environment producing a variety of scientifically interesting and commercially useful high value metabolites. The actinomycetes produce about 67 per cent of total 1200 microbial metabolites. The marine environment is a potential source for new actinomycetes and novel antibiotics (Mathew *et al.*, 1994). There are very few reports on antimicrobial activity from marine actinomycetes from west coast of India. In the last decades a few novel bioactive substances have been isolated from marine strains of Streptomycetes and other actinomycetes (Ravel *et al.*, 1998 and Canedo *et al.*, 2000). Work on diversity of marine actinomycetes is very much limited around the world. Only few institutions are working in this field as this type of research is more laborious and difficulties in isolating the actinomycetes. Keeping this in view, the review on actinomycetes has been given.

Biodiversity of Actinomycetes

Actinomycetes were first discovered by Cohn in 1875 and studies on actinomycetes have been numerous then. Erikson (1940) found that the anaerobic Actinomycetes *Antinomyces Israelli* settledout as compact bread 'crump' 'cauliflower' or puffball colonies in static culture. Laksman (1950) reported 21 recommended media in which four are frequently used for isolation of Actinomycetes. Kuster and Williams (1964) and Kriss *et al.* (1967) suggested that some actinomycetes might be terrestrial forms that have adapted to the salinity of seawater and sediments. Early work on marine actinomycetes was given by Baam *et al.*, 1966 and Okami and Okazaki (1972). Sambamurthy and Ellaiah (1974) isolated a new species called *Streptomycetes marinesis* from water of Palm Beach. Lakshmanaperumalsamy (1978) reported that out of 518 strains isolated from the marine sediments of Porto Novo, 85 per cent were *Streptomyces* species and 15 per cent were *Micromonospora* and *Nocardia*.

Postmaster and Freitas (1975) and Walker and Colwell (1975) reported that the *Streptomycetes* species showed high antibiotic activity. Vanajakumar (1979), Vanajakumar *et al.* (1981), Ellaiah and Reddy (1987) and Balagurunathan (1992) found that the isolated strains of actinomycetes (386 from different molluscs, 140 from marine sediments of Vishakapattinam coast and 51 from marine sediment of Parangipetti) were antagonistic against bacteria and fungi. Cross (1982), Wetzel (1983), Takizawa *et al.* (1993), Schallenberg and Kalff (1993) and Jiang and Xu (1996) reported that generally actinomycetes are abundance in the sediment.

Good fellow and Haynes (1984) and Ellaiah and Reddy (1987) reported that the marine sediments are potent sources for the isolation or actinomycetes with good antimicrobial activity. A high concentration of morphological form tends to produce a dispersed form of growth whilst a low concentration normally results in pellet formation (Lawton *et al.*, 1989). Jiang and Xu (1996) and Murphy and Hill (1998) reported that the actinomycetes are traditionally considered being soil bacteria but increasingly they are being isolated from sediments obtained from very deep waters and in the vicinity of hydrothermal vents.

Sivakumar (1998) isolated more than fifty strains of actinomycetes from mangrove sediments of Pitchavaram and more than 30 per cent were active against pathogenic microbes in which, more than 10 per cent showed broad spectrum of antagonistic activity and he also isolated some rare forms. Dhevendran and Annie (1999) and Balagurunathan (2000) found that in the marine and coastal ecosystems must be rich in gene pool possibly containing isolates, capable of producing useful metabolites. Mangroves, by virtue of fluctuating physico-chemical conditions can harbour actinomycetes of high genetic potential.

Gomathinayagam and Lakshmanaperumalsamy (2001) observed that the density of actinomycetes population in water and sediment varied monthly. A systematic and significant difference in population was observed in water and sediment. Higher density of actinomycetes population appeared in the period of pre-monsoon and relatively low population was recorded during post monsoon.

Whol and McArthur (2001) studied that the biomass and organic matter decomposition of actinomycetes-fungal interactions differ from those of individual isolates. Dhevendaran and Anithakumari (2002) isolated *Streptomyces* spp. from the gut of fish exhibiting L-asparaginase activity under growing conditions in a liquid broth. Similar trend was noticed with sodium chloride in the estuarine strains,

Mincer *et al.* (2002) and Kokare *et al.* (2004) studied that the sediment samples from Alibag, Janjira and Goa are potent sources for the isolation of bioactive actinomycetes. Most of the isolates from the sediment samples were identified as *Streptomyces.* Poorani (2005) isolated actinomycetes from the sediments of Southeast and Southwest coastal regions of India.

The followings are the attempts on the Indian marine actinomycetes

Authors	*Year*	*Places*
Chandramohan	1972	Sediments of Bay of Bengal
Postmaster and Freitas	1975	Marsh sediments of Bombay waters
Laksmanaperumalsamy	1978	Sediments of Parangipettai area
Vanajakumar	1979	Various parts of molluscs
Ellaiah and Reddy	1987	Sediments of Vizag
Balagurunathan	1992	Sediments of Parangipettai and mangrove area
Rathnakala	1995	Sediments of Cochin
Sivakumar	1998	Sediments of Pitchavaram
Devendaran and Annie	1999	Sediments of estuarine Veli lake
Dhevagi and Poorani	2006	Sediments of Pitchavaram

Isolation of Actinomycetes

Kanagawa and Kelly (1987) reported that the actinomycetes were isolated from sludge acclimate to thiophene-2-carboxylic (T2C) or 5-methyl-thiophene-2-carboxylic acid (T5M2C) and were identified as *Rhodococcus*.

Herron and Wellington (1990) reported a new method for the isolation and enumeration of *Streptomyces* spores from soil. Rathnakala and Chandrika (1993) observed that out of many recommended media for selective isolation of actinomycetes from soil, Glucose asparagine agar, Grein and Meyer's agar, Oat meal agar and Kuster's agar were found suitable for isolation of actinomycetes from mangrove sediments. In Kuster's agar when acetic acid was used, the colonies were countless showing selective enrichment of actinomycetes over control medium.

Whol and McArthur (1998) found that aquatic actinomycetes may have important differences from their terrestrial counterparts and a wide array of isolation techniques was most suitable for collection of aquatic actinomycetes (Figure 53.1). Lateef and Oloke (2003) observed that Lactose minimal agar is adjusted to be good for the isolation of actinomycetes, as it inhibited the growth of bacteria and fungi. Kuster and Williams (1964), Shirling and Gottlib (1966) and Parson *et al.* (1984) used sea water complex broth for culturing the actinomycetes.

Poorani (2005) isolated actinomycetes from the sediments of the Southeast and Southwest coastal regions of India and screened for their potential L-asparaginase activity using starch casein agar, Kuster's agar, seawater complex agar, Kenknight agar and glucose asparagine agar.

Importance of Actinomycetes

Actinomycetes have provided many important bioactive compounds of high commercial value and continued to be routinely screened for new bioactive compounds. Rainbow and Rose (1963) reported that the actinomycetes play a major role in producing antibiotics and other metabolites such as intracellular enzymes, pigments and growth promoting factors.

Okami and Okazaki (1972) and Chandramohan *et al.* (1972) reported that the actinomycetes may occur not only as a part of the marine ecosystem but also may contribute significantly to the economy of the sea. Young and Smith (1975) and Willoughby (1976), obtained many extra cellulase enzymes,

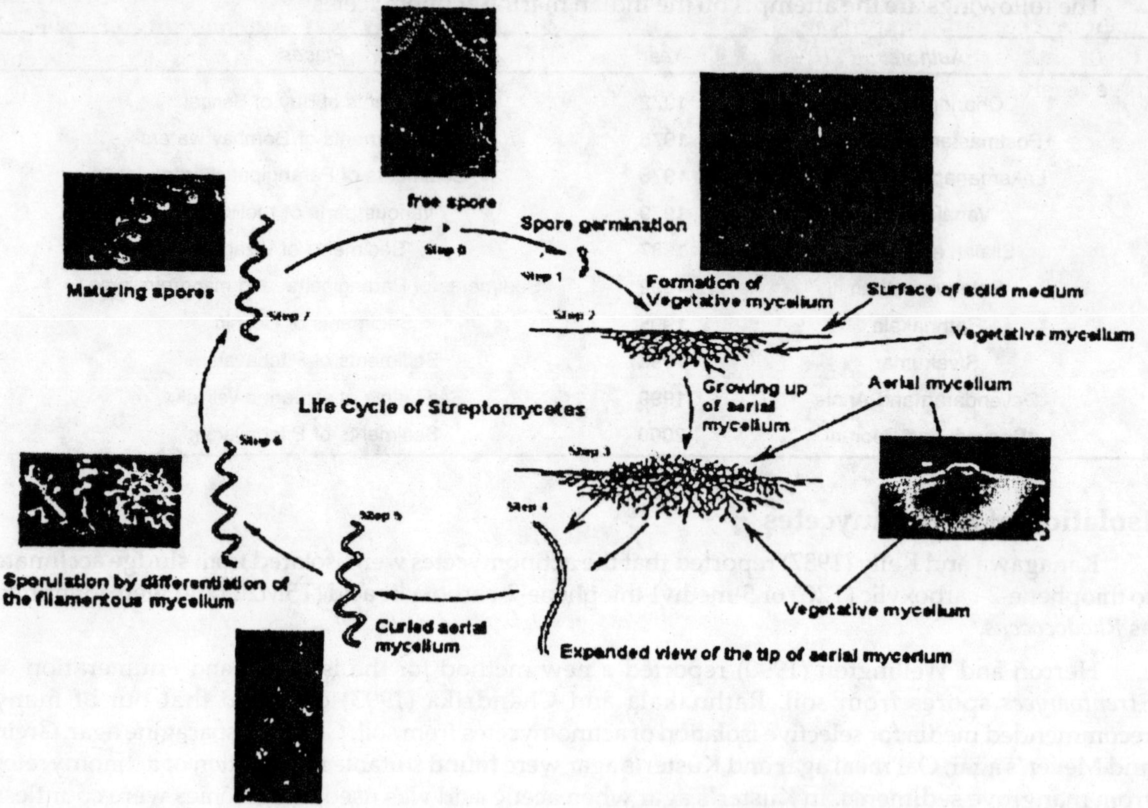

Figure 53.1: Complex Life Cycle of Streptomyces

which degrade cellulose, chitin, keratin and lignin from actinomycetes. They suggested that the marine actinomycetes might also take an active role in the deterioration of cellulosic substances in the marine environment.

Stastna *et al.* (1977), Metz *et al.* (1979) and Shomura *et al.* (1979) reported that production of antibiotic by some actinomycetes is affected by the morphological form of that organism. They found that the antibiotics are produced by filamentous growth forms, but not in fragmented mycelium. Balasubramanian *et al.* (1979) and Ishaque (1980) reported that the cellulase from the *Streptomycetes* isolates required 1–3 per cent sodium chloride for maximum activity and when the concentration was increased to 10–15 per cent there was no considerable loss in activity.

Kanagawa and Kelly (1987) reported that actinomycetes capable of complete degradation of thiophene derivatives and of their use by it as sole substrates for growth and they reported that thiophene-2- carboxylic acid and its 5-methyl derivative can be metabolized by some actinomycetes. Okami and Hotta (1988) reported that the actinomycetes have provided many important bioactive compounds successfully and approximately two-thirds of naturally occurring antibiotics of medical importance, which have high commercial value and continue to be routinely screened for new bioactive substances from Actinomycetes.

Ball *et al.* (1989) reported that native lignin degradation was associated with primary growth in actinomycetes and that the main activity was solubilization rather than mineralization. Ford *et al.* (1990) and Parthiana *et al.* (1991) and Takizawa *et al.* (1993) observed that the marine actinomycetes have received increasing attention as a source of novel bioactive compounds.

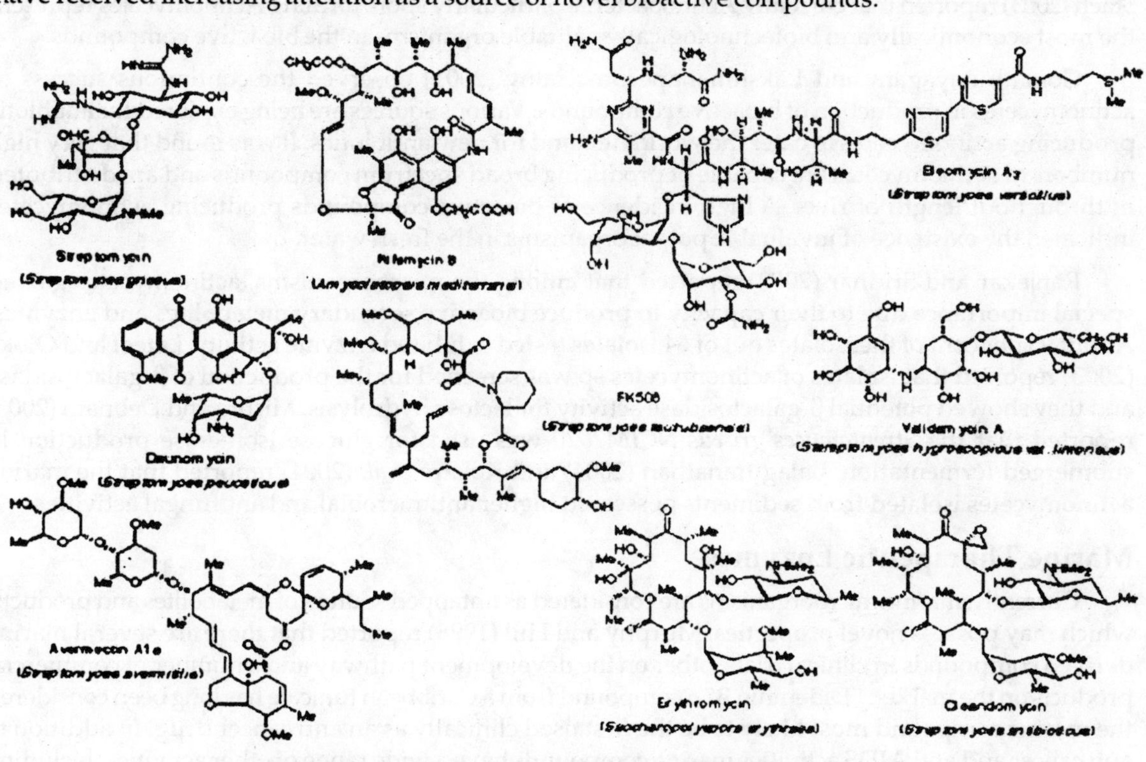

Figure 53.2: Antibiotics Produced by Streptomycetes

Jiang and Xu (1996) reported that the aquatic actinomycetes play a significant role in the decomposition of organic substances including some toxic compounds such as phenol in these lakes. Mercer *et al.* (1996) found that the actinomycetes are potentially rich sources of peroxidases for introduction into a market that was substantial and almost totally dominated Horse Radish Peroxidases (HRP) which was both well characterized and well active.

Murphy and Hill (1998) reported that a single group of bacteria produces over 70 per cent or all naturally occurring antibiotic drugs. Dhevenduran and Annie (1999) reported that the *Streptomycetes* associated with fish and shellfish have potential pharmaceutical value as they showed considerable antagonistic activity against fish and human pathogens.

Balagurunathan (2000) reported that the actinomycetes are well known for the production of high value metabolites such as nutrients, antibiotics, vitamins, enzymes, pigments etc. Today the value of antibiotic production from actinomycetes is > 1010 billion dollars/year. Novel strains of actinomycetes are registered as patents and sold at about 25–50 US dollar/strain from countries other than India. The biotechnological industries of developed countries are in need of potential strains of actinomycetes especially from the tropical areas that are blessed with rich microbial diversity.

Mieguelez *et al.* (2000) reported that the colony growth of the *Streptomycetes* provides an excellent prokaryotic experimental system for the study of the mechanism of cell death and its role in development. Zheng *et al.* (2000) reported novel antismoke agents from actinomycetes. Lazzarini *et al.* (2000) and Stach (2003) reported that currently Actinobacteria, particularly spore forming Actinomycetes represent the most economically and biotechnologically valuable organisms in the bioactive compounds.

Gomathinayagam and Lakshmanaperumalsamy (2001) observed the continuous success of actinomycetes in production of bioactive compounds. Various sources are being explored for antibiotic producing actinomycetes to cater the world demand for new antibiotics. It was found that very high numbers of actinomycetes are capable of producing broad spectrum compounds and are distributed in throughout length of river. A high incidence of bioactive compounds producing actinomycetes indicated the existence of invaluable potent organisms in the fresh water.

Ranjekar and Sridhar (2002) reported that among the microorganisms, actinomycetes gained special importance due to their capacity to produce bioactive secondary metabolites and enzymes. About 95 per cent of the isolates out of 64 isolates tested exhibited enzyme activity. Lateef and Oloke (2003) reported that isolates of actinomycetes sp was screened for the production of β-galactosidase and they showed potential β-galactosidase activity for lactose hydrolysis. Mishra and Debnath (2003) reported that the *Streptomyces griseus NCIM2020* was used for glucose isomerase production in submerged fermentation. Balaguranathan (2004) and Kokare *et al.* (2004) reported that the marine actinomycetes isolated from sediments possessed higher antimicrobial and antifungal activities.

Marine Therapeutic Enzymes

Currently, marine microorganisms are considered as untapped sources of metabolites and products, which may possess novel properties. Murphy and Hill (1998) reported that there are several marine derived compounds in clinical trials, other on the development pathway and a number of commercial products on the market. "Didemnin B" a compound from a Caribbean tunicate has long been considered the marine compound most likely to be the first used clinically as an anti cancer drug. In addition to anti cancer and anti AIDS activities marine compounds have a wide range of other activities, including neuro and immuno modulation, antibiotic activity and anti-inflammation.

Mellado and Ventosa (2003) and Sabu (2003) reported that the halophilic adaptation of enzymes from halophilic microorganism would permit a wide range of industrial applications. For these reasons, very recently extensive attempts to isolate different types of salt adapted enzymes from halophilic microorganisms have been performed. The use of salt tolerant enzymes from marine bacteria provides an interesting alternative for therapeutic purpose (Table 53.1). The major potential therapeutic application of enzymes is in the treatment of cancer (Table 53.2). Asparaginase has proved to be particularly promising for the treatment of acute lymphocytic leukaemia. Its ac;tion depends upon the fact that tumour cells are deficient in aspartate ammonia ligase activity, which restricts their ability to synthesize the normally non-essential amino acid, L-asparagine. Poorani (2005) reported that the partially purified L-asparaginase from marine actinomycetes showing potential growth inhibition on tumour cell lines.

L-asparaginase Enzyme from Microorganisms

Wide range of bacteria, fungi, yeast, actinomycetes and algae are very efficient producers of L-asparaginase. Mashbern and Wriston (1964) and Yellin and Wriston (1966) were successfully purified and demonstrate the tumourcidal activity of *E. coli* L-asparaginase. The partially purified two isoforms of L-asparaginase having the therapeutic potential by using salt precipitation, starch block electrophoresis and DEAE-Cellulose column chromatography. Oettegen *et al.* (1967) studied the efficacy of L-asparaginase in human beings with leukemia.

Table 53.1: Microbial Sources of Therapeutic Enzymes

Enzyme	*Source*
L-Glutaminase	*Beauveria bassiana, Vibrio coslicoia, Zygosaccharomyces rouxii,*
L-asparaginase	*Pseudomonas acidovorans, Acinetobacter* sp.
β-lactamase	*Citrobacter freundii, Serratia marcescens, Klebsiaiia neumonia*
Serratia peptidase	*Serratia marcescens*
Lipase	*Candida :iipolytica, C. rugosa, Aspergillus oryzae*
Alginate lyase	*Pseudomonas aeruginosa*
L-arabinofuranosidase	*Aspergillus niger*
Protease	*Bacillus polymyxa, Beauveria basssiana*
Superoxide dismutase	*Mycobacterium* sp., *Nocardia* sp.
Glucosidase	*Aspergillus niger*
Amylase	*Aspergillus oryzae, Bacillus* sp.
Serrapeptase	*Serratia marcescens*
Penicillin acylase	*Penicillium* sp.
Laccase	*Trametes versicolor*

Source: Sabu, 2003.

Table 53.2: Some Important Therapeutic Enzymes and their Applications

Enzyme	*Application*
L-asparaginase	Antitumour
L-Glutaminase	Antitumour
Superoxide dismutase	Anti-oxidant, anti-inflammatory
Serratia peptidase	Anti-inflammatory
Penicillin acylase	Synthetic antibiotic production
Collagenase	To treat Skin ulcers
Lipase	Digests lipids
Streptokinase	Anticoagulant
Urokinase	Anticoagulant
Laccase	Detoxifier
L-arginase	Antitumour
L-Tyrosinase	Antitumour
Glucosidase	Antitumour
Galactosidase	Antitumour
β-lactamase	Penicillin allergy
Ribonuclease	Antiviral

Source: Sabu, 2003.

Broome (1968) studied the effects of the L-asparaginase enzyme on asparagine levels in the blood, normal tissues and 6C3HED Lymphomas of mice. Heinemann and Howard (1969) reported that *Serrutia murcescens* ATCC 60 is a high potential strain to produce tumour inhibitory L-asparaginase in submerged fermentation for large scale production. Angeli *et al.* (1970) and Berenbaum (1970) found that the *A. terreus* asparaginase is not inactivated either at the pH of the peritoneal fluid or by the ascetic fluid. The weak activity on solid tumours may be correlated with the low diffusibility of this enzyme preparation from the peritoneal cavity into the blood stream.

Boyd and Phillips (1971) and Wade *et al.* (1971) reported that L-asparaginase and L-glutaminase are very common among bacteria. An automated modification of the test tube method was used to determine the asparaginase activities of about 200 strains from 78 species and the glutaminase activities of 46 strains from 13 species. Both the enzymes were widely distributed and at pH 8.5, asparaginase was generally the more active. Asparaginase was exceptionally active in some species of *Erwinia*. The tumour inhibitory L-asparaginase from *Serratia marcescens* showed a 365-fold purification and 15 per cent recovery of the enzyme, with specific activity of 250 IU per mg of protein.

Kafkewitz and Goodman (1974) and Wills and Woolfolk (1974) reported that asparagine requiring auxotroph of *Escherichia coli K-12* that have an active cytoplasmic asparaginase do not conserve asparagines supplements for use in protein synthesis. They reported that the rumen anaerobe *Vibrio succinogens* possesses a constitutive L-asparaginase. The compound supplied to the organism to generate the fumaric acid (affects the amount of enzyme produced.

Bascomb *et al.* (1975) and Gaffer and Shethna (1975) studied the purification, antitumour properties and large scale production in an inexpensive medium. The highest enzyme yield was obtained in corn-steep liquor medium (9.2 per cent w/v) at 37°C, the specific activity of the purified preparation was 45 IU/mg protein. The stimulation of L-asparaginase activity in *Azotobacter vinelandii* has been attempted using a variety of carbon and nitrogen sources, though the tested carbon sources failed to induce the enzyme, nitrogen sources such as ammonium salts, urea, L-aspartic acid and L-glutamic acid were found to be good inducers. The partially purified enzyme preparation possesses antitumour activity against Yoshida ascites sarcoma in rats.

Distasio *et al.* (1976) found that *V. succinogens* produced an L-asparaginase enzyme which showed higher L-asparaginase activities. Balakrish Nair *et al.* (1977), Barnes *et al.* (1977) and Selvakumar *et al.* (1977) reported that the details of the occurrence and activity of asparaginase in marine sediments. They suggests that marine sediments may be a good sources for active asparaginase producing fungi and found that the fungi isolated from the marine environment exhibit low asparaginase activity and their activity mainly restricted to the members of the genus *Aspergillus*.

Albanese and Katkewitz (1978) reported that the asparaginase synthesis by *Vibrio succinogens* is induced by ammonium ions. Kitto *et al.* (1979) described the isolation of two asparaginase enzyme from *P. geniculata* using a relatively straightforward purification procedure. The two enzymes were found to differ markedly both in their physicochemical properties and in their ability to affect the growth of murine neoplasms.

Mostafa (1979) reported that like bacteria, actinomycetes are also good source of L-asparaginase. Streptomyces in synthetic media with asparagine as a nitrogen source will stimulate more enzyme production than natural media. They found that the five genus of *Streptomycete* are capable of producing detectable amounts of L-asparaginase. Among them most potent L-asparaginase producers were identified as *Streptomyces collines, S. karnatakensis* and S. *venezulae* under different environmental and nutritional conditions. Cells grown on L-asparagine showed amidase activity with other amides but at a reduced rate.

Abuchowski *et al.*, 1979; Yashimoto *et al.*, 1986 and Jurgens *et al.* (1988) were reported that the biochemical properties of PEGylated asparaginase markedly differ from the native enzyme and its molecular weight is higher. The SDS is unable to separate the subunits of the enzyme due to cross-linking by PEG and having antitumour activity as clinically both in animals and humans. Arst and Bailey (1980) reported that the genetic evidence for a second asparaginase in *Aspergillus nidulans*.

Schemer and Holcenberg (1981) observed that L-asparaginase is the first enzyme with antitumour activity to be intensively studied in human beings. It is an enzyme drug of choice for acute lymphoblastic leukuemia in children used in combination therapy. Paul (1982) reported that L-asparaginase has been purified from a marine *Chlamydomonas* species, the first such enzyme to be purified from marine algae. The purified enzyme (mol.wt. 275000) possessed a km for asparagines of 1.34×10^{-4} M and showed limited antitumour activity in an antilymphoma assay *in vivo* and properties of the enzyme are contrasted with those of asparaginase from prokaryotic and eukaryotic sources.

Mostafa and Ali (1983) isolated the crude extract of L-asparaginase from *Thermoactinomycetes vulgaris* 13 MES by grinding the cells with sand and glass beads and followed by rapid freezing, thawing and exposure to ultrasonic waves. Wriston (1985) found that the purified asparaginase peak was pooled and usually found to be homogenous on analytical disc electrophoresis, although some preparations still show a second faint band. Gilbert *et al.* (1986) observed the expression of the cloned L-asparaginase gene was subject to glucose repression in *E. coli* but was not significantly repressed by glycerol. Recombinant plasmids, containing the asparaginase gene, when introduced into *Erwinia carotovora*, caused increased synthesis of the enzyme.

McEwen *et al.* (1987) and Leo *et al.* (1988) reported that the two L-asparaginase enzymes possess different immunological specifications. They confirmed that the L- asparaginase have immunogenicity in highly sensitized children *in vivo* with multiple acute lymphoblastic leukemia.

Asselin *et al.* (1989), Gallagher *et al.* (1989) and Masao (1989) reported that the L-asparaginase treatment for acute lympoblastic leukaemia is a major breakthrough in modern oncology because it induces complete emissions in over 90 per cent children with in weeks. The treatment with L-asparaginase as a single agent have kill the cells as both *in vitro* and *in vivo* in patients with acute lymphoblastic leukemia and besides these specific effects can exert immunosuppressive effects in general. This therapy got attention as a specific and favourable therapeutic idea which represents the inhibition of the leukemic cell growth without damaging the normal cells. Mesas *et al.* (1990) reported that the L-asparaginases are intracellular, the pH and the temperature for optima for L-asparginase production are the same as that for the growth of the enzyme producing organism.

Benny and Kurup (1991) and Selvakumar *et al.* (1991) attempted to screen L-asparaginase producing microorganisms from the estuarine sediments. The marine *Vibrio* isolates from the gut regions of *T. telescopium* showed the highest L-asparaginase activity and found that the bacterial isolates from different origins in the marine environment could yield potent strains for L-asparaginase production. The microbial population increased with increase of organic carbon content and decreased with increase of phosphate and nitrate. Percentage of L-asparaginase positive strains decreased with increase of organic carbon.

Benny and Ayyakkannu (1992) and Ramaiah and Chandramohan (1992) reported that the quantitative determination of L-asparaginase from the marine luminous bacteria and found the secretion of L-asparaginase by marine luminous bacteria did not seem to be influenced by the habitats. However, strains isolated from seawater particularly from the offshore regions appear to be potential sources of asparaginases. The total percentage of L-asparaginase positive population was slightly higher in sediments (66.6 per cent) than from fishes (62.8 per cent).

Maya *et al.* (1992), Maladkar *et al.* (1993) and Annie *et al.* (1994) reported that the, preferably NaCl are essential for the enzyme activity and growth of Streptomycetes. The purified L-asparaginase enzyme showed significant antitumor activity on experimental animal models. Kil *et al.* (1995) and Mohapatra *et al.* (1997) found that the halophilic nature of marine bacteria can be exploited industrially. The most of the microbial L-asparaginase is intracellular in nature except few.

Gulati *et al.* (1997) demonstrated that pH and dye-based procedure for screening L- asparaginase producing microorganisms. This procedure is suitable for bacterial and fungal screening and the results are obtained within 24 and 48 hours for bacteria and fungi respectively. Dhevendran and Annie (1999) screened *Streptomycetes* isolate's from fish, shellfish and sediment of estuarine Veli Lake, for antibiotic production and L-asparaginase activity. Shome and Shome (2001) reported that the random screening in the mangrove dominated areas of Andaman showed good proportion of L-asparaginase harbouring bacteria.

Dhevendaran and Anithakumari (2002) studied the L-asparaginase activity of antagonistic *Streptomycetes* associated with the gut of the fish, under different cultural conditions like temperature, pH, sodium chloride, carbon sources, amino acids, heavy metals, inorganic phosphate and sulfate and found that it is possible to optimize the growth conditions and maximum L-asparaginase enzyme synthesis under laboratory conditions.

Sabu (2003) and Savitri and Azmi (2003) reported that the L-asparaginase catalyzes the hydrolysis of L-asparagines into L-aspartic acid and ammonia. Certain tumour cells are deficient in their ability to synthesize the non-essential amino acid, L-asparagine and are forced to extract it from body fluids; by contrast, most normal cells can produce their own L-asparagines.

Poorani (2005) isolated the L-asparaginase from marine actinomycetes and it is partially purified for characterization of the enzyme. It was a molecular weight of 140 K Da and showed maximum activity at pH 8 and 60°C.

L-asparaginase Enzyme from Marine Actinomycetes

Marine actinomycetes have provided a number of biologically active substances, metabolites, antibiotics and enzymes for industries especially pharmaceutical industries. Imada *et al.* (1973) reported that 25 strains of *Streptomyces* sp. were cultivated in 40 ml of ST-2 medium in 200 ml Erlenmeyer flasks for 3 days with shaking, L-asparagines and L-glutamine deamidating activities were found in the sonicated preparations from such cultures. Selvakumar *et al.* (1977) observed that the asparaginase of *Streptomyces griseus* has a pH-optimum of 8.5 and it is interesting to note that the marine sediments harbour potential L-asparaginase enzyme producers.

Dhevendaran and Annie (1999) and Dhevendaran and Anithakumari (2002) reported that minimum concentration of salts, preferably NaCl are essential for the L-asparaginase enzyme activity and growth of *Streptomycetes*. Savitri and Azmi (2003) found that like bacteria, actinomycetes are also good source of L-asparaginase. Three most potent enzyme producers were identified as different strains of *Streptomyces collinus* and presence of L-asparagine in the culture medium induces the enzyme production but it is not essential for the enzyme biosynthesis.

Dhevagi and Poorani (2006) isolated the tumour inhibiting L-asparaginase from marine actinomycetes (Figures 53.3 and 53.4). They suggested that the enzyme from marine actinomycetes was more in starch casein broth.

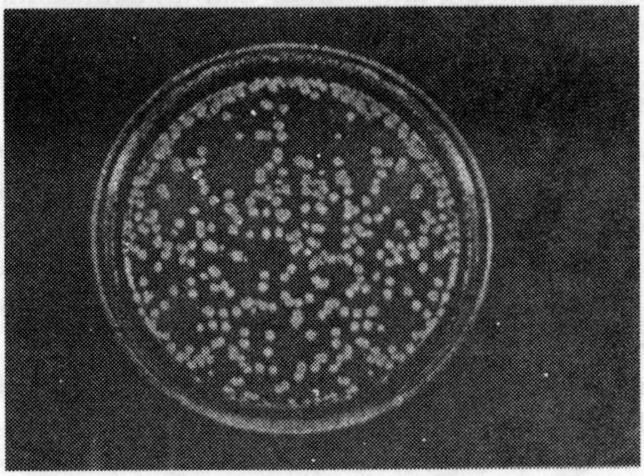

Figure 53.3: Actinomycetes Isolated from Pitchavaram Sediment

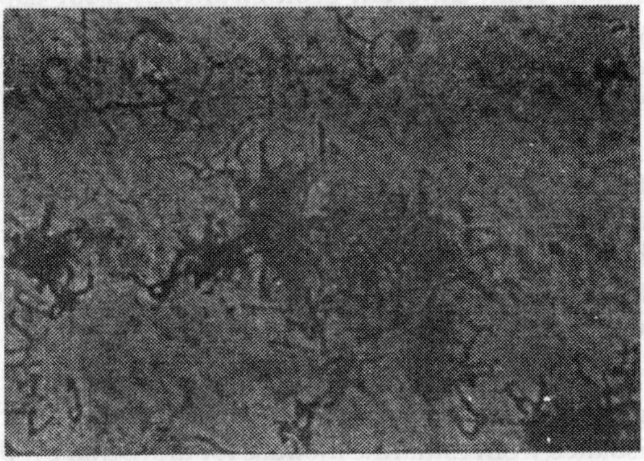

Figure 53.4: Microscopic View of Actinomycetes Isolated from Pitchavaram Sediment

Antitumour Activity of L-Asparaginase Enzyme

L-asparaginase is the first enzyme with antitumour activity to be intensively studied in human beings, rats and mice tumour cell lines. Angeli *et al.* (1970) and Berenbaum (1970) demonstrated the antitumour activity of L-asparaginase enzyme preparation on Walker 256 carcinosarcoma growing in ascetic form in the peritoneal cavity of the rat. They found that the preparation of *A. terreus* asparaginase has a greater careinostatic effect on the ascitic tumour and a smaller effect on those solid tumours that frequently develop in the subcutaneous at the inoculation site. The asparaginase effects the cells mediating the immune response and the effectiveness is lost once the cells begin to multiply rapidly, which may be relevant to the action of asparaginase in acute leukaemia.

Boyd and Phillips (1971) described the purification and properties of tumour inhibitory L-asparaginase enzyme from *Serratia marcescens.* It shows the inhibition of the Gardner lymphosarcoma 6C3HED in C3H mice. Complete regression of this tumour was obtained with a smaller dose of the enzyme from *S. marcescens* than with enzyme from *Escherichia coli.* The reason for this difference was not evident from a comparison of several properties of the two enzymes.

Bascomb *et al.* (1975) reported the properties and large scale production of an intracellular L-asparaginase with antitumour activity from a strain of *Citrobacter.* The anti-lymphoma activity of the enzyme was demonstrated with Gardner lymphosarcoma and was found only slightly less potent than Crasnitin, the most active asparaginase so far in this system. Gaffer and Shethna (1975) studied that the production, partial purification and antitumour activity of L-asparaginase from *Azotobacter vinelandii.* The partially purified enzyme preparation possesses antitumour activity against Yoshida ascites sarcoma in rats. The administration of 2,000 units/kg body weight, of L-asparaginase provided maximum protection against the tumour. It can also be seen that administration of increasing units of the enzyme results in a decrease in the survival period. This could be due to the toxicity of the enzyme preparation at high concentrations.

Distasio *et al.* (1976) and Barnes *et al.* (1977) reported the antitumour activity of purified L-asparaginase enzyme from *V. succinogens.* It possessed a potent antilymphoma activity in 6C3HED mice. The L-asparaginase therapy, alone or in combination with other drugs, is finding increased success in the management of acute lymphocytic leukemias. Selvakumar *et al.* (1977) screened for new strains from different ecological niches for asparaginase having antitumour and antileukaemia properties. Albanese and Kafkewitz (1978) found that the effect of medium composition on the growth and asparaginase production of *Vibrio succinogenes* and found that the properties of the *Vibrio* enzyme suggest that it may prove to be a clinically useful enzyme.

Kitto *et al.* (1979) reported that the tumour inhibitory and non-tumour inhibitory L-asparaginase from *Pseudomonas geniculata* and found that the asparaginase A was found to be devoid of antitumour activity in mice, whereas the asparaginase AG was effective in increasing the mean survival times of both C3H mice carrying the asparagines requiring Gardner 6C3HED tumour line and Swiss mice bearing the glutamine requiring Ehrlich ascites tumour line. These differences in antitumour activity were related to differences in the Km values for L-asparagine for the two enzymes.

Paul (1982) reported that the *Chlamydomonas* L-asparaginase possessed little ability to inhibit the growth of the Gardner lymphosarcoma in C3H mice. Benny and Kurup (1991) recommended *Aeromonas* isolated from estuarine mollusc for industrial production to meet the increasing demand of the enzyme in the treatment of lymphatic leukemia.

Selvakumar *et al.* (1991) reported that the L-asparaginase of marine *Vibrio,* isolated from the gut of *T. telescopium,* was found to be effective in suppressing Yoshida ascites sarcoma in Wistar rats. Benny and Ayyakkannu (1992) observed that the tumour cell requires an extra cellular source of L-asparagines and in its absence protein synthesis fails to occur in cancer. L-asparaginase hydrolyzes L-asparagine to aspartic acid and ammonia. When asparaginase from *Escherichia coli* was injected into mice suffering from certain types of leukaemia, marked important and often apparent cures occurred. Ueno *et al.* (1997) obtained the cell cycle arrest and apoptosis of leukemia cells induced by asparaginase. He found that the apoptotic cell death of murine leukemia cells was induced by *E. coli* L-asparaginase.

Savitri and Azmi (2003) and Sabu (2003) reported that the L-asparaginase (LA) can be effectively used for the treatment of lymphoblastic leukemia and tumour cells. Interest in this enzyme arose few decades ago when it was discovered that the antilymphoma activity of the guinea pig was due to LA. For pharmacologically and clinical tests, microbial sources are best for the bulk production of LA. Initially this drug failed to fall in antineopalstic category due to the immunogenic reactions caused by the foreign protein. With the development of protein engineering, modifications carried out in purified LA either reduced or completely eliminated the immunogenicity of LA in test model. The improved therapeutic activity and decreased immunogenicity of LA can be immensely used as an antineoplastic agent.

Poorani (2005) reported that the partially purified L-asparaginase isolated from marine actinomycetes showed maximum activity at 60°C and pH 8 and Stable upto 80°C. The enzyme has molecular weight of 140 K Da. The enzyme showed cytostatic and cytotoxic effects on tumour cell lines of American Type Culture Collections at 24 hours and 48 hours growth inhibition studies respectively.

Conclusion

It is interesting that the world's oceans, which cover 70 per cent of the earth's surface and include some of the most biodiverse ecosystems on the planet, have not been widely recognized as an important resource for novel actinomycetes. In fact, the distributions of actinomycetes in the sea remain largely undescribed and even today, conclusive evidence that these bacteria play important an ecological role in the marine environment has remained elusive. Speculation regarding the existence of indigenous populations of marine actinomycetes arises because these bacteria produce resistant spores that are known to be transported from land into the sea where they can remain viable but dormant for many years. Thus, it has been frequently assumed that actinomycetes isolated from marine samples are merely of terrestrial origin. The marine derived actinomycetes can be metabolically active and physiologically adapted to growth in seawater. Although some work has been done in the past 20 years on the marine actinomycetes not much work has been done in South Indian marine areas which themselves present a unique habitat and hence potentially new species and novel properties. Little information is available on the marine actinomycetes and its novel metabolites. Marine actinomycetes are new potential source of novel drugs and enzymes especially anti cancer enzyme and drugs. Therefore we report here the review on Marine actinomycetes and its potential anti leukeamic L-asparaginase.

References

Abuchowski, A. *et al.*, 1979. Treatment of L5178Y tumour bearing BDf mice with a non-immunogenic L-glutaminase-asparaginase. *Cancer Treat. Rep.*, 63: 1127–1129.

Asselin, B.L. *et al.*, 1989. *In vitro* and *in vivo* killing of all cells by L-asparaginase. *Cancer Res.*, 49: 4363–4369.

Albanese, E. and Kafkewitz, D., 1978. Effect of medium composition on the growth and asparaginase production of *V. succinogens. Appl. Env. Microbiol.*, 36(1): 25–30.

Angeli, L.C.D., Pocchiari, F., Russi, S., Tonolo, A. and Zurita, V.E., 1970. Effect of L-asparaginase from *Aspergillus terrus* on Ascites sarcoma in the rat. *Nature*, 225: 549–550.

Annie, K., Maya, J.S. and Devendaran, K., 1994. Antibiotic activity of *Streptomces* isolated from sediments. *Indian J. Mar. Sci.*, 17: 213–217.

Arst, H.N. and Bailey, C.R., 1980. Genetic evidence for a second *Aspergillus nidulans. J. of General Microbial.*, 121: 243–247.

Baam, R.B., Gandhi, N.M. and Freitas, Y.M., 1966. Antibiotic activity of marine microorganisms. *Helgoander Wiss. Meeresunters*, 13: 181–187.

Balagurunpathan, R., 1992. Antagonistic actinomycetes from shallow Sea sediments with references to alpha and beta unsaturated gamma-lactone type of antibiotic from *Streptomyces griseobrunneus* (p. 33). *Ph.D. Thesis*, Annamalai University, pp. 82.

Balagurunathan, R., 2000. *Actinomycetes* in mangrove ecosystems. In: *Flora and Fauna in Mangrove Ecosystems: A Manual for Identification*, pp. 148–152.

Balagurunathan, R., 2004. Marine actinomycetes: A promising future of pharmaceutical industry. *Advanced Biotech.*, 20(4): 16–21.

Balakrishna Nair, G., Selvakumar, N., Chandramohan, D. and Natarajan, R., 1977. Distribution and activity of L-asparaginase producing fungi in the marine environment of *Porto novo. Indian J. Mar. Sci.*, 6: 172–173.

Balasubramanian, T., Lakshmanaperumalsamy, P., Chandramohan, D. and Natarajan, R., 1979. Cellulolytic activity of *Streptomycetes* isolated from the digestive tract of a marine borer. *Indian J. Mar. Sci.*, 8: 111–114.

Ball, A.S., Betts, W.B. and McCarthy, A.J., 1989. Degradation of lignin-related compounds by *Actinomycetes. Appl. and Env. Microbiol.*, 55(6): 1642–1644.

Barnes, W.R., Dorn, G.L, and Vela, G.R., 1977. Effect of culture conditions on synthesis of L. asparaginase by *E. coli* A-1. *Appl. and Env. Microbiol.*, 33(2): 257–261.

Bascomb, S., Banks, G.T., Skarstedt, M.T., Fleming, A. and Bettelheim, K.A., 1975. The properties and large scale production of L-asparaginase from *citrobacter*. *J. General Microbiol.*, 91: 1–16.

Benny, A. and Ayyakkannu, 1992. L-asparaginase activity of benthic marine bacteria and of the muricid Gasteropod *chicoreus ramosus*, East coast of India. *Phuket. Mar. Biol. Spec. Publ.*, 11: 137–140.

Benny, P.J and Kurup, M., 1991. L-asparaginase activity in bacteria from estuarine sediments and mollusc. *Indian J. Mar. Sci.*, 20: 36–39.

Berenbaum, M.C., 1970. Immunosuppression of L-asparaginase. *Nature*, 225: 550–552.

Boyd, J.W. and Phillips, A.W., 1971. Purification and properties of L-asparaginase from *S. marcescens*. *J. Bacteriol.*, 106(2): 578–587.

Broome, J.D., 1968. Studies on the mechanism of tumour inhibition by L-asparaginase. *Br. J. Cancer*, pp. 1055–1072.

Canedo, L.M., Fernandez Puentes, J.L. and Baz, J.P., 2000. IB-G6212 a novel cytotoxic macrolide produced by a marine *Micramanospora* (II) physico-chemical properties and structure determination. *J. Antibiot.*, 53: 479–483.

Chandramohan, D., Ramu, S. and Natarajan, R., 1972. Cellulolytic activity of marine *Streptomycetes*. *Curr. Sci.*, 41(7): 245–246.

Cross, T., 1982. Aquatic actinomycetes: A critical survey of the occurrence, growth and role of actinomycetes in aquatic habitats. *J. Appl. Bacteriol.*, 50: 397–423.

Dhevagi, P. and Poorani, E., 2006. Isolation and characterization of actinomycetes from marine sediments. *J. Microb. World*, 8(1): 59–65.

Dhevendaran, K. and Annie, K., 1999. Antibiotic and L asparaginase of *Streptomycetes*.isolated from fish, Shellfish and sediments of veli estuarine along Kerala Coast. *Indian. J. Mar. Sci.*, 28: 335–337.

Dhevendaran, K. and Anithakumari, Y.K., 2002. L-asparaginase activity in growing conditions of *Streptomyces* spp. Associated with *Therapon jarbua* and *Villoritacyprinoids* of Veli lake, South India. *Fish. Technol.*, 39(2): 155–159.

Distasio, J.A., Niderman, R.A., Kafkewitz, D. and Goodman, D., 1976. Purification and characterization of L-asparaginase with anti-lymphoma activity from *Vibrio succinogens*. *The J. Biol. Chem.*, 251(22): 6929–6933.

Ellaiah, P. and Reddy, A.P.C., 1987. Isolation of *actinomycetes* from marine sediments of Vishakapatnam East coast of India. *Indian J. Mar. Sci.*, 16: 134–135.

Erikson, D., 1940. *Pathogenic Anaerobic Organisms of the Actinomyces Group*. Special report series, No. 240. Medical Research Council, London.

Ford, L.M., Eaton, T.E. and Godfrey, O.W., 1990. Screening of *St. ambofaciens* mutants that produce large quantities of spiramycin and determination of optimal conditions for spiramycin production. *Appl. and Env. Microbiol.*, 56(11): 3511– 3514.

Gaffar, S.A. and Shethna, Y.I., 1975. Partial purification and antitumour activity of L-asparaginase from *Azotobacter vinelandi*. *Curr. Sci.*, 44(20): 727–729.

Gallagher, M.P., Breuerjammli, M. and Mahro, B., 1989. L-asparaginase a drug for treatment of acute lymphoblastic leukemia. *Essays Biochem.*, 24: 1–40.

Gilbert, H.J., Blazek, R., Bullman, H.M.S. and Minton, N.P., 1986. Cloning and expression of *Erwinia chrysanthemi* asparaginase gene in *Escherichia coli* and *Erwinia carotovora. J. General Microbiol.*, 132: 151–160.

Gomathinayagam, P. and Laksmanaperumalsamy, P., 2001. Occurrence, distribution and antimicrobial activity of freshwater *actinomycetes. Asian J. Microbiol. Biotech. and Env.*, 3(3): 157–166.

Goodfellow, M. and J.A. Haynes, 1984. Actinomycetesin marine sediments. In: *Biological Biochemical Aspects of Actinim*, (Eds.) Oritz-Oritz, L., F. Bojatil and V. Yakoleff. Academic Press, London, pp. 453–472.

Gulati, R., Saxena, R.K. and Gupta, R., 1997. A rapid screening for L-asparaginase producing microorganisms. *Letters in Appl. Microbiol.*, 24: 23–26.

Heinemann, B. and Howard, A.J., 1969. Production of tumour inhibitory L–asparaginase by submerged growth of *Serratia marcescnes. Appl. Microbiol.*, 18(4): 550–554.

Herron, P.R. and Wellington, E.M.N., 1990. New method for extraction of *Streptomycetes* spores from soil and application to the study of lysogeny in sterile amended and nonsterile soil. *Appl. and Env. Microbiol.*, 56(5): 1406–1412.

Imada, A., Igarasi, S., Nakahama, K. and Isono, M., 1973. Asparaginase and glutaminase activities of microorganisms. *J. General Microbiol.*, 76: 85–99.

Ishaque, M. and Kluepfel, O., 1980. Cellulose complex of a mesophilic *Streptomyces* strain. *Can. J. Microbiol.*, 26: 183–189.

Jiang, C.L. and Xu, L.H., 1996. Diversity of aquatic *actinomycetes* in lakes of the middle plateau, Yunnan, China. *Appl. and Env, Microbiol.*, 62(1): 249–253.

Jurgens, H., Gibson, T.J., Plewniak, F. and Jeanmougin, F., 1988. Klinische erfahrungen mit polyathylengekoppelter *E. coli* asparaginase bei patuntenmit ALL-Mehrfachrezidiv klin. *Padiatr.*, 200: 184–191.

Kafkewitz, D. and Goodman, O., 1974. L-asparaginase production by the rumen anaerobe *V. succinogens. Appl. Microbiol.*, 27(1): 206–209.

Kanagawa, T. and Kelly, D.P., 1987. Degradation of substituted thiophenes by bacteria Isolated from activated sludge. *Microb. Ecol.*, 13: 47–57.

Kil, J.O., Church, D.M., Lash, A.E. and Leipe, D.D., 1995. Extraction of extracellular L-asparaginase from *Candida utilis. Biosci. Biotechnol. Biochem.*, 59: 749–750.

Kitto, J.B., Smith, G., Thiet, T.Q., Mason, M. and Davidson, L., 1979. Tumour inhibitory and non-tumour inhibitory L-asparaginase from *P. geniculata. J. Bacteriol.*, 137: 204–212.

Kokare, C.R., Mahadik, K.R., Kadam, S.S. and Chopade, B.A., 2004. Isolation of bioactive marine actinomycetes from sediments isolated from Goa and Maharashtra coastlines (West coast of India). *Indian J. Mar. Sci.*, 33(3): 248–256.

Kriss, A.E., Mishustina, J.E., Mitskevich, J.N. and Zemtsova, E.V., 1967. *Microbial Populations of Oceans and Seas.* Arnold, London, pp. 76–127.

Kuster, E. and Williams, S.T., 1964. Selection of media for the isolation of *Streptomycetes. Nature*, 202: 928–929.

Laksman, K., 1950. Isolation of marine microorganisms. *Appl. Environ. Microbiol.*, 36: 25–30.

Laksmanaperumalsamy, P., 1978. Studies on actinomycetes with special reference to antagonistic *Streptomycetes* from sediments of Porto Novo coastal zone. *Ph.D. Thesis*, Annamalai University, pp. 192.

Lateef, A. and Oloke, J.K., 2003. Enzymatic synthesis of oligosaccharides by *actinomycessp. Asian J. of Microbiol. Biotech. Env. Sci.*, 5(3): 291–295.

Lawton, P., Whitaker, A., Odell, D. and Stowell, J.D., 1989. *Actinomycetes* morphology in shaken culture. *Can. J. Microbiol.*, 26: 881–889.

Lclzzarini, A., Cavaletti, L., Toppo, G. and Marinelli, F., 2000. Rare genera of actinomycetes as potential producers of new antibiotics. *Antonie van Leeuwenhoek*, 78: 399–405.

Leo, H.T., Rappe, M.S., Hanada, Y. and Baden, D., 1988. Liposomal palmitoyl L-asparaginase; characterization and biological activity. *Cancer Chemother. Pharmacol.*, 34: 230–241.

Maladkar, N.K., Rainey, F.A. and Geobel, B.M., 1993. Fermentative production and isolation of L-asparaginase from *Erwinia carotovora*, EC-113. *Hindustan Antibiotic Bull.*, 35: 77–86.

Masao, N., 1986. Process for producing immobilized L-asparaginase preparations for the therapy of leukemia. *US Pat.*, 4: 617–627.

Mashbem, L.T. and Wriston, J.C., 1964. Tumour inhibitory effect of L-asparaginase from *E. coli. Arch. Biochem.*, 105: 450–458.

Mathew, A., Dhevendaran, K., Georgekutty, M.I. and Natarajan, P., 1994. L-asparaginase activity in antagonistic Streptomycetes associated with *Clam villorrita* cyprinoides (Hanley). *Indian J. Mar. Sci.*, 23: 204–208.

Maya, J.S., Devendaran, K. and Annie, K., 1992. L-asparaginase activity of *Streptomyces* sp. *Indian J. Mar. Sci.*, 12: 147–150.

McEwan, E.G., Lucon, C.M.M. and Rimmer, D.L., 1987. A preliminary study on the evaluation of asparaginase. *Cancer*, 59: 2011–2020.

Mellado M.E. and Ventosa, A., 2003. Biotechnological potential of moderately and extremely halophilic microorganisms. *Morgs for Healthcare, Food and Enz. Produc.*, pp. 233–256.

Mercer, D.K., Iqubal, M.P., Miller, G.G. and McCarthy, A.J., 1996. Screening *actinomycetes* for extra cellular peroxidase activity. *Appl. and Env. Microbiol.*, 62: 2186–2190.

Mesas, J.M., Ward, A.C. and Roux, K.H., 1990. Characterization and partial purification of L-asparaginase from *Cornebacterium glutamicum. J. Gen. Microbiol.*, pp. 136.

Metz, B., Kossen, N.W.F. and Van Suijdam, J.C., 1979. The morphology of actinomycetes. *Adv. Biochem. Eng.*, 11: 103–156.

Mieguelez, E.M., Hardisson, C. and Manzanal, M.B., 2000. *Streptomycetes:* A new model to study cell death. *Internatl. Microbiol.*, 3: 153–158.

Mincer, T.J., Jensen, P.R., Kauffman, C.A. and Fenical, W., 2002. Widespread and persistent populations of a new marine actinomycete taxon in Ocean sediments. *Appl. Environ. Microbiol.*, 68(10): 5005–5011.

Mishra, A. and Debnath, M., 2003. Studies on effect of nutrients on Glucose isomerase production by *St. griseus. NCIM* 2020. *Asian J. of Microbial. Biotech. Env. Sci.*, 5(1): 91–96.

Mohapatra, B.R., Rheims, H., Wolterink, A. and Stackebrandt, E., 1997. Production and properties of L-asparaginase from *Mucor* sp. associated with a marine sponge (*Spirastrella* sp.). *Cytobios*, 92: 165–173.

Mostafa, S.A., 1979a. Activity of L-asparaginase in cells of *Streptomyces karnatakensis. Zentralbl Bacteriol.*, (Naturwiss), 134: 343–351.

Mostafa, S.A., 1979b. Production of L-asparaginase by *Streptomyces karnatakensis* and *Streptomyces venezuelae. Zentralbl Bacteriol* (orig A), 134: 429–436.

Mostafa, S.A. and Ali, O.A., 1983. L-asparaginase activity in cell free extracts of *Thermoactinomyces vulgaris* 13 M.E.S. *Zbl. Microbiol.*, 5: 397–404.

Mostafa, S.A. and Salama, M.S., 1979. L-asparaginase producing *Streptomyces* from soil of Kuwait. *Zentralbl Bacteriol* (Naturwiss), 134: 325-334.

Murphy, P. and Hill, R.T., 1998. Marine vision becomes reality: Drugs from the sea. *Biofuture*, 119: 34–49.

Oettegen, H.F., P.A. Stephensen, M.K. Schwartz, R.D. Leeper, L. Tallal, C.C. Tan, B.D. Clarkson, R.B. Golbey, I.H. Krakoff, D.A. Karnofsky, M.L. Murphy and J.H. Burchenal, 1967. Toxicity of *E. coli* L-asparaginase in man. *Cancer*, 25: 253–278.

Okami, Y. and Hotta, 1988. Search and discovery of new antibiotics. In: *Actinomycetes in Biotechnology*. Academic Press, London, pp, 34–67.

Okami, Y. and Okazaki, T., 1972. Studies on marine microorganisms and its isolation from the sea. *J. Antibiot.*, 25: 456–460.

Parsons, T.R., Maita, Y. and Lalli, C.M., 1984. *A Manual of Chemical and Biological Methods for Seawater Analysis*. Pergamon Press (1st Edition), pp. 157–161.

Pathirana, C.D.M., Tapiolas, P.R., Jensen, R. Dwight and Fenical, 1991. Structure determination of madurakide: A new 24-membered ring macrolide glycoside produced by a marine bacterium (Actinomycetales). *Tetrahedron Lett.*, 32: 2323–2326.

Paul, J.H., 1982. Isolation and characterization of a *Chlamydomonas* L-asparaginase. *Biochem. J.*, 203: 109–115.

Poorani, E. and Dhevagi, P., 2005. Characterization of L-asparaginase from marine actinomycetes and checking its antitumour activity. *M.Phil. Thesis*, Bharathiar University, Coimbatore.

Postmaster, C. and Freitas, Y.M., 1975. An antibiotic producer from Marsh sediment. *Hindusthan Antibiot. Bull.*, 17: 118–120.

Rainbow, S.W. and Rose, H., 1963. Metabolites and antibiotics from Actinomycetes. *Microbiol Ecology*, 4: 24–27.

Ramaiah, N. and Chandramohan, D., 1992. Production of L-asparaginase by the marine luminous bacteria. *Indian J. Mar. Sci.*, 21: 212–214.

Rathnakala, R. and Chandrika, V., 1993. Effect of different media, for isolation growth and maintenance of *actinomycetes* from mangrove sediments. *Indian J. Mar. Sci.*, 22: 297–299.

Ravel, J. Schrempfu and Hill, R.T., 1998. Mercury resistance is encoded by transferable giant linear plasmids in two Chesapeake Bay Streptomycetes Strains. *Appl. Environ. Microbiol.*, 64: 3383–3388.

Renjekar, M.K. and Sridhar, K.R., 2002. Occurrence Find extracellular enzyme potential of *actinomycetes* of a thermal spring, Southern India. *Asian J. Microbial. Biotech. Env. Sci.*, 4(1): 59–64.

Sabu, A., 2003. Sources, properties and applications of microbial therapeutic enzymes. *Indian J. Biotechnol.*, 2: 334–341.

Sambamurthy, K. and Ellaiah, P., 1974. A new *Streptomycete* producing neomycin CB and W Complex. *S. marinenzis* (Part 1). *Hindustan. Antibiot. Bull.*, 17: 24–28.

Schemer, G. and Holcenberg, J.S., 1981. *Enzymes as Drugs*, (Eds.) J.S. Holcenberg and J. Roberts. Wiley Inter Science, New York, pp. 455–473.

Shomura, T., Yoshida, J., Amano, S., Kjima, M., Inouye, S. and Niida, T., 1979. Studies on actinomycetales producing antibiotics only on agar culture. I. Screening, taxonomy and morphology-productivity relationship of *S. halstedii* strain. *J. Antibiot.*, Tokyo, 52: 427–435.

Sivakumar, K., 1998. Research on marine *Actinomycetes.* In: *Flora and Fauna in Mangrove Ecosystems: A Manual for Identification*, pp. 47–51.

Savitri Asthana, N. and Azmi, W., 2003. Microbial L-asparaginase: A potent antitumour enzymes. *Indian J. Biotechnol.*, 2: 184–194.

Schallenberg, K.W. and Kalff, D.W., 1993. Ecology of actinomycetes. *Ann. Rev. Microbiol.*, 37: 189–216.

Selvakumar, N., Chandramohan, D. and Natarajan, R., 1977. L-asparaginase activity in marine sediments. *Curr. Sci.*, 46(9): 287–291.

Selvakumar, N., Vanajakumar and Natarajan, R., 1991. Partial purification, characterization a antitumour properties of L-asparaginase (Anti-leulkaemic agent) from *vibrio. Bioactive Compounds from Microorganisms*, pp. 289–300.

Shirling, E.B. and Gottlieb, D., 1966. Methods for characterization of *Streptomyces species. Internatl. J. Systematic Bacteriol.*, 6(3): 313–340.

Shome, R. and Shome, B.R., 2001. Microbial L-asparaginase from mangroves of Andaman Islands. *Indian J. Mar. Sci.*, 30: 183–184.

Sivakumar, K., 1998. Research on marine *Actinomycetes.* In: *Flora and Fauna in Mangrove Ecosystems: A Manual for Identification*, pp. 47–51.

Stach, J.E.M., Maldonado, L.A., Ward, A.C., Goodfellow, M. and Bull, A.T., 2003. New primers for the class actinobacteria application to marine and terrestrial environments. *Env. Microbiol.*, 5(10): 828–841.

Stastna, J., Caslavska, P., Wolf, A., Vinter, V. and Mikulik, K., 1977. Origin and morphology of a typical forms of *Streptomyces granaticolor. Folia Microbiol.*, 22: 339–345.

Takizawa, M., Colwell, R.R. and Hill, R.T., 1993. Isolation and diversity of *actinomycetes* in the Chesapeake Bay. *Appl. and Env. Microbiol.*, 59(4): 997–1002.

Ueno, T., Ohtawa, K., Mitsui, K., Kodera, Y., Hiroto, M., Matsushima, A.Y. and Nishimura, H., 1997. Cell cycle arrest and apoptosis of leukemia cells induced by asparaginase. *Leukemia*, 11(11): 1858–1861.

Vanajakumar, 1979. Studies on actinomycetes associated with molluscs from Porto novo coastal waters. *Ph.D. Thesis*, Annmalai University, pp. 236.

Vanajakumar, Selvakumar N. and Natarajan, R., 1981. Antagonistic Properties of *Actinomycetes* isolated from mollusks of Porto Novo Region, South India. In: *Bioactive Compounds from Marine Organisms: Emphasis on the Indian Ocean*, pp. 265–273.

Wade, H.E., Robinson, H.K. and Phillips, B.W., 1971. Asparaginase and Glutaminase activities of bacteria. *J. General Microbiol.*, 69: 299–312.

Walker, J.D and Colwell, R.R., 1975. Factors affecting enumeration and isolation of actinomycetes from Chesapeake Bay and southeastern Atlantic Ocean sediments. *Mar. Biol.*, 30: 193–201.

Wetzel, D.F., 1983. Enumeration, isolation and some physiological properties of actinomycetes from sediment and water. *Syst. Appl. Microbiol.*, 10: 85–91.

Whol, D.L. and McArthur, J.V., 1998. *Actinomycetes*-flora associated with submerged freshwater macrophytes. *FEMS Microbiol. Ecol.*, 26: 135–140.

Whol, D.L. and McArthur, J.L., 2001. Aquatic *actinomycetes:* Fungal interactions and their effects on organic matter decomoposition. *Microb. Ecol.*, 42: 446–457.

Wills, R.C. and Woolfolk, C.A., 1974. Asparagine utilization of *Escherichia coli*. *J. Bacteriol.*, 118(1): 231–241.

Willoughby, A.W., 1976. A continuing source of new metabolites. *Develop. Indust. Microbiol.*, 23: 1–7.

Wriston, J.C., 1985. *J. Methods in Enzymology,* 113, (Eds.) S.P Colowick and N.O. Kaplan. Academic Press, New York, pp. 608.

Yashimoto, T. *et al.*, 1986. Characterization of polyethylene glycol-modified L-asparaginase from *E. coli* and its application to therapy of leukemia. *Lap. Cancer. Res.*, 77: 1264–1270.

Yellin, T.Y. and Wriston, J.C., 1966. Antagonism of purified asparaginase from guinea pig serum towards Lymphoma. *Science*, 151: 998–1004.

Young, W.D. and Smith, P.M., 1975. Actinomycetes nature, distribution activities and importances. *Appl. Environ. Microbiol.*, 24: 17–22.

Zheng, W. and Crawford, D.L., Pometto, A.L. and Rafii, F., 2000. Survival and effects of wild-type, mutant and recombinant *Streptomyces* in a soil ecosystem. *Can. J. Microbiol.*, 35: 535–543.

Chapter 54

DGPS: Analysis of Differential Correction Patterns

D. Vijayalakshmi[1]*, I.V. Murali Krishna[2] and E.G. Rajan[3]

[1]Associate Professor, VNR Vignana Jyothi Institute of Engineering and Technology, Hyderabad

[2]Professor in Centre for Spatial Information Technology, JNTU, Hyderabad

[3]Professor in Department of ECE, Vasavi College of Engineering, Hyderabad

ABSTRACT

Differentially correcting GPS data by post processing uses a base GPS receiver that logs positions at a known location and a rover GPS receiver that collects positions in the field. Most of these errors can be "Calculated out" of the measurements by the process called differential correction. The specific objective of the study is to estimate the accuracies involved with DGPS analysis in post processing mode. The errors are reported using the factor 196σ. The deviation values (Co-ordinate Deltas) Δ northing, Δ easting and Δ elevations are all zeroes for all points. Base stations are set up with an elevation mask value of 10^0. A good rule of thumb is to increase the elevation mask for the rover by one degree depending upon the distance from the base station. This is found to be approximately equal to 75 km. Setting of 5 degrees works well for the rover in general if it is within 50 km from the base station. All the fixes with their obvious errors are now within the area of one small circle-that presumably includes the true location of the antennae. The ticks again are 10 meters apart, so the amount of error reduction and the accuracy you can expect from differentially corrected data is identified. The variations and accuracies of the received signals are to be better understood as operational models that could predict errors are still not available. The differential correction patterns are significantly influenced by a variety of parameters. The errors can be minimized through proper planning of the field data collection.

* E-mail: psridhar_babu_2k@yahoo.com; Phone: 9849267760

Proper understanding of the instrument performance is required for minimizing the errors. The point error ellipses suggest clearly the amount of deviation of signal. The horizontal and vertical precisions calculated in this study are good examples of quantifiable parameters that define the signal variation.

Introduction

Space environmental effects on the satellite communication can be considered as those effects related to space segment, those effects related to ground segment and those effects on the signals propagating through the earth's lower and upper atmosphere. The atmospheric structure has significant influence on signal propagation. This has definite influence on the data processing methodologies. The specific applications with which are concerned here are essentially related to navigation and positioning. Like any conventional navigational systems the GPS is subjected to errors that can degrade the precision of the system. Troposphere and ionosphere are the two regions, which approximately correspond to broad technical categorization into lower atmosphere and upper atmosphere zones respectively, have different properties of signal propagation. The structure of atmosphere can be described, for most practical purposes, through concentric layers of atmospheric domains with different physical and chemical properties. The signal propagation results in certain errors. These errors are ionospheric errors, tropospheric errors and multipath errors apart from the clock errors. In addition, the error in the signal can be significantly increased depending upon the geometry of the satellite used to determine a position. This can be estimated through the parameter PDOP (Position Dilution of Precision). The Dilution of precision factors (DOP) explain how geometry effects to yield position accuracy and scale ranging accuracy. The optimum geometry accuracy for four GPS satellites is achieved when three satellites are equally spaced on the horizon and one directly at zenith point of observation station.

DGPS, Differential GPS is the use of differential Correction data for the satellites in view, to remove the errors in the position measured by a GPS receiver due previously to SA (Selective availability), and to other effects. The Correction data is generated by a base station, which is a high quality GPS receiver with a good antenna sited at an accurately know location. The correction data is derived from the satellites pseudo range data and the base station's position, effectively correcting the timing errors artificially introduced into the atomic clock date by SA, or atmospheric effects. Differential correction techniques are used to enhance the quality of location data gathered using global positioning system (GPS) receivers. Differential correction can be applied in real-time directly in the field or when post processing data in the office. Although both methods are based on the same underlying principles, each accesses different data sources and achieves different levels of accuracy. Combining both methods provides flexibility during data collection and improves data integrity. With SA removed, a single GPS receiver from any manufacturer can achieve accuracies of approximately 100 meters. To achieve the accuracies needed for quality GIS records–from one to two meters up to a few centimetres requires, differential correction of the data. The majority of data collected using GPS for GIS is differentially corrected to improve accuracy.

The underlying premise of differential GPS (DGPS) as already mentioned is that any two receivers that are relatively close together will experience similar atmospheric errors. DGPS requires that a GPS receiver be set up on a precisely known location. This GPS receiver is the base or reference station. The base station receiver calculates its position based on satellite signals and compares this location to the known location. The difference is applied to the GPS data recorded by the second GPS receiver, which is known as the roving receiver. The corrected information can be applied to data from the roving receiver in real time in the field using radio signals or through post processing after data capture using special processing software.

DGPS

Differentially correcting GPS data by post processing uses abase GPS receiver that logs positions at a known location and a rover GPS receiver that collects positions in the field. The files from the base and rover are transferred to the office processing software, which computes corrected positions for the rover's file. This resulting corrected file can be viewed in or exported to a GIS.

There are many permanent GPS base stations currently operating throughout the world that provide the data necessary for differentially correcting GPS. Depending on the technology preferred by the base station owner, this data can be downloaded from the Internet or via a bulletin board system (BBS). Because base station data is consistent (with no gaps due to multipath errors) and very reliable because base stations usually run 24 hours, seven days a week, it is ideal for many GIS and mapping applications. Sources of base station data for post processing fall into four categories–public sources, commercial sources, web-based services, and base station ownership. Before purchasing a GPS receiver it is best to identify the source of base station data.

To attain accuracy levels on the order of one to 10 meters, differential correction is essential. The three main methods currently used for ensuring data accuracy are real-time differential correction, preprocessing real-time data, and post processing, Each method will achieve similar levels of accuracy, so the decision regarding which technique is appropriate will depend on factors such as project specifications, the end use of the data, and the sources available for differential correction.

In a nutshell, the differential correction process consists of setting a GPS receiver (Called a base station) at a precisely known geographic point. Since the base station knows exactly where its antenna is, it can analyze and record error in the received signals. These signal errors will be almost equivalent to the signal errors affecting other GPS receivers in the local area, so the accuracy of locations calculated by those other receivers may be improved, significantly integrating with base station data.

One measure of accuracy of GPS fixes is called circular error probable (CEP). It is the radius of a circle expressed in a linear unit, such as meters for a given situation, 50 per cent of the fixes will fall within the circle, and 50 per cent outside. Another measure of accuracy is based on two standard deviations of a normal distribution–called 2d RMS where RMS means root–mean–square, Ninety-five per cent of the fixes will lie within a circle with this radius.

Sources of GPS Error

The available data on GPS measurements in literature suggest that typical error sources and values for various types of receivers are different. There is no effort for rigorous estimate of error. However a broad estimate can be made as per data available in literature. These values could be mentioned as follows:

Satellite clocks < 1 meter

Ephemeris error < 1 meter

Receiver error < 2 meter

Ionospheric < 2 meter

Tropospheric < 2 meters

Multipath (depends on the situation and the receiver, may be large)

These values correspond to averages of many readings rather than the error that might be expected from a single reading. The field observations suggest that errors can be minimized by increasing the number of observable stations. The existence of systematic errors in a given situation make us to accept that the law of the large numbers is equally significant and relevant in the case of GPS observations also.

Table 54.1: Types of Errors

Types of Errors	*Significance*
Clock Errors	Synchronization of atomic clocks in the satellite is essential to evaluate the ability to determine the time taken for the signal to get received by the receiver from the satellite. Even a small amount of difference in the clocks can make a huge difference in the distance measurements, because the GPS signals travel at a speed 3×10^8 meters/sec
Ephemeris Errors	The exact position of a satellite is fixed through ephemeris representation. This representation is based on mean kepler ellipse. The time dependent deviations of the predicted orbit from this reference ellipse are transmitted to the receiver at regular intervals. Every hour or so, receiver gets position information at time "t". If this ephemeris prediction is incorrect the satellite will not be where it is expected to be, even by just a meter or two-then the measurement of the range from the receiver antenna to the satellite will be incorrect.
Receiver Errors	GPS receiver errors result from a number off actors related to receiver design, cost, and quality. These are many tunes difficult to assess without, proper instrument calibration.
Atmospheric Errors	The atmosphere has significant influence on transmission of signals. For most of its trip from the satellite to the receiver antenna, the GPS signal enjoys a trip through the virtual vacuum of "Empty space". Virtually all atmospheric influence is within about 120 km of the surface. Since the calculation of the range to the satellite depends on the speed of the signal, a change in signal speed due to atmospheric effect implies an error in distance, which produces an error in position finding. Significant changes in signal speed occur throughout the atmosphere, but the primary contributions to error come from the ionosphere, which contains charged particles under the influence of the earth's magnetic field

Field Data Collection

Most of these errors can be "Calculated out" of the measurements by the process called differential correction. The results of an experiment are as follows:

Table 54.2 shows the data collected during a field investigation taken with a Trimble GPS receiver placed in DGPS mode operated during two different time periods. It should be realized that each and every fix here is attempting to approximate the same true position. They show up in different positions because of errors. The true position of the antenna was not known, but almost certainly each point misses the exact, true position by some amount. The data given in the following section is actually collected by Trimble DGPS for 4 days during day time. The specific objective of the study is to estimate the accuracies involved with DGPS analysis in post processing mode. The network adjustment and adjustment style settings are shown in the following tables (Tables 54.2 and 54.3). The data is adjusted with reference to WGS84 as reference datum.

The adjusted grid co-ordinates are shown in Table 54.4. The geodetic co-ordinates in terms of latitude, longitude and heights along with errors are shown in Table 54.5. The errors are reported using the factor 1.96σ. The deviation values (Co-ordinate Deltas) Δ northing, Δ easting and Δ elevations are all zeroes for all points.

Table 54.2: Network Adjustment Report

User name	Administrator	*Date and Time*	1 : 13 : 09 PM 7/14/1005
Coordinate System	UTM	*Zone*	44 North
Project Datum	WGS 1984		
Vertical Datum		*Geoid Model*	EGM96 (Global)
Coordinate Units	Meters		
Distance Units	Meters		
Height Units	Meters		

Table 54.3: Adjustment Style Settings-95 per cent Confidence Limits

Residual Tolerances	
To End Iterations	0.000010m
Final Convergence Cutoff	0.005000 m
Covariance Display	
Horizontal	
Propagated Linear Error (E)	U.S.
Constant Term (C)	0.00000000 m
Scale on Linear Error (S)	1.96
Three-Dimensional	
Propagated Linear Error (E)	U.S.
Constant Term (C)	0.00000000 m
Scale on Linear Error (S)	1.96
Elevation Errors were used in the calculations.	
Adjustment Controls	
Compute Correlations for Geoid	False
Horizontal and Vertical adjustment performed	
Set-up Errors	
GPS	
Error in Height of Antenna	0.000 m
Centering Error	0.000 m

Table 54.4: Adjusted Coordinates

Adjustment Performed in WGS-84
Number of Points: 32
Numbers of Constrained Points: 0
Adjusted Grid Coordinates
Errors are reported using 1.96σ

Point Name	Northing	N Error	Easting	E Error	Elevation	E Error
1	1928712.764 m	0.012 m	216378.474 m	0.046 m	N/A	N/A
23	1926964.613 m	0.013 m	215470.731 m	0.046 m	N/A	N/A
24	1927234.048 m	0.012 m	215360.006 m	0.046 m	N/A	N/A
25	1927132.535 m	0.018 m	215102.747 m	0.047 m	N/A	N/A
26	1927219.160 m	0.311 m	214915.783 m	1.200 m	N/A	N/A
27	1926959.781 m	0.152 m	214780.009 m	0.684 m	N/A	N/A
28	1926874.901 m	0.013 m	214592.041 m	0.046 m	N/A	N/A
29	1926913.549 m	0.012 m	214466.250 m	0.046 m	N/A	N/A
30	1927392.761 m	0.014 m	213452.124 m	0.047 m	N/A	N/A
31	1927274.493 m	0.017 m	213283.905 m	0.048 m	N/A	N/A
32	1926847.800 m	0.015 m	213364.052 m	0.047 m	N/A	N/A
10	1929389.638 m	0.015 m	214080.107 m	0.047 m	N/A	N/A
11	1929509.183 m	0.016 m	213642.365 m	0.047 m	N/A	N/A
12	1929524.877 m	0.015 m	213614.240 m	0.047 m	N/A	N/A
13	1929565.012 m	0.025 m	213333.114 m	0.053 m	N/A	N/A
14	1929618.634 m	0.014 m	212655.503 m	0.047 m	N/A	N/A
15	1929463.154 m	0.015 m	212458.098 m	0.047 m	N/A	N/A
16	1929188.037 m	0.013 m	212038.254 m	0.046 m	N/A	N/A
17	1929043.280 m	0157 m	211855.872 m	0.359 m	N/A	N/A
18	1929419.754 m	0.015 m	210122.070 m	0047 m	N/A	N/A
19	1929126.837 m	0.014 m	210025.208 m	0.047 m	N/A	N/A
20	1927796.800 m	0.019 m	210243.863 m	0.048 m	N/A	N/A
21	1927647.922 m	0.013 m	213066.021 m	0.046 m	N/A	N/A
22	1928882.689 m	0.013 m	212694.270 m	0.046 m	N/A	N/A
2	1928197.391 m	0.015 m	216310.906 m	0.047 m	N/A	N/A
3	1927444.066 m	0.013 m	216305.024 m	0.046 m	N/A	N/A
4	1926648.851 m	0.018 m	215905.484 m	0.047 m	N/A	N/A
5	1926677.355 m	0.013 m	215597.701 m	0.046 m	N/A	N/A
6	1926829.184 m	0.014 m	215601.067 m	0.047 m	N/A	N/A
7	1926949.827 m	0.014 m	215521.521 m	0.047 m	N/A	N/A
8	1929082.109 m	0.015 m	215496.074 m	0.047 m	N/A	N/A
9	1929326.357 m	0.013 m	214524.356 m	0.046 m	N/A	N/A

Table 54.5: Adjusted Geodetic Coordinates

Errors are reported using 1.96σ

Point Name	Latitude	N Error	Longitude	E Error	Height	H Error
1	17°25'35.24265"N	0.012 m	78°19'48.54665"E	0.046 m	504.537 m	0.041 m
23	17°24'38.00578"N	0.013 m	78°19'18.63817"E	0.046 m	468.091 m	0.043 m
24	17°24'46.71345"N	0.012 m	78°19'14.76144"E	0.046 m	468.969 m	0.041 m
25	17°24'43.29661"N	0.018 m	78°19'06.09916"E	0.047 m	472.828 m	0.053 m
26	17°24'46.02715"N	0.311 m	78°18'59.72773"E	1.200 m	474.771 m	0.917 m
27	17°24'37.53409"N	0.152 m	78°18'55.25384"E	0.684 m	467.323 m	0.511 m
28	17°24'34.68930"N	0.013 m	78°18'48.92999"E	0.046 m	475.878 m	0.042 m
29	17°24'35.88815"N	0.012 m	78°18'44.65259"E	0.046 m	476.278 m	0.041 m
30	17°24'51.00100"N	0.014 m	78°18'10.08782"E	0.047 m	488.229 m	0.047 m
31	17°24'47.07963"N	0.017 m	78°18'04.44869"E	0.048 m	487.228 m	0.052 m
32	17°24'33.24682"N	0.015 m	78°18'07.36603"E	0.047 m	486.265 m	0.046 m
10	17°25'56.19680"N	0.015 m	78°18'30.39864"E	0.047 m	501.005 m	0.047 m
11	17°25'59.88204"N	0.016 m	78°18'15.51880"E	0.047 m	494.126 m	0.046 m
12	17°26'00.37927"N	0.015 m	78°18'14.55893"E	0.047 m	495.088 m	0.045 m
13	17°26'01.55489"N	0.025 m	78°18'05.02025"E	0.053 m	500.483 m	0.083 m
14	17°26'02.98642"N	0.014 m	78°17'42.04936"E	0.047 m	505.917 m	0.051 m
15	17°25'57.84179"N	0.015 m	78°17'35.43941"E	0.047 m	505.532 m	0.051 m
16	17°25'48.70584"N	0.013 m	78°17'21.35501"E	0.046 m	495.874 m	0.044 m
17	17°25'43.91651"N	0.157 m	78°17'15.24894"E	0.359 m	494.122 m	0.695 m
18	17°25'55.35125"N	0.015 m	78°16'10.35963"E	0.047 m	509.429 m	0.049 m
19	17°25'45.78537"N	0.014 m	78°16'13.22146"E	0.047 m	506.568 m	0.048 m
20	17°25'02.65539"N	0.019 m	78°16'21.26762"E	0.048 m	497.440 m	0.050 m
21	17°24'59.11789"N	0.013 m	78°17'56.89302"E	0.046 m	484.845 m	0.042 m
22	17°25'39.08273"N	0.013 m	78°17'43.71471"E	0.046 m	499.984 m	0.042 m
2	17°25'18.45954"N	0.015 m	78°19'46.50242"E	0.047 m	489.117 m	0.047 m
3	17°14'53.96966"N	0.013 m	78°19'46.65940"E	0.046 m	476.443 m	0.042 m
4	17°24'27.93947"N	0.018 m	78°19'33.50757"E	0.047 m	465.714 m	0.040 m
5	17°14'18.72613"N	0.013 m	78°19'23.07321"E	0.046 m	464.565 m	0.042 m
6	17°24'33.66291"N	0.014 m	78°19'23.11526"E	0.047 m	463.678 m	0.046 m
7	17°24'37.54825"N	0.014 m	78°19'20.36481"E	0.047 m	468.869 m	0.045 m
8	17°25'46.84713"N	0.015 m	78°19'18.49219"E	0.047 m	508.527 m	0.048 m
9	17°25'54.34306"N	0.013 m	78°18'45.47191"E	0.046 m	497.410 m	0.042 m

If the GPS receiver antenna, is placed precisely at that known point "T"–a point that has beer surveyed by exacting means and whose true position is known to be within a centimeter. Observatior "O" is taken by a base station receiver so three entities are to be considered.

1. The position of "T".

2. The reading "O" and
3. The "difference" between "O" and "T".

So an arrow is drawn from the measured point to the true point. This arrow which is shown in two dimensions but which would really exist in three, has both a length (called a magnitude) and a direction. An entity that has magnitude and direction is known as a vector Label the vector "E", for "error", because it represents the amount and direction by which the reading missed the true point. When GPS is used, using the reported co-ordinates, "O", as an approximation of "T'. The vector "E" was the (unknown) amount by which missing the determined position "T" equation could be written as T = O–E. Where we record "O" and we discard "E" to find an approximation of "T". That is, the true co-ordinates are the observed co-ordinates minus the error. It is better to estimate the magnitude–but not the direction of "E".

If exact value of T is known and with the measured value "O", then the above equation to find "E" is rewritten as

$$E = O - T.$$

The GPS observations and a posterior errors (calculated as 1.96σ) are given in Table 54.6. The Histogram of standardized results is shown in Figure 54.1.

Table 54.6: Adjusted Observations

Adjustment performed in WGS-84

GPS Observations

Number of Observations: 31

Number of Outliers: 23

Observation Adjustment (Critical Tau = 0.00)

Obs.	*From Pt.*	*To Pt*		*Observation*	*A-posterior Error (1.96σ)*	*Residual*	*Stand Residual*
B17	1	11	Az.	285°25'43.7023"	0°00'00.8402"	0°00'00.0000"	0.00
			ΔHt.	–10.411 m	0.022 m	0.000 m	0.00
			Dist.	2847.941 m	0.010 m	0.000 m	0.00
B14	1	31	Az.	244°16'20.8973"	0°00'00.6591"	0°00'00.0000"	0.00
			ΔHt.	–17.309 m	0.033 m	0.000 m	0.00
			Dist.	3410.409 m	0.013 m	0.000 m	0.00
B25	1	19	Az.	272°55'43.6080"	0°00'00.2121"	0°00'00.0000"	0.00
			ΔHt.	2.031 m	0.027 m	0.000 m	000
			Dist.	6362.819 m	0.009 m	0.000 m	0.00
B32	1	5	Az.	200°11'12.7727"	0°00'00.4074"	0°00'00.0000"	0.00
			ΔHt.	–39.972 m	0.011 m	0.000 m	0.00
			Dist.	2178.720 m	0.004 m	0.000 m	0.00
B28	1	22	Az.	271°50'26.2464"	0°00'00.2384"	0°00'00.0000"	0.00
			ΔHt.	–4.553 m	0.011 m	0.000 m	000
			Dist.	3685.881 m	0.006 m	0.000 m	000

Contd...

Table 54.6–Contd...

Obs.	From Pt.	To Pt		Observation	A-posterior Error (1.96σ)	Residual	Stand Residual
B33	1	6	Az.	201°37'39.4042"	0°00'00.7752"	0°00'00.0000"	0.00
			ΔHt.	–40.859 m	0.023 m	0.000 m	0.00
			Dist.	2036.486 m	0.008 m	0.000 m	0.00
B36	1	9	Az.	287°30'39.4542"	0°00'00.3806"	0°00'00.0000"	0.00
			ΔHt.	–7.127 m	0.008 m	0.000 m	0.00
			Dist.	1951.837 m	0.003 m	0.000 m	0.00
B12	1	29	Az.	225°56'39.5287"	0°00.00.1088"	0°00'00.0000"	0.00
			ΔHt.	–28.259 m	0.005 m	0.000 m	0.00
			Dist.	2624.022 m	0.002 m	0.000 m	0.00
B18	1	12	Az.	285°34'19.7592"	0°00'00.6677"	0°00'00.0000"	0.00
			ΔHt.	–9.449 m	0.019 m	0.000m	0.00
			Dist.	2879.321 m	0.008 m	0.000 m	0.00
B27	1	21	Az.	251°22'45.3649"	0°00'00.1439"	0°00;00.0000"	0.00
			ΔHt.	–19.692 m	0.009 m	0.000 m	0.00
			Dist.	3477.293 m	0.003 m	0.000 m	0.00
B13	1	30	Az.	244°55'16.3818"	0°00'00.4318"	0°00'00.0000"	0.00
			ΔHt.	–16.308 m	0.023 m	0.000 m	0.00
			Dist.	3208.344 m	0.008 m	0.000 m	0.00
B15	1	32	Az.	237°27'21.6307"	0°00'00.4378"	0°00'00.0000"	0.00
			ΔHt.	–18.272 m	0.022 m	0.000 m	0.00
			Dist.	3542.546 m	0.010 m	0.000 m	0.00
B7	1	24	Az.	213°45'26.6365"	0°00'00.2261"	0°00'00.0000"	0.00
			ΔHt.	–35.568 m	0.006 m	0.000 m	0.00
			Dist.	1794.443 m	0.002 m	0.000 m	0.00
B30	1	3	Az.	182°30'48.7669"	0°00'00.6253"	0°00'00.0000"	0.00
			ΔHt.	–28.094 m	0.009 m	0.000 m	0.00
			Dist.	1270.067 m	0.005 m	0.000 m	0.00
B16	1	10	Az.	285°36'34.4337"	0°00'00.8518"	0°00'00.0000"	0.00
			ΔHt.	–3.532 m	0.025 m	0.000 m	0.00
			Dist.	2394.522 m	0.009 m	0.000 m	0.00
B11	1	28	Az.	223°23'14"1852"	0°00'00.2790"	0°00'00.0000"	0.00
			ΔHt.	–28.659 m	0.011 m	0.000 m	0.00
			Dist.	2561"483 m	0.004 m	0.000 m	0.00
B29	1	2	Az.	186°40'09.2482"	0°00'03.0367"	0°00'00.0000"	0.00
			ΔHt.	–15.420 m	0.023 m	0.000 m	0.00
			Dist.	519.474 m	0.009 m	0.000 m	0.00
B35	1	8	Az.	291°54'44.9985"	0°00'01.9958"	0°00'00.0000"	0.00
			ΔHt.	3.990 m	0.027 m	0.000 m	0.00
			Dist.	956.009 m	0.008 m	0.000 m	0.00

Contd...

Table 54.6–Contd...

Obs.	*From Pt.*	*To Pt*		*Observation*	*A-posterior Error (1.96σ)*	*Residual*	*Stand Residual*
B26	1	20	Az.	260°42'28.5755"	0.00'00.4971"	0°00'00.0000"	0.00
			ΔHt.	–7.097 m	0.030 m	0.000 m	0.00
			Dist.	6198.795 m	0.013 m	0.000 m	0.00
B20	1	14	Az.	282°52'30.4219"	0.00'00.4384"	0°00'00.0000"	0.00
			ΔHt.	–1.380 m	0.032 m	0.000 m	0.00
			Dist.	3829.267 m	0.010 m	0.000 m	0.00
B22	1	16	Az.	275°26'56.5455"	0.00'00.2472"	0°00'00.0000"	0.00
			ΔHt.	–8.662 m	0.017 m	0.000 m	0.00
			Dist.	4363.503 m	0.005 m	0.000 m	0.00
B24	1	18	Az.	275°38'49.1020"	0.00'00.2927"	0°00'00.0000"	0.00
			ΔHt.	4.892 m	0.027 m	0.000 m	0.00
			Dist.	6292.342 m	0.010 m	0.000 m	0.00
B8	1	25	Az.	218°06'51.6818"	0°00'01.3843"	0°00'00.0000"	0.00
			ΔHt.	–31.709 m	0.035 m	0.000 m	0.00
			Dist.	2029.696 m	0.012 m	0.000 m	0.00
B34	1	7	Az.	205°07'27.7148"	0°00'00.6736"	0°00'00.0000"	0.00
			ΔHt.	–35.667 m	0.020 m	0.000 m	0.00
			Dist.	1959.012 m	0.007 m	0.000 m	0.00
B31	1	4	Az.	192°06'28.7513"	0°00'00.8669"	0°00'00.0000"	0.00
			ΔHt.	–38.823 m	0.022 m	0.000 m	0.00
			Dist.	2116.155 m	0.013 m	0.000 m	0.00
B23	1	17	Az.	273°22'46.6637"	0°00'07.1865"	0°00'00.0000"	0.00
			ΔHt.	–10.414 m	0.717 m	0.000 m	0.00
			Dist.	4531.896 m	0.369 m	0.000 m	0.00
B21	1	15	Az.	280°02'07.9366"	0°00'00.4198"	0°00'00.0000"	0.00
			ΔHt.	0.995 m	0.031 m	0.000 m	0.00
			Dist.	3989.119 m	0.010 m	0.000 m	0.00
B19	1	13	Az.	284°50'02.6757"	0°00'01.6003"	0°00'00.0000"	0.00
			ΔHt.	–4.054 m	0.074 m	0.000 m	0.00
			Dist.	3160.451 m	0.025 m	0.000 m	0.00
B10	1	27	Az.	221°33'37.8713"	0°00'47.6344"	0°00'00.0000"	0.00
			ΔHt.	–37.214 m	0.526 m	0.000 m	0.00
			Dist.	2370.926 m	0.471 m	0.000 m	0.00
B9	1	26	Az.	223°36'03.9868"	0°01'16.7901"	0°00'00.0000"	0.00
			ΔHt.	–29.766 m	0.946 m	0.000 m	0.00
			Dist.	2089.278 m	1.015 m	0.000 m	0.00
B6	1	23	Az.	206°38'28.4720"	0°00'00.4702"	0°00'00.0000"	0.00
			ΔHt.	–36.445 m	0.015 m	0.000 m	0.00
			Dist.	1968.602 m	0.003 m	0.000 m	0.00

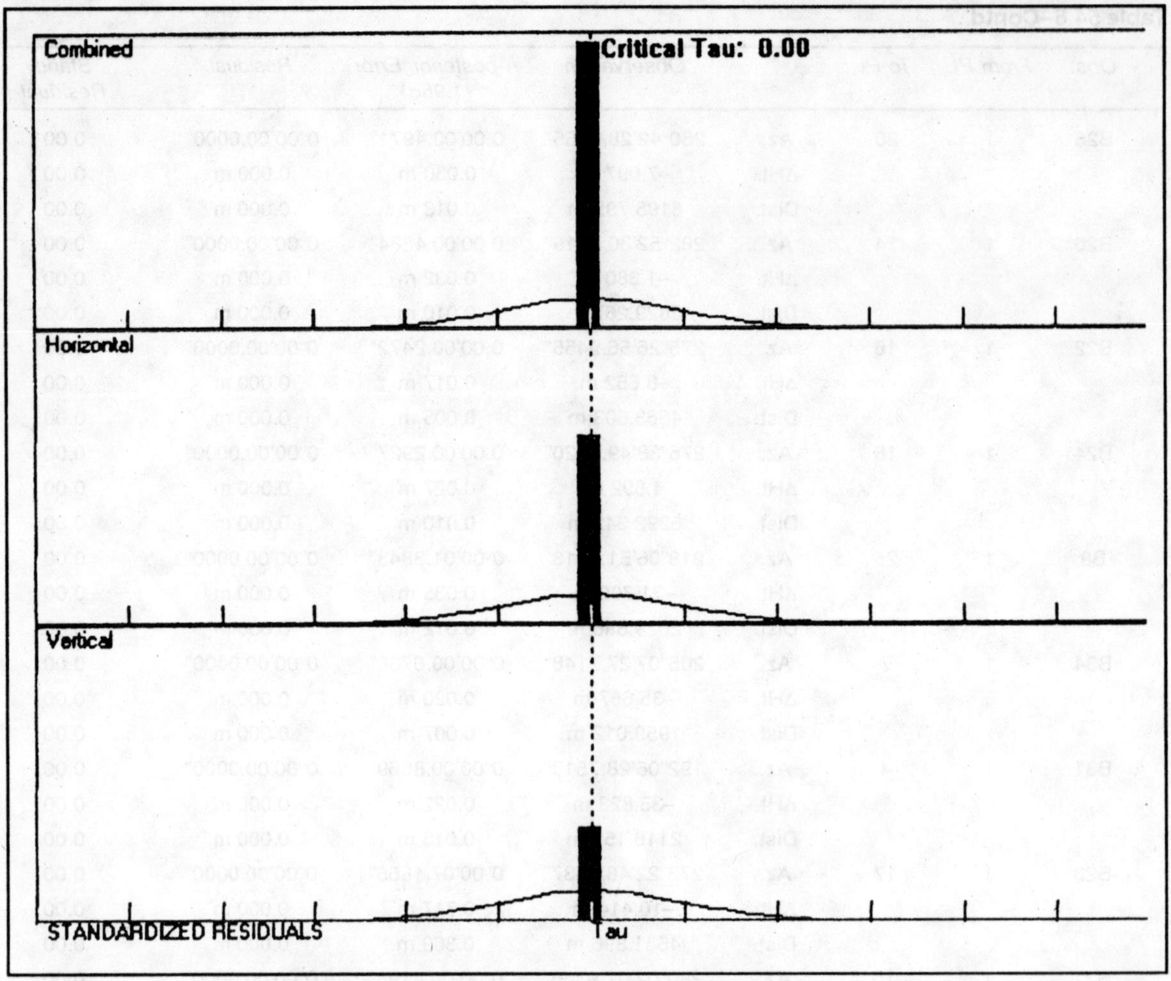

Figure 54.1: Histograms of Standardized Residuals

This technique provides an opportunity for canceling out most of the error in a GPS Position found by an antenna that is close to another antenna which is over a known point. The errors can be minimized through proper planning of the field data collection. Proper understanding of the instrument performance is required for minimizing the errors. The formula for the amount of error that might be expected with differentially corrected data is dependent on the distance between the base station antenna and the rover antenna. A rule of thumb is that the fix will be in error by one additional centimeter for each three kilometers between the two antennas. This relationship is approximately linear: 300 kilometers is found to produce error of about a meter.

Making Differential Correlation Work

Practically a number of conditions have to be satisfied for the process to work. The base and rover have to be taking data at the same time, and the base has to be taking data frequently. If the base station is to serve the rover, where ever it may be (with in the 500 km limit), regardless of when data are taken, the base station must take data from all satellites the rover might see. This can cause a problem; if the base and rover are widely separated and the rover might see a satellite that the base cannot view. The point error ellipses are shown in this case in Figure 54.2.

Base stations are set up with an elevation mask value of 10°. A good rule of thumb is to increase the elevation mask for the rover by one degree depending upon the distance from the base station. This is found to be approximately equal to 75 km. Setting of 5 degrees works well for the rover in general if it is within 50 km from the base station. About fifteen degrees is good for avoiding difficulties due to terrain, that elevation angle reduces errors that tend to occur when signals come from satellites low on the horizon, since those signals must pass through more atmosphere. In any event the rover elevation mask must be set high enough so that there is no possibility for the rover to record data from a satellite that the base is not recording as shown in Figure 54.2. Correcting errors by the differential correction method implies that the base station-receiver at the known point-can communicate with the roving receiver(s).

In post processing GPS, (more correctly called post mission-processing GPS), the data from both the base and the rover are brought together later in a computer, and the appropriate correction is applied to each fix created by the rover.

Here adding all the fixes that were displayed before, but here now each fix was differentially corrected. In other words, all the fixes with their obvious errors are now within the area of one small circle that presumably includes the true location of the antenna. The ticks again are 10 meters apart, so the amount of error reduction and the accuracy you can expect from differentially corrected data is identified. The variations and accuracies of the received signals are to be better understood as operational models that could predict errors are still not available. The differential correction patterns which are significantly influenced by a variety of parameters.

The point error ellipses suggest clearly the amount of deviation of signal. The horizontal and vertical precisions calculated in this study are good examples of quantifiable parameters that define the signal variation.

The covariant terms are shown in Table 54.7. The exact ratios for Horizontal precision and 3D precision are calculated. The exact errors in height measurement, distance and height are given in the table. The completed network of the cadastral map is shown in Figure 54.3.

Table 54.7: Covariant Terms

Adjustment performed in WGS-84

From Point	*To Point*		*Components*	*A-Posteliori Error (1.96σ)*	*Holiz. Precision (Ratio)*	*3D Precision (Ratio)*
1	23	Az.	206°38'28.4720"	0°00'00.4702"	1 : 613521	1 : 613521
		ΔHt.	–36.445 m	0.015 m		
		ΔElev.	?	?		
		Dist.	1968.602 m	0.003 m		
1	24	Az.	213°45'26.6365"	0°00'00.2261"	1 : 1165402	1 : 1165402
		ΔHt.	–35.568 m	0.006 m		
		ΔElev.	?	?		
		Dist.	1794.443 m	0.002 m		

Contd...

Table 54.7–Contd...

From Point	To Point	Components		A-Posteliori Error (1.96σ)	Holiz. Precision (Ratio)	3D Precision (Ratio)
1	25	Az.	218°06'51.6818"	0°00'01.3843"	1 : 165016	1 : 105016
		ΔHt.	–31.709 m	0.035 m		
		ΔElev.	?	?		
		Dist.	2029.696 m	0.012 m		
1	26	Az.	223°36'03.9868"	0°01'16.7901"	1 : 2058	1 : 2058
		ΔHt.	–29.766 m	0.946 m		
		ΔElev.	?	?		
		Dist.	2089.278 m	1.015 m		
1	27	Az.	221°33'37.8713"	0°00'47.6344"	1 : 5036	1 : 5036
		ΔHt.	–37.214 m	0.526 m		
		ΔElev.	?	?		
		Dist.	2370.926 m	0.471 m		
1	28	Az.	223°23'141852"	0°00'00.2790"	1 : 730758	1 : 730758
		ΔHt.	–28.659 m	0.011 m		
		ΔElev.	?	?		
		Dist.	2561.483 m	0.004 m		
1	29	Az.	225°56'39.5287"	0°00'00.1088"	1 : 1543071	1 : 1543071
		ΔHt.	– 28.259 m	0.005 m		
		ΔElev.	?	?		
		Dist.	2624.022 m	0.002 m		
1	30	Az.	244°55'16.3818"	0°00'00.4318"	1 : 387989	1 : 387989
		ΔHt.	–16.308 m	0.023 m		
		ΔElev.	?	?		
		Dist.	3208.344 m	0.008 m		
1	31	Az.	244°16'20.8973"	0°00'00.6591"	1 : 260172	1 : 260172
		ΔHt.	– 17.309 m	0.033 m		
		ΔElev.	?	?		
		Dist.	3410.409 m	0.013 m		
1	32	Az.	237°27'21.0307"	0°00'00.4378"	1 : 338667	1 : 338667
		ΔHt.	–18.272 m	0.022 m		
		ΔElev.	?	?		
		Dist.	3542.546 m	0.010 m		
1	10	Az.	285°36'34.4337"	0°00'00.8518"	1 : 267806	1 : 267806
		ΔHt.	–3.532 m	0.025 m		
		ΔElev.	?	?		
		Dist.	2394.522 m	0.009 m		
1	11	Az.	285°25'43.7023"	0°00'00.8402"	1 : 288317	1 : 288317
		ΔHt.	–10.411 m	0.022 m		
		ΔElev.	?	?		
		Dist.	2847.941 m	0.010 m		

Contd...

Table 54.7–Contd...

From Point	*To Point*		*Components*	*A-Posteliori Error (1.96σ)*	*Holiz. Precision (Ratio)*	*3D Precision (Ratio)*
1	12	Az.	285°34'19.7592"	0°00'00.6677"	1 : 351818	1 : 351818
		ΔHt.	–9.449 m	0.019 m		
		ΔElev.	?	?		
		Dist.	2879.321 m	0.008 m		
1	13	Az.	284°50'02.6757"	0°00'01.6003"	1 : 127790	1 : 127790
		ΔHt.	–4.054 m	0.074 m		
		ΔElev.	?	?		
		Dist.	3160.451 m	0.025 m		
1	14	Az.	282°52'30.4219"	0°00'00.4384"	1 : 366684	1 : 300684
		ΔHt.	1.380 m	0.032 m		
		ΔElev.	?	?		
		Dist.	3829.207 m	0.010 m		
1	15	Az.	280°02'07.9366"	0°00'00.4198"	1 : 404110	1 : 404110
		ΔHt.	0.995 m	0.031 m		
		ΔElev.	?	?		
		Dist.	3989.119 m	0.010 m		
1	16	Az.	275°26'56.5455"	0°00'00.2472"	1 : 814698	1 : 814698
		ΔHt.	–8.662 m	0.017 m		
		ΔElev.	?	?		
		Dist.	4363.503 m	0.005 m		
1	17	Az.	2.73°22'46.6637"	0°00'07.1865"	1 : 12287	1 : 12287
		ΔHt.	–10.414 m	0.717 m		
		ΔElev.	?	?		
		Dist.	4531.896 m	0.369 m		
1	18	Az.	275°38'49.1020"	0°00'00.2927"	1 : 648349	1 : 648349
		ΔHt.	4.892 m	0.027 m		
		ΔElev.	?	?		
		Dist.	6292.342 m	0.010 m		
1	19	Az.	272°55'43.6080"	0°00'00.2121"	1 : 729124	1 : 729124
		ΔHt.	2.031 m	0.027 m		
		ΔElev.	?	?		
		Dist.	6362.819 m	0.009 m		
1	20	Az.	260°42'28.5755"	0°00'00.4971"	1 : 488786	1 : 488786
		ΔHt.	–7.097 m	0.030 m		
		ΔElev.	?	?		
		Dist.	6198.795 m	0.013 m		
1	21	Az.	251°22'45.3649"	0°00'001439"	1 : 1171295	1 : 1171295
		ΔHt.	–19.692 m	0.009 m		
		ΔElev.	?	?		
		Dist.	3477.293 m	0.003 m		

Contd...

Table 54.7–Contd...

From Point	*To Point*		*Components*	*A-Posteliori Error (1.96σ)*	*Holiz. Precision (Ratio)*	*3D Precision (Ratio)*
1	22	Az.	271°50'26.2464"	0°00'00.2384"	1 : 655630	1 : 655030
		ΔHt.	–45.53 m	0.011 m		
		ΔElev.	?	?		
		Dist.	3685.881 m	0.006 m		
1	2	Az.	186°40'09.2482"	0°00'03.0367"	1 : 55113	1 : 55113
		ΔHt.	–15.420 m	0.023 m		
		ΔElev.	?	?		
		Dist.	519.474 m	0.009 m		
1	3	Az.	182°30'48.7669"	0°00'00.6253"	1 : 249404	1 : 249404
		ΔHt.	–28.094 m	0.009 m		
		ΔElev.	?	?		
		Dist.	1270.067 m	0.005 m		
1	4	Az.	192°06'28.7513"	0°00'00.8669"	1 : 163244	1 : 163244
		ΔHt.	–38.823 m	0.022 m		
		ΔElev.	?	?		
		Dist.	2116.155 m	0.013 m		
1	5	Az.	200°11'12.7727"	0°00'00.4074"	1 : 504349	1 : 504349
		ΔHt.	–39.971 m	0.011 m		
		ΔElev.	?	?		
		Dist.	2178.720 m	0.004 m		
1	6	Az.	201°37'39.4042"	0°00'00.7752"	1 : 263973	1 : 263973
		ΔHt.	–40.859 m	0.023 m		
		ΔElev.	?	?		
		Dist.	2036.486 m	0.008 m		
1	7	Az.	205°07'27.7148"	0°00'00.6736"	1 : 298200	1 : 298200
		ΔHt.	–35.667 m	0.020 m		
		ΔElev.	?	?		
		Dist.	1959.012 m	0.007 m		
1	8	Az.	291°54'44.9985"	0°00'01.9958"	1 : 113297	1 : 113297
		ΔHt.	3.990 m	0.027 m		
		ΔElev.	?	?		
		Dist.	956.009 m	0.008 m		
1	9	Az.	287°30'39.4542"	0°00'00.3806"	1 : 690946	1 : 690946
		ΔHt.	–7.127 m	0.008 m		
		ΔElev.	?	?		
		Dist.	1951.837 m	0.003 m		

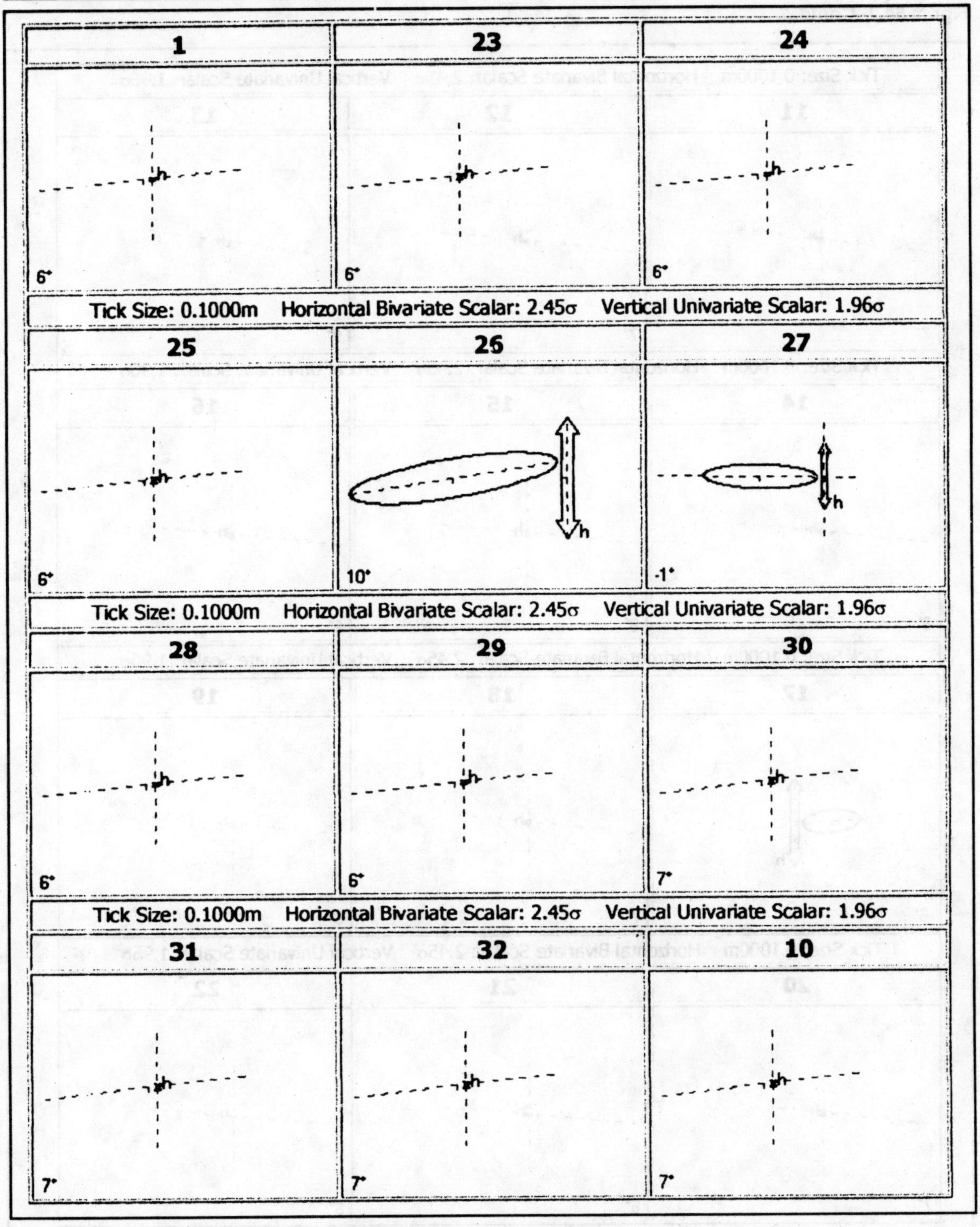

Figure 54.2: Point Error Ellipses

Figure 54.2–Contd...

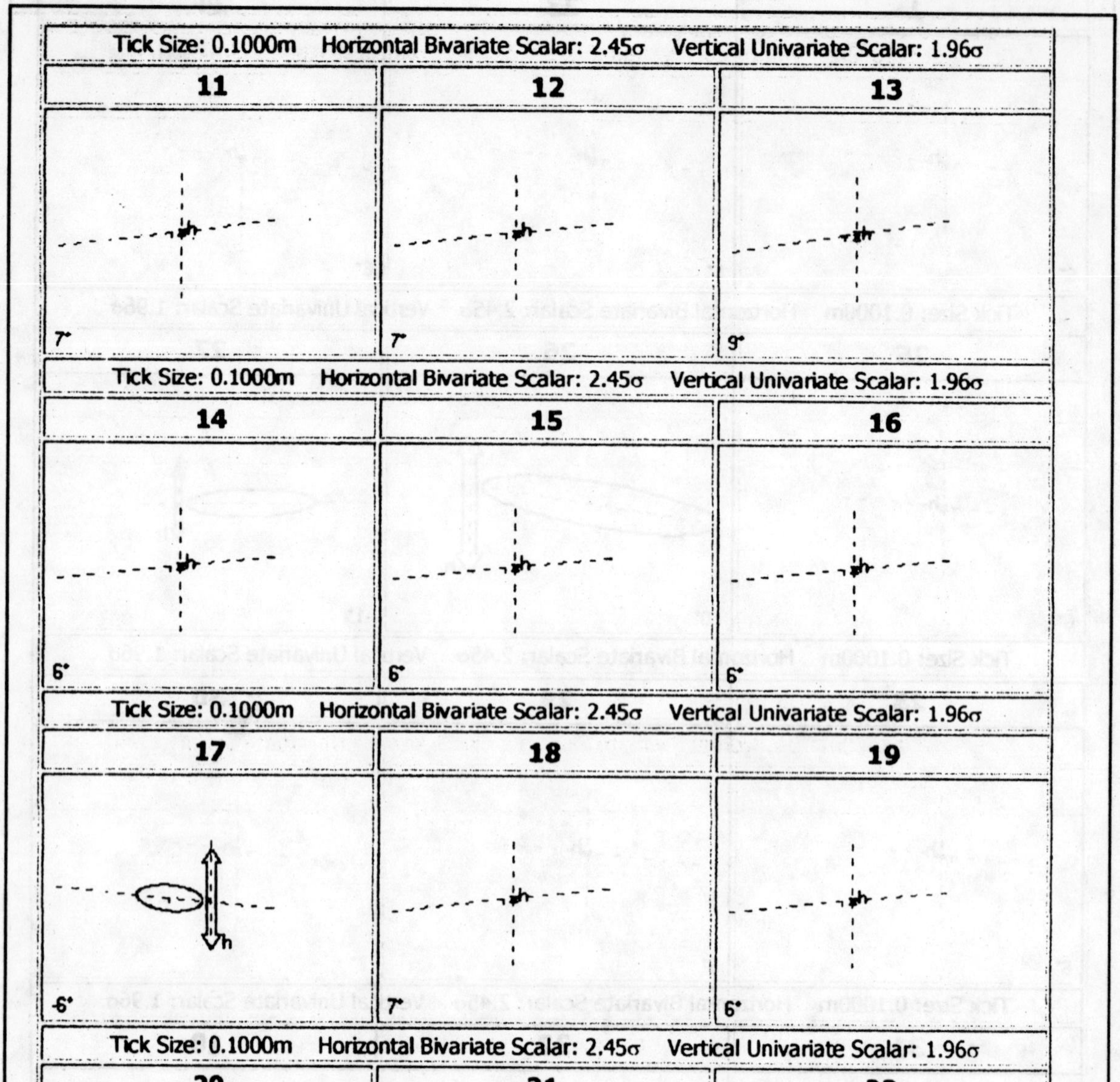

Figure 54.2–Contd...

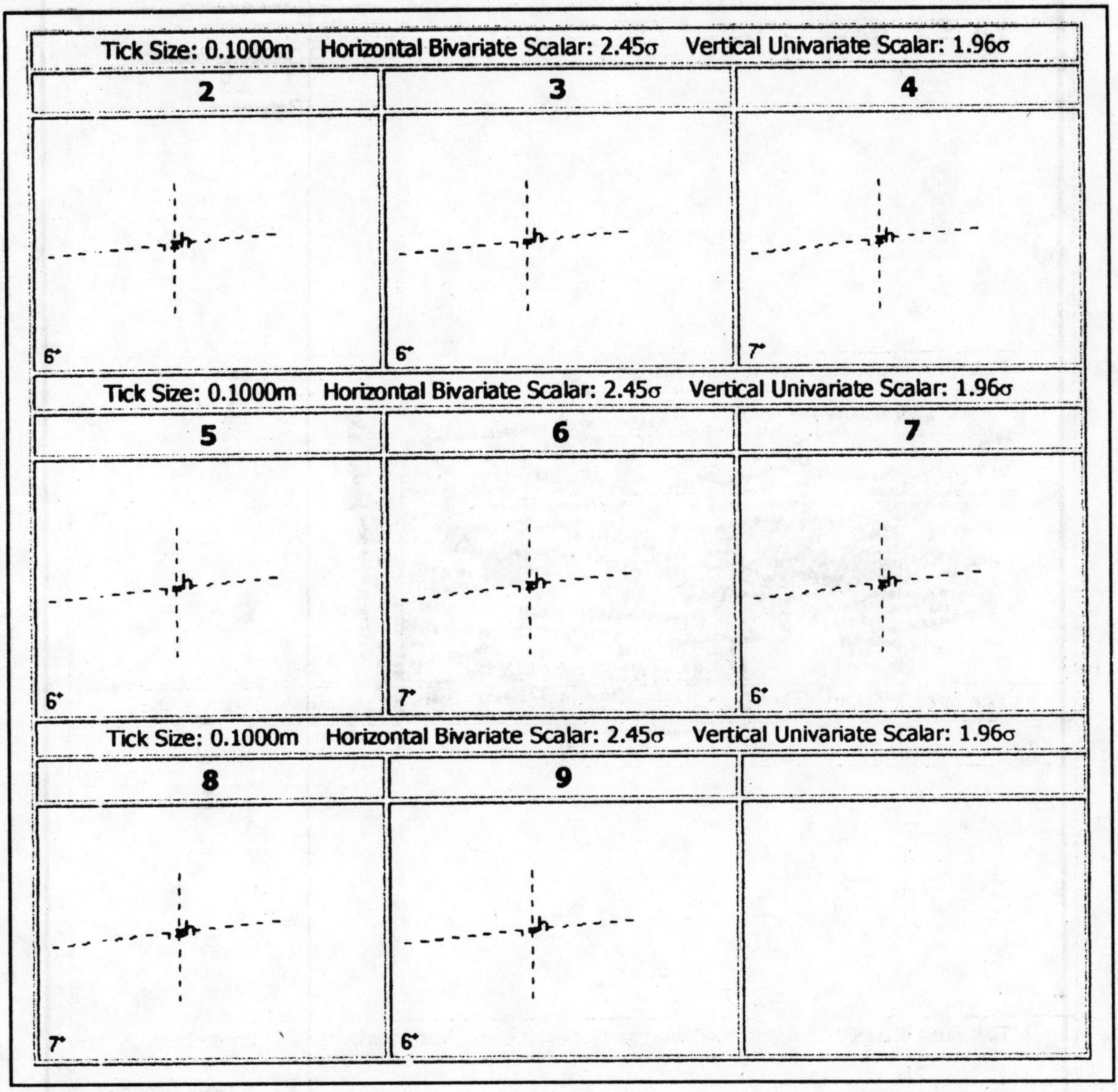

Conclusions

Extensive DGPS data has been collected and used in the present study. The variations and accuracies of the received signals are to be better understood as operational models that could predict errors are still not available. The differential correction patterns are significantly influenced by a variety of parameters. The errors can be minimized through proper planning of the field data collection. Proper understanding of the instrument performance is required for minimizing the errors. The point error ellipses suggest clearly the amount of deviation of signal. The horizontal and vertical precisions calculated in this study are good examples of quantifiable parameters that define the signal variation.

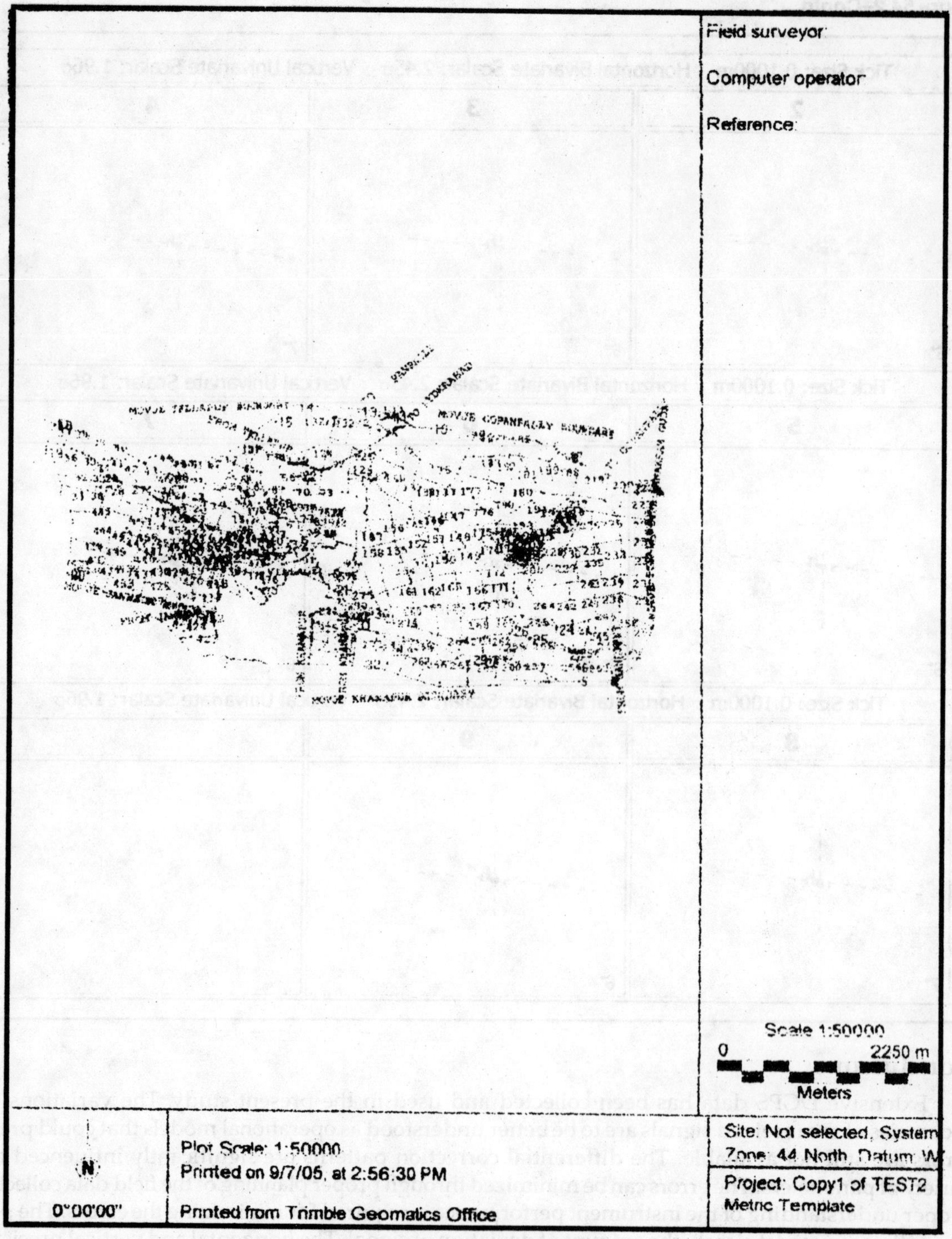

Figure 54.3: Cadastral Map

References

Adrados, C., Girard, I., Gendner, J. and Janeau, G., 2002. Global positioning system (GPS) location accuracy due to selective availability removal. *C.R. Biologies*, 325: 165–170.

Byun, S., G.A. Hajj and L.E. Young, 2002. Development and application of GPS signal. Multipath Simulator. *Radio Science*, 37(6): 1–23.

Misra, P. and P. Enge, 2001. *Global Positioning System Signals, Measurements, and Performance*. Ganga-Jamuna Press, Lincoln, MA.

Sardon, E., A. Rius and N. Zarraoa, 1994. Estimation of the transmitter and receiver differential biases and the ionospheric total electron content from global positioning system observations. *Radio Science*, 29(3): 577–586.

Chapter 55

Influence of Pollutants on Soil Bacteria

D. Vijayalakshmi[1], C. Krishna Kanth[2] and M.A. Singara Charya[3]
[1]Department of Microbiology, [2]Department of Physics
Kavitha Memorial P.G. College, Khammam – 507 001
[3]Department of Microbiology, Kakatiya University, Warangal – 506 009

ABSTRACT

The impact of pollutant generated from Dairy, Sewage, hospital and Automobile site on the soil bacteria was studied during 1996–1999. The minimum bacterial colonies (230×10^6) were recorded during rainy and maximum (606×10^6) during winter months. The dumps of automobiles caused for drastic reduction of bacteria in the soils (59×10^6). No significant relations were drawn between physico-chemical characteristics and bacterial densities in these polluted sites. pH correlated positively with bacterial densities of dairy soils ($r = 0.621$) at 5 per cent significant level. The date revealed the impact of toxic pollutants and the densities and diversities of bacteria and in turn the management of polluted soil systems.

Keywords: *Pollutants, Bacterial population, Degradative enzymes.*

Introduction

Soil is undoubtedly one of the most important need of the environment. Various physical, chemical and biological process are constantly changing soils over geologic time. Soil is a natural, three-dimensional array of vertically differentiated material at the surface of earths crust.

Pollution is often thought of as resulting from chemical contaminant generating from dairy industries, drug factories, sewage, automobiles etc. It is well known that microorganisms play the significant role in dynamics of organic matter by production of enzymes (Hattori, 1976).

Soil microorganisms in general bacteria in particular reproduce rapidly and adapt to new environmental situations is important to the decomposition and transformation of both natural and anthropogenic products. Some of the functions performed either entirely or in part by bacteria include nutrient cycling, decomposition of organic materials, Nitrogen fixation, Pesticide detoxification and oxidation-reduction reactions. Without bacteria to mediate these process is impossible. Soil conditions that affect the growth of bacteria are many. However, the most important factors are the O2, temperature, moisture, acidity, inorganic and organic nutrient status of the soil environment. Microbes also play an important role in mineralization process. The enzymatic activity of accumulated enzymes and from those in proliferating microorganisms. Accumulated enzymes in soils are regarded as enzymes present and active in a soil in which no microbial proliferation occurs (Kiss *et al.*, 1975). Sources of enzymes present and active in a soil in which no microbial proliferation occurs. Source of enzymes in soil are primarily the microbial biomass, although they can also originate from plant and animal residues. Enzymes can be immobilized on organic and inorganic constituents of soils. It is well known that microorganisms, the most significant sources of soil enzymes, are absorbed by soil organic and inorganic constituents. (Trivedi *et al.*, (2003), Amaud *et al.* (2004). In view of this to understand this rapid depilation of bacteria at pollution sites. A study was undertaken for three year in different seasons and the facts were established and presented in Tables 55.1–55.3. A correlation was drawn between the physical, chemical parameters and bacterial densities in the sites are increased pollution status were analyzed and presented in Tables 55.4–55.6.

Table 55.1: Bacterial Densities (1 × 10^6)/gm of Four Polluted Sites in Khammam During 1996–97

Polluted Sites	*Jun*	*Jul*	*Aug*	*Sept*	*Oct*	*Nov*	*Dec*	*Jan*	*Feb*	*Mar*	*Apr*	*May*
Dairy	402	500	320	230	570	606	272	398	406	270	368	255
Sewage	270	472	310	310	450	242	292	142	136	142	329	242
Hospital	392	266	276	175	272	190	169	252	190	284	284	271
Automobile	256	370	390	180	182	139	144	59	76	85	163	148

Table 55.2: Bacterial Densities (1 × 10^6)/gm of Four Polluted Sites in Khammam During 1997–98

Polluted Sites	*Jun*	*Jul*	*Aug*	*Sept*	*Oct*	*Nov*	*Dec*	*Jan*	*Feb*	*Mar*	*Apr*	*May*
Dairy	260	300	100	232	317	250	180	62	105	185	272	200
Sewage	205	180	166	300	472	362	309	180	286	192	250	360
Hospital	300	325	380	262	165	326	204	189	300	110	170	220
Automobile	145	172	180	175	170	168	108	72	142	140	148	189

Table 55.3: Bacterial Densities (1 × 10^6)/gm of Four Polluted Sites in Khammam During 1998–99

Polluted Sites	*Jun*	*Jul*	*Aug*	*Sept*	*Oct*	*Nov*	*Dec*	*Jan*	*Feb*	*Mar*	*Apr*	*May*
Dairy	150	320	108	125	310	216	172	187	165	207	216	117
Sewage	230	175	250	256	200	207	162	160	120	272	200	276
Hospital	85	100	95	162	161	150	171	106	162	106	101	106
Automobile	98	176	186	100	121	82	142	197	143	100	65	75

Table 55.4: Correlation Coefficient (r) Between Bacterial Densities and Physico-chemical Characteristics in Four Polluted Sites at Khammam During 1996–97

Parameters	*Polluted Sites*			
	Dairy	*Sewage*	*Hospital*	*Automobile*
pH	– 0.039	0.257	0.455	– 0.143
Temperature	– 0.069	0.218	0.781*	0.393
Conductivity	0.029	0.443	0.342	0.183
Turbidity	0.009	–0.176	–	–
Dissolved Oxygen	– 0.156	0.410	–	–
Biological Oxygen Demand	– 0.112	0.295	–	–
Chemical Oxygen Demand	0.284	0.230	–	–
Organic Matter	0.135	0.162	– 0.024	0.570
Phosphates	0.580	0.408	– 0.111	0.118
Sulphates	– 0.227	0.465	– 0.272	0.423
Nitrates	0.223	0.307	– 0.136	0.072

*: $P < 0.05$; –: Not Detected.

Table 55.5: Correlation Coefficient (r) Between Bacterial Densities and Physico-chemical Characteristics in Four Polluted Sites at Khammam During 1997–98

Parameters	*Polluted Sites*			
	Dairy	*Sewage*	*Hospital*	*Automobile*
pH	–0.039153	0.010	–0.097	–0.145
Temperature	–0.327	–0.307	–0.098	–0.102
Conductivity	–0.391	–0.164	–0.310	–0.757*
Turbidity	0.022	–0.309	–	–
Dissolved Oxygen	0.042	–0.332	–	–
Biological Oxygen Demand	0.386	0.044	–	–
Chemical Oxygen Demand	0.205	–0.006	–	–
Organic Matter	0.224	0.227	–0.413	–0.331
Phosphates	0.005	–0.073	–0.153	–0.576
Sulphates	0.435	0.113	–0.068	0.227
Nitrates	–0.453	–0.496	–0.217	–0.619

*: $P < 0.05$; –: Not Detected.

Materials and Methods

The physical Parameters (pH, Temperature, electrical conductivity and Turbidity) and chemical characteristics (Dissolved Oxygen, BOD, COD, Organic matter, Phosphates, Sulphates and Nitrates) of the soils collected from four polluted sites in, Khammam town were analyzed with monthly intervals for a period of three years. The analysis of these parameters were followed as suggested by Bradley

(1999). The serial dilution technique was used to isolate the bacteria from the polutted sites. From different habitates, each 10 gm of soil was suspended or agitated in a known volume of sterile water (90 ml) to make microbial suspension. Serial dilutions were made and desired amount of suspension transferred to culture plates and incubated at 37°C for 24–48 hrs and analyzed the plates for bacterial count. Standard method used for the analysis of desired samples and two way analysis of variance and correlation coefficient values between physico-chemical and bacterial densities were calculated and recorded in Table 55.1–55.6.

Table 55.6: Correlation Coefficient (r) Between Bacterial Densities and Physico-chemical Characteristics in Four Polluted Sites at Khammam During 1998–99

Parameters	*Polluted Sites*			
	Dairy	*Sewage*	*Hospital*	*Automobile*
pH	0.621	–0.384	–0.323	–0.041
Temperature	0.102	0.251	–0.589	–0.194
Conductivity	–0.299	0.131	0.393	–0.304
Turbidity	0.432	0.621*	–	–
Dissolved Oxygen	0.256	0.122	–	–
Biological Oxygen Demand	–0.207	–0.058	–	–
Chemical Oxygen Demand	–0.372	0.290	–	–
Organic Matter	–0.413	0.056	0.055	0.197
Phosphates	0.298	–0.334	–0.474	0.069
Sulphates	0.062	–0.056	–0.291	–0.049
Nitrates	0.437	–0.204	–0.179	–0.020

*: $P < 0.05$; –: Not Detected.

Results and Discussion

From the table, it was clear that the bacterial densities varied significantly with seasons and with the influence of effluents released from different industries. The maximum bacterial colonies were recorded in the month of November, 1996 (606×10^6) while minimum in September, 1996 (230×10^6) in the soils amended with dairy wastes. Remarkable reduction in the bacterial colonies were noticed in the same site during 1997–98, where the maximum bacterial colonies (370×10^6) were observed in the month of October 1997 and the minimum (62×10^6) in the month of January 1998. Subsequently, the bacterial colonies were recovered and stabilized during 1998–99 with maximum and minimum colony count as 320×10^6 and 125×10^6 respectively.

The enumeration of bacterial colonies in the sewage sites of Khammam town revealed that the constituents of the sewage promoted the growth of bacteria where, seasons were not showed their influence on the proliferation of the colonies.

The maximum colonies were 450×10^6 in the month of October while minimum 142×10^6 in the month of March 1997. The fluctuation of bacterial colonies in (1997–98) the site ranged from 166×10^6 (August 1997) to 472×10^6 (October 1997). The moderate densities of bacteria were noticed in all other months. The stability of the bacterial colonies were witnessed during 1998–99 with the range of densities from 120×10^6 (February 1999) to 272×10^6 (March 1999).

The surroundings of the hospital with higher possibilities of contamination with diversified bacterial genera were observed. The soil samples were collected around hospital site (where the waste dumped areas were selected) waste material showed high incidence of bacterial colonies. However, the densities varied with seasonal influence. The maximum colony count (392×10^6) was noticed in the month of June, 1996, while minimum (9×10^6) in December 1996. (The fluctuations in all other months of the year were in between these two values). The bacterial counts were varied from 110×10^6 (March 1998) to 380×10^6 (August 1997) significant decrease was observed in the colony counts of bacteria during 1998–99 (85×10^6 to 162×10^6). It was very difficult to identity the exact reasons and influence of different factors responsible for these bacterial decreases.

The fourth site is amended with used oils, bummed or partially bummed engine oils and varieties of hydrocarbon related compounds. These soils were shown a variety of bacterial genera with varied densities.

The maximum (390×10^6) bacterial colonies were recorded in the month of August, while minimum 59×10^6 in January. In general the months June to October showed moderate to high densities of bacterial population. Similar trend was continued in the enumeration of bacterial colonies. The highest densities of bacteria in the soils contaminated with oils and its related compounds was from May to October (1997–98). The minimum (72×10^6) number of colonies was noticed in the January 1998. The bacterial densities varied from 65×10^6 (April 1999) to 197×10^6 (January 1999). The gradual percolation and attachment of oil globules on the soil moieties could not show any remarkable fluctuations in densities of bacteria.

Similar to our observations many microbiologists (Border *et al.* (1985), Varada Raj and Ayyappan (1989), Douglas *et al.* (1990), Lindsay and Riley (1994), Jon *et al.* (1999) from different parts of world tried to isolate and identify the different bacterial genera in various polluted environments. They furtl _r signified the relationships between the bacteria and environmental conditions, Phyllis (1989).

Akira (1989) characterized the bacterial populations and structure in different degraded environments on the basis of respiratory quinine profiles (Mehta *et al.*, 2002; Datke *et al.*, 2003; Pawar *et al.*, 2004; Amaud *et al.*, 2004; Medina *et al.*, 2005). The effect of pollutant stress on the bacteria flora and relationships between nutritional requirements and presence of available bacteria aerobic and anaerobic systems.

Lalitha Kumari and Singara Charya (1996), they studied on the assessment of industrial and organic pollution and biological diversity of the soils in Warangal-India, observed the high incidence of bacteria in the soils amended by sewage waters.

Recently Baranyi and Carmen (1999), isolated variety of bacterial generation in UK with the help of numerical method. Greg *et al.* (1999) observed the dominance of *Pseudomonus aeruginosa* in different decayed environments and elaborated its physical characterization.

The correlation coefficient was determined between various physico-chemical parameters and bacterial densities of four polluted sites were presented (Tables 55.3–55.5). In general no major significant correlation was recorded between physico-chemical ad bacterial densities. In most case the significance was positive or negative above 5 per cent level. From the Tables 55.4–55.6, it was observed that the temperature showed a positive correlation with bacterial densities in soils near to the hospital site ($r = 0.781$) ($P < 0.01$). The correlation during (1997–98) were not in significant level in polluted sites of dairy, sewage and hospital but negative correlation was observed between conductivity and nitrates ($r = -0.757$) ($P < 0.01$) and ($r = -0.619$) ($P < 0.05$) with bacterial densities in the soils of automobile site. The significant correlations were observed in dairy and sewage sites (1998-99). pH correlated positively with bacterial densities of dairy soils (r = 0.621) at 5 per cent significant level. In the sewage soils the correlation was significant (r = 0.621) at 5 per cent significant level, and also turbidity and bacterial densities showed a negative correlation at sewage site (r = –0.621) ($P < 0.05$). Recently, Goovaerts (1999) Mehta *et al.* (2002), Datke *et al.* (2003), Pawar *et al.* (2004), Amaud *et al.* (2004), Medina *et al.* (2005), reviewed the statistics of soil science, its state of the art and perspective and concluded the close relations between various abiotic factor and their positive role in the bio-productivity of the soils.

Acknowledgement

The authors are thankful to Dr. S. Girisham Head Department of Micro Biology, Kakatiya University, Warangal for his encouragement. Two of us (DVL, CKK) are grateful to management, Sri Kavitha. Educational Society for providing Lab facilities.

References

Akira, H., Kiyoshi, M. and Kitamura, H., 1989.Characterisation of the Bacteria Population structure in an anaerobic-aerobic activated sludge system on the basis of Respiratory quinone profiles. *Appl. Env. Microbiol.*, 55: 897–901.

Arnaud, D., Mathews, R. and Jones, J.E., 2004. Perceptions and assessment of soil fertility by farmer in the mid hills of Nepal. *Agriculture Ecosystem and Environment*, 103: 191–206.

Border, M.M., Firehammer, B.D., Shoop, D.S. and Myer, 1985. Isolation of *Bacteriods fragilis* from the *Feces diarrheric* calves and lambs. *J. Clin. Micribiol.*, 21: 472–473.

Baranyi, J. and Carmen, P., 1999. Estimating bacterial growth parameters by means of detection time. *Appl. Environ. Microbiol.*, 65: 732–736.

Bradley, P.D., 1999. Microbial functional diversity can be influenced by the addition of simple organic substrates to soil. *Soil Biol. Biochem.*, 30: 1981–1988.

Datke, S.B., Minimol, J.S and Ujjainkar, V.V., 2003. Effect of Season on yield and yield components in ground nut PKV. *Res. J.*, 27: 57–60

Douglas, S., Myers, L. and Julie, B.L., 1990. Enumeration of Enbtero toxigenic *Bactriods fragilis* in Municipal sewage. *Appl. Environ. Microbiol.*, 56: 2243–2244.

Goovaerts, P., 1999. Geostatistics in soil science: State-of-the-art and perspectives. *Goederema*, 89: 1–45.

Jon, H., Gatlin, L.E., K. Charles, G.H. Choi and M.S. Hanson, 1999. Identification and characterization of Sir A, an Iron regulated protein from *Staphylococcus aureus*. *J. Bacteriol.*, 181: 1436–1443.

Kiss, S., Bularda, M.D. and Radulescn, D., 1975. Nutrient cycling in soil. *Adv. Agron.*, 27: 25–87.

Lalitha Kumari, B., 1996. Assessment of industrial and organic pollution biological diversity and enzyme productions of the soils in warangal, A.P. *Ph.D. Thesis*, Kakatiya University.

Lindsay, J.A. and Riley, T.V., 1994. Staphylococcus iron requirements siderophore production and iron regulated protein expression. *In fact Immun.*, 62: 2309–2314.

Kumar, A.S., 2005. Effect of salt stress on germination of three varieties of *Vigna radiata*. *Ind. J. Env. and Eco-planning*, 8: 201–203.

Madhuya, D.A., S. Adak and Purohit, M.K., 2003. Assessment of effluents from cement industries: A pollution problem. *Indian J. Environ. and Ecoplan.*, 7: 497–500.

Medina, A., Probanza, A. and Azcon, R., 2005. Interactions of arbuscular mycorrhizal fungi and bacillestrains and their effects on plant growth microbial rhizosphere activity. *Applied Soil Ecology*, 22: 15–28.

Pawar, S.N., Divekar, S.P., Glule, S.B. and Kadale, A.S., 2004. Effect of mulching on moisture conservation and yield of summer groundnut crop. *J. Sci. Crops*, 14: 410–413.

Phyllis, A., Martin, W. and Rusel, S.T., 1989. World wide abundance and distribution of *Bacillus thuringenues* isolates. *Appl. Environ. Microbiol.*, 55: 2437–2442.

Hattori, J., 1976. The physical environment in soil microbiology. An attempt to extend principles of micro-biology to soil microorganisms. *CRC Crit. Rev. Microbiol.*, 4: 423–461.

Trivedi, B.S., 2003. Use of farm residues for integrated nutrient management Indian. *J. Agric. Environ and Biotech.*, 1:75–93.

Varadaraj, P. and Ayyappan, S., 1989. Population and distribution of Z heterotrophic bacterial flora in freshwater bodies of Bangalore. *Ind. J. Env. Hlth.*, 31: 162–170.

Chapter 56

Assessment of Biodegradation Potential of Linear Alkyl Benzene Sulphonate by *Anabaena doliolum*

Dilip K. Rathore[1], Preeti Sharma[2], Praveen K. Chandel[3] and Sandhya Tyagi[1]

[1]Laboratory of Anatomy and Eco-physiology of Medicinal Plants and Cyanobacterial Research; [2]Microbial Research Laboratory; [3]Laboratory of Biomolecular Technology

Department of Botany, University College of Science, Mohanlal Sukhadia University, Udaipur – 313 001, Rajasthan

ABSTRACT

Cyanobacteria, commonly known as blue-green algae, are largest and most unique group of photosynthetic prokaryotes inhabiting diverse habitats. They are endowed with property of photosynthesis as well as aerobic nitrogen fixation. Many of these organisms respond to stressful condition in presence of water polluted with LABS (Linear Alkyl Benzene Sulphonate). Cyanobacteria are known for their ability to degrade many pollutants present in water, out of which LABS is very important detergent compound used worldwide. As its presence is much common in altered aquatic habitats and few studies are available in terms of chemical nature, degradation and analytical studies. But not much work has been done regarding its effects on algal growth and its biodegradation by these organisms. In this article, we examined biodegradation potential and growth rate percentage of a cyanobacterium *Anabaena doliolum* Bharadwaja using 5 µg/ml and 10 µg/ml LABS concentration at regular time interval (5, 10, 15, 20, 25, 30 days). The per cent growth rate and biodegradation potential were estimated using pigment analysis method and Longwell and Maniece Methylene Blue method (1955) respectively. *Anabaena doliolum* was found to degrade LABS efficiently up to 95.14 per cent and 94.73 per cent at 5 and 10 μg/ml concentration, respectively after 25 days.

Keywords: *Anabaena doliolum, LABS, Detergent, Biodegradation, Water self-purification.*

Introduction

Cyanobacteria are one of the largest and most diverse group of photosynthetic prokaryotes. They are cosmopolitan occurring in fresh and marine water to terrestrial environments. The filamentous heterocystous cyanobacteria are endowed with property of being photosynthetic as well as aerobic nitrogen-fixers. These organisms respond to stress conditions such as waters polluted with linear alkyl benzene sulphonate (LABS) (Cook, 1998). Interest centres on cyanobacterial degradation of environmental pollutants. The conditions that favour cyanobacterial degradation in nature or in some future controlled process will reflect particular needs of the important species. This report summarizes findings with cyanobacteria. Our research has sought, therefore, to identify the species capable of degrading detergent compounds. There is lot of literature available on LABS in terms of their degradation, analytical studies, chemical nature, etc. (Huddleston *et al.*, 1963; Goodnow and Harrison, 1972; Schleheck *et al.*, 2004) but there is scanty work and literature available on the effect and biodegradation of LABS on algal (oxygenic) organisms (Kerterz *et al.*, 1994; Mampel, 1998). Surfactants produce negative and sometimes also stimulatory effects on cyanobacteria, green algae, diatoms, and plant seedlings (Ostroumov and Maertz-Wente, 1991; Goryunova and Ostroumov, 1986; Waterburg and Ostroumov, 1994). We have examined biodegradation potential and growth rate percentage of *Anabaena doliolum* bharadwaja on LABS at 3, 10, 15 μg/ml concentration. So, interest centres on algal degradation of environmental pollutants and detergent compounds of any water body near urban dwelling.

Materials and Methods

In the present study sodium dioctyl sulpho succinate (Manoxol OT) has been selected as the reference material for LABS to observe its effect on growth of test organism *Anabaena doliolum* under culture conditions. Initially growth experiments were set using various concentrations of Manoxol OT and it was found that concentrations higher than 20 μg/ml were inhibitory to the growth of test organism but concentration lower than this supported measurable growth which was also used to calculate biodegradation potential of test organism in terms of its growth rate percentage and total chlorophyll-a. The cultures were grown in Allen and Arnon's nitrate free medium (Allen and Arnon, 1955) (Kaushik, 1987). Prior to experiments, the cultures were made bacteria-free by following standard manuals and axenic cultures of the test organism were maintained and were used as inoculums for further experiments. The average temperature maintained was 28±2°C and white fluorescent light at a distance of 30 cm and intensity of 1800 lux was provided to growing cultures. Culture media and glass wares were autoclaved every time before use at 1.1 kg/cm^2 for 15 minutes at least. The growth period of the cyanobacteria was taken as 18 days. Exponentially growing cultures were homogenized by sonication and used for inoculation in culture tubes (125 × 25 mm). The cyanobacterial sample was centrifuged keeping the supernatant for determination of LABS concentration using the process as indicated and pellet was kept for pigment extraction at 4°C for 24 hours and then chlorophyll absorbance was recorded at 630 nm, 645 nm, and 665 nm by UV-VIS spectrophotometer systronics-117 using following formula (Kaushik, 1987).

$$\text{Total growth (Chl.a)} = C_a \cdot V_a / V_c \text{ mg/l}$$

where,

C_a: Amount of Chl. a

V_a: Volume of 80 per cent acetone

V_c: Volume of culture concentrate

(C_a: $11.6\ OD_{665} - 1.31\ OD_{645} - 0.14\ OD_{630}$)

And the growth rate percentage was calculated using this formula (Kaushik, 1987)

$$\text{Growth Rate Percentage} = dx/dt.1/x.100$$

where,

dx: Change in growth (final – initial growth)

dt: Time (days)

x: Initial growth

For determination of concentration of Manoxol OT, Longwell and Maniece Methylene Blue Method (Manivasakarn, 1986) was used which is a quite sensitive test for quantitative measurement and gives blue colour by reacting with anionic surfactants. Absorbances of the samples were recorded at 652 nm using control as blank. The amounts were determined using standard curves and the results were expressed in mg surfactant µg/ml as Manoxol OT into the culture medium, obtained as supernatant liquid after centrifugation.

Results and Discussion

Initial growth experiments were set to observe the effect of different concentrations of LABS in the range of 0.1 µg/ml to 20 µg/ml using test organism *Anabaena doliolum* for growth period of 18 days. It was observed that there was negative correlation between growth and concentration of LABS used. At 3 µg/ml, growth rate increased at a rapid pace and it exhibited the same trend up to 5 µg/ml of LABS concentration, which is also the point for maximum growth rate and decline thereafter regularly and at highest level of LABS tested, *i.e.*, 20 µg/ml least growth was recorded due to toxicity of LABS as shown in Figure 56.1. At second stage taking 5 µg/ml and 10 µg/ml as sub-lethal concentration of LABS, growth rate percentage and biodegradation potential of test organism was observed at regular

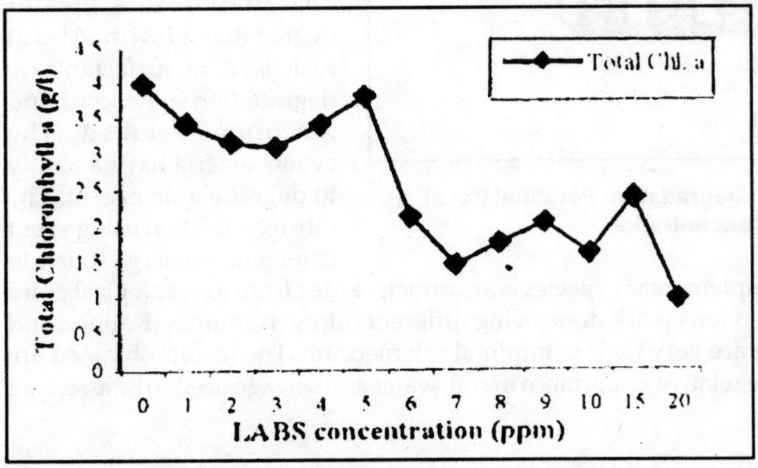

Figure 56.1: LABS Concentration and Algal Growth

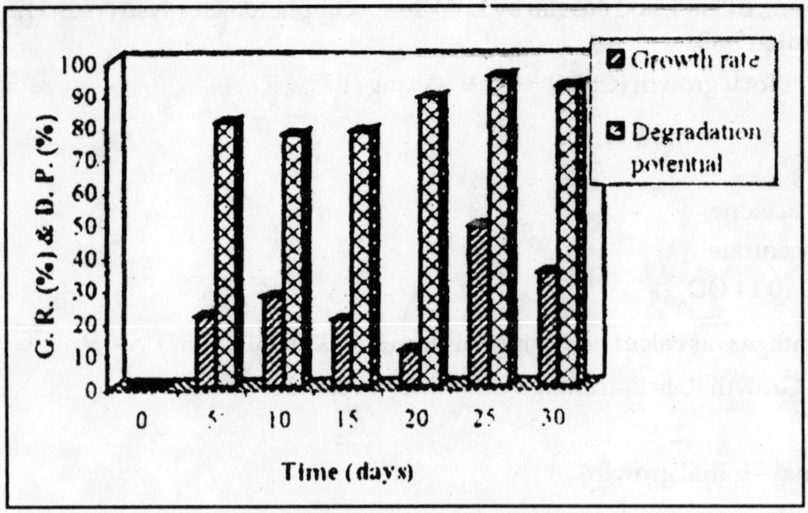

Figure 56.2: Growth Rate (%) and Degradation Potential (%) at 5 ppm LABS Concentration

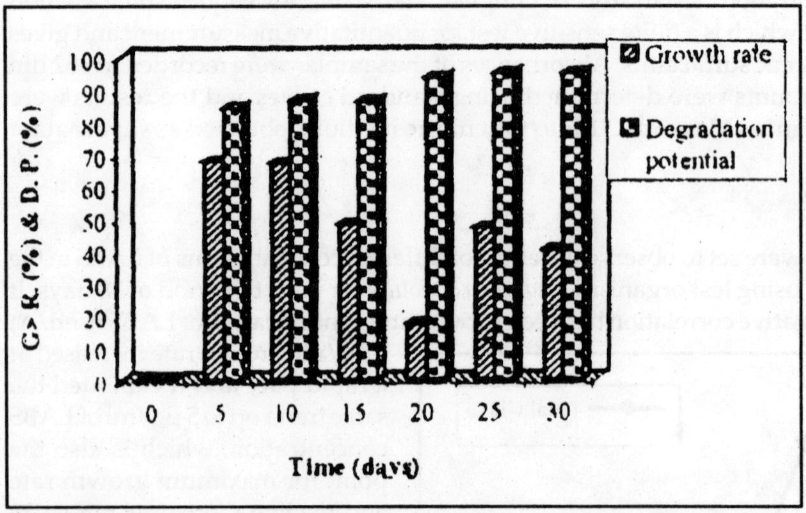

Figure 56.3: Growth Rate (%) and Degradation Potential (%) at 108 ppm LABS Concentration

interval of 5 days starting from 0 to 30 days. It was found that at the concentration of 5 and 10 µg/ml, the growth rate percentage increased significantly up to 10 days and later decreased gradually up to 20 days and then increases again up to 25 days and then again decreases as shown in Figure 56.2 and 56.3.

In terms of LABS degradation, the amount of remaining LABS was least after 25 days in 5 and 10 µg/ml concentration that was indicative of degradation potential of *Anabaena doliolum* with reference to time as shown in Figures 56.2 and 56.3. Differences in ability to degrade surfactants may occur within a particular genus as evidenced by our results. Also high concentration of surfactant may increase degradation to a greater degree than growth. Also at toxic level of surfactants no degradation may occur and inoculum is killed. The cyanobacteria have a ability to degrade a detergent. The nitrogen fixation using spent detergent, as energy source in conjunction with bio-chemically complementary species is an attractive idea from the ecological point of view that will add more to the previous work done using different nitrogen sources (Kapoor, and Arora, 1996). Sometimes detergents are very toxic in minimal salt medium. The results obtained are under laboratory conditions but there relative importance in soil, water and sewage need to be assessed.

Conclusion

In the present investigation it has been concluded that *Anabaena doliolum* can effectively degrade LABS with 95.14 per cent and 94.73 per cent biodegradation at 5 and 10 µg/ml, respectively. Therefore it can be used as an agent for water self-purification processes which would help towards pollution monitoring of natural water bodies around urban and rural areas by the *Anabaena doliolum* growing therein which also help LABS mineralization and top-down control in trophic chain and maintenance of water reservoirs specially which are in use for municipal water supply. Also, the biodegradation potential of *Anabaena doliolum* growing in local wetlands was ascertained. This will further strengthened water self-purification process, control and utilization of algal blooms.

Acknowledgements

The authors are grateful to CSIR for their financial help.

References

Cook, A.M., 1998. Sulfonated surfactants and related compounds; facets of their desulfonation by aerobic and anaerobic bacteria. *Tenside Surfactants Detergents*, 35: 52–56.

Goodnow, R.A. and Harrison, A.P., 1972. Bacterial degradation of detergent compounds. *Applied Microbiology*, 24(4): 555–560.

Goryunova, S.V. and Ostroumov, S.A., 1986. Effect of an anionic surfactant on green algae and seedlings of some angiosperms. *Biological Sciences*, 7: 84–86.

Huddleston, R.L. and Allred, R.C., 1963. Microbial oxidation of sulfonated alkyl benzenes. *Develop. Ind. Micobiol.*, 4: 24–37.

Kapoor, K. and Arora, Leenta, 1996. Ecophysiological and developmental growth behaviour of cyanobacteria in different nitrogen sources. *Ecol. Env. and Cons.*, 2: 133–141.

Kaushik, B.D., 1987. *Laboratory Methods for Blue-green Algae.*

Kerterz, M.A., Kalbener, P., Stockinger, H., Beil, S. and Cook, A.M., 1994. Desulfonation of linear alkyl benzene sulfonate surfactants and related compounds by bacteria. *Appl. Environ. Microbiol.*, 60 : 2296–2303.

Mampel, J., Hitzler, T., Ritter, A. and Cook, A.M., 1998. Desulfonation of biotransformation products from commercial linear alkyl benzene sulfonate (LABS). *Environ. Toxicol. Chem.*, 17: 1960–1963.

Manivasakam, N., 1986. *Physico-chemical Examination of Water, Sewage and Industrial Effluents.* Pragati Prakashan, Meerut, India, pp. 201–203.

Ostroumov, S.A. and M. Maertz-Wente, 1991. Effect of the non-ionic surfactants on marine diatoms. In: *National Meeting of American Chemical Society*, 31: 18–19.

Schleheck, D., Knepper, T.P., Fischer, K. and Cook, A.M., 2004. Mineralization of Individual congeners of linear alkyl benzene sulfonate by defined pairs of heterotrophic bacteria. *Applied and Environmental Microbiology*, 70(7): 4053–4063.

Waterburg, J. and Ostroumov, S.A., 1994. Effect of a non-ionic surfactant on marine cyanobacteria. *Microbiology*, 63: 259–263.

Chapter 57

Effect of Different Concentrations of Sewage on the Haematological Parameters of *C. mirghala*

S. Muthalagi
Department of Zoology, Govindammal Aditanar College, Tiruchendur – 628 215, Tamil Nadu, India

ABSTRACT

The impact of different concentrations of domestic sewage on the haematological parameters of *C. mirghala* was studied. Dose and time dependant changes were observed on all the blood parameters in the test animal. Exposure to sub-lethal concentration of domestic sewage causes significant reductions in RBC count, haemoglobin content and haematocrit values leading to anemia in the fish. Sewage exposure stimulates the immune response of test animal as indicated by the enhanced values of lymphocytes.

***Keywords:** Sewage, C. mirghala, Haematological parameters.*

Introduction

Haematological procedure has been used for a long time by fishery biologists as a screening method to assess the state of health of fish stocks in fish biology research (Larson *et al.*, 1985). They have been applied to investigate the condition and to detect and diagnose disease and parasite attacks in fish (Hesser 1960; Blaxhall, 1972). Haematological parameters can be altered by both environmental and endogeneous factors. Review of the past literature clearly states that, many works have been done on the effect of pesticides and heavy metals on haematological parameters of the fish. The effect of diluted sewage on fish has been studied but to a small degree.

Large quantities of treated and non-treated sewage are usually discarded into river water and they flow into the sea. Raw sewage contains high concentrations of ammonia, nutritive salts, bacteria and other pollutants and they give a harmful effect on the health of fish. Very little information is available about the effects of sewage on physiological status in fish (Bucher and Hofer, 1993; Kakuta *et al.*, 1994), though a lot of studies have been conducted to assess the sub-lethal effects of environmental stressors on fish. So, in the present study, an attempt has been made to find out the effect of different concentration of domestic sewage on the haematology of *C. mirghala*.

Materials and Methods

Domestic sewage was collected from the Ladies hostel of Govindammal Aditanar College for Women at Tiruchendur, Thoothukudi District. The physico-chemical characters of raw sewage were analysed as per APHA (1985). Juveniles of *C. mirghala* (2.5±0.5 g live weight) were procured from Manimuthar dam and acclimated to laboratory conditions for a period of one month in de-chlorinated tap water. During the acclimation period, the fish were fed on pelletted feeds with 38 per cent protein level. After characterization and static bioassay of the domestic sewage, the test fishes were exposed to chosen sub-lethal levels (50 to 100 per cent) of domestic sewage which was taken in the plastic troughs of 30 liters capacity. Ten fishes were introduced into each trough and triplicates were maintained in each concentration. The test fish reared in de-chlorinated tap water served as control.

Haematological studies were carried out at the 30th, 60th and 90th day of exposure. Methods described by Hesser (1960) for routine fish haematology were followed to estimate all blood parameters in this study. Red cell indices were worked out using the formula suggested by Johannson-Sjobeck and Larson (1978).

Results and Discussion

The physico-chemical characteristics of domestic sewage were given in Table 57.1. The different blood parameters of control and sewage treated fishes are shown in Table 57.2. The total red cell counts, Hb content, PCV and MCHC were found to be significantly reduced in the test animal, on exposure to different sub-lethal concentrations of domestic sewage. Dose and time-dependant changes were observed on all the blood parameters. Similar observation were made by Sambasiva Rao *et al.* (1985) in fishes treated with pesticides, by Subramaniam *et al.* (1988) in fishes exposed to tannery effluent and by Izwou Kakuta (1997) in fishes exposed to raw and treated sewage. On the other hand, the mean corpuscular volume (MCV) showed gradual increase in relation to sewage concentration and exposure period.

Exposure of the test animals for 30,60 and 90 days to sub-lethal concentrations of domestic sewage caused significant reduction in total RBC count, haemoglobin content and haematocrit value leading to anemia. Since the red cell indices like MCHC show gradual reduction during exposure period, the anaemic condition can be classified as microcytic and hypochromic anemia (Johannson-Sjobeck and Larson, 1978). The anemic effects in fishes are variously attributed to the inhibition of erythrocyte production (Larson *et al.*, 1976) or haemodilution (Mcleay 1973) or coagulation of blood following destruction of intestinal cells (Johannson-Sjobeck and Larson, 1978) or destruction of liver cells (Waluga, 1966). In the present study, the anemic effects may be due to both reduction in the number of erythrocytes and haemoglobin content of each cell as indicated by a marked decrease in MCHC value. The significant reduction in total RBC count may be attributed to the destruction of red cells by pollutants, because significant increase in MCV is considered to be the indices of red cell destruction leading to anemia (Johannson-Sjobeck and Larson, 1978).

Table 57.1: Physico-chemical and Biological Parameters of Raw Sewage at the Inlet Point (Average of Weekly Samplings for One Year)

Parameters	*Raw Sewage*	*Parameters*	*Raw Sewage*
pH	8.25	Temperature °C	29
Total dissolved solids	865	Total suspended solids	162
Dissolved oxygen	1.02	Dissolved carbon-dioxide	62.75
BOD_5 (as O_2)	285	Ammonia nitrogen	1.89
Nitrite nitrogen	0.0025	Nitrate nitrogen	0.57
Total nitrogen	2.68	Total phosphorus (as P)	4.35
Dissolved phosphorus (as P)	2.67	Zn	3.52
Pb	0.21	Cu	0.64
Cr	0.12	Total bacteria 10^6 ml^{-1}	6.42
Coliforms 10^4 ml^{-1}	7.38	Salmonella × 10^3 l^{-1}	32.00
Faecal streptococci × 10^2 l^{-1}	5.72		

All values in mg l^{-1}, otherwise stated.

Table 57.2: Blood Parameters of *C. mirghala*

Blood Parameters	*Exposure Days*	*Control*	*50 per cent*	*100 per cent*
RBC (Total) × 10^6 mm^{-3}	0	4.58	4.4	4.1
	30	4.49	4.36	4.06
	60	4.38	4.25	3.94
	90	4.55	4.18	3.88
Haemoglobin (%)	0	13.29	12.94	12.11
	30	13.26	12.91	12.08
	60	12.78	12.52	11.65
	90	13.04	12.41	11.54
Haematocrit (%)	0	66.31	64.87	60.87
	30	66.25	64.6	61.35
	60	65.15	63.9	59.6
	90	66.55	63.63	59.49
MCV (ft)	0	14.48	14.74	14.85
	30	14.76	14.82	15.11
	60	14.87	15.04	15.13
	90	14.63	15.22	15.33
MCH (pg)	0	29.02	29.41	29.54
	30	29.53	29.61	29.75
	60	29.18	29.46	29.57
	90	28.66	29.69	29.74
MCHC (g l^{-1})	0	20.04	19.95	19.89
	30	20.02	19.98	19.69
	60	19.62	19.59	19.55
	90	19.59	19.5	19.4

An increase in the number of immature RBC in sewage treated individuals precludes the possibility of inhibition of erythropoietin. The increase in the number of circulating immature erythropoiesis by elevated demands of O_2 and CO_2 transport in sewage exposed fish as a result of physical and metabolic activity.

Thus, the present study clearly reveals that the domestic sewage affects the blood parameters of the test animals at higher concentration.

References

APHA, 1985. *Standard Methods for the Examinations of Water and Wastewater*, 18th edn. Inc. New York.

Blaxhall, P.C., 1972. The haematological assessment of the health of freshwater fish: A review of selected literature. *J. Fish. Biol.*, 4: 593–604.

Bucher, F. and R. Hofer, 1993. The effects of treated domestic sewage on three organs (gills, kidney, liver) of brown trout (*Salmo trutta*). *Water Res.*, 27: 255–261.

Hesser, E.F., 1960. Methods for routine fish haematology prog. *Fish Cult.*, 22: 164–171.

Izuru Kakuta, 1997. *Environmental toxicology and water quality*. John Willey and Sons, ISSN 1053–4725, 12 (1) 43–51.

Johannson-Sjobeck, M.K. and Larson, A., 1978. The effect of cadmium on the haematology and on the activity of amino-levulinic acid dehydralase (ALA-11) in blood and haemopoetic tissues of the flounder *Plueronectes flesus. L. Environ. Res.*, 17: 191–204.

Kakuta, I., K. Ishii and S. Murachiet, 1994. Effects of diluted sewage on biological parameters of carp. *Cyprinus carpio. Comp. Biochem. Physiol.*, 107: 289–294.

Larsson, A., *et al.*, 1976. Some haematological and biochemical effects of cadmium on fish. In: *Effects of Pollutants on Aquatic Organisms*, (Ed.) A.P.M. Lockwood. Soc. Exp. Bio. Sem. Ser. Cambridge University Press, Cambridge, 2: 35–45.

Larsson, A. *et al.*, 1985. Fish physiology of metal pollution results and experiments from laboratory and field studies. *Ecotoxicol. Environ. Saj.*, 9: 250–281.

Mcleay, D.J., 1973. Effects of 12th and 25 days exposure to kraft pulp mill effluent on the blood and tissues of juvenile *Coho salmon. J. Fish. Res. Bd., Can.*, 30: 395–400.

Sambasiva Rao, K.R., K.S. Prasada Rao, I.K. Ahaned Sahib and K.V. Raman Rao, 1985. Combined action of carbaryl and phenthoate on a freshwater fish, *C. puncatus* (Bloch). *Ecotoxicol and Environ. Saj.*, 10: 209–217.

Subramaniam, M.A., M. Sahul Hameed, and G. Varadaraj, 1988. Haematological changes in the fish *Macrones keletius* (Gurther) after exposure to the tannery effluent. *The Indian Zoologist*, 12: 71–74.

Waluga, D., 1966. Phenol effects on the anatomical and histopathological changes in bream (*Abramis brama.* L) Acta. *Hydrobiol.*, 8: 55–78.

Chapter 58

Identification of Antifungal Activity in Different Species of *Cassia*

Chatterjee Padma and Dutta Tuhina
Plant Physiology and Biochemistry Laboratory, Department of Botany, University of Kalyani, Kalyani – 741 235, West Bengal, India

ABSTRACT

The bioactivity of three different species of *Cassia viz., Cassia fistula, Cassia sophera* and *Cassia tora* were experimented in terms of their antifungal action. The crude 50 per cent aqueous ethanolic leaf extract of the three species were tested for antifungal effect by the standard Agar cup method. Antifungal activity was located in the crude extracts of all the tested plant extracts against *Fusarium oxysporum* and *Phytophthora* sp. The antifungal property was also evaluated by studying the effects of the leaf extracts on healthy and fungal infected seeds of *Phaseolus aureus* in different doses (2 mg/ml, 2.4 mg/ml, 2.8 mg/ml and 3.2 mg/ml). Maximum inhibitory effect was exhibited by *Cassia sophera* (3.2 mg/ml dose).

***Keywords**: Antifungal effect, Cassia fistula, Cassia sophera, Cassia tora, Phaseolus aureus, germination percentage.*

Introduction

The plant *Cassia* belongs to the family Leguminosae, sub family Caesalpiniaceae. *Cassia fistula* Linn. is a widely growing tree whereas *Cassia sophera* L. and *Cassia tora* L. are widely growing shrubs. It was reported that *Cassia sophera* and *Cassia fora* completely inhibited conidial germination of *Phyllactinia coryllea* (Maji, Chattopadhay, Kumar and Saratchandra, 2005). Phongrpaichit, Pujenjob, Rukachaisirikul, and Ongsakul (2004) reported that *Cassia fistula* leaf extract had most potent inhibitory effect against *Penicilium merneffi*. Anthraquinones isolated from *Cassia tora* seed also showed an antifungal property against phytopathogenic fungi (Kim, Lee, Kim and Lee, 2004).

Materials and Methods

Matured leaves of *Cassia fistula* were collected from Bolpur, Birbhum whereas *Cassia sophera* and *Cassia tora* were collected from Sonarpur-Harinavi-Kamalgazi region, Kolkata, during the month of May-July. Fungal strains of *Fusarium oxysporum* and *Phytophthora* sp. were taken from the Department of Botany, Kalyani University. Viable seeds of *Phaseolus aureus* were collected from Mohanpur, B.C.K.V., Kanchrapara, 24 Pgs (N) West Bengal.

10 gms of sundried powdered leaves of each species were soaked in 50 per cent aqueous-ethanol at room temperature for 72 hrs. The extract was charcolised and filtered and dark brown residual solid was obtained.

Antifungal Assay

The antifungal assay was performed following the Agar cup method (Pumbit *et al.*, 1995). 1 ml of fungal suspension of *Fusarium oxysporum* and *Phytophthora* sp. were uniformly separately over agar plate. In each case on the agar plate the wells of 0.8 cm diameter was bored with sterile cork borer and on next day 1 ml of respective crude extract was poured into well. The petridishes were incubated at 30ºC and after 48 hrs the inhibition zone surrounding the point of application was noticed in each species.

Studies on Germination Percentage

The residual solid of each species was made powdery and their effect on germination percentage of seeds of *Phaseolus aureus* were observed. Seeds of *Phaseolus aureus* infected with fungal association of *Fusarium* sp., *Mucor* sp. etc. were pre-soaked for 6 hrs in different doses–2 mg/ml, 2.4 mg/ml, 2.8 mg/ml and 3.2 mg/ml respectively of each test compound. Parallely control sets as well as comparison with standard antifungal compound Griseofulvin were also maintained. Readings were taken at the interval of 24 hrs and 48 hrs respectively.

Results and Discussion

From Table 58.1, it is clear that maximum antifungal activity was shown by *Cassia sophera* crude leaf extract.

Table 58.1: Comparative Account of the Effect of the Antifungal Activity Shown by the Crude Leaf Extracts of 3 Different Species of *Cassia*

Sl.No.	*Name of the Plant*	*Effect of Inhibition of Fungal Growth Against*	
		Fusarium oxysporum	*Phytophthora sp.*
1.	*Cassia fistula* (1 ml)	+	+
2.	*Cassia sophera* (1 ml)	++	++
3.	*Cassia tora* (1 ml)	+	+

Conclusion

From Table 58.2 it is clear that maximum antifungal activity was located in *Cassia sophera* crude leaf extract at 3.2 mg/ml dose. The test compound was non toxic to healthy seeds and behaved almost similarly with the synthetic fungicide Griseofulvin.

Table 58.2: Comparative Account of the Antifungal Activity Shown by the Crude Leaf Extract of Different *Cassia* sp. on the Germination Percentage of Seeds of *Phaseolus aureus*

Treatment with Different Doses	*Germination Percentage after Hours Interval*					
	Cassia fistula		*Cassia sophera*		*Cassi fora*	
	24 hrs	*48 hrs*	*24 hrs*	*48 hrs*	*24 hrs*	*48 hrs*
Healthy (H)	100	100	100	100	100	100
H+2 mg/ml	100	100	90	90	90	90
H+2.4 mg/ml	100	100	100	100	90	92
H+2.8 mg/ml	90	90	100	100	100	100
H+3.2 mg/ml	100	100	100	100	100	100
H+Griseofulvin	100	100	80	90	90	100
Infected (I)	0	10	0	10	10	10
I+2 mg/ml	10	10	60	60	10	10
I+2.4 mg/ml	93	93	80	80	66	66
1+2.8 mg/ml	35.7	35.7	90	90	66	66
1+3.2 mg/ml	21	42	90	90	83	83
I+Griseofulvin	90	90	90	100	83	83
F-Value	10.6	101.38	123.74	16.63	0.88	10.62
Significance Level	At 1%	At 0.1%	At 0.1%	At 1%	Non Sig.	At 0.1

References

Dev Maji, M., Chattopadhay, S., Kumar, P. and Saratchandra, B., 2004. *In vitro* screening of some plant extracts against fungal pathogens of Mulberry (*Moms* spp.). *Archives of Phytopathology and Plant Protection*. Taylor and Frances, 38: 157–164.

Phongrpaichit, S., Pujenjob, N., Rukachaisirikul, Y. and Ongsakul, M., 2004. Antifungal activity from leaf extract of *Cassia alata* L., *Cassia fistula* L. and *Cassia tora* L. *Songklanakarin J. Sci. Technol.*, 26(5): 741–748.

Kim, Y.M., Lee, C.H., Kiln, Y.O. and Lee, H.S., 2004. Anthraquinones isolated from *Cassia fora* (Leguminosae) seed show an antifungal property against phytopathogenic fungi. *J. Agric Food Chem.*, 52(20): 6096–6100.

Chapter 59

A Model Approach on Groundwater Quality of Anekal Taluk, Bangalore Urban District

K.L. Prakash, V. Raghu, and R.K. Somashekar
Department of Environmental Science, Bangalore University, Bangalore – 560 056

ABSTRACT

Application of linear regression analysis is very much significant to construct the predictive model using the present groundwater quality data from which, helps to predict the futuristic quality of groundwater of particular area. 1026 groundwater samples of 272 villages were collected and analyzed for 14 parameters for Physico-chemical and biological quality of groundwater. Karl-Pearson Correlation coefficients, r, has been used and attempts have been made to develop the linear regression equations (y = Ax + B) for significantly correlated parameters. The results of the linear model emphasises a positive and significant correlation among the water quality parameters.

Keywords: *Linear regression model, Groundwater quality, Water pollution, Anekal taluk.*

Introduction

Today Groundwater is the most precious and significant for drinking purpose, is one among the most widely distributed water resources of the earth. It is known that 97 per cent of freshwater present on the earth surface is in the form of groundwater. The freshwaters of streams, lakes and rivers hold only 3 per cent of the total available water. Groundwater is considered as a reliable and safe source of water, protected from surface contamination by geological filters (Venkata Mohan and Jayarama Reddy, 1995), hence serving as a major source for drinking, irrigation and industrial uses (Kausik *et al.*, 2002). Hence the freshwater is a finite and limited resource (Bouwer, 2000). Recently many studies have been shown that groundwater is highly susceptible to pollution from natural as well as anthropogenic activities such as industrial effluents, by leakage and spills occurring during transport of chemicals (Srikanth, *et al.*, 1995), municipal sewage, application of chemicals to agricultural field (Rao and Reddy, 1995 and Mandeep Singh *et al.*, 2003), runoff, leaching, land fills, refuse dumps, and septic tanks (Oloniya and Saxena, 1977) besides changes in land use pattern climate and infiltration rates have affected the quality of groundwater. Groundwater quality monitoring is essential for assessing the harmful concentration of toxic materials (Miṣhra Ravi, 1999). Therefore there is a need to a monitor as well provide a contamination free water for the end users. Application of linear regression analysis is very much significant to construct the predictive model using the present groundwater quality data from which, one can predict the futuristic quality of groundwater of particular area. Consequently, linear regression equations greatly facilitate rapid monitoring of parameters. Hence models are necessary for the groundwater pollution studies and management.

Study Area

Anekal Taluk lies geographically between latitude 12°39′ and 13°18′ and longitude 77°22′ and 77°52′ spanning over an area of 535 km^2 (Figure 59.1). The topography is undulated terrain/landscape

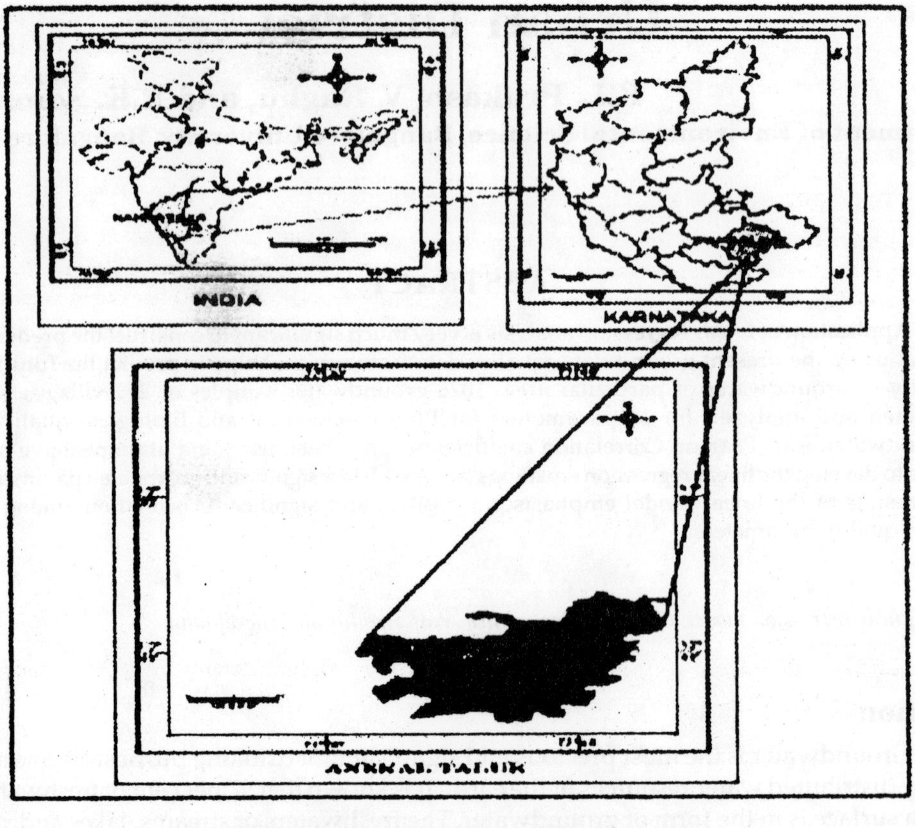

Figure 59.1: Location Map of the Study Area

with intermingling of hills and valleys and bare rocky outcrops facilitating to rapid erosion. The eastern portion of taluk forms a plain area and western portion is covered by jungle and marked by a continuous chain of hills through which several rivulets combine together and drain into river Arkavathi. Geologically, the area consists of gneissic granites and dyke rocks belonging to Precambrian age. Groundwater in the district occurs under water table conditions in the weathered mantle of granitic gneisses and joints, cracks, and crevices of the basement rock. It is estimated that nearly 80 per cent rainfalls is lost through surface runoff and evapo-transpiration and remaining 20 per cent becomes part of groundwater (Kamataka Gazetteer, 2001).

Materials and Methods

Sample Collection

The grab or snap sampling method was adopted to collect the groundwater samples considered as a representative samples. Water sources as such existing hand pumps, Mini water supply and Pipeline water supply were selected, which were being used for drinking as well as other domestic purposes. The samples were collected in polythene containers of 2 lts capacity and sterilized glass bottles of 1 lts capacity for physico-chemical analysis and bacteriological analysis respectively. Samples after collection were transported to the laboratory under freezer condition (4°C).

Analysis of Groundwater Samples

The groundwater quality was assessed, while selecting the parameters, only important drinking water quality of 14 physico-chemical and bacteriological parameters such as pH, Colour, Turbidity, Electrical Conductivity, Total Dissolved Solids, Alkalinity, Chlorides, Total hardness, Calcium hardness, Nitrates, Sulphates, Iron, Fluorides and *E. coli* using standard methods (APHA, 1995 and Manivasakam, 2000).

Linear Regression

Linear regression is an estimation of one independent variable in terms of the other. Co-variation of two independent magnitudes is known as correlation. If two variables x and y are related in such away that increase or decrease in one of them corresponds to increase or decrease in the other, we say that the variables are positively correlated. If increase or decrease in one of them corresponds to decrease or increase in the other, the variables are said to be negatively (inversely) correlated (Garg *et al.*, 2003; Singh and Singh 2001; Draper *et al.*, 1981) have applied linear regression equations in their study.

The statistical analysis was carried out using the SPSS Package.

The correlation coefficient, r, was calculated using the equation:

$$r = \frac{N\Sigma X_1 Y_1 - \Sigma X_1 \Sigma Y_1}{\{[N\Sigma X_1^2 - (\Sigma X_1)^2] . [N\Sigma Y_1^2 - (\Sigma Y_1)^2]\}^{1/2}} \quad (1)$$

where, x and y are any 2 variables, N is the number of observations. Standardization of linear regression equation was made for the groundwater quality data.

where,

$$A = \frac{N\Sigma X_1 Y_1 - \Sigma X_1 \Sigma Y_1}{N\Sigma X_1^2 - (LX_1)^2} \quad (2)$$

$$B = \frac{1}{N(\Sigma Y_1 - A\Sigma X_1)} \quad (3)$$

From the values of A and B, values of Y were calculated using the following equation.

$$Y = Ax + B \quad (4)$$

If x and y are correlated, the best fitting straight line in the least square sense gives reasonably a good relation between x and y. Higher the value of r, clearly shows that a good linear relationship exists between the two variables. If r = 1, then it shows an indicator of perfect relation ship between variables. Percentage of variance is explained by $R^2 \times 100$, more percentage of variance gives perfect relationship between variables.

Results and Discussion

The intercorrelations among 14 variables have been shown in Table 59.1. The intercorrelations between explanatory variables distort the parametric estimates and significance tests. Statistically, substantial conclusions could be drawn when the estimates are distorted.

Table 59.1: Correlation among Fourteen Variables

	Alkalinity	*Calcium*	*Chloride*	*Colour*	*E. coli*	*EC*	*Fluoride*	*Iron*	*Nitrate*	*pH*	*Sulphate*	*TDS*	*TH*	*Turbidity*
Alkalinity	–													
Calcium	0.441	–												
Chloride	0.435	0.927	–											
Colour	0.061	0.139	0.132	–										
E. coli	0.015	–0.028	–0.028	–0.140	–									
EC	0.564	0.908	0.927	0.132	–0.030	–								
Fluoride	0.149	0.011	0.017	0.042	0.031	0.042	–							
Iron	–0.026	0.128	0.132	0.368	–0.051	0.109	–0.011	–						
Nitrate	0.078	0.106	0.086	–0.071	–0.022	0.138	–0.090	–0.032	–					
pH	–0.004	–0.107	–0.108	–0.092	0.029	–0.122	0.051	0.091	–0.075	–				
Sulphate	0.185	0.129	0.097	0.037	0.061	0.127	0.024	–0.049	–0.001	–0.139	–			
TDS	0.680	0.643	0.686	0.076	–0.024	0.811	0.040	0.037	0.149	–0.125	0.154	–		
TH	0.431	0.948	0.923	0.125	–0.009	0.896	0.018	0.123	0.098	–0.087	0.083	0.621	–	
Turbidity	0.042	0.216	0.206	0.618	–0.066	0.181	0.017	0.420	–0.005	–0.064	0.033	0.127	0.189	–

From the point of view of analysis, only those variables that explain a significant portion of variance are considered important. For parameters with r values greater than 0.6, is taken for developing the linear regression models (y = Ax + B) for Anekal Taluk and equations is shown in the Table 59.2.

A correlation observed among the parameters and having a correlation more than +6.0 were considered for the study. The highest positive correlation, r = 0.948 is found between total hardness and calcium. This observation indicates that total hardness has significantly contributed by calcium. Similarly, high relation is also noticed between total dissolved solids, chloride, total hardness and calcium with electrical conductivity, total hardness and calcium, which indicate the presence of calcium chloride, sulphate and iron in these samples. Correlation of Electrical conductivity and Total dissolved solids was observed to be +0.811, which indicated the presence of dissolved ions in water is capable of conducting current, therefore conductivity is a good and rapid measure of Total dissolved solids (Shrivastava *et al.*, 2002). However TDS is also correlated among the parameters such as alkalinity, Total hardness, Chloride and Calcium, the correlation values were observed and values ranged between +0.621 and +0.686. The correlation is found between colour and turbidity (+0.618), which is due to suspended matter, iron, humus materials etc (Clair *et al.*, 2000). Hence these equations are more significant in calculating the variables without experimental determination, by substituting the parameters or variables. So with minimum errors, it is possible to assess the water quality with the minimum experimental variables through which it can reduce the time as well as the cost during water-monitoring program.

Table 59.2: Linear Regression Model Developed for Groundwater Quality Data of Anekal Taluk

X	*Y*	*r*	*Equations*
Colour	Turbidity	0.618	Y = 2.30x – 0.70
EC	TDS	0.811	Y = 0.378x + 257.41
EC	Chloride	0.927	Y = 0.155x – 103.46
EC	Total Hardness	0.896	Y = 0.109x – 1.52
EC	Calcium	0.908	Y = 0.083x – 25.08
TDS	Alkalinity	0.680	Y = 0.1 07x + 164.26
TDS	Chloride	0.686	Y = 0.24x – 39.00
TDS	Total Hardness	0.621	Y = 0.164x + 55.65
TDS	Calcium	0.643	Y = 0.128x + 0.41
Chloride	Total Hardness	0.923	Y = 0.674x + 78.54
Chloride	Calcium	0.927	Y = 0.511x + 37.23
Total Hardness	Calcium	0.948	Y = 0.716x – 12.85

Application of the linear regression model was very much significant in the present study, while assessing the groundwater quality. From the model result it can be concluded that groundwater mainly contain calcium, chloride, iron and sulphates. Among analyzed parameters positive correlation (Table 59.2) was noticed. The standardized linear regression model (Table 59.2) facilitates the future prediction of water quality parameters without experimental observations. This standardized model can be successfully applied for the model verification and evaluation. Hence this model would facilitate the task of rapid monitoring of water quality.

References

APHA, AWWA and WPCF, 1995. *Standard Methods for the Examination of Water and Wastewater*, 19th Ed. APHA, New York.

Bouwer, H., 2000. Integrated water management: Emerging issues and challenges. *Agricultural Water Management*, pp. 45.

Clair, N., Sawyer, Perry L. McCarty, and Gene F. Parkin, 2000. *Chemistry for Environmental Engineering*, 6th Ed. Tata McGraw-Hill Publishing Company Limited, New Delhi, pp. 439–485.

Draper, N.R. and Smith, H., 1981. *Applied Regression Analysis*, 2nd Ed. John Wiley and Sons, Inc., New York.

Garg, V.K., Sharma, B.P. and Rakesh K. Hooda, 2003. Groundwater contamination in an urban area. *Journal of the IPHE, India*, (2): 22–28.

Gazetteer, 2001. *Karnataka Gazetteer*, pp. 30–49.

Kumar, A. Kaushik, Kanchan, Taruna and Sharma, H.R., 2002. Water quality index and suitability assessment of urban groundwater of Hisar and Panipat in Haryana. *J. Environ. Biol.*, 23(3): 325–333.

Manivasakam, N., 2000. *Physico-chemical Examination of Water, Sewage and Industrial Effluents*, 6th Ed. Pragati Prakashan, Meerut.

Mishra, Ravi and Richariya, L.K., 1999. Nitrate concentration in groundwater of Beehar basin, Rewa (M.P) India. *J. Env. Poll.*, 6(4): 261–264.

Oloniya, M.S. and Saxena, K.K. 1977. Groundwater pollution by open refuse dumps in Jaipur. *Indian J. Env. Health*, 19: 177–188.

Rao, P.L.K.M. and Reddy, L.V.R., 1995. Distribution of heterotrophic and coliform bacterial flora in shallow groundwaters in Hyderabad metropolitan city (Andra Pradesh). *Poll. Res.*, 14(2): 165–170.

Sharma, Moti Ram, 2003. Assessment of groundwater quality of Bilaspur area in Himachal Pradesh. *Journal of the IFHE, India*, (2): 60–62.

Shrivastava, V.S. and Patil, P.R., 2002. Tapti river water pollution by industrial wastes: A statistical approach. *Nature Environment and Pollution Technology*, 1(3): 279–283.

Singh, Mandeep, Samanpreet Kaur and Sooch, S.S., 2003. Groundwater pollution: An overview. *Journal of the IFHE, India*, (2): 29–31.

Singh, T.N. and Singh, S.N., 2001. Statistical treatment of data on river Varuna at Varanasi with emphasis on chromium and cadmium. *J. Env. Poll.*, 8(2): 149–153.

Srikanth, R. and Anees Khanam, A. Madhu Mohan Rao and Vasanth Rao, 1995. Nitrate levels in groundwater in 20 villages of Medak district (Andra Pradesh) as affected by industrial pollution. *Poll. Res.*, 14(2): 195–197.

Venkata Mohan, S. and Jayarama Reddy, S., 1995. Assessment of overall water quality of Tirupati. *Poll. Res.*, 14(3): 275–282.

Chapter 60

Ferns of Banswara District (Wagad Region) of Rajasthan, India

B.L. Chaudhary and S. Rana*

Bryology Laboratory, Department of Botany, College of Science, Mohanlal Sukhadia University, Udaipur – 313 001, Rajasthan, India

ABSTRACT

Eight ferns have been collected from Banswara Rajasthan. These are *Actiniopteris radiata* (Swartz) Link, *Adiantum incisum* Forsk., *A. lunulatum* Burm. *Adiantum capillus-veneris* Linn. *Cheilanthes albomarginata* Clarke, *Cheilanthes farinose* (Forsk) Kaulf. *Christella dentata* (Forsk) Holttum, *Pteris vittata* Linn.

Keywords: *Ferns, Banswara, Peepalkhunt, Kagdifall, Mahi river, Rajasthan.*

Introduction

Rajasthan is the largest state in India with an area of 3,42,274 sq.km. and lies between latitude 23°3′N and 30°12′N and longitude 69°31′E and 78°17′E. The climate of the Rajasthan is one of the great extremes with sudden change in temperature, precipitation and wind, which restricts the pteridophyte flora. As such only a limited number of species, which have a wide range of tolerance, occur in the state.

Banswara (Figure 60.1) district is located in the south part of Rajasthan. The district having an area of about 5037 square km is situated between 23°11′ and 23°56′ north latitude and 73°58′ and

* E-mail: s.rana@rediffmail.com

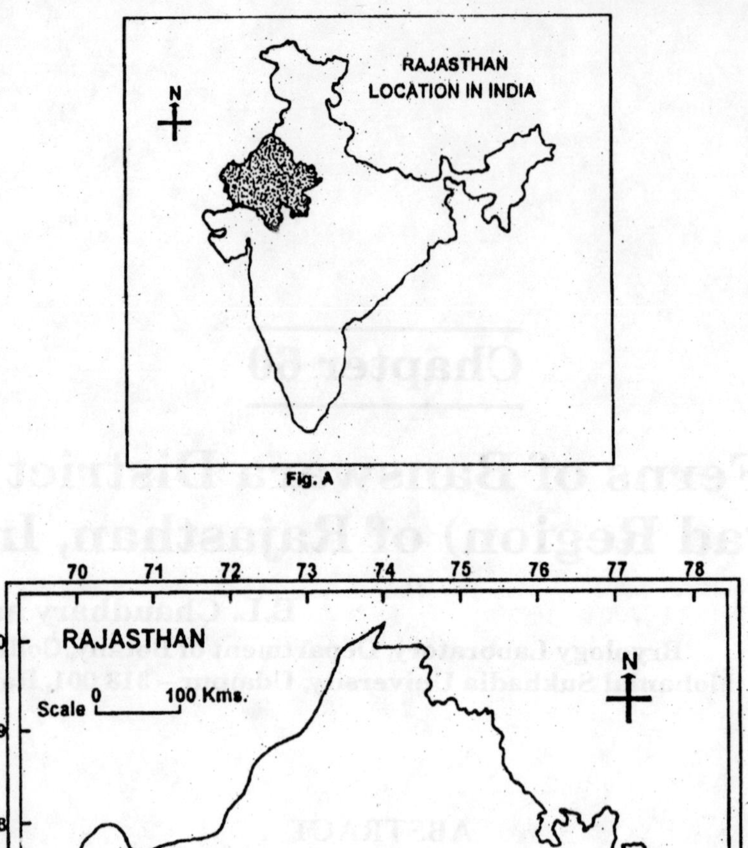

Figure 60.1: (A) Location of Rajasthan; (B) Location of Site of Collection (Banswara)

74°49′ east longitudes. The district has average rainfall 950 mm and on the whole has a salubrious climate. Further it has a highly varied physiography, from plateau lands to hilly tracts. Due to high rainfall and humidity, it is one of the richest spot of Angiospermic vegetation in general and pteridophyte in particular of Rajasthan. It includes variety of trees, shrubs, herbs, climbers and grasses. The subtropical evergreen forest of Banswara consists of mixed tree growth with *Tectona grandis* L. f.nom cons. and *Mangifera indica*. Linn. as a predominant species followed by *Anogeissus, latifolia* (Rxb. ex DC.) Wall. ex Guill and Perr., *Diospyros melanoxylon* L. Roxb., *Madhuca indica* J.F. Gmelin, *Dendrocalamus strictus* (Roxb.) Nees. *Ficus religiosa* L. No reports are available regarding the occurrence of these ferns from this region of Rajasthan, therefore the present study was carried out.

Materials and Methods

The present study was undertaken to identify the pteridophyte flora of the Banswara. The area was visited several times during different seasons of the year, especially rainy season, when pteridophytes flourish well due to high humidity and optimum temperature. Field notes were taken at the time of collection to observe habit, habitat and localities during period under study and brought to the laboratory in the sealed bags and taken photographs of plants and herbarium mounts with their ventral side bearing sori and pressed in standard herbarium sheets and all measurements were brought in record in metric scale. The specimens are incorporated in the herbarium of Bryology Laboratory, Department of Botany, University College of Science for further study.

For the description of genera, standard books and publication by Bir and Verma (1963), Gena (1998), Sharma, Bohra (1977) and Khullar (1994, 2000), advanced pteridophytes (2002) were consulted. For morphological studies, material was dissected and observed under Olympus Binocular Microscope. Camera Lucida drawings were made under the microscope.

Results and Discussion

The following ferns were identified:

Pteridales: Actiniopteridaceae

Actiniopteris radiata (Swartz) Link.

The plant is 8 to 25 cm. long, roots in the rock cervices. The rhizome is oblique to horizontal 1.5 to 2.0 cm length, the young leaves show circinate vernation, the stripe is 3 to 20 cm. long. The laminae are stiff and rough, composed of radiating segments. The sori are produced in linear row on the abaxial surface of the mature laminae, sporangia are leptosporangiate. Spores triangular, trilete, exine reticulated, tuberculated or irregular.

Locality

It was found distributed all over the study area.

Distribution

Pakistan, Punjab, Haryana, Uttar Pradesh and most widely distributed xerophytic fern found almost throughout Rajasthan and India.

Pteridales: Adiantaceae

Adiantum incisum Forsk.

Rhizome short, erect, tufted, dense scaly, dark brown, concolorous or bicolorous with darker center, linear, base cordate, margins entire, apices hair pointed; lamina, linear, lanceolate, 1-pinnate, sub-coriaceous, hairy throughout; rachis hairy elongated and often rooting at the apex; pinnae sub-opposite, sessile, lobed on upper margin, basal pair of pinnae generally reduced and deflexed, both surfaces hairy; veins forked, free; sori 3–8 in each pinna on the upper margins, rounded to elongate, indusiate, indusia thick, oblong or sub-reniform hairy on margins; spores trilete tetrahedral, yellow, non-perisporiate, smooth.

Locality

Near the Kagdi Fall alt. 500 m.

Distribution

Kashmir, Himachal Pradesh, Mount Abu, Punjab, Sikkim, Darjeeling hills, South-central India, Rajasthan.

Adiantum lunulatum Burm.

Rhizome short, erect or suberect bearing a tuft of fronds and dense tufts of fibrous roots, fronds simply pinnate, rachis often elongated and rooting at the apex, pinnae, semilunar shaped distinctly stalked, texture herbaceous, stipes clustered towards rhizome apex, black, glabrous, glossy; rachis with a board terminal pinna or sometime extended and then with an apical vegetative bud, lamina pinnate, lanceolate or oblong-lanceolate, texture herbaceous, glabrous; pinnae 5–12 pairs, alternate, long petiolate, smaller ones on the extended rachis, semi-orbicular, lunate, outer margin rounded, almost entire or with very shallow and remote incisions, in lower pinnae the two margins meet at the petiole to form an almost continuous crescent-shaped curve, the lower basiscopic margin recurved or straight. Sori brown margin entire and spores dark-brown.

Locality

Near Peepalkhunt valley, the plants grow on slopes of rocks and moist soil particles alt. 700 m.

Distribution

Kashmir, Himachal Pradesh, Dehra Dun, and Eastern, southern and central India.

Adiantum capillus-veneris Linn.

Rhizome short, stout, ascending, scaly, scales 3 mm long, linear-lanceolate, margins entire, apices acuminate; stipe chestnut coloured, polished, scaly at base, glabrous above, scales as on rhizome; lamina 10–28 × 4–8 cm, linear-lanceolate, unipinnate, herbaceous, shining green, glabrous, occasionally rooting at the apex; pinnae lunate, opposite or alternate, stalked, base truncate, margins lobato-incised, apices round, basal linna largest; distal pinnae dimidiate, irregular; rachis glabrous; veins flabellately forked, free; sori usually interrupted on the upper edge of pinnae, rarely continuous; indusia brown, crescent shaped; spores, trilete tetrahedral, yellow, non-perisporiate, smooth.

Locality

Widely distributed on moist soil and shady rocks around the study area alt. 500 to 650 m.

Distribution

Throughout the plains of India, Pakistan, Punjab, Haryana, Uttar Pradesh.

Pteridales: Sinopteridaceae

Cheilanthes albomarginata Clarke.

Rhizome short erect and covered by persistent leaf bases and dense tufts of highly branched fibrous Roots, paleae are narrowly lanceolate, with acuminate gland-tipped apices and smooth margins bearing glandular hairs when young. A broad band of cells towards the middle of place are dark brown in colour while the other are hyaline, non-glandular, uniseriate acicular hairs occur mixed with paleae, leaves spreading, tufted, almost horizontally placed and forming tosettes. The stipe is brittle, cylindrical, reddish brown in colour and more or less densely paleae but glossy beneath the paleae. The lamina is compact, ovate-lanceolate to deltoid-lanceolate in outline, with an acute apex. The upper surface of lamina is glossy, the lower bear paleae as well as, acicular hair and is covered more or less densely by a white silvery crust. Rachis is cylindrical and similar to stipe. Sporangia are few per sorus.

Locality

Near Peepalkhunt Ghat Road alt. 450 to 600 m growing on calcareous moist soil.

Distribution

Pakistan, Uttar Pradesh, Punjab, Haryana, Mt. Abu.

Cheilanthes farinosa (Forsk) Kaulf.

Rhizome short, stout, erect, stipe dark-chestnut coloured, slightly scaly at the very base, scales as on the rhizome; lamina delicate, apex acute, board at the base, tripinnatifid, coraceous, upper surface green, lower-surface farinose, farina snow white; pinnae up to 10 pairs, opposite, basiscopic pinnules of the basal pair of pinnae large and deeply lobed, basal pair deltate, upper pinnate ovate-lanceolate, veins forked; sofi indusiate, pseudoindusium continuous, margins undulate; spores brown, globose, trilete, perispore with filiform appendices. This small tufted fern popularly know as the 'Silver Fern' due to the underside of the lamina covered with white indumentum of small waxy papillae.

Locality

Widely distributed on moist, soil and shady rocks around the study area alt. 500 to 650 m.

Distribution

Sikkim, Darjeeling hills, South central India and widely distributed throughout Aravallis and Haroti region.

Thelypteriales: Thelypteridaceae

Christella dentata (Forssk) Holttum.

Rhizome creeping, black, densely covered by adventitious roots and old leaf stipes. Fronds 0.3 to 0.9 m long, stipe and main rachis pubescent, scales present on young rachis and stipes, rachis grooved adaxially throughout, pinnate compound. The lower pinna generally reduced. Each pinna generally cut down about half way to the mid rib into blunt subfalcate lobes, oblong lanceolate texture herbaceous. Margin serrate and pubescent throughout apex acute, pinnate and free except the lowest pair which fuses with the vein of the adjacent pinna, vein and venules pubescent throughout on both the surface but denser on adaxial surface. Sori involucrate, involucres, densely hairy, reniform, located on the lateral venules near the midrib or throughout, 25 to 30 sporangia per sorus. Sporangia having thick single celled stalk which is quite long fibrillar, uniseriate, annulus more than three-fourth. Spores radial, monolate, perineous.

Locality

Near bank of Mahi river alt. 550 to 650 m on moist calcareous soil and rock.

Distribution

Sitamata Wild Life Sanctuary, Maharashtra. Throughout tropics and sub tropics of the world.

Pteridales: Pteridaceae

Pteris vittata L.

Rhizome sub-erect, scales linear, apex sharply acute; thin, light brown in colour. Fronds about 45 to 75 cm long, rachis adaxially grooved throughout, hairs present (scales being replaced by hairs) green when young and dark brown when mature. Simple pinnate with a terminal pinna like the lateral ones. Pinnae numerous, middle ones longest form 5 to 12 cm long, basal ones gradually reduced, upper ones slightly reduced and much shorter than the terminal pinna. Pinnae sessile, oblong, lanceolate, glabrous, base cordate, margin slightly serrtate throughout in sterile pinnae, entire and reflexed (revoluted) in fertile pinnae. Apex acuminate, midrib distinct on both surfaces, hair present on lower of midrib, veins forked, not anastomosing, sori continuous indusium membranous; sporangia, with brown annulus, paraphyses present, spores monolate, with a reticulate exine.

Locality

Near Peepalkhunt forest alt. 500 to 600 m.

Distribution

Western Himalaya, Pakistan, Kashmir, Gujarat, Uttar Pradesh, Sikkim, Darjeeling hills.

References

Bhardwaja, T.N, Yadav, A.K. and Gena, C.B., 1977. New localities for leptosporangiate ferns in Rajasthan. *J. Bomaby Nat. His. Soc.*, 75: 533–535.

Bhardwaja, T.N., Gena, C.B. Takker, T., Kaur, R. and Wadhwani, C., 1981. Newly recorded and rediscovered ferns at Mt. Abu. *J. Bombay Nat. His. Soc.*, 87: 639–640.

Bhardwaja, T.N., Gena, C.B. and Varma, S., 1987. Status survey of pteridophytic flora of Rajasthan with special reference to endangered ferns and fern allies. *Indian Ferns J.*, 4: 47–50.

Bohra, D.R., Singh, R. and Sharma, B.D., 1980. Survey of the pteridophytic flora of Rajasthan. *Geobios*, 7: 334–336.

Bir, S.S. and Verma, S.C., 1963. Ferns of Mt. Abu. Res. Bull. (N.S.) Punjab University, 14: 187–202.

Sharma, B.D. and Bohra, D.R., 1977. Year round pteridophytic flora of Mt. Abu. *Geobios*, 4: 102–103.

Chapter 61

Seasonal Changes in the Abundance and Diversity of Benthic Invertebrates in Two Ponds of Varied Nature

R. Sivakami[1], S. Thangaraj[1], P. Sugi[1], M. Suganya[1] and G. Premkishore[2]

[1]P.G. and Research Department of Microbiology, Kurinji College of Arts and Science, Tiruchirappalli – 620 002, Tamil Nadu

[2]Alan Feldman Public School and Junior College, Kazhakuttum, Trivandrum, Kerala

ABSTRACT

Benthic animals form an important constituent of the nutrient cycle in an ecosystem and also the major food items of the bottom dwelling fishes. An analysis of the benthic fauna can give us idea about the pisciculture productivity of an ecosystem. With this in mind, a study was carried out to examine the seasonal changes in the abundance and diversity of two water bodies with different water sources, but situated close together. Results indicate that though rotifers were the dominant groups in both the ponds, there were differences in the composition of zoobenthos. Nevertheless as rotifers are excellent live feeds for fishes, both the ponds appear to have a high degree of aquaculture potential.

Keywords: *Benthic invertebrates, Water sources, Seasonal variation.*

Introduction

Benthic animals are extremely diverse and are represented by nearly all phyla from protozoans to chordates; however, their quantitative analysis are very difficult and taxonomy confusing (Jana and

Manna, 1995). Nevertheless, they form an important constituent of the nutrient cycle in an ecosytem (Mir, 1995) as also the major food items of the bottom swelling fishes (Sivakami, 1996). The parameters and characteristics of the bottom, its changes and properties exert a major influence on the fish component of the biocoenosis of water system (Jana and Manna, 1995). A knowledge of the benthic fauna and bottom conditions thus helps to evaluate the potentiality of fish production in water body. Literature indicates that studies on benthic organisms of freshwater bodies especially of Tamil Nadu (India) are meagre. This study was carried out to examine the seasonal changes in the abundance and diversity of benthic invertebrates of two ponds of varied water source. One perennial and the other, astatic but located close by from 1 km distance and assessed this pond in aquaculture point of view.

Materials and Methods

After a preliminary survey of various freshwater lentic systems in the outskirts of the city of Tiruchirappalli (lat. 10°44′N, in long. 78°43′E) in central Tamil Nadu (India), two water bodies were chosen. Of these one is a rain fed seasonal pond (Ragavendra pond: 0.4 ha and 2 m depth when full), and the other is a river fed perennial pond (Gundur pond: 0.5 ha and 3 m depth when full) and located about 10 km from Tiruchirappalli junction (Figure 61.1). The distance between the two ponds is about a kilometer and hence the general weather conditions prevailing in the ponds are almost similar. The study was made for two years during 2004 to 2006.

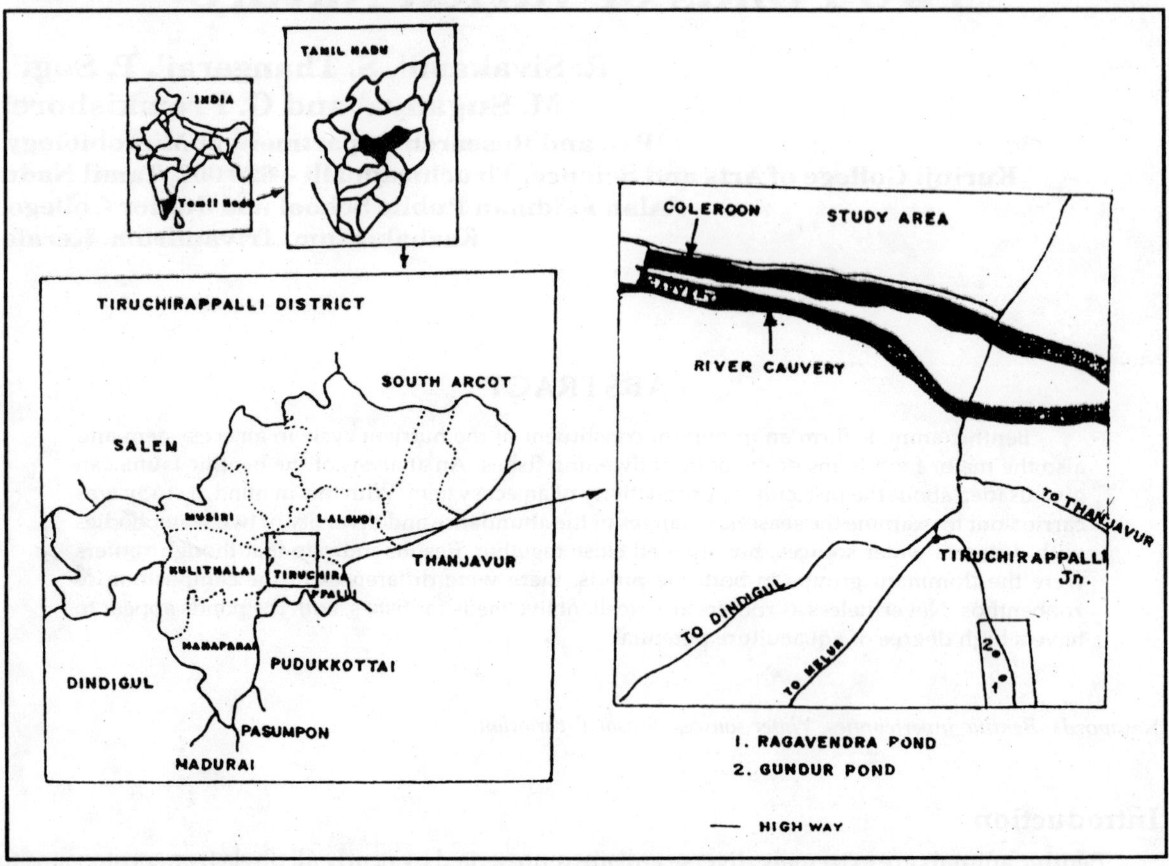

Figure 61.1: Map Showing the General Location of the Target Areas in Tiruchirappalli

Sampling was done for three seasons of an year (rainy/winter season, pre summer season and post summer season). While samples for the rainy/winter season was taken in November, samples for pre summer and post summer were taken in February and June respectively. The experiment was repeated again, the next year at the same date and time to confirm the pattern of occurrence.

Samples were collected from two different sites of each pond separately at 9.00 a.m. during the sampling day for both the year. An Ekman's grab (size 15 cm^2) was employed for collecting samples (sediments) for analysis of benthic organisms. Soon after lifting the grab from the bottom, a polyethylene bag was placed beneath the grab to prevent loss of samples.

The samples were immediately transferred to an enamel basin and then sieved through a standard sieve (BSS No 40; Pore size: 0.425 mm) with washing of water and the contents preserved in 10 per cent formaldehyde for detailed analysis. Both the counts and gravimetric methods were used for the analysis of the bottom biota. The forms were identified up to generic level using Edmondson (1959), Needham and Needham (1974), Pennak (1978) Tonapi (1980) and expressed as numbers $meter^{-2}$.

Results

The characteristics of the soil of both the ponds are presented in Table 61.1. Benthic invertebrates of Ragavendra pond for the three seasons during the two year study recorded a total of 19 species belonging to five groups–Rotifera, Oligochaeta, Ostracoda, Insecta and Mollusca (Table 61.2, Figure 61.2). Gundur pond during the same period recorded 30 species (Table 61.3, Figure 61.3) the organisms recorded belonged to the same five groups as observed in Ragavendra pond.

Table 61.1: Soil Characteristics of the Two Different Ponds

Parameter	*Unit*	*Ragavendra Pond*		*Gundur Pond*	
		I Year	*II Year*	*I Year*	*II Year*
Sand	(per cent)	40	42	48	45
Silt	(per cent)	29	28	18	19
Clay	(per cent)	26	23	30	31
Course sand	(per cent)	5	7	4	5
Water holding capacity	(per cent)	34.5	42.6	48.6	45.3
pH		8.1	8.4	8.3	8.5
Phosphate	(mg/g dw)	0.038	0.041	0.046	0.042
Silicate	(mg/g dw)	0.27	0.31	0.33	0.4
Nitrate-N	(mg/g dw)	0.0415	0.0243	0.051	0.074
Nitrite-N	(mg/g dw)	0.012	0.016	0.007	0.013
Ammonia-N	(mg/g dw)	0.0235	0.0471	0.036	0.043
Sulphate	(mg/g dw)	0.502	0.621	0.592	0.741
Calcium	(mg/g dw)	4.75	4.98	6.85	7.21
Magnesium	(mg/g dw)	0.036	0.042	1.83	1.44
Chloride	(mg/g dw)	0.28	0.32	0.76	0.84
Organic matter	(per cent)	1.03	1.06	1.11	1.24

Table 61.2: Variation Benthic Invertebrates (Average) for the Three Different Seasons of Ragavendra Pond (no/m²) (2004–06)

	Genus	Season Month	Rainy/Winter (November)		Pre-Summer (February)		Post-Summer (June)	
		Year	I	II	I	II	I	II
I	**Rotifera**							
1.	*Brachionus*		130	110	220	80	130	120
2.	*Keratella*		–	–	1600	1410	860	62
3.	*Filinia*		–	130	1910	1540	120	130
4.	*Rotaria*		620	530	2040	1820	740	690
	Class Count		750	770	5770	4850	1850	1560
	Percentage		27.67	33.33	67.32	63.23	58.73	60.93
II	**Oligochaeta**							
1.	*Branchura*		–	10	40	20	10	20
2.	*Nais*		10	10	30	10	30	20
3.	*Tubifex*		40	20	70	40	20	40
	Class Count		50	40	140	70	60	80
	Percentage		1.84	1.73	1.63	0.91	1.90	3.12
III	**Ostracoda**							
1.	*Cypris*		140	130	260	240	160	170
	Class Count		140	130	260	240	160	170
	Percentage		5.16	5.62	3.03	3.12	5.07	6.64
IV	**Insecta**							
	Hemiptera							
1.	*Anisops*		200	200	200	100	–	–
2.	*Belostoma*		–	–	300	100	200	100
3.	*Notonecta*		500	360	2000	1200	300	250
	Ephemeroptera							
4.	*Calnis*		100	200	300	200	–	–
	Coleoptera							
5.	*Hydaticus*		–	–	100	–	200	100
	Diptera							
6.	Chironomus larvae		800	500	700	400	200	160
7.	Dragonfly larvae		–	–	500	300	150	100
8.	*Pentaneura*		–	–	200	90	–	–
	Class Count		1,600	1260	2300	2390	1050	750
	Percentage		55.04	54.54	26.83	31.16	33.3	27.73
V.	**Molluscs**							
1.	*Lymnaea*		80	60	40	40	–	–
2.	*Planorbis*		10	10	–	20	10	20
3.	*Livipara*		18	40	60	60	20	20
	Class Count		170	110	100	120	30	40
	Percentage		6.27	4.76	1.16	1.56	0.95	1.56
	Total Count		2710	2310	8570	7670	3150	2560

Table 61.3: Variation Benthic Invertebrates (Average) for the Three Different Seasons of Gundur Pond (no/m²) (2004–06)

	Genus	*Season Month*	*Rainy/Winter (November)*		*Pre-Summer (February)*		*Post-Summer (June)*	
		Year	*I*	*II*	*I*	*II*	*I*	*II*
I	**Rotifera**							
	Brachionus		310	390	1300	1020	60	40
	Callidina		320	310	200	140	–	–
	Keratella		260	240	–	–	120	60
	Filinia		740	760	1210	1160	1460	600
	Rotaria		760	770	1280	1060	1860	800
	Trichocerca		840	800	120	100	910	600
	Class count		3230	3270	4110	34.80	4410	2100
	Percentage		41.83	43.36	46.02	47.54	52.43	43.20
II	**Oligochaeta**							
	Branchura		10	20	–	–	30	10
	Nais		10	10	20	30	40	30
	Lambriculus		20	30	10	20	30	10
	Tanypus		40	30	–	–	20	20
	Tubifex		160	170	120	100	260	100
	Class Count		240	260	150	150	380	170
	Percentage		3.10	3.44	1.67	2.04	4.51	3.49
III	**Ostracoda**							
	Cypris		940	900	1200	800	400	200
	Class Count		940	900	1200	800	400	200
	Percentage		12.17	11.93	13.43	10.92	4.75	4.11
IV	**Insecta**							
	Hempitera							
1.	*Anisops*		–	–	140	100	170	140
2.	*Belostoma*		160	170	110	90	210	200
3.	*Corixa*		140	150	–	–	120	90
4.	*Notonecta*		170	140	240	200	190	110
	Ephemeroptera							
5.	*Beatis*		110	100	–	–	140	180
6.	*Calnis*		120	100	120	100	–	–
	Coleoptera							
7.	*Cybister*		120	120	120	90	–	–
8.	*Hydaticus*		110	120	130	90	150	160

Contd...

Table 61.3–Contd...

	Genus	Season Month	Rainy/Winter (November)		Pre-Summer (February)		Post-Summer (June)	
		Year	I	II	I	II	I	II
	Diptera							
9.	Chironomus larvae		1060	980	1200	1010	1040	400
10.	Damselfly larvae		–	–	160	100	–	–
11.	Dragonfly larvae		110	60	280	200	140	120
12.	*Pentanura*		120	110	140	100	–	–
13.	*Tanypus*		160	150	–	–	140	120
	Class Count		2380	2200	2640	2080	2300	1520
	Percentage		30.82	29.17	29.56	28.41	27.34	31.27
V.	**Mollusca**							
1.	*Paludomus*		150	150	–	–	220	210
2.	*Cyraulus*		140	130	160	150	230	220
3.	*Melaria*		230	220	240	220	150	130
4.	*Parreysia*		170	180	140	120	320	310
5.	*Vivipara*		240	230	320	310	–	–
	Class Count		930	910	830	810	920	870
	Percentage		12.04	12.06	9.29	11.06	10.93	17.90
	Total Count		7720	7540	8930	7320	8410	4860

During the rainy/winter season, Ragavendra pond recorded 12 species in the first year which increased to 14 species in the second year. Combining both the years on an average, rotifers recorded 30.5 per cent oligochaetes 1.9 per cent, ostracods 5.39 per cent, insects 56.5 per cent and molluscs 5.52 per cent of the benthic fauna for this season. Gundur pond for the same period recorded 28 species for both the years with no change in the species composition. On the whole rotifers contribute 42.56 per cent, oligochaetes 3.27 per cent, ostracods and molluscs 12.05 per cent each and insects 29.99 per cent of the benthic fauna.

In the presummer season, Ragavendra pond recorded 18 species for both the years. However, there was a change in species composition with *Planobis* sp. being documented in the 2nd year in absence of Hydaticus sp. Taking both the years into consideration, rotifers contributed 65.25 per cent, oligochaetes 1.27 per cent, ostracods 3.08 per cent, insects 28.99 per cent and molluscs 1.36 per cent of the benthic fauna. Gundur pond for this season recorded a total of 23 species with the same species occurring for both the years. On an average, rotifers contributed 46.78 per cent, oligochaetes 1.86 per cent, ostracods 12.18 per cent, insects 28.98 per cent and molluscs 10.18 per cent of the benthic fauna.

For the post summer season, Ragavendra pond recorded 15 species for both the years, with both the years showing the same composition. On an average, rotifers represented 59.83 per cent, oligochaetes 2.5 per cent, insects 30.63 per cent, molluscs 1.26 per cent and ostracods 5.86 per cent of the total benthic fauna. Gundur pond during the same period recorded 24 species for both the years with a similar composition rotifers contributed 47.82 per cent, oligochaetes 4 per cent, ostracods 4.43 per cent, insects 29.3 per cent and molluscs 14.42 per cent of the benthic fauna.

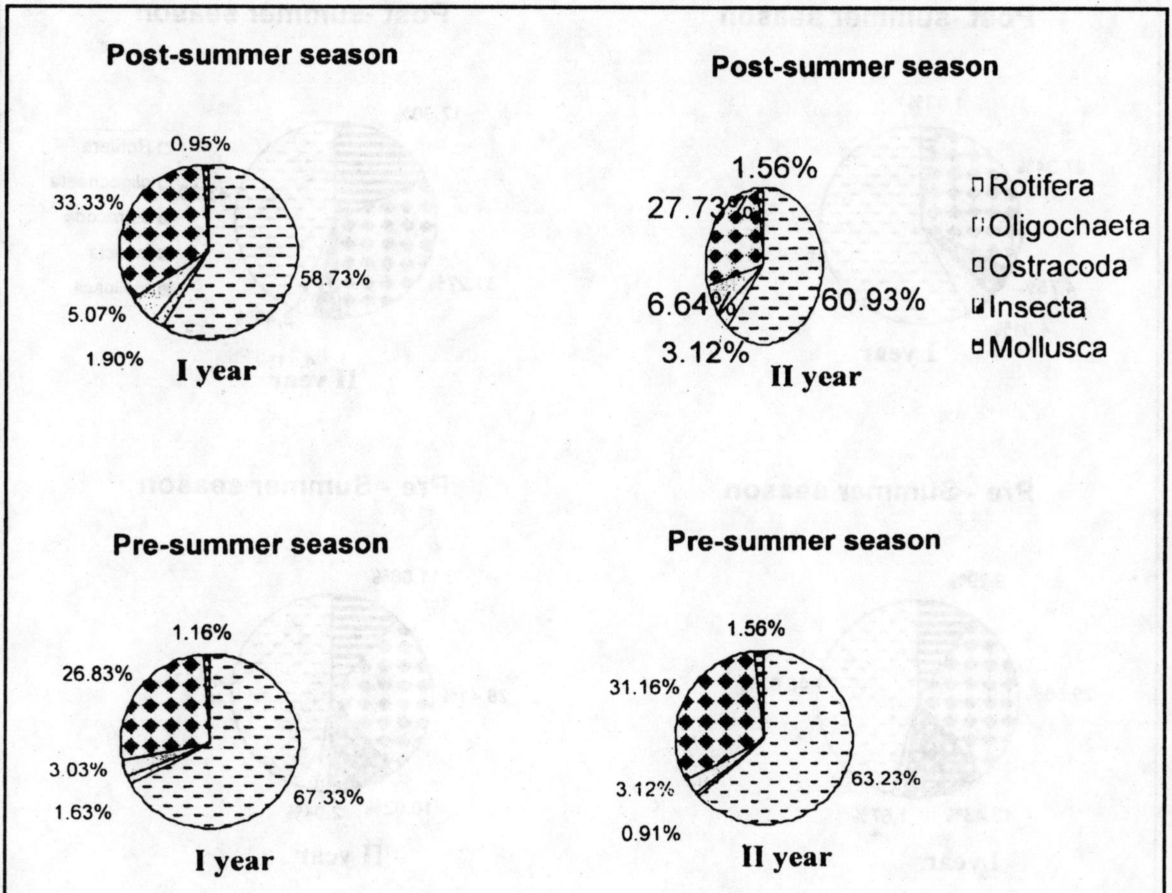

Figure 61.2: Percentage Composition of Various Groups of Benthic Invertebrates during Three Different Seasons in Ragavendra Pond

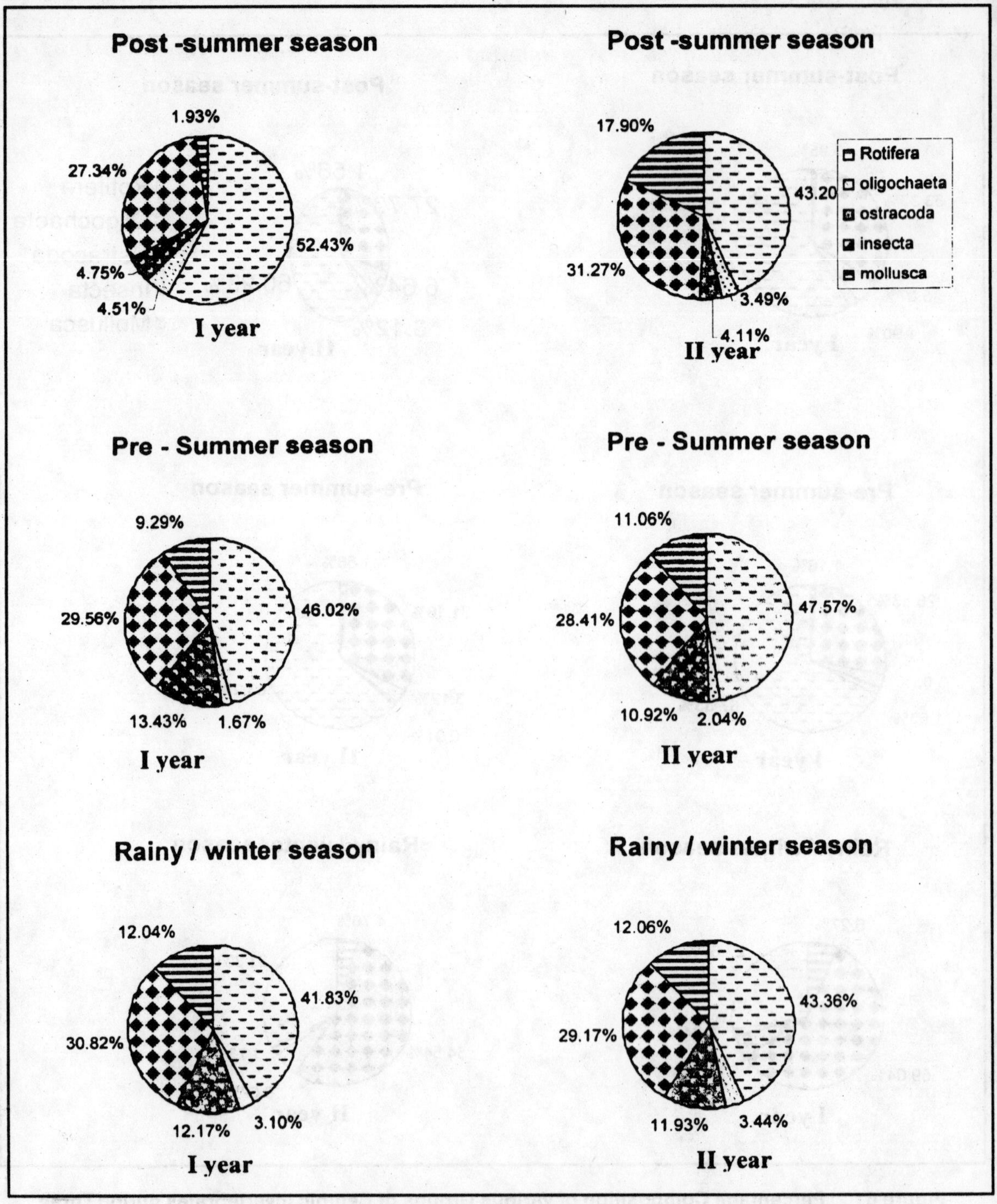

Figure 61.3: Percentage Composition of Various Groups of Benthic Invertebrates during Three Different Seasons in Gundur Pond

Pooling data for the three seasons of both the years reveal that for Ragavendra pond, the dominant group appeared to be rotifers followed by insects, ostracods, molluscs and oligochaetes with diversity being highest in the pre summer season followed by post summer and rainy/winter season. For Gundur pond again rotifers appeared to be the dominant group followed by insects, molluscs, ostracods, and oligochaetes but with rainy season showing the maximum diversity followed by post summer and pre summer.

Discussion

Results indicate that for the both the ponds the highest rotifers population was recorded either in the pre summer or the post summer seasons while the lowest uniformity in winter/rainy season. The higher counts in the pre summer and post summer seasons are attributed to the increase presence of *Keratella rotaria* and *Filinia* species during these seasons when compared to the rainy/winter season. In addition, the intensity of light which will be least during the rainy/winter season will also inhibit the production of young ones (Minkoff *et al.*, 1986) particularly during hatching of resting eggs. This will further decreased the rotifer population during this season Barbhuyan and Khan (1992) while studying the structure and function of benthic ecosystem in a perennial pond also reported the lowest rotifer values in winter season. Further selective predation by cladocerans and copepods (Williamson, 1983) will also result in the reduction of density in the rotifer population in winter.

Comparing both the ponds, Gundur pond recorded higher diversity than Ragavendra pond for all the three seasons for both the years. Rotifer density also was higher in Gundur pond for all the seasons, except for the pre summer season when Ragavendra pond recorded higher densities. This is because of *Filinia, Rotaraia* and *Keratella* sp of Ragavendra pond recording the highest population during this season. In addition, the absence of *Keratella* sp in Gundur pond during this season also decreased the rotifer density.

Oligochaetes did not so much variation in the percentage composition during the different seasons. Cowell and Vodopich (1981) and Barabhuyan and Khan (1992) also reported that the temporal distributions in the abundance of oligochaetes was not much pronounced and there was uniformity in their abundance through out the year as reported in the present study. Nevertheless comparing the three seasons the highest percentage of oligochaetes was noticed in post summer followed by rainy/ winter season and pre summer for the both the ponds. Bhowmick (1968) and Jhingran (1991) while studying the zoobenthos of freshwater ponds also reported that oligochaetes constituted the bulk especially during the summer and the monsoon months. The maximum number and oligochaetes noticed in the summer season might be due to the presence of higher quantities of feed in the form of organic matter during this season (Brinkhursts, 1974). While its decreased presence in pre summer to the consumption as food by bottom dwelling predators. Boss and Lakra (1994) reported the occurrence of large amount of oligochaetes to the presence of clay. In the present study, also among the two ponds Gundur pond (Table 61.1) which had a higher amount of clay reported higher percentage of oligochaetes, thus adding credibility to this statement.

Ostracoda of Ragavendra and Gundur ponds was recorded by a single similar species (*Cypris*) which did not show much variations for the three seasons. Barbhuyan and Khan (1992) also reported that the percentage composition of this group did not show much variation in an eutrophic pond. Comparing the three season, Ostracoda of Ragavendra pond recorded maximum percentage during the post summer followed by rainy/winter and pre summer season, while Gundur pond recorded maximum percentage in rainy/winter season followed by pre and post summer season. The differences in the pattern of occurrence between the two ponds may be attributed to predations in addition to changes in the physio-chemical nature of the soil and nutrients.

The group Insecta recorded their maximal percentage in the rainy/winter season followed by post summer for both the ponds. The maximum number of insects noticed during the winter/rainy season might be due to the lesser decomposing rate and congenial bottom climate as suggested by Gopinathan (1985), Anwar and Siddiqui (1988), and Bais *et al.* (1992). While the minimum percentage noticed during pre summer season to the unfavorable bottom environment which would have compelled the insects either to go deep into the sediments or even perish in addition to predation.

Mollusca of Ragavendra pond recorded highest percentage during rainy/winter season followed by pre summer and post summer while for Gundur pond the maximum percentage was recorded in post summer followed by rainy/winter and pre summer season. The maximum percentage of molluscs recorded during the summer season can be related to optimum temperature, high percentage of organic matter coupled with appreciable amount of calcium and magnesium in the sediment (Satyamurthi, 1960; Vasisht and Bhandal, 1981). The maximum levels noticed in winter in Ragavendra pond can be attributed to trapping of large amounts of hibernating animals and optimal amount of nutrients. One of the main seasons for the insects of Ragavendra pond not recording their highest percentage in summer was that the pond was getting dried up. This would have resulted in a high degree of predation besides even death due to the unfavorable conditions even though the percentage of organic matter available was high.

Result of the present study indicates that though the ponds were situated in close proximity both exhibited certain similarities as also uniqueness in that diversity appeared to be more for Ragavendra pond during the pre summer season while for Gundur pond highest diversity was noticed in the rainy/winter season. The differences in the abundance of benthic organisms in both the ponds can be attributed to the combination of the favorable conditions such as temperature, alkalinity, depth, chemical nature of soil sediment and growth of vegetation. Differences in the nature of the above variables could account for the differences in the species diversity noticed in two ponds. Wetzel (1983) reported that the distribution of zoobenthos in lakes is extremely heterogeneous as a result of variable requirements for feeding, growth and reproduction. These requirements are in turn influenced by the changes in water and sediment characteristics with passage of time. A perusal of the sediment characteristics of both the ponds revealed that there are differences between the two and probably this could be the reason of the differences in the composition of benthic fauna in both the ponds.

Regarding the order of abundance of zoobenthos, both the ponds recorded rotifers as the dominant group followed by insects, ostracods/molluscs and oligochaetes. Barbhuyan and Khan (1992) while studying zoobenthos of sewage ponds recorded Oligochaeta, Nematoda, Ostracoda and Rotifera in the order of abundance while Bose and Lakra (1994) reported the dominance of chironomites in one pond and oligochaetes in another and Bais *et al.* (1992) reported oligochaetes as the main group. Recently Jana and Mana (1995) reported the dominance of Ciliophora in one and Rotifera in another water body. The differences in the abundance of benthic groups noticed in the present study as well as the differences noticed with other water bodies can again be attributed to the differences in the nature of the soil, nutrient availability, physiochemical variable and the degree of predation. Nevertheless the point which emerges from the present study was that both the ponds appear to be suitable for aquaculture in as much as the zoobenthos was dominated by rotifers. It is a well known fact that rotifers are extensively fed by both the young and adult fish. Goldman and Horne (1983) reported that all fish even large predators like pike and Lake trout feed on tiny rotifers. Sarma (1991) reported that of the 2000 rotifer species identified so far *Brachionus* is particularly the most suitable for feeding fish larvae. In this context again both the ponds appears to be suitable for aquaculture. Theinemann (1925) considered a lake bed producing 1000 animals or less per square meter as poorly productive. Taking this into account, both the ponds recorded higher values and hence may ranked as productive. This study emphasis, rainwater pond in the context of conservation also ranked as productive can utilize for aquaculture.

References

Anwar, S. and Siddiqui, M.S., 1998. On the distribution and dynamics of macro invertebrate fauna of the River Kali in Northern India. *J. Environ. Biol.* (Suppl.), 9(3): 333–341.

Bais, V.S., Yatheesh, S. and Agarwal, N.C., 1992. Benthic macro-invertebrate in relation to water and sediment chemistry. *J. Freshwater Biol.*, 4(3): 183–199.

Barbhuyan, S.I. and Khan, A.A., 1992. Studies on the structure and function of benthic ecosystem in an eutrophic body of water: Temporal and spatial distribution of benthos. *J. Freshwater Biol.*, 4(4): 239–247.

Bhowmick, M.L., 1968. Environmental factors effecting fish food in freshwater fisheries, Kalyani, West Bengal. *Ph.D. Thesis*, University of Kalyani, Kalyani, West Bengal.

Bose, S.K. and Lakra, M.P., 1994. Studies on the macro 200 benthos of two freshwater ponds of Ranchi. *J. Freshwater Biol.*, 6(2): 135–142.

Brinkhurst, R.O., 1974. Factors ameoliating interspecific aggregation of tubificid oligochaetes. *J. Fish Res. Bd., Can.*, 31: 460.

Cowell, B.C. and Vodopich, D.S., 1981. Distribution and abundance of benthic macroinvertebrates in a subtropical Florida lake. *Hydrobiologia*, 78: 97–105.

Edmondson, W.T., 1959. *Freshwater Biology*, 2nd edition. John Wiley and Sons, Inc., New York, pp. 1248.

Goldman, C.R and Horne, A.J., 1983. *Limnology*. McGraw Hill International Book Company, New Delhi, 466 pp.

Gopinathan, K., 1985. Ecology and fisheries of certain inland water bodies of Kerala. *Ph.D, Thesis*, Kerala University, Kerala.

Jana, B.B. and Manna, A.K., 1995. Seasonal changes of Benthic invertebrates in two tropical fish ponds. *J. Freshwater Biol.*, 7(2): 129–136.

Jhingran, V.G., 1991. *Fish and Fisheries of India*, 3rd edition. Hindustan Publishing Corporation, New Delhi.

Minkoff, G., Lubzens, E. and Kahan, D., 1986. Environmental factors affecting hatching of rotifers (*Brachionus Pilicatilis*) resting eggs. *Hydrobiol.*, 104: 61–69.

Mir, M.F., 1985. Ecological studies on zoobenthic community of Dal lake, Kashmir. *Ph.D Thesis*, The University of Kashmir, Kashmir, India.

Needham, J.G. and Needham, P.R., 1974. *A Guide to the Study Freshwater Biology*. Holden Bay, San Francisco, U.S.A.

Pennak, R.W., 1978. *Freshwater Invertebrates of the United States*, 2nd edition. Wiley-Interscience, New York, pp. 803.

Satyamurthi, T.S., 1960. Land and freshwater mollusca. *Bull. Madras Govt. Museum*, p. 177.

Sharma, S.S.S., 1991. Rotifers and aquaculture. *Environ. Ecol.*, 9(2): 414–428.

Sivakami, R., Premikishore, G. and Chandran, M.R., 1996. Diel migratory pattern of plankton in Ayilapettai pond at Tiruchirappalli, India. *J. Freshwater Boil.* (In Press).

Thienamann, A., 1925. Die Binnengewasser, Mittelleuro pass Ein Limnologis the Eiuferhrug Binnengewasser, 1: 1–255.

Tonapi, G.T., 1980. *Freshwater Animals of India: An Ecological Approach.* Oxford and IBH, New Delhi, p. 1–317.

Vasisht, H.S. and Bhandal, R.S., 1981. Seasonal variation of benthic fauna in some north Indian lake and ponds. *J. Ecol.*, 69(1–3).

Williamason, C., 183. Invertebrate predation on planktonic rotifers. *Hydrobiol.*, 104: 385–396.

Chapter 62

Cyanobacteria of Two Contrasting Lentic System at Tiruchirappalli, Tamil Nadu, India

R. Sivakami[1], P. Jijendra Kumar[2], J. Vijay Swarna Rani[1], K. Renuka[1] and G. Prem Kishore[2]

[1]P.G. and Research Department of Microbiology, Kurinji College of Arts and Science, Tiruchirappalli – 620 002, Tamil Nadu

[2]Alan Feldman Public School and Junior College, Kazhakuttum, Trivandrum, Kerala

ABSTRACT

Tiruchirappalli located in the central Tamil Nadu has a large number of aquatic bodies with varied sources of water. A study was undertaken over a period of twelve months, during 2005 at Ayilapettai region (outskirt of Tiruchirappalli city) to find the cyanobacterian population in two water bodies with different nature. While perennial pond dependent river water and other seasonal pond dependent rain water. Among the two bodies, the perennial river fed pond recorded 9 genera consisting of 14 species while the seasonal water body recorded 7 genera with 9 species. Result indicates that the perennial body had a higher amount of cyanobacterian when compared to the seasonal body. This is attributed to the difference in the amount of nutrients in both the water bodies.

***Keywords**: Cyanobacteria, Perennial pond, Seasonal pond, Seasonal abundance.*

Introduction

One of the most interesting geomophological features of tropical India is the presence of innumerable temporary to semi-permanent freshwater bodies, 'Kuttais'. These are similar in nature to the 'playa' lakes of southern north America and the 'pans' of South Africa (Rajendran, 1981). Slightly bigger manmade multipurpose water bodies near human settlement occur in this part of the country and are generally known as 'Kulams'. These are comparable to the 'ponds' sensu Welch (1952). The importance of sustained hydrobiological studies on Indian waters is now recognized in water resources management mainly because of the rapid rate of indiscriminate exploitation of freshwater resources. In India, there are about 60,000 inland water bodies, *i.e.* ponds, lakes, etc., with about 3.0×10^6 ha area (Mahajan, 1985). Investigation of biological diversity and primary productivity rates are of great importance to determine the potential for organic production and subsequent exploitation. The need for knowledge of Cyanobacteria diversity is often stressed as a reliable indicator of the integrity of aquatic ecosystem (Barbosa *et al.*, 1995). Recognition of freshwater as a limited resource and an understanding of its dynamics is necessary for its conservation and management. To that end, two representative ponds from Ayilapettai region on Tamil Nadu of peninsular India were subjected to a fortnightly Cyanobacterial study for one year (2005). One pond is perennial, another one is temporary / seasonal.

Materials and Methods

The planktons net was used to collect the cyanobacteria from samples of water at different sites of the pond throughout the year for both ponds at 8.00 am during 2005. The preservation of collected Cyanobacteria was done in 4 per cent of formalein and observation was made in both fresh and in preserved samples. The (+) was used to express the degree of relative abundance of Cyanobacteria according (Kanunga *et al.*, 2005).

Results and Discussion

Results indicate that among the two bodies, the perennial river fed pond recorded 9 genera consisting of 14 species while the seasonal water body recorded 7 genera with 7 species. In the river fed water body, four species *viz.*, *Anabaena circinalis, Microcystis aerigenosa, Oscillatoria acuta, Synechocystis aquatilis* were found to be perennial. Among these *M. aeruginosa* was found to be the most dominant. In this perennial water body, *Chroococcus turgidus* was found to be occurring only for the least period. As far as the seasonal pond was concerned *Chroococcus giganteus, C. turgidus* were recorded the lowest count it appears that a significant amount of Cyanobacteria occurred in the perennial water body which contained a lesser amount organic matter as compared to the seasonal body. As similar result was also reported by Kanungo *et al.* (2005). Literature also reveals that most of the Cyanobacteria reported in these bodies have also been reported by Hosmani and Bharati (1975), Bose and Gorai (1993) and Mendhe (1996).

In the river fed perennial pond *Aphanocapsa* genera had the maximum number of species (4) followed by *Anabaena* and *Gloeocapsa* while in rain fed seasonal body the genera *Chroococcus* and *Synchocystis* was recorded by two species each in general most of the species of Cyanobacterians was found to have occurred in November/December in both the water bodies this is probably due to decreased amount of nutrients due to dilution because of the increased level if water in these two ponds. The maximum Cyanobacterians was reported when the nutrients were low reported by (Pearsall,

1932; Seenayya, 1980; Sinha and Chandrawati, 1996; Smith, 1983; Kanungo *et al.*, 2006). The present study indicates that the perennial pond because of flowing water had decreased nutrients than the seasonal pond and consequently higher Cyanobacteria in perennial pond than the seasonal pond.

Table 62.1: Seasonal Abundance of Cyanobacteria in Rain Fed Pond of Ayilapettai at Tiruchirappalli, Tamil Nadu

Name of the Cyanobacteria	Month											
	Jan	Feb	Mar	Apr	May	Jun	Jul	Aug	Sep	Oct	Nov	Dec
Anabaena circinalis	+				Dry	Dry	Dry	+	+	+	+	+
Chroococcus giganteus	+	+			Dry	Dry	Dry	+	+	+	+	+
C. turgidus	+				Dry	Dry	Dry				+	+
Lyngbya limnetica					Dry	Dry	Dry		+	+	+	
Microcystis aeruginosa	+	+	+	+	Dry	Dry	Dry					
Oscillatoria acuta	+	+	+	+	Dry	Dry	Dry					
Phormidium tenue	+				Dry	Dry	Dry	+	+	+	+	+
Synechocystis aquatilis	+	+	+	+	Dry	Dry	Dry	+	+	+	+	+
Synechocystis pevalekii	+	+			Dry	Dry	Dry			+	+	+

+, ++, +++ are degree of abundance of Cyanobacteria.

Table 62.2: Seasonal Abundance of Cyanobacteria in River Fed Perennial Pond of Ayilapettai Region Tiruchirappalli, Tamil Nadu

Name of the Cyanobacteria	Month												
	Jan	Feb	Mar	Apr	May	Jun	Jul	Aug	Sep	Oct	Nov	Dec	
Anabaena circinalis	+	+	+	+	+	+	+	+	+	+	+	+	
Anabaena utermohlii				+					+	+	+		
Aphanocapsa grevillei	+	+									+	+	
A. koordesi	+								+	+	+	+	
A. montana			+	+	+								
A. pulchra						+	+	+					
Chroococcus turgidus			+	+									
Gloeocapsa atrata	+	+									+	+	
G. punctata				+	+	+				+	+	+	
Lyngbya limnetica				+	+			+	+	+	+		
Microcystis aeruginosa	+	+	+	+	+	+	+	+	+	+	+	+	
Oscillatoria acuta	+	+	+	+	+	+	+	+	+	+	+	+	
Phormidium tenue	+	+									+	+	+
Synechocystis aquatilis	+	+	+	+	+	+	+	+	+	+	+	+	

+, ++, +++ are degree of abundance of Cyanobacteria.

References

Barbosa, F.A.R., Bicudo, C.E.M. and Huszar, V.L.M., 1995. Phytoplankton studies in Brazil: Community structure variation and diversity. In: *Limnology in Brazil*, (Eds.) Tundisi, J.S., Bicudo, C.E.M. and Matsumura-Tundisi, T. Brazilian Limnological Society, Sao Paulo, pp. 18–36.

Bose, S.K. and Gorai, A.C., 1993. Seasonal fluctuation of plankton in relation to physico-chemical parameters of a freshwater tank of Dhanbad, India. *J. Freshwater Biol.*, 5: 133–140.

Hosmani, S.P. and Bharati, S.G., 1980. Limnological studies in ponds and lakes of Dharwar: Comparative phytoplankton ecology of four water bodies. *Phykos*, 19: 27–43.

Kanungo, V.K. and Verma, J.N. and Patel, D.K., 2005. Cyanobacteria of a temple tank at Raipur Chhattisgarh. *Indian J. Environ. and Ecoplan.*, 10(3): 743–745.

Mahajan, K.K., 1985. Water resources, evaluation and optimum utilization in India. In: *Environment and Natural Resources*, (Eds.) Agarwal, V.P. and Rana, S.V.S. Jagmandar Book Agency, New Delhi, pp. 27–42.

Mendhe, K., 1996. Study phytoplanktonic population of Bilwali Talab Indore. *Indian J. Appl. Pure Biol.*, 11: 83–88.

Pearsall, W.H., 1932. Phytoplankton in the English lakes. II. The composition of the phytoplankton in relation to dissolved substances. *J. Ecol.*, 20: 241–262.

Rajendran, M., 1986. Studies on South Indian astatic ponds. In: *Proc. All India Symposium on Freshwater Biology*, Tamil Nadu, India, 164 pp.

Seenayya, G., 1980. Water pollution and algal blooms in some of the ponds of Hyderabad. In: *Progress in Ecology*, (Eds.) V.P. Agarwal and Sharma, V.K., 4: 37–52.

Shannon, C.E. and Wiener, W., 1963. *The Mathematical Theory of Communication*. Univ. Illinois Press, Urbana, 117 pp.

Sinha, R. and Chandrawati, J., 1996. Physio-chemical studies and phytoplankton population in four ponds of Patna. *Biol. J.*, 8: 83–89.

Welch, P.S., 1952. *Limnology*, 2nd edition. McGraw-Hill Book Company, New York, 538 pp.

Chapter 63

Response of Potato to Different Depths and Methods

H.M. Patil[1], B.T. Shete[1] and S.S. Ilhe[2]

[1]Junior Research Assistant, [2] Assistant Professor, Department of Agronomy, Mahatma Phule Krishi Vidyapeeth, Rahuri – 413 722, Dist. Ahmednagar, M.S.

ABSTRACT

The investigation entitled, "Response of Potato to Different Depths and Methods" was undertaken during *Rabi* 2002–03 with potato cv. Kufri Jyoti at Water Management Project, Mahatma Phule Krishi Vidyapeeth, Rahuri. The field experiment was laid out in split plot design with three replications and twenty treatments comprising of four main plot treatments *viz.*, sprinkler, micro-sprinkler, drip, surface and five sub plot treatments *viz.*, depth of 3.53, 3.03, 2.43, 1.98, 1.50 cm. water at 25 mm CPE by sprinkler and micro-sprinkler, wetted area of 1.30, 1.10, 0.90, 0.70, 0.50 for drip, depth of 7.06, 6.05, 4.85, 3.95 and 3.00 cm water at 50 mm CPE by surface and method. The soil was clayey in texture, low in available nitrogen (182.68 kg ha^{-1}), medium in available phosphorus (16.80 kg ha^{-1}) and rich in available potassium (562 kg ha^{-1}) with slightly alkaline in reaction (pH 8.2).

The growth and yield contributing characters attained maximum value under sprinkler method than other methods. The highest tuber yield (23.39 t ha^{-1}) and haulms yield (1.68 t ha^{-1}) was recorded due to sprinkler method. There was an increase of 64.55 per cent in tuber yield over surface method. The big size tubers were maximum under sprinkler methods, while the total number of tubers per plant was higher under micro-sprinkler method. The highest tuber yield (20.85 t ha^{-1}) was obtained from regime which was at par with regime h (20.24 t ha^{-1}). The water requirement of regime I_1 (43.90 cm) was highest, but WOE (474.9 kg ha^{-1} mm) was lowest. The gross and net returns were maximum for sprinkler and micro-sprinkler methods with regime I_1 with B : C ratio of 4.80 and 3.27 respectively, while the drip system was not economical for potato.

Keywords: *Potato, Irrigation depth methods.*

Introduction

Potato provides vitamins, minerals, proteins and valuable food for those, who suffers from stomach acidity and low blood pressure. In the year 1999–2000 production of potato was 24.2 million tones from area of 1.24 million hectares, with the productivity 19.4 tonnes per hectare. (Anonymous, 2001). About 82 per cent area and 88 per cent production are contributed by its winter crop alone (Sangwan, 1991).

Maharashtra accounts 15000 hectares are *i.e.* 1 to 2 per cent of all India acreage with productivity of4.8 t ha-l (Anonymous, 1999). Irrigation water may be applied to the crops either by surface method or by pressurized irrigation methods such as drip, sprinkler and minor sprinkler. In order to overcome the losses which takes place by surface methods and to use available water efficiently, the modern pressurized methods are adopted. Optimum use of water as per crop requirement is the times need. The water production function is the quantitative expression relating yield output to irrigation input. It is a valuable tool to decide the level of water use. The formation of crop response to quantity of water applied will be of practical use in deciding upon the application of irrigation water to maximize the total production per unit input of irrigation water.

As the potato crop is susceptible to the excess and shortage of irrigation water the optimum levels of irrigation water should be applied at particular time with modern methods of irrigation like drip, sprinkler and micro-sprinkler. A systematic attempt on the basis of IW/CPE ratio for potato has not been studied. For recommending the most appropriate irrigation method and the optimum irrigation depth, so that the present investigation was planned by following objectives:

1. To study the effect of different irrigation methods on growth and yield of potato.
2. To study the growth and yield response of potato to varying water application depth in relation to irrigation methods.

Materials and Methods

The investigation entitled, "Response of Potato to Different Depths and Methods" was undertaken during *Rabi* 2002–03 with potato cv. Kufri Jyoti at Water Management Project, Mahatma Phule Krishi Vidyapeeth, Rahuri. The field experiment was laid out in split plot design with three replications and twenty treatments comprising of four main plot treatments *viz.*, sprinkler, micro-sprinkler, drip, surface and five sub plot treatments *viz.*, depth of 3.53, 3.03, 2.43, 1.98, 1.50 cm. Water at 25 mm CPE by sprinkler and micro-sprinkler, wetted area of 1.30, 1.10, 0.90, 0.70, 0.50 for drip, depth of 7.06, 6.05, 4.85, 3.95 and 3.00 cm water at 50 mm CPE by surface and method. The soil was clayey in texture, low in available nitrogen (182.68 kg ha^{-1}), medium in available phosphorus (16.80 kg ha^{-1}) and rich in available potassium (562 kg ha^{-1}) with slightly alkaline in reaction (pH 8.2). Uniformity Coefficient (UC)

$$UC = 100\left[1 - \frac{\sum X}{m + n}\right]$$

UC: Uniformity coefficient per cent.

m: Average depth of all observations, mm

n: Total number of observation points.

For micro-sprinkler UC was 82 per cent and for overhead sprinkler 85 per cent. Water requirement, of the crop

$$WR = IR + ER + S$$

WR: Water requirement, mm
IR: Total irrigation water applied mm
ER: Effective rainfall
S: Groundwater Contribution

Field Water Use Efficiency (kg ha^{-1} mm)

$$WUE = \frac{Y\ (kg\ ha^{-1})}{WR\ (mm)}$$

WUE: Water use efficiency (kg ha^{-1} mm)
y: Yield of crop (kg ha^{-1})
WR: Seasonal water requirement of crop (mm)

Treatment Details

A. Irrigation methods (4)

1. Over head sprinkler
2. Micro-sprinkler
3. Drip
4. Surface

B. Irrigation Regimes (5)

1. Over head sprinkler: 3.53, 3.03, 2.43, 1.98 and 1.50 cm Depth of water at 25 mm CPE.
2. Micro-sprinkler: 3.53, 3.03, 2.43, 1.98 and 1.50 cm Depth of water at 25 mm CPE.
3. Drip method: 1.30, 1.10, 0.90, 0.70, 0.50 wetted area factor.
4. Surface method: 7.06, 6.05, 4.85, 3.95 and 3.00 cm. Depth of water at 50 mm CPE.

Table 63.1: Potato Tuber Yield and Haulms Yield (t ha^{-1}) as Influenced by Different Irrigation Methods and Irrigation Regimes

	Irrigation Methods	*Potato Tuber Yield (t ha^{-1})*	*Haulms Yield (t ha^{-1})*
A.	Main plot treatments		
	Sprinkler	23.39	1.68
	Micro sprinkler	18.70	1.25
	Drip	15.98	1.08
	Surface	14.11	1.16
	SEm ±	0.49	0.02
	C.D. at 5%	1.71	0.06

Contd....

Table 63.1–Contd...

	Irrigation Methods	*Potato Tuber Yield (t ha^{-1})*	*Haulms Yield (t ha^{-1})*
B.	Sub plot treatments		
	I_1	20.85	1.42
	I_2	20.24	1.39
	I_3	18.35	1.40
	I_4	16.76	1.13
	I_5	14.05	1.11
	SEm ±	0.36	0.03
	C.D. at 5%	1.05	0.08
	Interaction	Significant	Significant
	General Mean	18.05	1.29

I_1: Depth of 3.53 cm of 25 mm CPE by sprinkler, micro-sprinkler, depth of 7.06 cm of 50 mm CPE by Surface and 1.30 wetted area for drip.

I_2: Depth of 3.03 cm of 25 mm CPE by sprinkler, micro-sprinkler, depth of 6.05 cm of 50 mm CPE by Surface and 1.10 wetted area for drip.

I_3: Depth of 2.43 cm of 25 mm CPE by sprinkler, micro-sprinkler, depth of 4.85 cm of 50 mm CPE by Surface and 0.90 wetted area for drip.

I_4: Depth of 1.98 cm of 25 mm CPE by sprinkler, micro-sprinkler, depth of 3.94 cm of 50 mm CPE by Surface and 0.70 wetted area for drip.

I_5: Depth of 1.50 cm of 25 mm CPE by sprinkler, micro-sprinkler, depth of 3.00 cm of 50 mm CPE by Surface and 0.50 wetted area for drip.

Table 63.2: Potato Tuber Size Distribution Under Irrigation System

Sl.No.	*Treatment*	*Small Size (< 50 gm)*		*Medium Size (50–100 gm)*		*Large Size (> 100 gm)*		*Greening Potato %*	
		% No.	*% Wt.*	*% No.*	*% Wt.*	*% No.*	*% Wt.*	*% No.*	*% Wt.*
(A)	Irrigation System								
	Sprinkler	23.05	11.63	33.34	22.47	43.80	63.97	4.48	2.64
	Micro-sprinkler	21.81	12.39	38.20	25.20	43.19	62.66	10.43	5.08
	Drip	33.36	20.72	33.79	39.97	30.49	40.22	4.16	2.34
	Surface	34.52	25.31	30.82	32.58	33.10	42.57	9.40	5.57
	SEm ±	3.02	0.73	2.06	2.72	2.67	3.74	1.03	0.98
	C.D. at %	10.46	2.54	NS	9.44	9.25	12.96	3.57	NS
(B)	Irrigation Regimes								
	I_1	25.20	11.91	30.70	26.26	43.60	61.49	10.68	5.74
	I_2	30.55	17.46	32.26	27.84	39.19	59.36	7.98	4.17
	I_3	28.44	20.45	33.92	30.38	39.55	51.34	7.52	3.77
	I_4	23.42	17.26	35.89	31.29	34.49	47.28	6.69	3.18
	I_5	33.33	20.50	37.41	34.51	31.40	42.31	2.72	2.68
	SEm ±	2.65	1.13	2.95	2.84	2.14	3.23	1.35	0.86
	C.D. at %	NS	3.24	NS	NS	6.17	9.31	3.89	NS

Results and Discussion

Tuber Yield (t ha^{-1})

Effect of Irrigation Methods

The sprinkler method recorded the highest yield (23.39 q ha^{-1}) which was significantly superior over other methods, followed by micro-sprinkler and drip method. The lowest yield was (14.11 t ha^{-1}) obtained at surface method. There was an increase of 64, 32 and 13 per cent for sprinkler, micro-sprinkler and drip respectively over surface method of irrigation. Yadava *et al.* (1988), Saggu and Kaushal (1993) obtained similar results, Patel and Patel (2001).

Effect of Irrigation Regimes

The highest tuber yield was obtained at regime I_1 (20.85 t ha^{-1}) which was at par with regime I_2 (20.24 t ha^{-1}). This was significantly superior over rest of regimes, followed by regime I_3, I_4 regime I_5 recorded the lowest tuber yield (14.05 t ha^{-1}), which was 32.61 per cent less than regime I_1 similar results were reported by Patil (1993), Anonymous (2002).

Dry Haulm's Yield (t ha^{-1})

Effect of Irrigation Methods

The dry haulms yield (t ha^{-1}) was significantly influenced due to different irrigation methods. Sprinkler (1.68 t ha^{-1}) recorded significantly superior haulms yield over all other methods, followed by micro-sprinkler and surface method. The lowest haulms yield was recorded (1.08 t ha^{-1}) for drip irrigation. These results are in conformity with those obtained by Anonymous (2001), Anonymous (2002).

Effect of Irrigation Regimes

The highest haulms yield (1.42 t ha^{-1}) was recorded by regime I_1 which was at par with regime I_3 and regime I_2. The lowest haulms yield (1.11 t ha^{-1}) was recorded regime I_5 which was at par with regime I_4.

Similar results were reported Anonymous (2001), Anonymous (2002).

References

Anonymous, 1999. *Economic Survey of Maharashtra* (1998–99).

Anonymous, 2001. *A Report of Research Work Done on Irrigation Water Management* at DMPKV, Rahuri, 2000–01.

Anonymous, 2001. *Economic Survey of India* (2000–01).

Anonymous, 2002. *A Report of Research Work Done in Irrigation Water Management* at MPKV, Rahuri, 2001–02.

Patel, J.C. and Patel, B.K., 2001. Response of potato to nitrogen under drip and furrow methods of irrigation. *J. Indian Potato Assoc.*, 28(2–4): 293–295.

Patel, J.C. and Patel, L.R., 2001. Effect of irrigation and nitrogen on yield attributes in potato. *J. Indian Potato Assoc.*, 28(2–4): 285–287.

Patil, R.B., 1993. Irrigation scheduling and water requirement of potato. *M.Tech. Thesis*, MPKV, Rahuri.

Saggu, S.S. and Kaushal, M.P., 1993. Drip irrigation is better than furrow irrigation for potato crop. *Indian Farming*, 43(9) : 11–12.

Sangwan, S.S., 1991. *Production and Marketing of Potato in India*. Mittal Publication, New Delhi, pp. 258.

Yadava, A.C., Pandit, M.L. and Yadav, S.K. 1988. Water management in solanaceass vegetables: A review. *Agril. Reviews*, Karnal, Haryana. Agril. Univ. Hissar, India 9(3): 111–124.

Chapter 64

Mycelial Growth Response to 2,4-D Exposure

Namita Joshi and Deep Gupta
Department of Environmental Science, Kanya Gurukul Mahavidyalaya, Gurukul Kangri University, Haridwar, Uttaranchal, India

ABSTRACT

The effect of different concentrations of 2,4-D *viz.*, EC_{50}, ½ EC_{50} and ¼ EC_{50} on mycelial growth of four fungi namely *Mucor circinelloides, Penicillium lanosum, Fusarium, roseum* and *Aspergillus flavus* was studied on PDA culture plates. All the three concentrations of 2,4-D drastically inhibited *Mucor circinelloides* while moderate suppression was observed in *Penicillium lanosum* and *Fusarium roseum. Aspergillus flavus* showed increase in colony zone with ascospores during exposure of 5 to 7 days.

Keywords: *2,4-D, Mycelial growth, Mucor circinelloides, Penicillium lanosum, Fusarium roseum and Aspergillus flavus.*

Introduction

Herbicides play an important role in weed control. When such chemicals are applied in agricultural systems, the possibility exists that the chemicals may affect the soil microflora. These may involve influences on soil processes such as, nutrient cycling which intern influencing the fertility of soil system (Goring and Laskowski, 1981). Generally soil is regarded as nutrient deficient for microbial growth, which requires carbon. Thus a unique characteristic of microorganisms is their utilization of pesticide chemicals as a carbon source. Some chemicals have been found to serve as the sole carbon source for microbial cultures, indicating that at least some of the pesticide substrates are true energy source (Matsumura, 1975). Effect of different pesticides on soil microbial population has been studied by several authors (Wain Wright, 1978; Reddy *et al.*, 1989). 2,4-Dichlorophenoxyacetic acid (2,4-D) is a member of phenoxy acid herbicides, which is used because of its effectiveness against certain weed species. Garbito *et al.* (1991) observed that soil microorganisms get affected by the phenoxy acid herbicides. The purpose of our study to find the effects of 3 different concentrations of 2,4-D on mycelial growth of *Mucor circinelloides, Penicillium lanosum, Fusarium roseum* and *Aspergillus flavus* through mycelial growth and isolation of 2,4-D degrading fungi by enrichment culture technique.

Materials and Methods

To determine the effect of 2,4-D on the mycelial growth of *Mucor circinelloides, Penicillium lanosum, Fusarium roseum* and *Aspergillus flavus* PDA was amended with EC_{50} (138), ½ EC_{50} (69) and ¼ EC_{50} (34) mg l^{-1} doses of 2,4-D after autoclaving the, medium. Plates were inoculated centrally with a 5 mm mycelial plug obtained from a two-day old culture of *Mucor circinelloides* and four to five day old cultures of *Penicillium lanosum, Aspergillus flavus* and *Fusarium roseum.* All treatments were replicated three times. Replicate plates were incubated at 28±2°C and mycelial growth was recorded after 48 h for *Mucor circinelloides* and 96 h for *Penicillium lanosum, Aspergillus flavus* and *Fusarium lanosum.*

Fungi were also isolated by enrichment culture technique, involving sequential transfer of an inoculate consisting of different concentrations of 2,4-D and glucose in KM medium (Kirk *et al.*, 1978). The method was based on the premise that elevated concentration of 2,4-D (increasing from 150 to 900 mgl^{-1}) may select 2,4-D degrading fungi, while steady reduction of the glucose concentration (from 10 to 0/g) will select those fungi that can utilize cellulose or 2,4-D. After 4–5 weeks of exposure to the final medium with highest concentration, the fungi in the medium were streaked on PDA and grown at 28±2°C for 7 days.

Results and Discussion

Table 64.1 shows the per cent inhibition or stimulation of four fungal genera *i.e. Mucor circinelloides, Penicillium lanosum, Aspergillus flavus* and *Fusarium roseum* in 2,4-D doses (EC_{50}, ½ EC_{50} and ¼ EC_{50}). Results show that inhibition of *Mucor circinelloides* (60 per cent) in EC_{50} dose was higher as compared to the ¼ EC_{50} (58 per cent) and ¼ EC_{50} (53 per cent). Whereas the growth of *Penicillium lanosum* was moderately inhibited by all the three doses of 2,4-D, it was found 21 per cent (EC_{50}), 14 per cent (½ EC_{50}) and 7 per cent (¼ EC_{50}) respectively. Inhibition of *Penicillium lanosum* was directly proportional to the concentrations used. Same trend was observed in the case of *Fusarium roseum* where 22 per cent inhibition with EC_{50} then 17 per cent (½ EC_{50}) and 11 per cent (¼ EC_{50}). In contrast to the three fungal species *Aspergillus flavus* showed the stimulation in all the treatments. Stimulation, was directly proportional to the concentrations. per cent stimulation was 28 per cent (EC_{50}), 7 per cent (½ EC_{50}) and 7 per cent (¼ EC_{50}).

Colony Morphology

Morphologically the colonies of *Mucor circinelloides* and *Fusarium roseum* were not different from their controls in all the three treatments. Whereas the colonies of *Penicillium lanosum* in EC_{50} dose were light green in color. Similarly, *Aspergillus flavus* also showed the light green color with ascospores of black color in EC_{50} dose. While all the four species showed formation of chlamydospores. Decreases in colony zonation and sporulation rings were observed in *Penicillium lanosum, Mucor circinelloides* and

Fusarium roseum. Only *Aspergillus flavus* showed enhancement in zone of colony following 2,4-D exposure. In this case *Aspergillus flavus* used 2,4-D as a carbon source that was more favourable to the mycelial growth and formed ascospores. The formation of ascospores could have suggested the ability of the fungi to utilized 2,4-D as nutrient. It also influences the general physiology of certain microbes, where resistant spores are formed under less favourable conditions (Tan and Chua 1987). Kaufman and Blake (1973) reported that some fungi *viz*., *Aspergillus* sp., *Penicillium* sp. and *Fusarium* sp. easily degraded herbicides like Propham, Propanil, Solan, and Swep in soil enrichment cultures as well as in pure cultures. During present study degradation capacity of *Aspergillus sp.* is more pronounced than other fungi. Abdel Fattah *et al*. (1983) also found same results with two Triazine herbicides on Egyptian soil fungi. Triazine promoted the total counts of *Aspergillus* sp. fungi as *A. terrus* and *A. flavus* when supplemented in agar medium. While the mycelium growth of *Alternaria alternata, Trichoderma viride, Myrothecium verrucaria, Cunninghamella echinulata, Gliocladium roseum* and *Penicillium verruculosum* was significantly suppressed. During the present study suppression of *Mucor circinelloides* was quite pronounced as compared to other species, which indicates adverse effect of 2,4-D on multiple development stages of *Mucor circinelloides* (Michailides and Sports 1991).

Table 64.1: Per cent Inhibition or Stimulation of Mycelial Growth (mm/day Mean Growth Rate) in 2,4-D Treatments

Herbicide Concentrations	*Mucor circinelloides*	*Penicillium lanosum*	*Aspergillus flavus*	*Fusarium roseum*
EC_{50}	–60	–21	18	–22
½ EC_{50}	–58	–14	7	–17
¼ EC_{50}	–53	–7	7	–11

Negative values are per cent inhibition in mycelial growth.

References

Abdel-Fattah, H.M., Abdel-Kader, M.I.A. and Hamida, S., 1983. Selective effects of two triazine herbicides on Egyptian soil fungi. *Mycopathologia (Historical Archive)*, 82(3): 143–151.

Garabito, G., Alejandra, E. and Trujillo, A.H.S., 1991. *Review of Latinoam Microbiology*, 33(2): 191–201.

Goring, C.A.I. and Laskowski, D.A., 1980. The effect of pesticides on nitrogen transformation in soils. In: *Nitrogen in Agricultural Soils*, (Eds.) Stevenson, F.J., A.S.A. Madison, WI. pp. 689–720.

Kaufman, D.D. and Blake, J., 1973. Microbial degradation of several acetanide, acylanilide, carbamate, toluidine and urea pesticides. *Soil Biol. Biochem.*, 5: 297–308.

Kirk, T.K., Schultz, E., Connors, W.J., Lorenz, L.F. and Zeikus, J.G., 1978. Influence of culture parameters on lignin metabolism by Phanerochete chrysorposium. *Archives of Microbiology*, 21: 104–107.

Matsumura, F., 1975. Environmental alteration of insecticide residues. In: *Toxicology of Insecticides*, Plenum Press, New York.

Michailides, T.J. and Spotts, R.A., 1991. Effect of certain herbicides on the fate of sporangiospores of *Mucor piriformis* and conidia of *Botrytis cinerea* and *Penicillium expansum*. *Pesticide Sci.*, 33: 11–22.

Reddy, D.R., Purushotam, K.R. and Ramamurthy, R., 1989. Influence of fenitrothion on carbohydrate constitutes and some enzyme levels in the earthworm. Insecticide pollution and soil fertility. In:

Soil Pollution and Soil Organisms, (Ed.) Mishra, P.C. Ashish Publishing House, New Delhi, pp. 63–85.

Tan, T.K. and Chua, L.H., 1987. The response of three soil fungi to dalapon and Diuron treatments. *Pesticide Sci.*, 20: 233–240.

Wainwright, M., 1978. A review of the effects of pesticides on soil microbial activity in soil. *Soil Sci.*, 29: 287–298.

Chapter 65

Effects of N, P and K on Productivity and Soil Fertility in Maize (*Zea maize*)–Wheat (*Triticum aestivum*) Cropping System

R.B. Hile[1], H.M. Patil[2], Y.J. Patil[2] and S.S. Bhosale[1]

[1]Assistant Professors, [2]Junior Research Assistants,
Cropping System Research Project, Department of Agronomy,
M.P.K.V., Rahuri – 413 722, M.S.

ABSTRACT

A field experiment was conducted on cultivators' field for two years *i.e.* during *kharif* and *Rabi* seasons of 2004 to 2006 on medium black soils in the plain zone area of Nasik district of Maharashtra in order to study the effect of fertilizer levels on yield and uptake of nutrients in Maize (*Zea mays*) + Wheat (*Triticum aestivum*) cropping sequence. The field experiment was laid out on permanent site in Randomized Block Design with six replications and five treatments as detailed in Table 65.2. The treatment consisted equal levels, each of N (120 kg/ha.) and P_2O_5 (60 kg/ha) and K_2O (40 kg/ha.) along with control half dose of N and full dose of P and K were applied as basal to both the crop at the time of sowing. The remaining half dose of N was applied as a top dressing at 25–30 days after sowing to both the crops. Maize variety cv. 'Super900M' was sown at the onset of monsoon in the second week of July followed by wheat variety cv. 'NIAW301' in the third week of November during each season.

The fertility level increased the yield and up take of nutrients by Maize-Wheat cropping sequence. The grain yields of Maize and wheat with combined use of fertilizers were high during both years. The maximum productivity and net returns, improvement in fertility status and chemical properties of soil could be possible from Maize + Wheat cropping with application of

100 per cent RDF in the respective crops. A significant increase in the grain yield of Maize-Wheat cropping sequence was observed with increase in nitrogen level from 0 to 120 kg/ha., whereas an increase was significant up to 120 kg N/ha. Level in maize + wheat yield. The highest net return was brought about with 120, 60 and 40 N, P_2O_5, and K_2O kg/ha respectively with adequate supply of irrigation water.

The productivity and net monetary returns were recorded highest with the application of recommended dose of nutrients (N, P and K) to Maize-wheat-cropping sequence.

***Keywords**: Soil fertility, Productivity, Levels of NPK.*

Maize-wheat cropping sequence has gained popularity in semi-arid agro climatic condition of plain zone area of Maharashtra with high yielding and fertilizer responsive cultivators of these two staple food crops. There has been growing interest in cropping sequence as a potential tool in improving and sustaining soil health as well as productivity. Cropping sequence is traditionally a low cost input agriculture system. Information on nutrient management on individual crops is available, while cropping system, it is lacking. Moreover, the single nutrient approach has been replaced by multi-nutrient to proved balanced nutrients to boost up crop productivity and nutrient use efficiency. Beside nutrient management in cropping system is more efficient and judicious than individual crop, as following crop take care of the residual effects of nutrients, N, P and to some extent K. Keeping these considerations in view, the present field investigation was undertaken.

Materials and Methods

A field experiment was conducted for three consecutive seasons (2004–05 to 2005–06) on medium black soil in plain zone area of Maharashtra. The soil low in organic carbon (0.59) and nitrogen (223.60 kg/ha) low to medium in available phosphorus (14.56 kg/ha) and high in available potassium (548. 60 kg/ha). The field experiment was laid out on permanent site in Randomized Block Design with six replications and five treatments as detailed in Table 65.2. The treatment consisted equal levels, each of N (120 kg/ha) and P_2O_5 (60 kg/ha) and K_2O (40 kg/ha) along with control, half dose of N and full dose of P and K were applied as basal to both the crop at the time of sowing. The remaining half dose of N was applied at top dressing 25–30 days after sowing to both the crops. Maize variety cv. "Super 900 M" was sown at the onset of monsoon in the second week of July followed by Wheat variety cv. "NIAW-301" in the third week of November during each season. Dibbling was carried out at 75 × 25 cm row spacing of maize and wheat of 22.5 cm by drilling. The recommended dose for maize and wheat was 120.60.40 and 120.60.40 kg N : P : K/ha, respectively. 'Biofertilizers *viz.* Azotobactor and PSB were used for seed treatment wherever applicable as per treatments.

Results and Discussion

Response of Maize to N, P and K

The grain yield of maize increased significantly with increasing levels of N up to 120 kg/ha. In 2004–05 and 2005–06 on an average, application of 120 kg/ha N increased the grain yield by 9.65 q/ha. over absolute control. These findings were confirmed with the results of Singh and Singh (1979), Patel *et al.* (1991) and Varma *et al.* (1999). Application of P was effective however marked response was observed up to 60 kg P_2O_5/ha only (Table 65.1). This might be attributed to more P fixing capacity of soil of experimental plot. These results are in conformity with the findings of maize remained unaffected by potash application in all the 3 seasons. This might be due to sufficient potash reserve and high potash buffering capacity of experimental plot. Similar results were also reported by Patel *et al.* (1991).

Table 65.1: Nutrient Uptake Pooled Data for the Year 2004–06

Sl.No.	*Maize*			*Wheat*		
	N	*P*	*K*	*N*	*P*	*K*
T_1	74.77	11.23	73.26	80.19	11.94	76.02
T_2	112.00	17.08	109.80	109.60	16.62	104.30
T_3	137.70	21.59	132.60	126.90	20.01	121.00
T_4	156.40	25.88	155.4	134.30	22.76	127.70
T_5	183.10	30.74	177.80	165.90	28.97	158.70
SE.m±	2.54	0.86	3.21	3.36	0.96	3.49
CD 5%	6.97	2.16	8.91	9.60	2.42	9.79

Table 65.2: Pooled Mean of Yield, Gross and Net Monetary Returns and B : C Ratio of Maize-Wheat Cropping Systems

Treatment Details	*Kharif Maize Yield (q/ha)*		*Rabi Wheat Grain Yield (q/ha)*		*Gross Monetary Returns (Rs. 000/ha)*	*Net Monetary Returns (Rs. 000/ha)*	*B : C Ratio*
	Grain	*Straw*	*Grain*	*Straw*			
T_1 Control (No fertilizers)	21.18	30.47	21.23	30.65	33.18	7.67	1.29
T_2, N	30.83	44.38	29.53	40.55	47.16	18.00	1.61
T_3, N and P	38.32	55.13	35.26	46.65	56.59	25.22	1.79
T_4, N and K	42.32	61.88	37.56	47.74	61.63	30.73	1.98
T_5, N, P and K	49.03	69.89	44.09	57.70	71.64	39.09	2.19
SE.m±	0.69	1.34	0.80	1.71	0.94	1.81	0.29
C.D.	2.04	3.96	2.36	3.79	2.78	5.08	0.13

Rates: Rs./Qtls. Maize: 500/-, Wheat: 800/-, Straw: 40/-.

Response of Wheat to N, P and K

The yield of wheat increased significantly with increasing levels of N (Table 65.1), and a linear response to N was observed up to 120 kg/ha in all the seasons. On an average N kg/ha. increased yields by 8.30q/ha respectively over absolute control. These findings were confirmed with the results of Patel *et al.* (1991) and Varma *et al.* (1999).

Grain and Fodder Yield

The application of N, P and K, RDF + biofertilizers recorded significantly higher grain yield of maize (49.03) and wheat (44.09 q/ha) followed by the treatment involving application of N&K RDF + biofertilizers to both the crops. (Table 65.1) However, these two treatments were at par with each other for both the crops. Similar results were obtained in case of stalk/fodder yield of maize and wheat, respectively.

Soil Fertility Status

The highest available nitrogen (159.70 kg/ha), phosphorus (33.43 kg/ha) and organic carbon (0.61 per cent) content was observed due to application of 100 per cent RDF + biofertilizers followed by N and K RDF + biofertilizers. The treatment effects were significant for available N, K and organic carbon content, while, non-significant for available P, pH and EC. In general, wherever there was inclusion of organic source either biofertilizer, the improvement in the chemical properties of soil was observed more pronouncely as compared to use of only inorganic source. This trend is in conformity with the findings of Patel (1994). It is interesting to note that even in treatment 'absolute control', the EC as well as available nitrogen and potassium content of soil was improved as compared to initial status. This was because of beneficial effects of FYM and biofertilizers. This is in conformity with the research findings of Varma *et al.* (1999).

Economics

The pooled data on economic studies of maize-wheat cropping system under all recommended package of practices (Table 65.2) revealed that gross monetary returns were significantly higher due to application of 100 per cent RDF + biofertilizers (Rs. 71,640/ha) and followed by treatment T_4. However, Treatment T_5 has highest net returns and B: C ratio occurred due to 100 per cent RDF + to both the crops were (Rs. 39,090 and 2.19) followed by other treatment.

Thus, it could be possible to achieve maximum productivity and monetary returns from maize + wheat cropping system with the application of RDF + biofertilizers to both the crops.

References

Patel, A.M., 1994. Response of nitrogen through manures and fertilizers alone and their combinations on pigeon pea. *M.Sc. Thesis*, MPKV, Rahuri, India.

Sharma, R.A. and Dixit, B.K., 1987. Effect of nutrient application on rainfed soybean. *J. Indian Soc. Soil Sci.*, 35: 456–459.

Sharma, P.B., Jain, P.B. and Saran, R.N., 1990. Effect of land shaping on productivity of different cropping schedule with varying sources of nutrients. *Indian J. Agri. Sci.*, 60(9): 592–605.

Verma, O.P.S., 1996. Integrated nutrient management in pearl millet under rainfed conditions. *Indian J. Agron.*, 41(1): 58–60.

Verma, Upendra and Rajput, O.P., 1999. Effect of N, P and K productivity and soil fertility in maize (*Zea maize*)-wheat (*Triticum aestivum*) cropping system. *Indian J. Agron.*, 44(3): 483–487.

Yadav, R.P., Sharma, R.K. and Shrivaṣtava, V.K., 1997. Fertility management in pigeon pea based intercropping system under rainfed conditions. *Indian J. Agron.*, 42(1): 46–49.

Chapter 66

Diel Variation in Zooplankton and their Environment of Kadinamkulam Lake with Special Reference to Rotifer Species, Southern Kerala

V. Sobha[1], Anuradha Rammohan, S. Santhosh[2] and J. Chandra Thara

[1]Dean, Professor and Head,

Department of Environmental Sciences, University of Kerala

[2]Technology Assistant, Department of Environmental Sciences,

Kariavattom Campus, University of Kerala, Thiruvananthapuram – 695 581, India

ABSTRACT

The present study mainly concentrated on the diel variation of Zooplankton and the community characteristics of rotifer species in the Kadinamkulam lake of Kerala during 11th and 12th of October, 1992 (full moon). Zooplankton composed of Tintinnids, Medusa, Rotifers, Nematodes, Polychaete larvae, Cladocerans, Ostracods, Copepods, Amphipods, Decapods larvae, Insects, Insects larvae, Gastropods larvae, Bivalve larvae, Fish eggs and Fish larvae. Rotifers species diversity, species richness, eveness and dominance indices were worked out. Their population dynamics were mainly influenced by saline water intrusion and husk retting activity.

***Keywords**: Zooplankton, Plankters, Rotifers, Diatoms, Husk retting, Intrusion.*

Introduction

The Kadinamkulam backwater (Latitude 8°35′N; Longitude 76°45′ to 76°52′E) is one of the largest backwater systems of Trivandrum in southern Kerala. It lies almost parallel to the adjoining Arabian Sea for some distance and maintains a temporary connection with the sea at the north western region. The Vamanapuram river originating from the Western Ghats, after traversing for about 80 km and carrying a drainage area of 540 km^2, empties into the backwater through the Anchuthenghu lake in the vicinity of the temporary barmouth. The formation and closure of the sand bar at the mouth leads to pronounced changes in the hydrobiological conditions of the backwater. Not only are they very productive in the tropical level or biological energy flow sense, but are productive also by virtue of their essential role as spawning and nursery grounds for many migratory species of marine fish and as feeding and resting areas for many species of water birds (Milne and Dunnet, 1972). Like all ecosystems, estuaries are dependent on the functions of primary production, primary consumption, predation and decomposition (Correl, 1978). Diel studies in the Kerala coast have so far been restricted mainly in the Cochin backwater area. Shynama and Balakrishnan (1973) reported the diurnal variation of some physico-chemical factors during the south-west monsoon. Pillai and Pillai (1973) and Madhupratap and Rao (1979) investigated the influence of tides on the diel variation of zooplankton population. Diel variability in the zooplankton occurrence, abundance and composition in the water column of Minicoy lagoon and surrounding area outside the coral reef at Kavaratti, Lakshadweep have also been reported by Goswami (1979), Goswami and Goswami (1990) and Madhupratap *et al.* (1991). It is obvious that no attempt has been hitherto made to elucidate the diel variation in the planktons especially rotifers features in the Kadinamkulam backwater system. Hence, the present study was taken up to evaluate the magnitude of the short-term variation of zooplankton population of this dynamic back water system. Taking into perspective the importance of rotifers in aquaculture as live feed and as a bioindicators pioneering attempt was also made to study the diel changes in the species diversity of this interesting group of zooplankton.

Materials and Methods

For the present investigation, samples were collected from only two stations *viz.*, Station–I (Murukumpuzha-retting area) and Station–III (Thalamukku-non-retting area), at intervals of four hours on 11th and 12th October 1992 (Full Moon), covering the entire diel cycle. Zooplankton are collected monthly from three stations using plankton net made up of bolting silk (mesh size 75 μm) and preserved in 15 per cent formalin. Counting of zooplankton was done by using Sedgwick-Rafter cell and density is represented in percentage. Zooplanktons were identified as per the method followed by Neyar (1968), Nayar and Nair (1969), Battish (1992) and Ward and Whipple (1992).

Results

The tides were of a mixed, semi-diurnal type with two successive high and low waters. During the time of sampling the high water was at 11.0 hrs and 0100 hr when the sampling was completed low tide was in progress. At station–I, 1.51 m was the least depth measured at 2200 hrs and the highest of 2.23 m was observed at 0600 hrs, during the beginning of the sampling. At station–III also a similar trend was observed. The zooplankton biomass in the area varied from 1.04 to 4.16 ml $(100^3)^{-1}$ at station I and from 4.16 to 10.04 ml $(100^3)^{-1}$ at station–III (Figure 66.1). Average values for day and night samples were 2.24 and 2.5 ml $(100^3)^{-1}$ respectively at station–I, and 8.72 ml $(100^3)^{-1}$ respectively at station–III. Zooplankton biomass and population were quite high for the samples collected during early morning and decreased gradually with lowest values at 1400 hrs at station land 1430 hrs at station–III. The values increased again at dusk, attaining a peak at 0200 hrs at station–I and 2230 hrs at station–III. Total zooplankton population also showed similar fluctuations as that of biomass. The minimum population [12848 No. $(100^3)^{-1}$] was attained at 1400 hrs and maximum [111074 No. $(100^3)^{-1}$] at 0200 hrs at station–I, whereas at station–III the minimum population [58692 No. $(100^3)^{-1}$] was recorded at 1430 hrs and maximum population [133845 No. $(100^3)^{-1}$] at 2230 hrs. The highest population density was represented by 10 faunal groups at station–I (0200 hrs to 0600 hrs) and 14 at station–III (2230 hrs). Diel variation in the zooplankton composition, the quantitative abundance and the percentage composition are shown the Table 66.1.

The occurrence of all the rotifer species encountered throughout the diel cycle in the two stations, their relative abundance represented in No. $(100^3)^{-1}$ and their percentage composition are illustrated in the Table 66.2. Totally ten families of rotifers belonging to three orders, and 29 species belonging to 14 genera and two sub-genera were encountered during the present study. Brachionidae constituted the largest family during the present study. Ten species belonging to four genera were observed in this family and except for *Brachionus patulus* and *Platyias quadricornis*, which were completely absent at station I, the other species were all well distributed. Euchlanidae included two species, *viz., Euchlanis dilatata* and *Dipleuchlanis propatula*. The former was encountered only at 0600 hrs at station–I and was completely absent at station–III, while the latter was absent at station land quite well distributed during the diel cycle at station–III. Lecanidae constituted the second largest family during the present investigation, having nine species belonging to one genera and two subgenera. Lecane (Monostyla) bulla and L (Lecane) papuana were the two most common species found in this family. The percentage composition of L. (M.) bulla fluctuated between 0 to 1.78 per cent at station–I and from 0.19 per cent to 1.76 per cent at station–III. Family Trichocercidae was represented only by *Trichocerca* sp. At station I it was completely absent throughout the period of study, while at station–III a large number of them were observed only at 2300 hrs and were absent during the rest of the cycle. *Asplanchna brightwelli* was the only species that represented the family Asplachnidae. At both the stations they were observed only in the last sample, at the end of the cycle the following day (0600 hrs at station land 0630 hrs at station–III). Family Filiniidae represented by two species and both were commonly found and well distributed. Total species abundance showed that the minimum was at 1400 hrs and the maximum at 0200 hrs at station–I, while at station–III, the minimum was at 1830 hrs and the maximum at 2230 hrs. In general, total numbers of species were at station–III than at station–I.

The rotifer community during the present study was composed of 23 species at station–I and 27 species at station–III. The variation in the indices such as diversity (H), richness (d), evenness (e) and dominance(c) of the rotifers species at the two stations during the diel cycle is shown in the Table 66.3. The diversity index (H) varied from 0.7649 (0200 hrs) to 1.4088 (1400 hrs) at station I and an opposite trend was observed at station–III, wherein the diversity fluctuated between 0.8724 (1430 hrs) and 1.5796 (0630 hrs). At station–I the maximum was during the afternoon, while at station–III it was the minimum during the same time. At station–I the number of species ranged from 10 at 1400 hrs to 15 at 0200 hrs to 0600 hrs and at station–III, it ranged from 13 at 1430 hrs to 1830 hrs to a maximum of 20 species at 0600 hrs. When the two stations were compared, station–III had more number of species than station–I. The index of species richness (d) ranged between 0.9996 and 1.3484 at station–I and between 1.2013 and 1.7996 at station–III. In both the cases the minimum was during the afternoon to evening hours (1400 hrs at station–I and 1830 hrs at station–III) and the maximum was at dawn. Station III was more species rich than station– I. Eveneness or equitability index (e) fluctuated between 0.2825 and 0.6118 at station–I and between 0.3401 and 0.5273 at station–III. Here the maximum was observed at 1400 hrs at station–I, while it was the minimum that was observed at the same time at

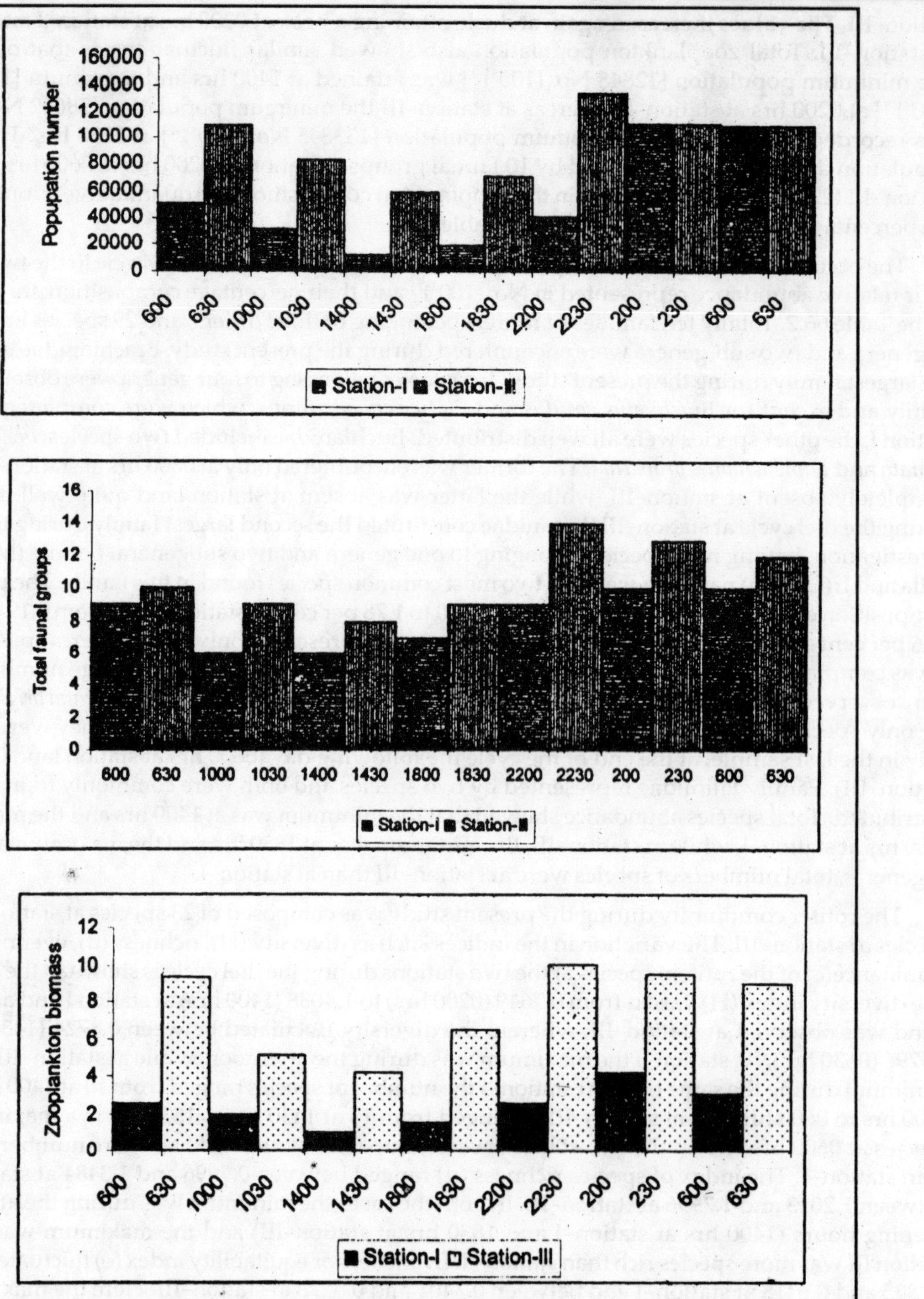

Figure 66.1: Diel Variation in Total Zooplankton Population [No. $(100m^3)^{-1}$], Total Faunal Groups (No) and Zooplankton Biomass [ml $(100m^3)^{-1}$] at Station–I and Station–III

Table 66.1: Diel Variation in the Quantitative Abundance (No. 100m^{-3}) of the Various Zooplankton Groups at Stations–I and III (The percentage composition given in parenthesis)

Zooplankton	*Station–I*							*Station–III*						
	Time of Sampling (hrs)							*Time of Sampling (hrs)*						
	0600	*1000*	*1400*	*1800*	*2200*	*0200*	*0600*	*0630*	*1030*	*1430*	*1830*	*2230*	*0230*	*0630*
Tintinnids	–	–	104 (0.81)	42 (0.21)	187 (0.46)	73 (0.07)	62 (0.06)	–	125 (0.15)	62 (0.11)	250 (0.40)	458 (0.34)	52 (0.05)	208 (0.19)
Rotifers	31075 (62.10)	16370 (51.98)	8133 (63.32)	14258 (72.28)	12542 (30.84)	33124 (29.82)	32291 (29.86)	22214 (19.55)	35183 (41.96)	28299 (48.22)	21781 (34.96)	38938 (29.09)	29016 (25.15)	38470 (34.78)
Nematodes	–	42 (0.13)	–	–	21 (0.05)	–	–	62 (0.05)	73 (0.09)	–	94 (0.15)	62 (0.05)	–	62 (0.06)
Polychaete Larvae	42 (0.08)	–	–	21 (0.11)	–	10 (0.01)	42 (0.04)	62 (0.05)	–	10 (0.02)	10 (0.02)	31 (0.02)	62 (0.05)	42 (0.04)
Cladocerans	114 (0.23)	–	–	–	–	21 (0.02)	135 (0.12)	31 (0.03)	31 (0.04)	10 (0.02)	10 (0.02)	31 (0.02)	62 (0.02)	42 (0.06)
Ostracods	21 (0.04)	–	–	–	–	–	10 (0.01)	–	–	–	–	10 (0.01)	21 (0.02)	10 (0.01)
Copepods/ Copepodites	7894 (15.77)	1560 (4.95)	333 (2.59)	298 (1.51)	12480 (30.68)	32305 (29.08)	25334 (23.43)	11856 (10.43)	2371 (2.83)	1997 (3.40)	2622 (4.21)	30701 (22.94)	25376 (21.99)	31824 (28.77)
Amphipods	–	–	–	–	–	–	–	–	–	–	–	21 (0.02)	10 (0.01)	–
Decapods Larvae	5418 (10.83)	4586 (14.56)	2330 (18.14)	1924 (9.75)	9911 (24.37)	33509 (30.17)	32323 (29.89)	19250 (16.94)	11149 (13.30)	13089 (22.30)	24173 (38.80)	36402 (27.20)	33488 (29.02)	22183 (20.05)
Insects	–	–	–	–	–	–	–	10 (0.01)	10 (0.01)	–	–	–	–	–
Insects Larvae	83 (0.17)	– –	– –	– –	21 (0.05)	31 (0.03)	52 (0.05)	– –	– –	– –	– –	31 (0.02)	52 (0.05)	114 (0.10)
Gastropod Larvae	374 (0.75)	260 (0.83)	114 (0.89)	302 (1.53)	1778 (4.37)	2340 (2.11)	9360 (8.66)	1290 (1.14)	250 (0.30)	166 (0.28)	812 (1.30)	4025 (3.01)	2808 (2.43)	1685 (1.52)
Bivalve Larvae	5023 (10.04)	8674 (27.54)	1830 (14.25)	2882 (14.61)	3723 (9.15)	9651 (8.69)	8528 (7.89)	58843 (51.79)	34653 (41.33)	15059 (25.66)	12544 (20.14)	23104 (17.26)	24440 (21.18)	15912 (14.38)
Fish eggs	–	–	–	–	–	–	–	–	–	–	–	21 (0.02)	10 (0.01)	–
Fish larvae	–	–	–	–	10 (0.02	10 (0.01)	–	10 (0.01)	–	–	–	10 (0.01)	21 (0.02)	52 (0.05)

Table 66.2: Diel Variation in the Quantitative Abundance (No. $100m^{-3}$) of the Rotifer Species Groups at Stations–I and III (The percentage composition given in parenthesis)

Rotifer Species	*Station–I* Time of Sampling (hrs)							*Station–III* Time of Sampling (hrs)						
	0600	*1000*	*1400*	*1800*	*2200*	*0200*	*0600*	*0630*	*1030*	*1430*	*1830*	*2230*	*0230*	*0630*
Anuraeopsis fissa	52 (0.17)	–	–	–	62 (0.49)	11 (0.03)	–	42 (0.19)	–	21 (0.07)	–	343 (0.88)	52 (0.18)	104 (0.27)
Brachionus angularis	561 (1.81)	1394 (8.52)	312 (3.84)	208 (1.46)	125 (1.00)	1539 (4.65)	925 (2.86)	302 (1.36)	260 (0.74)	863 (3.05)	707 (3.25)	562 (1.44)	624 (2.15)	364 (0.95)
B. calyciflorus	312 (1.00)	301 (1.84)	1508 (18.54)	291 (2.04)	375 (2.99)	458 (1.38)	489 (1.51)	188 (0.85)	52 (0.15)	62 (0.22)	468 (2.15)	562 (1.44)	520 (1.79)	624 (1.62)
B. caudatus	208 (0.67)	–	–	–	62 (0.49)	–	–	135 (0.61)	–	–	–	–	–	624 (1.62)
B. patulus	–	–	–	–	–	–	–	–	–	–	–	–	–	104 (0.27)
B. plicatilis	8923 (28.71)	10109 (61.75)	3588 (44.12)	11160 (78.27)	9163 (73.06)	27601 (83.33)	24533 (75.971)	12241 (51.10)	24835 70.59)	18076 (63.88)	13853 (63.6)	22923 (58.87)	19656 (67.74)	20322 (52.83)
B. quadridentatus	145 (0.47)	353 (2.16)	21 (0.26)	94 (0.66)	270 (2.15)	624 (1.88)	2486 (7.70)	135 (0.61)	281 (0.80)	21 (0.07)	333 (1.53)	343 (0.88)	156 (0.54)	312 (0.81)
B. rubens	–	94 (0.57)	–	208 (1.46)	125 (1.00)	42 (0.13)	62 (0.19)	322 (1.45)	780 (2.22)	62 (0.22)	125 (0.57)	562 (1.44)	520 (1.79)	416 (1.08)
Keratella tropica	20093 (64.66)	2694 (16.46)	2236 (27.49)	2090 (14.66)	1737 (13.85)	1747 (5.27)	1924 (5.96)	8008 (36.05)	7550 (21.46)	8820 (31.17)	5669 (26.03)	8497 (21.82)	4368 (15.06)	9568 (24.87)
Platyias quadricornis	–	–	–	–	–	–	–	42 (0.19)	31 (0.09)	–	–	–	–	–
Euchlanis dilatata	104 (0.33)	–	–	–	–	–	–	–	–	–	–	–	–	–
Dipleuchlanis propatula	–	–	–	–	–	–	–	62 (0.28)	42 (0.12)	–	42 (0.19)	–	52 (0.18)	–
Colurella obtuse	–	–	–	–	–	–	125 (0.39)	–	–	–	–	458 (1.18)	104 (0.36)	624 (1.62)
Lecane (Lecane) *arcula*	–	–	21 (0.26)	–	–	–	–	–	31 (0.09)	–	–	–	–	–

Contd...

Table 66.2–Contd...

Rotifer Species	*Station–I*							*Station–III*						
	Time of Sampling (hrs)							*Time of Sampling (hrs)*						
	0600	*1000*	*1400*	*1800*	*2200*	*0200*	*0600*	*0630*	*1030*	*1430*	*1830*	*2230*	*0230*	*0630*
L. (L.) *crepida*	–	–	–	–	–	–	–	42 (0.19)						
L. (L) *curvicornis*	–	94 (0.57)	–	31 (0.22)	–	–	–	31 (0.14)	–	–	–	–	–	104 (0.27)
L. (L) *nana*	–	–	–	–	–	11 (0.03)	–	–	–	–	–	–	–	–
L. (L) *papuana*	52 (0.17)	135 (0.82)	21 (0.26)	31 (0.22)	62 (0.46)	–	–	114 (0.51)	187 (0.53)	21 (0.07)	42 (0.19)	114 (0.29)	–208	(0.54)
L. (Monostyla) *bulla*	146 (0.47)	–	145 (1.78)	31 (0.22)	125 (1.00)	62 (0.19)	312 (0.97)	42 (0.19)	94 (0.27)	62 (0.22)	42 (0.19)	686 (1.76)	312 (1.08)	624 (1.62)
L. (M.) *furcata*	–	–	–	31 (0.22)	–	–	125 (0.39)	–	–	–	–	–	–	104 (0.27)
L. (M.) *hamata*	–	–	–	–	–	–	–	–	–	–	–	114 (0.29)	104 (0.36)	–
L. (M.) *lunaris*	–	–	–	–	–	62 (0.19)	62 (0.19)	–	–	–	42 (0.19)	–	–	208 (0.54)
Trichocera sp.	–	–	–	–	–	–	–	–	–	–	–	686 (1.76)	–	–
Asplanchna brightwelli	–	–	–	–	–	–	125 (0.39)	–	–	–	–	–	–	208 (0.54)
Hexartha sp.	–	94 (0.57)	–	–	–	62 (0.19)	–	62 (0.28)	146 (0.41)	104 (0.41)	–	686 (1.76)	156 (0.54)	104 (0.27)
Filinia longiseta	385 (1.24)	873 (5.33)	198 (2.43)	–	–	395 (1.19)	686 (2.12)	6218 (0.98)	572 (1.63)	104 (0.37)	333 (1.53)	1009 (2.59)	1040 (3.58)	1768 (4.60)
Filinia sp.	–	135 (0.82)	83 (1.02)	83 (0.58)	187 (1.49)	250 (0.75)	125 (0.39)	62 (0.28)	124 (0.35)	–	83 (0.38)	686 (1.76)	832 (2.87)	624 (1.62)
Testudinella patina	94 (0.30)	–	–	–	62 (0.49)	62 (0.19)	62 (0.19)	83 (0.37)	42 (0.12)	21 (0.07)	–	–	–	–
Rotaria sp.	–	94 (0.57)		–	187 (1.49)	198 (0.60)	250 (0.77)	83 (0.37)	156 (0.44)	62 (0.22)	42 (0.19)	707 (1.82)	520 (1.82)	1456 (13.78)

station–III, depicting an inverse relation between the two stations. The dominance index (c) ranged from 0.3071 to 0.7001 at station–I and from 0.3462 to 0.5453 at station–III. Here also an inverse relation between the two stations were observed. At station–I, the minimum was observed at 1400 hrs and the maximum at 0200 hrs, while at station–III, the minimum was at 0630 hrs and the maximum at 1030 hrs. In general, the dominant index was more at station–I than at station–III, unlike the other indices.

Table 65.3: Diel Variation in Diversity, Richness, Eveness and Dominance Indices of Rotifer Species at Stations–I and III

	Station–I						
	Time of Sampling (hrs)						
	0600	*1000*	*1400*	*1800*	*2200*	*0200*	*0600*
Diversity Index (N)	0.9551	1.3146	1.4088	0.7924	1.0589	0.7649	1.0258
Richness Index (d)	1.0634	1.1336	0.9996	1.0455	1.2716	1.3451	1.3484
Evenness Index (e)	0.3843	0.5290	0.6118	0.3304	0.4128	0.2825	0.3788
Dominance Index (c)	0.5012	0.4196	0.3071	0.6351	0.5551	0.7001	0.5884
	Station–III						
	Time of Sampling (hrs)						
	0630	*1030*	*1430*	*1830*	*2230*	*0230*	*0630*
Diversity Index (N)	1.1385	0.9508	0.8724	1.0709	1.4495	1.2368	1.5796
Richness Index (d)	1.7985	1.4329	1.170	1.2013	1.4191	1.3625	1.7996
Evenness Index (e)	0.3866	0.3429	0.3401	0.4175	0.5228	0.4567	0.5273
Dominance Index (c)	0.4343	0.5453	0.5061	0.4743	0.3974	0.4853	0.3462

Discussion

The present study revealed the existence of marked variations in the biological features of the Kadinamkulam lake. The tides were of a mixed, semi-diurnal, semi diurnal type, with two successive high and low tide waters. During the time of sampling the high water was at 1100 hrs and 0100 hrs. Similar tidal system was observed at Cochin backwaters (Pillai and Pillai, 1973; Shynamma and Balakrishnan, 1973). In the present investigation, tidal influence on the various features of the estuary was not obvious. This may be due to the position of the two stations which were situated rather far away from the barmouth *i.e.*, the stations were situated near the lower reaches of the backwater, and thus the two stations were far away from the direct influence of the tidal inflow.

Zooplankton biomass was high during early morning, decreased gradually with lowest values at 1400 hrs/1430 hrs, and the biomass increased again at dusk gradually reaching a peak at 0200 hrs and 0230 hrs at stations–I and III respectively. Total zooplankton population and diversity of various groups of zooplankton were also high during the night collection. It is seen that both the stations were very shallow and the high illumination of the water column probably made the zooplankton to take refuge very close to the bottom, thus the day time collection had less faunal diversity. Similar conditions have been reported in the shallow regions of Narakkal, Cochin backwater (Madhupratap and Rao, 1979) and also at Kavaratti atoll (Goswami, 1979 and Goswami and Goswami, 1990). Increase abundance of zooplankton during the night has also been observed in the Zuari estuary (Goswami *et al.*, 1979), in the river Potengi, Brazil (Sankarankutty *et al.*, 1979), in Malad Creek (Gajbhiye *et at.*, 1984a), off Versova (Gajbhiye *et al.*, 1984b). Despite the fact that light is the most important factor which governs the diel rhythm in the incidence and abundance of the plankters, the role of the other ecological and biological factors cannot be ignored. In this context the effect of tides on zooplankton variation can be mentioned. However, in the present instance the zooplankton population in general. In this context the effect of tides on zooplankton variation can be mentioned. However, in the present instance the zooplankton population in general was apparently more abundant during the second high water period, and this was more conspicuous as it was during the late night hours. Apart from this, no more significant effect of tides on zooplankton population was observed. Madhupratap and Rao (1979) are of the opinion that although tidal exchanges contribute much to the exchange of water in the estuary, it seems that the zooplankton are able to adjust and maintain their position within the estuary during study where no significant tidal influence on the variation of zooplankton population was observed. But reports have shown that the zooplankton biomass appear to be influenced by the incoming tides in the Mandovi estuary (Selvakumar *et al.*, 1986). According to Sankarankutty *et al.* (1979) both tide and illumination are responsible for the diel variation in the abundance of zooplankton in a tidal estuary. Temperature also have affected the zooplankton abundance *i.e.*, greater zooplankton abundance was observed at low temperature, and a similar observation was also made by Goswami (1979) in the Kavaratti atoll. Mathew *et al.* (1977) have opined that a high production rate alone is not enough to attract the zooplankton to a particular area, and this was so in the present instance, because at night when the zooplankters were in abundance, the productivity was considerably low.

The faunistic composition showed more diversity at station–III than at station–I. This can be attributed to the faunal depletion in retting zones as has been observed earlier (Abdul Aziz and Nair, 1986; Bijoy Nandan *et al.*, 1989). The rotifers constituted the largest zooplankton group and their abundance was relatively low during the morning hours and high during the night. The other major zooplankton groups, *viz.*, copepods, decapods larvae, gastropod larvae, Cladocerans and tintinnids, were also found predominantly in the night samples. However, the percentage composition of bivalve larvae followed an inverse trend, *i.e.*, it increased during the day and showed an almost gradual decline during the nights hours. Gajbhiye *et al.* (1984a) have also observed the incidence of bivalve larvae in the Malad Creek being limited during to the samples collected during the day. While Ayyakkannu (1989) has reported nocturnal maxima of bivalve larvae in the Coleroon estuarine system. Zooplankton groups like Ostracods, amphipods, insects, insect larvae, fish eggs and fish larvae were encountered irregularly and in lesser numbers. However, apart from insects, the incidence of the other lesser groups was generally during the night to dawn hours.

The rotifers species showed marked diel rhythm in their occurrence. The species which were observed during the night and early morning hours include *Brachionus patulus, Euchlanis dilatata, Colurella obtuse, Lecane* (Lecane) *nana*, L. (Monostyla) *furcata*, L. (M) *hamata*, (M) *Lunaris, Trichocerca* sp. and *Asplanchna* sp. While *Platyias quadricornis*, L. (L) *arcula* and L. (L.) *crepida* were observed only during the day. *Testudinella patina* was found only at night in station–I and during the day at station–III. While the more common species like *B. angularis, B. calyciflorus, B. plicatilis, B. quadridentatus, Keratella tropica* and L. (M.) *bulla* were observed throughout the diel cycle, but their numbers being more generally during dawn and also during the night hours. Rest of the species were scattered and showed no distinct diel rhythm in their occurrence. Species diversity of rotifers was maximum during the afternoon at station–I, whereas it was minimum during the same time at station–III. Thus species diversity showed an inverse relation between the two stations. Highest numbers of species of species were observed during the late night to early morning hours. Species richness of rotifer was maximum at dawn and minimum during the afternoon to evening hours, showing the effect of light on the species richness of rotifers. Evenness index was maximum at 1400 hrs at station–I, whereas at station–III it showed a minimum during the same time, thus showing similar variation as that of species diversity. Dominance index was maximum at 0200 hrs and 1030 hrs at stations–I and III respectively, and minimum at 1400 hrs and 0630 hrs at the two stations respectively. Station–III was more species rich, while the dominance index was greater at station I is due to the pollution load in the particular area.

Thus in the present study, the physico-chemical features did not reveal a distinct diel variation in their fluctuations, but the overall pattern of plankton variation (with special emphasis to rotifers) followed a diel rhythm and was not directly linked to tides. However, in a polluted environment as in station I, the adverse conditions will obviously affect the water quality, leading to variations in the composition of plankton.

Acknowledgements

The second author is thankful to University of Kerala, Trivandrum for awarding senior Research Fellowship during the tenure of the study.

References

Abdul Aziz, P.K. and Nair, N.B., 1986. Ecology of the coconut husk retting grounds in Kerala. *Symp. Ser. Mar. Biol. Assoc., India*, 6: 1115–1130.

Ayyakkannu, K., 1989. Diel variation and relative abundance of planktonic larvae in Coleroon estuarine complex, south east coast of India. *J. Mar. Biol. Ass., India*, 31(1–2): 276–286.

Battish, S.K., 1992. *Freshwater Zooplankton of India*. Oxford Publ. Co., New Delhi, 235 pp.

Bijoy Nandan, S., P.K. Abdul Azis and Natarajan, P., 1989. Water quality and faunal depletion in the retting zones of the back waters of Kerala. *Proceedings of the Kerala Science Congress, Cochin*, pp. 167–175.

Correll, D.L., 1978. Estuarine productivity. *Bioscience*, 28(10): 646–650.

Gajbhiye, S.N., Nair, V.R. and Desai, B.N., 1984a. Diurnal variation zooplankton in Malad Creek (Bombay). *Indian J. Mar. Sci.*, 13(2): 75–79.

Gajbhiye, S.N., Nair, V.R. and Desai, B.N., 1984b. Diurnal variation zooplankton off Versova (Bombay). *Mahasagar*, 17(4): 203–210.

Goswami, S.C., Selvakumar, R.A. and Goswami, U., 1979. Diel and tidal variations in zooplankton populations in the Zuari Estuary, Goa. *Mahasagar*, 12(4): 247–258.

Goswami, S.C. and Goswami, U., 1990. Diel variation in zooplankton in Minicoy Lagoon and Kavaratti Atoll (Lakshadweep Islands). *Indian J. Mar. Sci.*, 19(2): 120–124.

Goswami, S.C., 1979. Diel variation of zooplankton at Kavaratti Atoll (Lakshadweep). *Indian J. Mar. Sci.*, 8(4): 247–251.

Madhupratap, M. and Rao, T.S.S., 1979. Tidal and diurnal influence on estuarine zooplankton. *Indian J. Mar. Sci.*, 8(1): 9–11.

Madhupratap, M., Achuthankutty, C.T. and Nair, S.R.S., 1991. Zooplankton of the lagoons of the laccadives: Diel patterns and emergence. *J. Plankton Res.*, 13(5): 947–958.

Mathew, K.J., Gopinathan, C.P., Rao, D.S., Regunathan, A. and Murty, A.V.S., 1977. Diurnal variations in the distribution of zooplankton in relation to currents and other ecological parameters of the Mud Bank of Alleppey, Kerala. In: *Proc. Symp. Warm Water Zoopl.*, Spl Publ., UNESCO/NIO, Goa, p. 250–263.

Milne, H. and Dunnet, G.M., 1972. Standing crop, productivity and trophic relation of the fauna of the Ythan estuary. In: *The Estuarine Environment*, (Eds.) R.S.K. Barnes and J. Green. Applied Science Publications, London, p. 86–106.

Nayar, C.K.G., 1968. Rotifer fauna of Rajasthan. *Indian J. Fish.*, 31: 168–185.

Nayar, C.K.G. and Nair, K.K.N., 1969. A collection of Brachionid Rotifers from Kerala. *Proc. Indian Acad. Sci.*, 69B(4): 223–233.

Pillai, V.K. and Pillai, M.A., 1973. Tidal influence on the diel variation of zooplankton with special reference to copepods in the Cochin backwater. *J. Mar. Biol. Ass.*, India, 15(1): 411–417.

Sankarankutty, C., De Medeiros, G.F. and De Santos, N., 1979. On diurnal variations of zooplanktons in a tidal estuary. *J. Mar. Biol. Ass.*, India, 21(1–2): 187–190.

Santhosh, S., 2002. Hydro-chemistry of Paravur-Kappil backwater with special reference to plankton, Kollam district, Kerala, India. *Ph.D. Thesis*, University of Kerala, Trivandrum.

Selvakumar, R.A., Goswami, S.C. and Goswami, V., 1986. Tidal and diel influence on zooplankton occurrence in the Mandovi estuary, Goa. *J. Indian Fish. Assoc.*, 16–17: 39–49.

Shynamma, C.S. and Balakrishnan, K.P., 1973. Diurnal variation of some physico-chemical factors in Cochin backwater during south-west monsoon. *J. Mar. Biol. Ass., India*, 15(1): 391–398.

Ward, H.B. and Whipple, G.C., 1992. *Freshwater Biology*, (Ed.) W.T. Edmondson, 1247 pp.

Chapter 67

Qualitative and Quantitative Study of Zooplankton in Kadinamkulam Lake with Special Reference to Retting Activity, Southern Kerala

V. Sobha[1], Anuradha Rammohan, S. Santhosh[2] and J. Chandra Thara

[1]Dean, Professor and Head,
Department of Environmental Sciences, University of Kerala
[2]Technology Assistant, Department of Environmental Sciences,
Kariavattom Campus, University of Kerala, Thiruvananthapuram – 695 581, India

ABSTRACT

Qualitative and quantitative study of zooplankton population in the Kadinamkulam lake of Kerala during October 1991 to September 1992 was studied. Zooplankton composed of Tintinnids, Medusa, Rotifers, Nematodes, Polychaete larvae, Cladocerans, Ostracods, Copepods, Amphipods, Decapods larvae, Insects, Insects larvae, Gastropods larvae, Bivalve larvae, Fish eggs and Fish larvae. Their population dynamics.were mainly influenced by saline water intrusion and husk retting activity. Among zooplankton, Rotifer were mainly dominant group followed Copepods/Copepodites, decapods larvae and gastropod larvae.

Keywords: *Kadinamkulam, Zooplankton, Rotifer, Copepods, Husk retting, Intrusion.*

Introduction

The zooplanktons occupy a central position between the autotrophs and other hetertrophs and form an important link in food web of the freshwater ecosystems. Zooplankton is the intermediate link between phytoplankton and fish and which are the secondary producers in the aquatic environment. It contains both herbivores and carnivores, the latter belonging to the tertiary producers, or even to some higher level of production. A knowledge of their abundance, composition and seasonal variation, therefore, is an essential pre-requisite for any successful aquaculture programmes. Zooplankton is a good indicator of changes in water quality because it is strongly affected by environment conditions and responds quickly to changes in environmental quality. Among the zooplankton, rotifers are apparently the most sensitive indicators of the water quality. The rotifer fauna of Kerala was brought to light by Nayar and Nair (1969) and Nair (1972). Shibu (1991) gave an account on the rotifers of Paravur lake. Bijoy Nandan (1991) and Bijoy Nandan and Abdul Azis (1994) have observed a few rotifers species and those were only from the retting areas of Kadinamkulam backwaters and no attempt has been hitherto made to study the ecological aspects of this interesting group of zooplankton. Hence qualitative and quantitative studies of zooplankton are of great importance. In the present study of zooplankton studies of Kadinamkulam backwater is very important because this may help to assess the environmental degradation due to intensive husk retting, salt water intrusion from adjacent sea through open barmouth and other anthropogenic activities.

Materials and Methods

The Kadinamkulam backwater (Latitude 8°35′N; Longitude 76°45′to 76°52′E) is, one of the largest backwater systems of Trivandrum in southern Kerala. It lies almost parallel to the adjoining Arabian Sea for some distance and maintains a temporary connection with the sea at the north western region. The Vamanapuram river originating from the Western Ghats, after traversing for about 80 km and carrying a drainage area of 540 km^2, empties into the backwater through the Anchuthenghu lake in the vicinity of the temporary barmouth. The formation and closure of the sand bar at the mouth leads to pronounced changes in the hydro biological conditions of the backwater. For the present investigation, three station are selected and samples were collected seasonally from the Kadinamkulam backwater. The study was conducted during the period from October 1991 to September 1992. Zooplankton are collected monthly from three stations using plankton net made up of bolting silk (mesh size 751.1 μm) and preserved in 15 per cent formalin. Counting of zooplankton was done by using Sedgwick-Rafter cell and density is represented in percentage. Zooplanktons were identified as per the method followed by Nayar (1968), Nayar and Nair (1969), Battish (1992) and Ward and Whipple (1992).

Results

The total zooplankton population ranged between 32687 No. $(100m^3)^{-1}$ to 3085150 No. $(100m^3)^{-1}$. Station–II recorded the lowest total annual population, while the highest was recorded at station III. Seasonal variations in the zooplankton population were rather pronounced. In all the three stations the lowest population was recorded during postmonsoon and the highest during premonsoon at station–II and during monsoon at station–I and III, where the population showed a peak during the month of August (Table 67.1). Zooplankton in the Kadinamkulam lake is composed of tintinnids, Medusae, rotifers, nematodes, polychaete larvae, Cladocerans, Ostracods, Copepods/copepodites, Arnphipods, decapods, decapods larvae, insects, insect larvae, gastropod larvae, bivalve larvae, fish eggs and fish larvae. In all the stations, rotifer constituted the dominant group (69.24 per cent, 54.96 per cent and 73.27 per cent at stations–I, II and III respectively). The number of faunal groups encountered during the period of study ranged from 5 to 15. At Station–I the number of faunal groups ranged from a total of 5 (May) to 12 (April, September, December), at Station–II the number varied from 9 (April, July, September) to 15 (December) and at Station–III from a total of 9 (April, May) to 14 (September, December) Table 67.2. In general, the number of faunal groups were more during the premonsoon period and among the Stations, Station–II had the highest number of faunal groups. Community characteristics of rotifers such as Species diversity, species richness, evenness and dominance indices were also tabulated at Table 67.3. Analysis of variance also recorded between stations, seasons and months Table 67.4.

Table 67.1: Seasonal Variation of Quantitative Abundance (No. 100 m^{-3}) of the Various Zooplankton Groups at Kadinamkulam Back Water

Zooplankton Groups	*Station–I*			*Station–II*			*Station–III*		
	Pre-monsoon	*Monsoon*	*Post-monsoon*	*Pre-monsoon*	*Monsoon*	*Post-monsoon*	*Pre-monsoon*	*Monsoon*	*Post-monsoon*
Tintinnids	10439	Nil	29	1612	13	2049	4342	143	330
Medusae	60	Nil	57	26		130	39	Nil	213
Rotifers	426598	759270	37614	207483	215163	53009	312866	849732	35827
Nematodes	Nil	Nil	Nil	Nil		88	Nil	3	55
Polychaete larvae	16	31	76	29	13	335	63	107	398
Cladocerans	180	1735	81	29	1199	67	8	1885	42
Ostracods	Nil	11	21	18	68	44	16	29	5
Copepods/copedites	190310	66401	44562	128752	20642	83879	157082	36510	58817
Amphipods	Nil	3	5	31	8	5	11	26	Nil
Decapods (Lucifer)	Nil	Nil	Nil	3		Nil	3	Nil	Nil
Decapods larvae	54270	120573	19191	23728	35007	36720	34536	269682	22753
Insects	Nil	3	24	3	13	5	Nil	Nil	3
Insects larvae	21	55	23	18	34	81	68	83	8
Gastropod larvae	3768	8050	1698	5236	2967	8807	24302	6672	11796
Bivalve larvae	463	26762	2252	424	55671	2899	1407	21008	4756
Fish eggs	3	8	8	8	3	21	11	13	18
Fish larvae	11	16	Nil	16	Nil	44	29	8	31

Discussion

The plankton composition is known to change over the years. Previous studies in the Kadinamkulam lake biotopes have revealed striking variation in zooplankton compositions within this region. Difference in the seasonal maxima and density of plankton are dependent on a number of physical, chemical and biological factors. The zooplankton density was found to decrease during the Northeast monsoon (Post monsoon season) and the lowest values were noticed in November in all the three stations. The zooplankton population increased during the monsoon at stations–I and II, and during premonsoon at station–II the highest being in March. A similar observation was made in the Cochin backwater by Nair and Tranter (1971), where the distribution and abundance of zooplankton was found to be high in April and low in November and according to them such an increase and decrease was due to the conditions during the premonsoon, being more favorable than those during the post-monsoon.

Table 67.2: Seasonal and Annual Mean (± S.E) of Major Zooplankton Groups in Kadinamkulam Lake

Zooplankton Groups	*Stations*	*Seasonal Mean*			*Annual Mean*
		Premonsoon	*Monsoon*	*Postmonsoon*	
Rotifers	I	62.64±9.98	61.75±18.02	46.37±9.57	57.08±6.88
	II	54.78±8.16	61.56±15.57	34.25±14.17	50.20±7.33
	III	68.16±13.71	62.43±21.07	41.00±15.87	57.2±9.22
Copepods	I	27.59±6.97	11.63±6.24	31.89±9.91	23.70±4.67
	II	35.93±6.44	8.60±4.47	37.05±15.83	27.19±6.36
	III	24.27±11.26	2.91±0.62	31.64±16.07	19.60±6.67
Decapods larvae	I	6.18±0.89	22.02±11.09	13.55±2.58	13.92±3.79
	II	7.41±3.28	16.93±11.25	20.07±4.02	14.80±3.90
	III	4.36±1.90	30.84±21.33	13.31±3.72	16.17±7.03
Gastropod larvae	I	1.04±0.70	1.21±1.00	2.94±1.58	1.73±0.63
	II	1.03±0.62	0.65±0.35	4.66±2.76	2.11±0.98
	III	1.83±1.04	0.47±0.25	6.50±3.33	2.94±1.26
Bivalve larvae	I	0.15±0.07	2.99±2.27	4.38±2.39	2.51±1.08
	II	0.16±0.08	11.42±10.20	2.99±2.03	4.86±3.31
	III	0.25±0.19	2.99±2.48	6.83±4.62	3.36±1.70

Table 67.3: Variation in Diversity, Richness, Evenness and Dominance Indices of Rotifer Species at Stations–I, II and III from October'91 to September'92

Seasons	*Months*	*Diversity Index (H)*			*Richness Index (d)*			*Evenness Index (e)*			*Dominance Index (c)*		
		Stn. I	*Stn. II*	*Stn. III*	*Stn. I*	*Stn. II*	*Stn. III*	*Stn. I*	*Stn. II*	*Stn. III*	*Stn. I*	*Stn. II*	*Stn. III*
Post-monsoon	Oct.	1.0765	1.1761	1.4209	1.1561	0.7670	1.3045	0.4079	0.5108	0.5247	0.5041	0.3827	0.3088
	Nov.	1.4187	0.9574	1.3191	1.4264	1.2180	1.4305	0.5239	0.3733	0.4871	0.3784	0.5500	0.4839
	Dec.	0.1909	0.3398	0.4360	0.6040	1.0066	1.0806	0.0918	0.1367	0.1700	0.9344	0.8720	0.8015
	Jan.	0.1918	0.3235	0.2814	0.4119	0.3176	0.5270	0.1192	0.2333	0.1571	0.9335	0.8670	0.8966
Pre-monsoon	Feb.	0.0776	0.0970	0.1167	03346	0.3462	0.4903	0.0482	0.0603	0.0600	0.9767	0.9703	0.9641
	Mar.	0.3600	0.1569	0.0875	0.5366	0.5371	0.5405	0.1850	0.0754	0.0398	0.8342	0.9439	0.9693
	Apr.	0.0579	0.2854	0.6362	0.2121	0.2489	0.6636	0.0418	0.2059	0.2763	0.9810	0.8695	0.6661
	May	0.1777	0.1233	0.1511	0.2621	0.2612	0.3413	0.1282	0.0890	0.0939	0.9268	0.9539	0.9408
Monsoon	Jun.	0.1418	0.2492	0.2690	0.6172	0.5754	0.8825	0.0645	0.1198	0.1083	0.9479	0.8920	0.8804
	Jul.	1.1386	1.1678	1.1755	0.8790	1.0663	1.2600	0.4748	0.4700	0.4454	0.4105	0.4645	0.4644
	Aug.	0.2121	0.4473	0.4680	0.4780	0.5444	0.7413	0.1020	0.2151	0.1883	0.9276	0.8241	0.7921
	Sep.	0.8070	0.7218	0.9261	1.0509	1.1248	1.260	0.3058	0.2665	0.3269	0.6674	0.6721	0.5784

Table 67.4: Analysis of Variance of Major Zooplankton Groups in Kadinamkulam Lake

Source	*df*	*F Ratio*				
		Rotifers	*Copepods*	*Decapods*	*Gastropod Larvae*	*Bivalve Larvae*
Stations	2	1.753	2.0831	0.1637	0.6600	0.4867
Seasons	2	16.298*	27.6362**	9.5653**	7.8664**	3.0587
Months	9	20.9843**	11.9475**	8.8617**	2.5538*	3.7533**

*: Significant ($P < 0.05$); **: Significant ($P < 0.01$).

Unlike the total zooplankton population distribution, the zooplankton biomass was at its minimum during monsoon and maximum during postmonsoon, when three stations were compared, station–III recorded the highest population and biomass, this can be attributed to the unpolluted nature of this station, when with the other two. Nair *et al.* (1983) have reported that in the near shore waters of Thal, in accordance with the increase in biomass, consequent increase in the total population of zooplankton was not observed. This is in conformity with the present findings. An earlier study in Kadinamkulam lake has shown low zooplankton density during monsoon (Nair *et al.*, 1984). But present study the zooplankton density was at its highest during the monsoon and not lowest as observed a decade ago. In the present study, rotifers formed the predominant group followed by copepods/copepodites, decapod larvae, bivalve larvae and gastropod larvae. Among these common groups of zooplankton, rotifers, decapods larvae and gastropod larvae were most abundant at station–III, while copepods and bivalve larvae were predominant at station–II. Due to the heavy pollution load at station–I the total faunal groups represented here were rather low. The percentage of incidence of rotifers in relation to other groups observed in the present study was higher than that reported from earlier works conducted in the Kadinamkulam lake (Sobha and Miranda, 1987; Bijoy Nandan, 1991). This shows the singularly important position occupied by this group in the zooplankton population of this dynamic, brackish water lake.

Rotifers population was low during postmonsoon and high during premonsoon and monsoon, whereas copepods were low in number during monsoon and high during postmonsoon, thereby showing a clear cut inverse relation between the two groups. During the present study, the rotifer and copepod population maxima succeeded each other and never occurred together in large numbers. However, the larval groups (decapod, gastropod and bivalve) were low in number during premonsoon and high during postmonsoon and monsoon, thus depicting an increase in larval population during both the south-west and north-east monsoons, in the lake. Silas and Pillai (1975) showed that the decapods larvae formed a major component of zooplankton I the Cochin backwater. They observed a peak during monsoon and a minor peak was also observed during the postmonsoon. Studies have shown that temperature, salinity, dissolved oxygen, pollution and even phytoplankton and production have their effect on the distribution and abundance of the various zooplankton groups. When the temperature was low during postmonsoon the zooplankton population also showed a decline and during warmer months (premonsoon months and to some extent during the monsoon as well), the zooplankton population attained its maxima thereby showing that temperature had a direct effect on the seasonal abundance of zooplankton population. Similarly Lopes (1994) reported that temperature fluctuations influenced the seasonal occurrence of the zooplankton with higher densities occurring during the warmer months in the Guaran river estuary. Santhosh (2002) has shown that pH, is a limiting factor for a set of species in an aquatic environment. In the present study pH showed an inverse relation with rotifer population, low pH recorded during the monsoon season coincided with rather high rotifer population during the same season.

The extent of salt water intrusion depends on a number of factors such as tidal regime, quantity of rainfall and freshwater influx. The Kadinamkulam lake is periodically exposed to the sea and the condition of the barmouth has a crucial role in this context. The zooplankton population in the present study was generally abundant during the low saline, monsoonal period and least during the postmonsoon, when the salinity was rather high. Lesser groups like tintinnids and polychaete larvae also showed their maxima when the salinity was generally high. Similarly, in the present investigation, the copepods were observed to endure low concentration of oxygen and they were abundant during the postmonsoon when the dissolved oxygen was rather low. However, their number dwindled in monsoon at the time of high dissolved oxygen concentration. Low group diversity at station I was observed and this was due to the deterioration in the water quality, associated with long tears of coconut husk retting in this region. Station–II was also polluted due to retting activity but a diverse faunal group were observed in this station and this may be due to the dominance by few indicator (opportunistic) groups such as the insect larvae, nematods and polychaete larvae, apart from the presence of other faunal groups. Nair *et al.* (1981) have stated that in polluted estuaries, the cumulative effect of various factors is responsible for an eventual stress on aquatic life. Abundance and scarcity of more sensitive groups and species can be considered as biological indices, to forecast the polluted and unpolluted waters.

During the monsoon season, the retting areas showed considerably high zooplankton density and according to Abdul Azis and Nair (1978), this may be because of the flushing out of the retting byproducts from the retting grounds due to some extent by monsoonal floods. Apart from them five dominant zooplankton groups, others such as polychaete larvae, nematodes, insect larvae and Cladocerans were the predominant groups in the retting areas and thus these can be considered as pollution tolerant groups. Abundance of similar zooplankton groups were observed by Abdul Azis and Nair (1978, 1986) in the retting areas of Edava-Nadayara back water system. A stable environment can sustain a diverse faunal assemblage, indicating healthy condition. Thus in the present study, station III showed a stability in its physico-chemical features and thereby sustained a diverse faunal assemblage and this features supports the aforementioned observation made by Gajbhiye and Desai (1981). Rotifers formed the most important group of zooplankton in all the stations (they were most abundant at station–III) and were distributed in abundance throughout the period of investigation (October'91 to September'92), seasonal trend in their distribution showed that the premonsoon and monsoon were the most favorable time for the incidence of this group. Although the species diversity was less during premonsoon, the biomass in number was however high and during monsoon, the species diversity was maximum.

During the present study, species diversity was minimum during the premonsoon, more so at Station–I and maximum during the postmonsoon and also at Station–III. Diversity is also used as in index of environmental pollution and Desai *et al.* (1983) have shown that diversity is lowest in polluted waters. Hence this explains the low diversity during the premonsoon and also at station–I, because during this season and at Station–I, the water is turbid with ret liquor and highly polluted due to the presence of high hydrogen sulphide concentration and less dissolved oxygen concentration. Thus the maximum diversity during postmonsoon and at station–III can be due to the presence of high salinity, moderate dissolved oxygen concentration and high phytoplankton productivity, coupled with astable and unpolluted environment. Diversity is also a function of two elements, one being the number of species (richness) and the other its equality (evenness). Species richness and evenness also showed a similar variation as that of species diversity, that is least during premonsoon and at station–I and highest during postmonsoon and at station–III. It is thus obvious that more the number of species (richness) and its eveness, more will be its diversity. Winner *et al.* (1980) also observed an increase in diversity index as a result of increase in evenness and richness in the zooplankton community in the lake St. Clair, the low richness and evenness indices observed during the premonsoon and also at Station–I can also be attributed to the environmental degradation, instability and pollution suffered by this region due to the presence of hydrogen sulphide as a result of retting.

All organisms in a community are not important in determining the nature of the whole community. Certain species or species groups normally control the energy flow and strongly influence the prevailing environments of all other species and they are known as 'ecological dominants', a dominant species in a community may exert a powerful control over the occurrence of other species. Odum (1971) has explained this relation by stating that it is the large number of the rare species that determine the species diversity of the trophic groups and whole communities, while the dominants represented by a few common species largely account for the energy flow in each species group.

Green (1993) has also observed that dominance is generally inversely related to diversity and according to him an abnormally high dominance for a number of species can be indication of pollution or some other form of environmental stress. This supports the observation made during the present study, where in the high dominance index recorded during the premonsoon and also at station–I is in consequence to the high pollution load due to sulphide pollution caused by the retting of coconut husk in the lake. So it is very urgent to take necessary actions to mitigate the problems.

Acknowledgements

The second author is thankful to University of Kerala, Trivandrum for awarding senior Research Fellowship during the tenure of the study.

References

Azis, A.P.K. and Nair, N.B., 1986. Ecologyz of the coconut husk retting grounds in Kerala. *Symp. Ser. Mar. Biol. Assoc.*, India, 6: 115–1130.

Azis, A.P.K. and Nair, N.B., 1978. The nature of pollution in the retting zones of backwaters of Kerala. *Aquatic Biology*, 3: 41–62.

Battish, S.K., 1992. *Freshwater Zooplankton of India*. Oxford Publ. Co., New Delhi, 235 pp.

Desai, B.N., Gajbhiye, S.N., Jiyalal Ram, M. and Nair, V.R., 1983. Comparative account on zooplankton in polluted and unpolluted estuaries of Gujarat. *Mahasagar*, 16(3): 281–291.

Gajbhiye, S.N. and Desai, B.N., 1981. Zooplankton variability in polluted and unpolluted waters off Bombay. *Mahasagar*, 14(3): 173–182.

Green, J., 1993. Diversity and dominance in planktonic Rotifers. *Hydrobiologia*, 255: 345–352.

Lopes, R.M., 1994. Zooplankton distribution in the Guaran River Estuary (South-eastern Brazil). *Estuar. Coast. Shelf Sci.*, 39: 287–302.

Nair, K.K.C. and Tranter, D.J., 1971. Zooplankton distribution along salinity gradients in the Cochin Backwater before and after the monsoon. *J. Mar. Biol. Ass., India*, 13(2): 203–210.

Nair, K.K.N., 1972. Sessile rotifers of Kerala. *J. Ker. Acad. Biol.*, 3(2): 30–36.

Nair, N.B., Krishnakumar, K., Arunachalam, M., Abdul Azis, P.K. and Dharmaraj, K., 1984. Ecology of Indian estuaries: Studies on the zooplankton ecology of Kadinamkulam Backwater. *Proc. Indian. Acad. Sci. (Anim. Sci.)*, 93(6): 573–584.

Nair, N.B., Krishnakumar, K., Abdul Azis, P.K., Dharmaraj, K., Arunachalam, M. and Balasubramanium, N.K., 1983. Ecology of Indian estuaries, Part III: Physico-chemical features of Akathumuri-Anchuthengu-Kadinamkulam back water systems, S.W. Coast of India. *Indian J. Mar. Sci.*, 12: 154–159.

Nair, N.B., Arunachalam, M., Abdul Azis, P.K., Krishnakumar, K., and Dharmaraj, K., 1984. Ecology of Indian estuaries: Distribution and seasonal variation of zooplankton in the Ashtamudi estuary. *Proc. Nat. Acad Sci.*, India, 55B: 272–289.

Nair, V.R., Gajbhiye, S.N., Jiyalal Ram, M. and Desai, B.N., 1981. Biomass and composition of zooplankton in Auranga, Ambika, Purna and Mindola estuaries of South Gujarat. *Indian J. Mar. Sci.*, 10(2): 116–122.

Nandan, B.S., 1991. Effect of coconut husk retting on the water quality and biota of an aquatic biotope in Kerala. *Ph.D. Thesis*, University of Kerala, Trivandrum.

Nandan, B.S. and Abdul Azis, P.K., 1994. Zooplankton of the retting zones in the Kadinamkulam estuary, Kerala. *Mahasagar*, 27(1): 59–65.

Nayar, C.K.G., 1968. Rotifer fauna of Rajasthan. *Indian J. Fish.*, 31: 168–185.

Nayar, C.K.G. and Nair, K.K.N., 1969. A collection of Brachionid Rotifers from, Kerala. *Proc. Indian Acad. Sci.*, 69B(4): 223–233.

Odum, E.P., 1971. *Fundamentals of Ecology*, 3rd Edition. W.B. Saunders Company, London, 573 pp.

Santhosh, S., 2002. Hydro-chemistry of Paravur-Kappil backwater with special reference to plankton, Kollam district, Kerala, India. *Ph.D. Thesis*, University of Kerala, Trivandrum.

Shibu, S., 1991. Ecology of the Paravur lake. *Ph.D. Thesis*, University of Kerala, Trivandrum.

Silas, E.G. and Pillai, P.P., 1975. Dynamic of zooplankton in a tropical estuary (Cochin backwater), with a review on the plankton fauna of the environment. *Bull. Dept. Mar. Sci. Univ. Cochin*, 7(2): 329–355.

Sobha, V. and Miranda, I., 1987. Nature of plankton production in Kadinamkulam Lake and Ashtamudi estuary of Kerala. *Proc. Nat. Sem. Estuar. Manag*, Trivandrum, pp. 307–317.

Ward, H.B. and Whipple, G.C., 1992. *Freshwater Biology*, W.T. Edmondson (ed.), 1247 pp.

Winner, J.M., Patrick, H. and Wallen, D.G., 1980. Zooplankton species diversity in the lake St. Clair, Ontario, Canada. *Hydrobiol. Bull.*, 75: 57–63.

Chapter 68

Histological Alterations Under Tannery Effluent Stress During Development of Oocytes in the Ovary of *Pantala flavescens* (Fabricius) (Libellulidae : Anisoptera)

A. Parithabhanu and M.A. Subramanian
P.G. and Research Department of Zoology, C.N. College, Erode – 638 004, Tamil Nadu

ABSTRACT

The last three larval instars and the adults of various stages of female *Pantala flavecens* and the composite tannery effluent collected from a local tannery were used in the present study. Static bioassays were carried out to determine LC_{50} 96 hour value of the effluent to each larval instar. Then the individual larvae were reared in sublethal concentrations of the effluent for 30 days. The ovaries from the control and effluent-treated insects were processed for histological preparations. The tannery effluent was found to bring remarkable alterations in the oocytes of the ovary both in nucleus and follicle cells. There was no indication of vitellogenesis in the oocytes under the tannery effluent stress. Thus the effluent can be considered to have chaemo sterilent property.

Keywords: *Pantala flavescens, Tannery effluent, Vitellogenesis follicle cells.*

Introduction

In the recent past, number of industries have been established in most of the developing countries in India. The tannery industry is the oldest industry and forms the basis for foreign exchange as it is export oriented but the discharge of effluents from tannery units into the water bodies leads to adverse effects of the flora and fauna. Arora (1981) has shown that the tannery effluent consists of a number of potentially harmful substances among all types of industrial effluents. Though much information are made available on harmful effects of industrial effluents on various animals groups (Ambrose *et al.*, 1994;Varadaraj and Subramanian, 1991; Sakthivel, 1994) similar investigations on odonate larvae are scanty.

In insects, the major source of nutrients for the development of embryo is the egg yolk. The growth of the oocyte during the maturation phase is due to vitellogensis. The yolk materials are lipo-phospho-glyco-proteins which are stored in the oocytes during oogenesis. Investigations have been made to find the impact of pollutants on the development of ovary in many fishes (Sudha, 1998; Ghosh *et al.*, 1999). Schneider *et al.* (1995) and Anitha *et al.* (1999) have made studies to show the effect of plant extracts on the ovarian development in many insects. The diagnosis of the impact of pollutants by means of behavioral, physiological and biochemical parameters alone can not cope up with to understand the harmful effects of the pollutants. Therefore the histomorphological studies are highly useful to assess the pathological conditions of tissues in animals. The literature survey indicates that there is no emphasis on the effect of pollutants on the development of oocytes in insects. Therefore the present study has been carried out to show the impact of tannery effluent on histological alterations during the development of oocytes in the ovary of dragonfly *Pantala flavescens.*

Materials and Methods

The dragonfly larvae were manually collected from their local breeding places, brought to the laboratory and maintained as per the method given by Varadaraj 1978. The last three larval instars and the adults of various stages of female were selected.

The composite tannery effluent collected from a local tannery and dechlorinated tap water as the diluent medium were used. The effluents having the same physicochemical properties were selected every time. In the test water, variables such as temperature, pH, salinity and dissolved oxygen were controlled.

Static bioassays were carried out to determine LC_{50} 96 hr value of the tannery effluent to each larval instar. Then individual larval instars were reared in 5 different sublethal concentrations of the effluent (0.5 per cent, 1.0 per cent, 1.5 per cent, 2.0 per cent and 2.5 per cent) for 30 days with appropriate controls. For histological preparations, ovaries from penultimate and final instar lavae as well as adults of various stages of development were selected. The ovaries from control and effluent-treated insects were processed for routine microtechnique method (Smith, 1940). The paraffin sections of 8 μ thickness were cut, stained and microphotographed.

Results and Discussion

In the control penultimate larvae, the vitellarium consisted of a number of growing oocytes of which the younger immature oocytes were seen near the. germarium whereas the mature oocytes were near the lateral oviduct (Plate 68.1; Ph.m.1.) Histologically, the ovarioles of the final instar larvae and adult-1 stage exhibited more or less similar structure like those of the penultimate larvae except increase in size of oocytes especially the basal oocytes (previtellogenic) (Plate 68.1; Ph.m.2) During

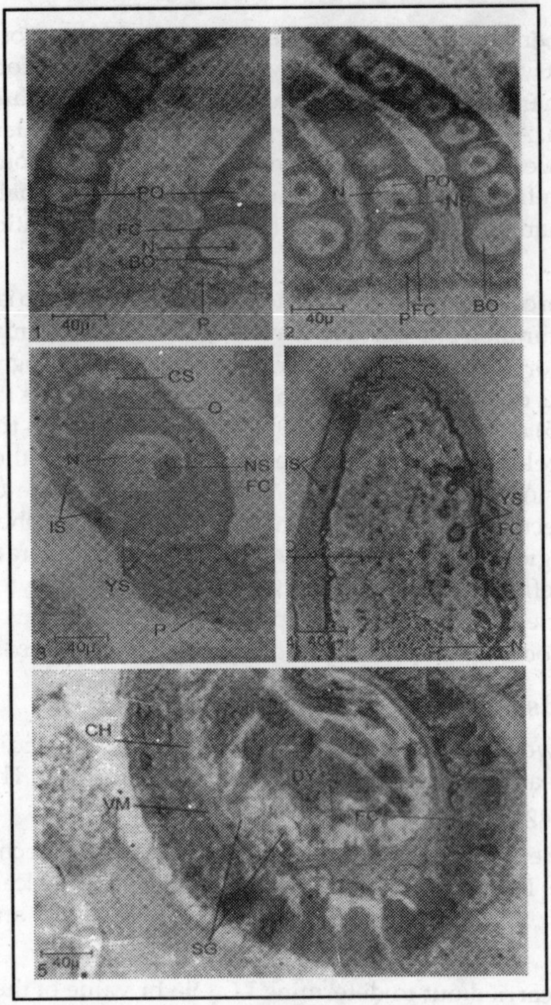

Plate 68.1

Ph.m.1: L.S. of a portion of ovary showing enlarged ovarioles in the control final instar larva of female *P. flavesens.* (400x)
(BO: Basal oocyte; FC: Follicular cell; N: Nucleus; P: Pedicel; PO: Penultimate oocyte)

Ph.m.2: L.S. of a portion of ovary showing enlarged ovarioles in adult stage I (control) with previtellogenic basal oocytes of female *P. flavescens.* (400x)
(BO: Basal oocyte; FC: Follicular cell; P: Pedicel; PO: Penultimate oocyte; N: Nucleus)

Ph.m.3: L.S of an early vitellogenic oocyte in adult stage II (control) of female *P. flavescens.* (100x)
(CS: Clear space; FC: Follicular cell; IS: Intercellular space; N: Nucleus; NS: Nucleolus; O: Ooplasm; P: Pedicel; YS: Yolk spheres)

Ph.m.4: L.S. of a vitellogenic basal oocyte in adult stage III (control) of female *P. flavescens* (with yolk bodies at the periphery). (400x)
(FC: Follicular cell; IS: Intercellular space; N: Nucleus; O: Ooplasm; YS: Yolk spheres).

Ph.m.5: L.S of a post vitellogenic oocyte in adult stage IV (control) of female *P. flavescens* (with yolk bodies surrounded by vitelline membrane and chorion). (400x)
(CH: Chorion; DY: Dense yolk; FC: Follicular cell; SG: Secretory granules; VM: Vitelline membrane)

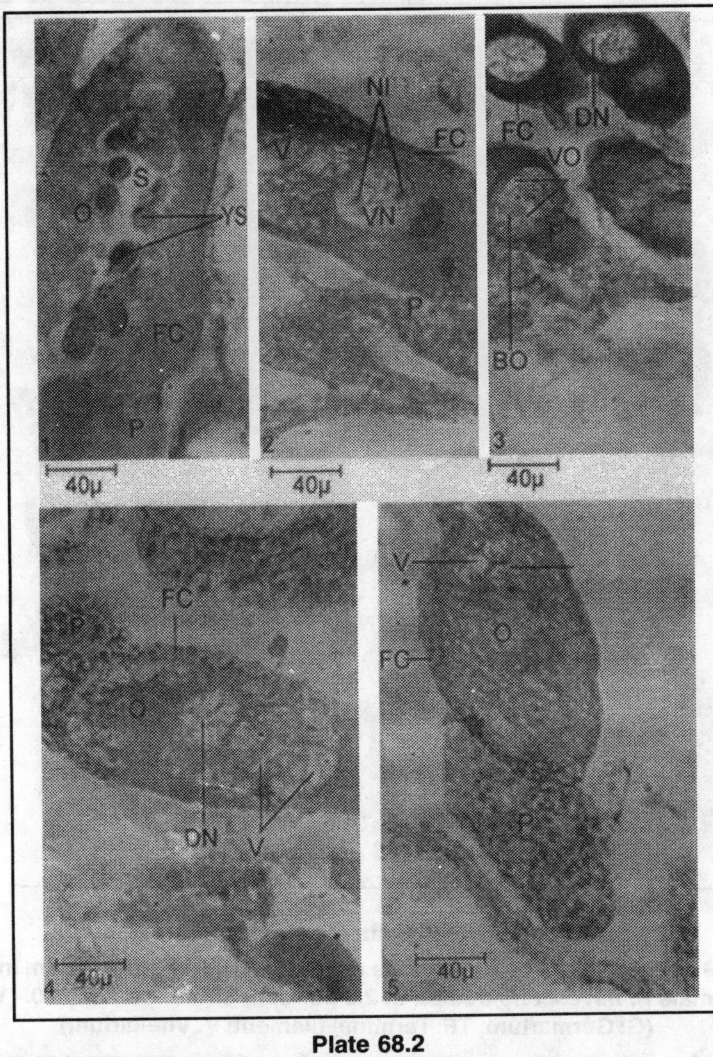

Plate 68.2

Ph.m.1: L.S of a basal oocyte from the ovary in the larvae of female *P. flavescens* treated in 0.5 per cent effluent on day 30. (400x)
(FC: Follicular cell; O: Ooplasm; P: Pedicel; S: Space; YS: Yolk spheres)

Ph.m.2: L.S of a basal oocyte from the ovary in the larvae of female *P. flavescens* treated in 1.0 per cent effluent on day 30. (400x)
(FC: Follicular cell; NI: Nucleoli; P: Pedicel; V: Vacuole; VN: Vacuolate nucleoplasm)

Ph.m.3: L.S of a basal oocyte from the ovary in the larvae of female *P. flavescens* treated in 1.5 per cent effluent on day 30. (400x)
(BO: Basal oocyte; DN: Disintegrated nucleus; FC: Follicular cell; P: Pedicel; VN: Vacuolated nucleoplasm)

Ph.m.4: L.S of a basal oocyte from the ovary in the larvae of female *P. flavescens* treated in 2.0 per cent effluent on day 30. (400x)
(DN: Disintegrated nucleus; FC: Follicular cell; O: Ooplasm; P: Pedicel; V: Vacuole)

Ph.m.5: L.S of a basal oocyte from the ovary in the larvae of female *P. flavescens* treated in 2.5 per cent effluent on day 30. (400x)
(DN: Disintegrated nucleus; FC: Follicular cell; O: Ooplasm; P: Pedicel; V: Vacuole)

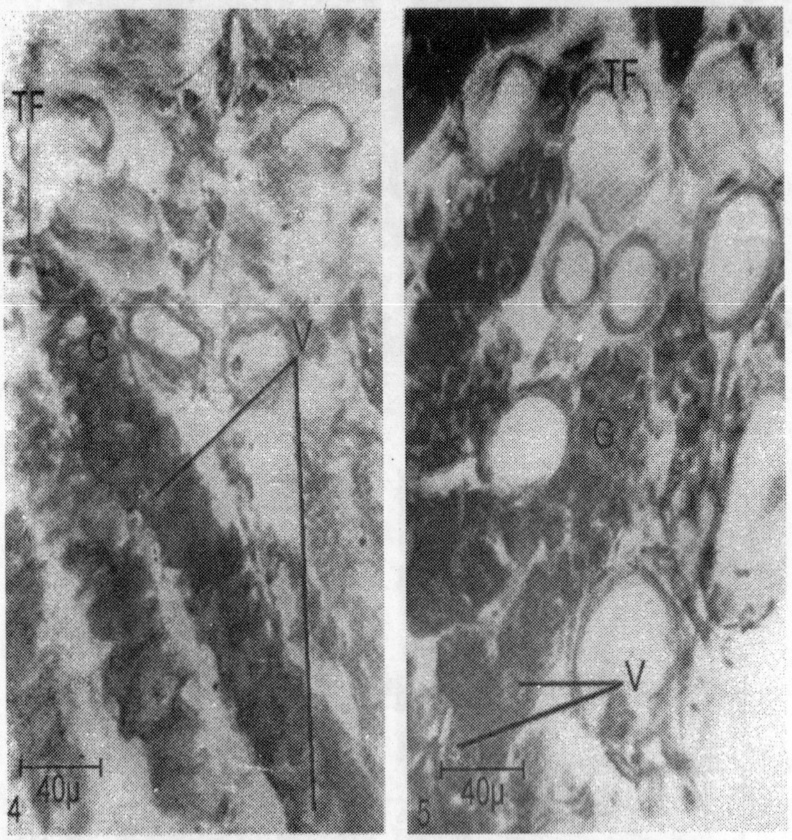

Plate 68.3

Ph.m.1: L.S of a portion of ovary with the ovarioles showing germarium in the larvae of female *P. flavescens* treated in 2.0 per cent effluent on day 30. (100x)
(G: Germarium; TF: Terminal filament; V: Vitellarium)

Ph.m.2: L.S of a portion of ovary with the ovario1es showing germarium in the larvae of female *P. flavescens* treated in 2.5 per cent effluent on day 30. (100x)
(G: Germarium; TF: Terminal filament; V: Vitellarium)

adult stage II the basal oocytes exhibited early vitellogenic phase (Plate 68.1; Ph.m.3). In adult stage III, the vitellogenic ovary showed prominent histological changes connected with vitellogenesis. The ooplasm showed small yolk spheres along the periphery of the oocytes (Plate 68.1; Ph.m.4). A number of secretory granules were also noticed at the peripheral region of the ooplasm. In adult IV stage, the oocytes contained fully packed yolk bodies in their ooplasm (post vitellogenic) (Plate 68.1; Ph.m.5).

Quite remarkable alterations were noticed in the basal oocytes of the ovary treated in tannery effluent for 30 days. In 0.5 per cent effluent, the ooplasm of the basal oocyte contained clear spaces with very few yolk spheres indicating the vitellogenic effort by the oocyte (Plate 68.2; Ph.m. 1). On the other hand, in other concentrations of the effluent, there was no indication of yolk formation in the oocytes. In these oocytes, the follicle cells were found to assume less columnar. In lower concentrations of the effluent, nucleus was found to be less stained with vacuolated nucleoplasm. The nucleus was disorganized without nucleous and nuclear membrane in higher concentrations of the effluent (Plate 68.2; Ph.m.2-5). Changes were also noticed in the germarium of the ovarioles in the ovary. On 30th day of exposure, the germinal epithelial cells were found to undergo active cell division but produced immature cells only in lower concentrations of the effluent. On the contrary, the germarium of the ovarioles appeared to be disintegrated with vacuolated areas due to necrosis of cells in the higher concentrations of the effluent (Plate 68.3; Ph.m. 1 and 2).

In *P. flavescens*, the ovary is panoistic type as it contains no trophic cells and the basal oocytes increase in size and become spherical in shape at the beginning of vitellogenesis. The ooplasm contains central vacuolated region without yolk granules as also noticed by Elliott and Gillott (1976) in the ooplasm of *Melenoplus.* At initial adult stages, the yolk spheres appear in the cortical region of the ooplasm and large yolk spheres, in the central region of ooplasm. At postvitellogenic phase, the oocytes are found to develop an inner vitelline membrane and an outer chorion.

In the present investigation, the tannery effluent is found to cause alterations in the developing oocytes. The egg cytoplasm as well as nucleoplasm become vacuolated. In higher concentrations of the effluent, the periphery of the ooplasm is clear with large vacuoles. The nucleus become disorganized without nucleous and nuclear membrane. There is no indication of yolk deposition in the basal oocytes. In short, there is no occurrence of vitellogenesis, thus showing the adverse effect of tannery effluent on the normal development and maturation of oocytes in *P. flavescens.* The inhibition of follicular growth and maturation of oocytes during ovarian development has been demonstrated in insects treated with apholate, antifertility agents, diflubenzuron and trehalose inhibitors (Mathur, 1998; Rao and Patnaik, 1999). In inhibiting the development of oocytes in the ovary of dragonfly, the tannery effluent can be considered to have chemosterilent property for which further investigations are required.

References

Ambrose, T., Cyrill Arunkumar, L., Vincet, S., and Lambert, R., 1994. Biochemical responses of *Cyprinus carpio* Communis to toxicity of tannery effiuent. *J. Ecobiol.*, 3(b): 213–216.

Anitha, R.H., Sabitha Raja, S., Manjula, C. and Raman, C.V., 1999. Effect of Precocene on the nuclicacid contents in the ovaries of *Chilo artellus* Swinhoe. *Entomon.*, 24(4): 307–313.

Arora, H.C., 1981. *Standard Methods of Examination Water and Wastewater*, 15th edn. Washington, DC.

Elliott, R.H. and Gillott, C., 1976. Histological changes in the ovary in relation to yolk deposition, allatectomy and destruction of the median neurosecretary cells in *Melanoplus sanguinipes. Can. J. Zool.*, 54: 185–192.

Ghosh, D., Chel, S.G. and Pal, G., 1999. RNA-synthesis in the oocytes of *Gesonula puncifrons* (Orthoptera: Acrididae): Autoradiographic and amanitin inhibition studies. *Entomon.*, 24(1): 1–8.

Mathur, G.K., 1998. Laboratory evaluation of diflubenzuron: Deleterious effects on ovaries of *Poekilocerus pictus* (Fabr.) *Entomon.*, 23(1): 67–68.

Rao, L.M. and Patnaik, R.M.S., 1999. Heavy metal accumulation in the cat fish *Mystus vittatus* (Bloch) from Mehadrigedda stream of Visakhapatnam, India. *Poll. Res.*, 19: 325–329.

Sakthivel, M., 1994. Sublethal effects of tannery and textile mill effluents on gill structure, respiration and bioconcentrations of chromium in *Cyprinus carpio* (Linnaeus). *J. Environ. Poll.*, 1(324): 83–88.

Schneider, M., Wisel, G., and Dom, A., 1995. Effect of JH III and JH Analogues on phase related growth, egg maturation and lipid metabolism in *Schistocera gregaria* females. *J. Insect Physiol.*, 41(1): 31–32.

Smith, S.G., 1940. A new embedding schedule for insect cytology. *S.T.*, 15: 175–176.

Sudha, H.R., 1998. Morpho-histology of the pineal gland and the effect of altered environmental factors on the pineal gland and ovary of *Mystus vittatus* (Bloch). *J. Ecol. Physiol.*, 1(2): 57–74.

Varadaraj, G., 1978. Studies on the haemolymph and cuticle of a dragonfly *Anax immaculifrons* (Anisoptera : Aeshnidae). *Ph.D. Thesis*, Madras Univ., Madras, India.

Varadaraj, G. and Subramanian, M.A., 1991. The impact of tannery effluent on feeding energetics and oxygen consumption in the fingerlings of *Oreochromis mossambicus*. *J. Ecotoxicol. Environ. Monit.*, 1(4): 270–276.

Chapter 69

Analyses of Total Count of Hemocytes of Estuarine Crab, *Scylla serrata* Under Acute Arsenic Exposure

Sanjib Saha, Mitali Ray and Sajal Ray
Department of Zoology, University of Calcutta,
35 Ballygunge, Circular Road, Kolkata – 700 019, West Bengal, India

ABSTRACT

Hemocyte of *Scylla serrata*, an economically important estuarine crab plays diverse immunological functions. Kinetics of total homocyte count (THC) was determined under the exposure of 1, 2 and 3 ppm of sodium arsenite in diverse span of exposure *i.e.* 24, 48, 72, and 96 hours in controlled laboratory condition. Shift in THC was recorded against all the concentration screened. The pattern of change was uniform. Arsenic induced alteration in THC indicated a state of immunological stress in estuarine crab. Data is suggestive of cellular stress which may lead to decline in endemic population of *Scylla serrata* in arsenic affected districts of West Bengal.

Keywords: Sodium arsenite, Scylla serrata, Hemocytes, Total count.

Introduction

Estuarine mud crab, *Scylla serrata* (Crustacea: Decapoda) is an economically important edible species. The animal is widely distributed in the mud flat of the intertidal zone of Sunderban delta of West Bengal (Ali *et al.*, 2004). Different toxicants, heavy metals and pesticides have a major role in the contamination of the aquatic ecosystem (Eisler, 1988; Selvakumar *et al.*, 1996; Shibu Vardhanan and

Radhakrishnan, 2002). All the pollutants released on land will eventually find their way to the rivers and oceans as the final repository. Since the estuaries are the links to the freshwater and marine systems, they contain a variety of anthropogenic chemicals referred to as xenobiotics, *viz.* heavy metals, pesticides, fertilizers etc. (Vijayavel and Balasubramanian, 2005). The arsenic contamination of aquatic ecosystem is through the use of pesticides, non-ferrous smelters, coal-fired and geothermal power plant and also from the earth crust (Wang and Hang, 1994; Gomez-Caminero *et al.*, 2001). Natural habitat of *Scylla serrata* faces environmental threat due to exposure of such diverse toxins including metalloid toxin arsenic (Krishnaja *at al.*, 1987; Acharya, 2002; Saha and Ray, 2005). Hemocytes, the chief effector cells of *Scylla serrata* are considered as prime immuno-effector cells. Diverse immunological responses including phagocytosis, non-self recognition, nodule formation, surface adhesion, cellular aggregation, super oxide anion generation etc. are carried out by the immuno-reactive hemocyte subpopulations (Sindermann, 1971; Hose *et al.*, 1990; Soderhall and Cerenius, 1992; Johansson *et al.*, 2000; Yildiz and Atar, 2002). In this present study adult *Scylla serrata* were exposed to different sub-lethal concentrations (1, 2 and 3 ppm) of sodium arsenite in controlled laboratory condition for analysis of alteration of total count of hemocytes in varied span of exposure (24, 48, 72 and 96 hours). Arsĕnic induced modulation of total count of hemocyte of *Scylla serrata* may lead to increase in susceptibility of the animal for parasitic infection and affect normal physiology of the habitat in its natural habitat. A shift in total count under sodium arsenite exposure is in *Scylla serrata* is reported for the first time.

Materials and Methods

Live adult *Scylla serrata* were collected (N = 120) from selected habitat of South 24 pgs. district of West Bengal and transported to the laboratory. The length and breadth of the carapace were 6 ± 0.5 c.m. and 9 ± 0.5 c.m. respectively. Animals were maintained in standard glass aquaria in batches and fed with fresh flesh of mollusks. The water of the static water environment was replenished and animals were acclimatized for one week (Heasman and Fielder, 1983). The temperature of water kept constant at 28±2°C. *Scylla serrata* were exposed to 1, 2 and 3 ppm for sodium arsenite in batches along with the control for a span of 24, 48, 72 and 96 hours duration. During treatment animal received uniform ration of daylight and checked for mortality and morbidity. Fresh blood was collected aseptically from heart using anticoagulant (1 per cent EDTA), 23 gauge sterile needle and volume of collected blood never exceeded 0.5 ml (Yildiz and Atar, 2002) from post treatment *Scylla serrata.* Total hemocyte count (THC) was analyzed using improved Neubauer hemocytometer (Jussila *et al.*, 1997; Manjula *et al.*, 1997).

Results and Discussion

Total count of hemocyte of *Scylla serrata* was recorded as 47.31±34.05 × 10^5 cells/ml. An increase in hemocyte density was recorded against 1 ppm, 2 ppm, and 3 ppm of sodium arsenite for 24 hours and 96 hours. Increased total count were recorded as 105.2±1.12 × 10^5 cells/ml, 108.73±1.92 × 10^5 cells/ml and 129.6±1.31 × 10^5 cells/ml against 1 ppm, 2 ppm and 3 ppm per 24 hours and 95.04±1.82 × 10^5 cells/ml, 72.96±1.63 × 10^5 cells/ml and 214.48±1.73 × 10^5 cells/ml against 1 ppm, 2 ppm and 3 ppm per 96 hours of exposure respectively (Figures 69.1–69.3). Intermediate value of total count were recorded as 42.32±1.56 × 10^5 cells/ml and 33.28±1.93 × 10^5 cells/ml; 47.2±1.39 × 10^5 cells/ml and 46.08±1.53 × 10^5 cells/ml, 61.68±1.72 × 10^5 cells/ml and 50.32±1.12 × 10^5 cells/ml against 1 ppm, 2 ppm and 3 ppm per 48 and 72 hours of sodium arsenite exposure respectively (Figures 69.1–69.3). An exposure of 3ppm of arsenite for 96 hours of span resulted a highest increase in total count in this experiment (Figure 69.3). Initial increase (24 hours) followed by a decline in cell count (48 and 72 hours) and subsequent fluctuation of THC against 96 hours of exposure is indicative of destabilization of cellular homeostasis of *Scylla serrata* under the exposure of sodium arsenite. *Scylla serrata* is distributed in habitat, which bears the risk of arsenic contamination from various sources. Industrial activities and biological leaching of arsenic might affect diverse aquatic organisms including *Scylla serrata* adversely (Gomez-Caminero *et al.*, 2001). Xenobiotics induced alteration of immunological state may lead to decline of the population of this economically importance endemic species from Sundarbans delta leading to a possible loss of biodiversity (Tamaras and Michael, 2001). Report of arsenic tox:icity in aquatic organism is scanty. Present information would help to formulate a sustainable strategy of aquaculture of mud crab in Sunderbans. Moreover present data is suggestive to establish *Scylla serrata* as a biomonitor species against toxic metal and metalloid in estuarine environment.

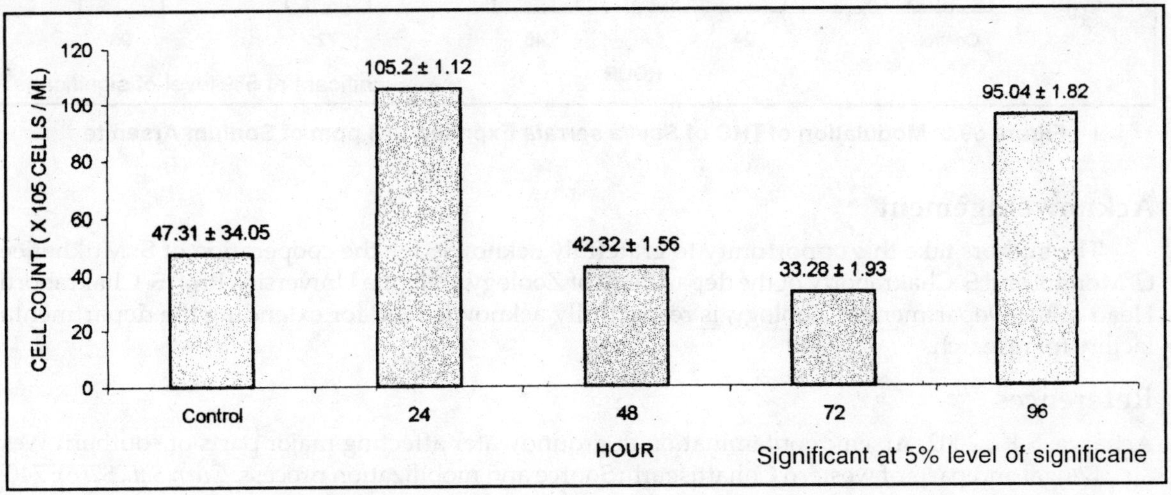

Figure 69.1: Modulation of THC of *Scylla serrata* Exposed to 1 ppm of Sodium Arsenite

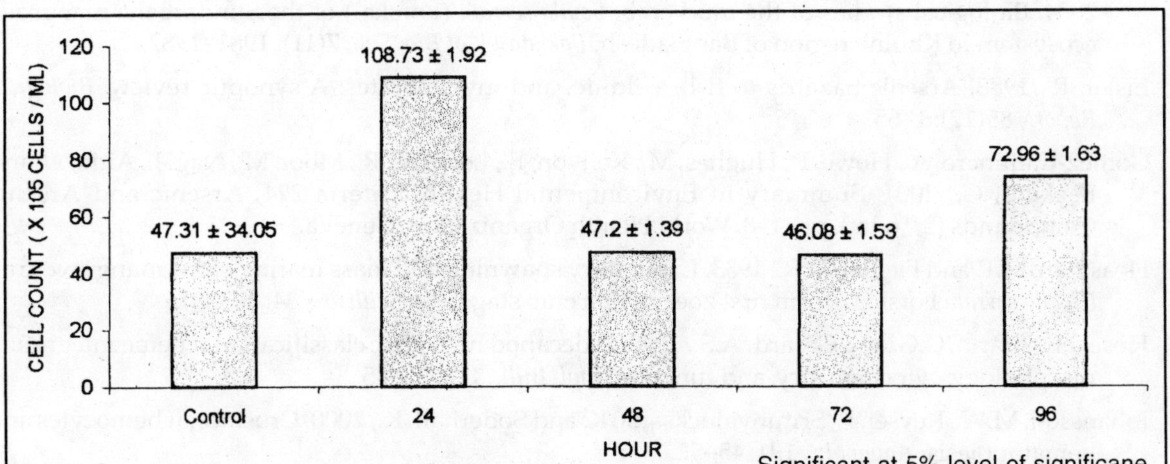

Figure 69.2: Modulation of THC of *Scylla serrata* Exposed to 2 ppm of Sodium Arsenite

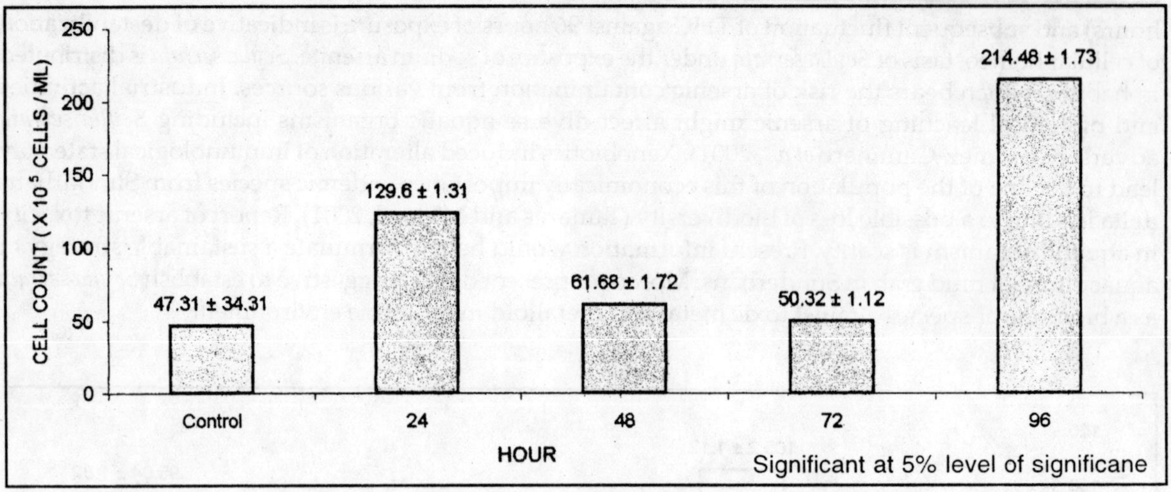

Figure 69.3: Modulation of THC of *Scylla serrata* Exposed to 3 ppm of Sodium Arsenite

Acknoweldgement

The authors take this opportunity to gratefully acknowledge the cooperation of S. Mukherjee, C. Mondol and S. Chakraborty of the department of Zoology, Calcutta University. Prof. S. Chakraborti, Head of the Department of Zoology, is respectfully acknowledged for extending the departmental facility for research.

References

Acharya, S.K., 2002. Arsenic contamination in groundwater affecting major parts of southern West Bengal and parts of western Chhattisgarh: Source and mobilization process. *Curr. Sci.*, 82(6): 740–744.

Ali, M.Y., Kamal, D., Hossain, S.M.M., Azam, M.A., Sabbir, W., Murshida, A., Ahmed, B. and Azam, K., 2004. Biological studies of the mud crab, *Scylla serrata* (Forskal) of the Sundarbans mangrove ecosystem in Khulna region of Bangladesh. *Pakistan J. of Bio. Sci.*, 7(11): 1981–1987.

Eisler, R., 1988. Arsenic hazards to fish, wildlife, and invertebrates: A synoptic review. *Biological Report*, 85(12): 1–65.

Gomez-Caminero, A., Howe, P., Hughes, M., Kenyon, E., Lewis, D.R., Moor, M., Nag. J., Aitio, A. and Becking, G., 2001. Summary in Environmental Health Criteria 224, Arsenic and Arsenic Compounds (2nd edn.). pp. 1–8. World Health Organization, Geneva.

Heasman, M.P. and Fielder, D.R., 1983. Laboratory spawning and mass rearing of the mangrove crab (*Scylla serrata* Forskal), from first zoea to first crab stage. *Aquaculture*, 34: 303–316.

Hose, J.E., Marin, G.G. and Gerard, A.S., 1990. A decapod hemocyte classification scheme integrating morphology, cytochemistry, and function. *Biol. Bull.*, 178: 33–45.

Johansson, M.W., Keyser, P., Sritunyalucksana, K. and Soderhall, K., 2000. Crustacean hemocytes and hematopoiesis. *Aquacult.*, 191: 45–52.

Jussila, J., Jago, J., Tsvetnenko, E., Dunstan, B. and Evans, L.H., 1997. Total and differential hemocyte counts in western rock lobsters (*Panulirus cygnus* George) under post-harvest stress. *Mar. and Freshwater Res.*, 48(8): 863–867.

Krishnaja, A.P., Rege, M.S. and Joshi, A.G., 1987. Toxic effects of certain heavy metals (Hg, Cd, As, Se) on the intertidal crab *Scylla serrata*. *Mar. Environ. Res.*, 21(2): 109–119.

Manjula, P.L., Rahman, K. and Abraham, T. J., 1997. Hemocyte classification and differential counts in Indian spiny lobster, *Panulirus homarus* (linneaus). *J. Aquacult. The Tropics*, 12(2): 113–121.

Saha, S. and Ray, S., 2005. Arsenic induced alteration of hemocyte count in estuarine mud crab, *Scylla serrata*. In: *International Conference on Toxicology, Environmental and Occupational Health*, Lucknow U.P. Abstract volume. p. 7.

Selvakumar, S., Khan, A.S. and Kumaraguru, 1996. Acute toxicity of some heavy metals, pesticides and water soluble fractions of diesel oil to the larvae of some brachyuran crabs. *J. Environ. Biol.*, 17(3): 221–226.

Shibu Vardhannan, Y. and Raadhakrishnan, 2002. Acute toxicity of copper, arsenic and HCH to paddy field crab, *Paratelphusa hydrodromus* (Herb.). *J. Environ. Biol.*, 23(4): 387–392.

Sindermann, C.I., 1971. Internal defenses of Crustacea: A review. *Fish. Bull.*, 69: 455–489.

Tamaras, S.G. and Michael, H.D., 2001. Immunotoxicity in Invertebrates: Measurement and ecotoxicological relevance. *Ecotoxicology*, 10: 5–23.

Vijayavel, K. and Balasubramanian, M.P., 2005. Fluctuations of biochemical constituent and marker enzymes as a consequence of naphthalene toxicity in the edible estuarine crab, *Scylla serrata*. *Ecotoxi. and Environ. Saf.*, pp. 1– 7.

Yildiz, H.Y. and Atar, H.H., 2002. Hemocyte classification and differential counts in freshwater crab (*Potamon fluviatilis*). *Turk. J. Vet. Anim.*, 26: 403–406.

Chapter 70

Pseudomonas fluorescens as an Ideal Candidate for Azoe Dye Decolourisation

C.S. Shobana* and P. Amsaveni

Faculty of Biotechnology, International University College of Technology Twintech, Bandar Sri Damansara, 52200, Kula Lumpur, Malaysia

ABSTRACT

Pseudomonas fluorescens strains were isolated from textile dye effluent contaminated soil samples collected from different parts Tirupur town in Coimbatore district of Tamil Nadu, India. The ability of these strains to decolourise Congo red was studied in detail. The decolourisation parameters *viz.*, dye concentration, temperature, pH and incubation time were optimised. Presence of plasmid was confirmed by obtaining plasmid profile on agarose gel electrophoresis. Plasmid curing was carried out using sodium dodecyl sulphate (SDS) and elevated temperature. The lose of plasmid was confirmed by antibiotic sensitivity tests. The optimal dye concentration, temperature, pH and incubation time for decolourisation of Congo red by uncured strains were found to be 0.7–1.75 mg/50 ml, 40ºC, 7.0 and 3 days respectively. The effect of plasmid curing on decolourisation of Congo red was also analysed. The results indicated the involvement of plasmid in decolourisation activity.

Keywords: Pseudomonas fluorescens, Azo dyes, Congo red, Decolourisation.

Introduction

Population explosion and industrial development during the 20th century has not only increased the conventional solid and waste pollutants to critical levels but also produced a range of pollution

* Corresponding Author: E-mail: shobanasenthilkumar@gmail.com

problems. A major share of India's export, hosiery and knitwear products are produced in Tirupur, a small town in Coimbatore district of Tamil Nadu. There are about 650 to 700 bleaching and dyeing units located in Tirupur (Kannabiran *et al.*, 1993).

Triphenylmethane and azo dyes are used extensively in textile industry for dying nylon, polyacrylonitrile modified nylon, wool, silk and cotton. Biodegradation of dyes has received a considerable attention because conventional wastewater treatment processes does not work well with recalcitrant xenobiotic compounds (Walmik *et al.*, 1998). Environmental regulations in most of the countries have made it mandatory to decolourise the wastewater prior to its discharge into water bodies. These highly coloured compounds also lead to the reduction of sunlight penetration in rivers, lakes or lagoons which in turn decreases photosynthetic activity and dissolved oxygen concentration causing detrimental effect on aquatic ecosystem (Kannabiran *et al.*, 1993; Agarwal and Kumar, 1990). The release of azo dyes into the environment is a concern due to its toxicity, mutagenicity and carcinogenicity, colouration of natural waters and their biotransformation products. Azo dyes constitute a major class of environmental pollutants accounting for 60 to 70 per cent of all dyes and pigments used. These dyes are characterized by aromatic moieties linked together with azo groups (-N=N-) and Congo red is one such dye. At present, considerable attention has been given for evaluating the fate of azo dyes during wastewater treatment (Vaidya and Datye, 1982). The efficiency of *P. pseudomallei* in biodegradation of four triphenylmethane dyes *viz.*, basic fuchsin, methyl violet, crystal violet, and Victoria blue have been studied in detail (Yatome *et al.*, 1991). Ogawa and Yatome (1990), Puvaneswari *et al.* (2002) and Bhatt *et al.* (2005) have experimented with *Pseudomonas* for the decolourisation of azo dyes. This research was conducted with the objective of investigating the ability of *P. fluorescens* to decolourise Congo red, the model azo dye, analyse various parameters *viz.*, dye concentration, temperature, pH and incubation time and evaluate the effect of plasmid curing on decolourisation.

Materials and Methods

Soil Samples

107 dye effluent contaminated soil samples were collected from different parts of Tirupur town in Coimbatore district ofTarnil Nadu and processed for the isolation of *P. fluorescens*.

Bacterial Cultures and Maintenance

Isolation and enrichment of dye degrading microorganisms were done by employing the soil embedation technique (Paszczynski *et al.*, 1992 and Pasti-Grigsby *et al.*, 1992). Standard biochemical tests were performed for the confirmation of strains (Buchanan and Gibbons, 1974). The isolated strains of *P. fluorescens* were maintained on Nutrient agar slants containing 0.2 per cent of Congo red dye.

Congo Red Dye Solution

Congo red solution was prepared by mixing 0.35 g of Congo red dye (HiMedia) in 100 ml of sterile distilled, water and stored at 10°C for further use.

Inoculum Preparation and Sampling

The *P. fluorescens* culture was inoculated in 50 ml of nitrogen and sulfur free broth containing Congo red at various concentrations of 0.70, 1.75, 3.50, 5.25 and 7.00 mg/50 ml of distilled water in 100 ml conical flasks. Uninoculated control was also maintained. The flasks were incubated in shakers at 37°C for 1–5 days. After incubation, the broth was centrifuged at 10,000 rpm for 10 minutes. Absorbance of the supernatant was measured at 419 nm periodically for 5 days and expressed as optical density (O.D). The rate of colour reduction was assessed by comparing the O.D. of the test samples with that of the control (Table 70.1).

Table 70.1: Absorbance of Congo Red Solution at Various Concentrations

Sl.No.	Congo Red Concentration mg/50 ml	OD at 419 nm
1.	0.70	0.618
2.	1.75	0.840
3.	3.50	1.206
4.	5.25	1.410
5.	7.00	1.782

Decolourisation Assessment

The percentage of colour reduction was measured as per the method described by Kumar *et al.* (1997).

$$\text{Colour removal (\%)} = \frac{\text{(A)} - \text{Absorbance of residual broth}}{\text{(A)}} \times 100$$

(A): Absorbance of uninoculated broth.

Optimization of Decolourisation Parameters

The parameters *viz.*, dye concentration, temperature, pH and incubation time were optimized for the cultures of *P. fluorescens* based on the percentage of decolourisation of Congo red at various dye concentration, temperature, pH and incubation time.

Plasmid Studies

The experimental strains were subjected to plasmid isolation (Bimboim and Doly, 1979 and Ish-Horowicz and Burke, 1981). The strains were a cured of plasmid using SDS and elevated temperature (El-Mansi *et al.*, 2000). The presence and absence of plasmid was confirmed by agarose gel electrophoresis (Sambrook and Russel, 2001). Plasmid profiles were obtained by electrophoresis in a submerged horizontal gel apparatus on 0.8 per cent agarose in tris acetate at 80V for 3 hours. The gel was stained for 15 minutes with ethidium bromide at the concentration of 10mg/rnl of staining solution, de stained and then visualised under UV transilluminator (Maniatis *et al.*, 1982). In order to study the resistant pattern, antibiogram of the cured and uncured strains was performed using standard antibiotics (Bauer *et al.*, 1966).

Decolourisation of Congo Red by Plasmid Cured Strains of *P. fluorescens*

The inoculum was prepared in cetrimide broth. The broth containing Congo red at various concentrations of 0.70, 1.75, 3.50, 5.25 and 7.00 mg/50 ml at the optimal pH of 7.0 was prepared and inoculated with plasmid cured strain of *P. fluorescens* and incubated at optimized temperature and time. After incubation, the percentage of decolourisation was calculated (Table 70.3).

Results and Discussion

This investigation was aimed at the exploration of the decolourisation ability of the indigenous *P. fluorescens*. The colonies showing green pigmentation and zone of clearance were selected for further studies. Out of 107 soil samples collected, 73 samples were positive for the presence of *P. fluorescens*. Apart from *P. fluorescens*, the dominant microbial flora of the effluent were *Bacillus subtilis*, *Aspergillus flavus* and *A. niger*. The strains were further confirmed for their identity by standard biochemical tests.

The isolated strains of *P. fluorescens* were gram-negative motile rods producing catalase, oxidase and gelatinase, reduced nitrate and positive for alkaline reaction in litmus milk test.

Table 70.2: Percentage of Decolourisation of Congo Red by *P. fluorescens* at Various Temperature

Sl.No.	*Congo Red concentration mg/50 ml*	*Percentage of Decolourisation of Congo red at*			
		10°C	*20°C*	*30°C*	*40°C*
1.	0.70	2.3	9.5	21.7	39.7
2.	1.75	5.5	16.5	34.2	41.5
3.	3.50	8.4	12.6	17.5	28.7
4.	5.25	7.5	14.5	21.9	24.5
5.	7.00	4.7	9.3	15.5	20.2

Table 70.3: Percentage of Decolourisation of Congo Red by *P. fluorescens* at Various pH

Sl.No.	*Congo Red concentration mg/50 ml*	*Percentage of Decolourisation of Congo red at pH*			
		5.0	*6.0*	*7.0*	*8.0*
1.	0.70	10.84	16.99	36.89	26.86
2.	1.75	10.71	20.24	39.29	23.92
3.	3.50	10.17	16.21	19.02	16.54
4.	5.25	10.39	14.98	19.22	17.52
5.	7.00	5.50	10.05	15.27	12.54

Microorganisms play an important role in reducing the colour intensity of the azo dye solution (Ogawa and Yatome, 1990; Kumar *et al.*, 1997; Walmik *et al.*, 1998 and Mali *et al.*, 1999). Hu (1994) reported that azoreductase of *P. luteola* decolourised Red G, RBB, RP2B, and V2RP at a concentration of 100 mg/l. The decolourisation percentage was observed to be 37.4 per cent, 93.2 per cent, 92.4 per cent, and 88 per cent within 2 days. Azoreductase enzyme is responsible for the initiation of dye degradation by bacterial strains (Zirnmerman *et al.*, 1982).

Absorbance of Congo red at various concentrations of 0.70, 1.75, 3.50, 5.25 and 7.00 mg/50 ml of distilled water was measured at 419 nm (Table 70.1). The bacterial isolate *P. aeruginosa* NBAR12 is capable of decolourising 12 different dyes tested for decolourisation efficiency varied in the range of 80 to 95 per cent (Bhatt *et al.*, 2005).

The optimal temperature for the decolourisation of Congo red was found to be 40°C (Table 70.2). A bacterial consortium RVM 11.1 exhibited 94 per cent decolourisation of Reactive Violet 5 (RV5) under temperature ranging from 25°C to 40°C (Moosvi *et al.*, 2005). Temperature is an important determinant in biodegradation reactions. In the context of effluent treatment, temperature controls the rate of metabolism and microbial growth. All organisms have their own optimum, maximum and minimum temperature for growth and degradative reactions. The optimal parameters necessary for the bacterial consortium SV5 to decolorise Ranocid Fast Blue dye were temperature of 37°C in less than 24 h with an initial dye concentration of 100 ppm (Mathew and Madamwar, 2004).

In the persent investigation, the test strains of *P. fluorescens* recorded the maximum percentage of decolourisation of Congo red at pH 7.0 (Table 70.3). Direct Red 81 (DR 81) decolourizing activity of a novel bacterial consortium was observed at pH 7.0 by Junnarkar *et al.* (2006). The diazo-dye Reactive Blue 172 was decolourised by *P. aeruginosa* NBAR12 at optimal pH of 7.0 (Bhatt *et al.*, 2005).

The maximum decolourisation (34.47 per cent) of Congo red was observed during the third day of incubation (Table 70.4). *Phlebia brevispora, P. fascicularia* and *P. floridensis* have been evaluated for their potential to decolourise eight industrial dyes including reactive yellow, reactive orange, reactive red, rathidol scarlet, coracryl black, coracryl pink, coracryl violet and coracryl red. The extracts obtained from the cultures grown for six days significantly decolourised the tested dyes (Arora and Chander, 2004).

Table 70.4: Percentage of Decolourisation of Congo Red by *P. fluorescens* at Various Incubation Time

Sl.No.	*Concentration*	*Percentage of Decolourisation of Congo Red at Different Incubation Time (in days)*				
	mg/50 ml	*1*	*2*	*3*	*4*	*5*
1.	0.70	3.20	9.06	34.47	12.60	2.70
2.	1.75	8.50	18.90	30.95	17.86	11.7
3.	3.50	8.70	13.50	17.60	4.89	0.90
4.	5.25	4.96	11.50	15.20	4.25	0.71
5.	7.00	3.36	5.80	12.20	3.30	0.39

The genes that encode for biodegradation of many recalcitrant organic chemicals are often encoded on plasmids. Sometimes, only one or two enzymes are encoded on a plasmid while the rest on the bacterial chromosome. In some other bacteria an entire pathway is plasmid encoded. There is a range of causes for their instability but often a plasmid will be lost from an organism if it is cultivated in the absence of the substrate whose degradation it encodes. Plasmids tend to remain stable in a population when their presence confers a selective advantage on individual cells which carry them. When no advantage accrues to the organism, plasmids can be lost because their maintenance in the population represents a drain of resources which can impose a selective disadvantage. As stated in the reports of El-Mansi *et al.* (2000), the effect of SDS proved to be very effective in plasmid with relatively higher frequency and this agent showed the same effect on the plasmid of *P. fluorescens.* At the concentration of 2 per cent w/v, the curing efficiency of SDS was 3 per cent. Plasmid profile analysis clearly indicated the loss of plasmid by the cured strains (Figure 70.1).

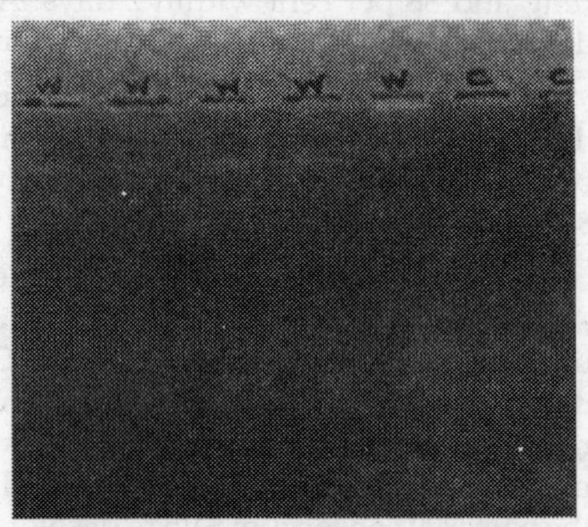

Figure 70.1: Plasmid Profile of Wild and Cured Strains of *P. fluorescens* w: uncured; c: cured.

The uncured strain produced bluish green, smooth, opaque and large colonies whereas the cured strains produced white, flat, rough, small and irregular colonies. The margin of the cured strain colonies was irregular whereas the wild strain showed a smooth margin. When inoculated on King's medium, the cured strains produced brown coloured colonies whereas the wild strains produced bluish green colonies. A similar kind of change in colony. morphology was also reported by Sujoy Saha *et al.* (2000).

An essential trait of any plasmid is the selection marker associated with replicon. Antibiotic resistance is one of the important plasmid borne characters. Many genetic determinants of antibiotic resistance in Enterobacteriaceae and Staphylococci are on extrachromosomal elements (Mitsuhashi, 1965). The antibiotic sensitivity pattern of the *Pseudomonas* strains was determined using Kirby-Bauer method (Bauer *et al.*, 1966). The antibiotic susceptibility tests showed the results as in Table 70.5.

Table 70.5: Antibiogram of Uncured and Cured Strains of *P. fluorescens*

Sl.No.	Antibiotics	Concentration	Sensitivity Pattern	
			Uncured	Cured
1.	Ampicillin (A)	25 μg/disc	R	R
2.	Chloramphenicol	25 μg/disc	R	S
3.	Gentamycin (G)	50 μg/disc	R	S
4.	Tetracycline (O)	30 μg/disc	R	R
5.	Streptomycin (S)	10 μg/disc	R	I
6.	Kanamycin (K)	30 μg/disc	R	R
7.	Nalidixic acid (Na)	30 μg/disc	R	R
8.	Erythrornicin (E)	15 μg/disc	R	R
9.	Rifampicin (R)	5 μg/disc	R	R

R: Resistant; S: Sensitive; I: Intermediate susceptibility.

Both the cured and uncured cells showed resistance to nalidixic acid, erythromycin, rifampicin, kanamycin and tetracycline. The uncured cells were resistant to gentamycin and chloramphenicol whereas the cured cells were sensitive to the same. This clearly indicated the plasmid borne resistance to the antibiotics, chloramphenicol and gentamycin. The uncured cells were resistant to the antibiotic streptomycin whereas the cured cells showed intermediate response to the same. In the current analysis, it was found that the cured strains of *P. fluorescens* lost its ability to decolourise Congo red (Table 70.6). The maximum percentage of decolourisation by cured strains of *P. fluorescens* was only 11.31 at 0.70 mg/50 ml concentration of Congo red. The results clearly indicated that these plasmids have some catabolic function.

Table 70.6: Decolourisation of Congo Red at Various Concentrations by Plasmid Cured Strains of *P. fluorescens*

Sl.No.	Congo Red Concentration mg/50 ml	Percentage of Decolourisation
1.	0.70	11.31
2.	1.75	9.69
3.	3.50	7.41
4.	5.25	4.75
5.	7.00	2.82

Whatever may be the mechanism responsible for the phenomenon, we believe that the ability to cure plasmid should prove useful in the study of this organism and its plasmids. Further, the transformation and expression of this potential plasmid in other degrading bacteria may yield a novel strain to the environmental biotechnologist.

References

Agarwal, R.K. and Kumar, P., 1990. Textile industrial effluents: Implication and possible use for aforestation in western Rajasthan. *Annals of Arid Zone*, 29: 295–302.

Arora, D.S. and Chander, M., 2004. Decolourisation of diverse industrial dyes by some *Phlebia* spp. and their comparison with *Phanerochaete chrysosporium. J. Basic Microbiol.*, 44(5): 331–338.

Bauer, A.W., Kirby, W.M., Sherris, J.C. and Truck, M., 1966. Antibiotic susceptibility testing by a standardized single disk method. *Am J. Clin. Path.*, 45: 494.

Bhatt, N:, Patel, K.C., Keharia, H. and Madamwar, D., 2005. Decolourization of diazo-dye Reactive Blue 172 by *Pseudomonas aeruginosa* NBAR12. *J. Basic Microbiol.*, 45(6): 407–418.

Bimboim, H.C. and Doly, J., 1979. A rapid extraction procedure for screening recombinant plasmid DNA. *Nucl. Acids Res.*, 7: 1513–1523.

Buchanan, R.E. and Gibbons, N.E. (eds), 1974. *Bergey's Manual of Determinative Bacteriology*, 8th Edition. The Williams and Wilkins Co., Baltimore.

El-Mansi, J.A., Karen, A.I., Craig, K.K. Linda and David, J.P., 2000. *FEMS Microbial.*, 151: 201–208.

Hu, T.L., 1994. Decolourization of reactive azo dyes by transformation with *Pseudomonas luteola. Bioresource Technol.*, 49(1): 47–51.

Ish-Horowicz, D. and Burke, J.F., 1981. Rapid and efficient cosmid cloning. *Nucl. Acids Res.*, 9: 2989–2998.

Junnarkar, N., Murthy, D.S., Bhatt, N.S. and Madamwar, D., 2006. Decolourization of diazo dye Direct Red 81 by a novel bacterial consortium. *World J. Microbiol. Biotechnol.*, 22(2): 163–168.

Kannabiran, B. and Prasasan, A., 1993. Wastewater treatment from textile industry. *Geobios*, 20: 108.

Kumar, V.L., Wati, F.F., Fitz Gibboa, F., Nigam, P., Banal, I.M., Sindh, D. and Marchant, R., 1997. Biodecolourisation of Azodyes. *Biotechnol. Lett.*, 19: 311.

Mali, P.L. Mahajan, M.M., Patil, D.P. and Kulkarni, M.V., 1999. Biodecolourisation of Members of Triphenylmethane and Azo Group of Dyes. *J. Scientific and Ind. Res.*, 59: 221–224.

Maniatis, T., Fritsch, E. and Sambrook, J., 1982. *Molecular Cloning: A Laboratory Manual*. Cold Spring Harbor Laboratory, Cold Spring Harbor, NY.

Mathew, S. and Madamwar, D., 2004. Decolourization of Ranocid Fast Blue dye by bacterial consortium SV5. *Appl. Biochem Biotechnol.*, 118(1–3): 371–381.

Mitsuhashi, S., 1965. Transmissible drug-resistance factor R. *Gunma J. Med. Sci.*, 14: 169–209.

Moosvi, S., Keharia, H. and Madamwal, D., 2005. Decolourization of textile dye Reactive Violet 5 by a newly isolated bacterial consortium RVM 11.1. *World J. Microbiol. Biotechnol.*, 21(5): 667–672.

Ogawa, T. and Yatome, C., 1990. Biodegradationofazo dyes in multistage rotating biological contactor immobilized by assimilating bacteria. *Bull. Environ. Contam. Toxicol.*, 44(4): 561–566.

Pasti-Grigsby, M.B., Paszczynski, A., Goszczynski, S., Crawford, D.L. and Crawford, R.L., 1992. Influence of aromatic substitution patterns on azo dye degradability by *Streptomyces* spp. and *Phanerochaete chrysosporium. Appl. Environ. Microbiol.*, 58(11): 3605–3613.

Paszczynski, A., Pasti-Grigsby, M.B., Goszczynski, S., Crawford, R.L. and Crawford, D.L., 1992. Mineralization of sulfonated azo dyes and sulfanilic acid by *Phanerochaete chrysosporium* and *Streptomyces chromofuscus. Appl. Environ. Microbiol.*, 58: 3598–3604.

Puvaneswari, N., Muthukrishnan, J. and Gunasekaran, P., 2002. Biodegradation of benzidine based azodyes direct red and direct blue by the immobilized cells of *Pseudomonas fluorescens* D41. *Indian J. Exp. Biol.*, 40(10): 1131–1136.

Sambrook, J. and Russel, D.W., 2001. *Molecular Cloning: A Laboratory Manual,* 3rd edn. Cold Spring Harbor Laboratory Press, Cold Spring Harbor, NY.

Sujoy Saha, Singh, R.P., Verma, J.P. and Jayaraman, J., 2000. Plasmid-borne determinants of colony morphology, pigmentation, antibiotic resistance and antibiosis in *Pseudomonas* species antagonistic to bacterial blight of cotton. *Current Science,* 79(9): 1384–1385.

Vaidya, A.A. and Datye, K.V., 1982. Environmental pollution during chemical processing of synthetic fibers. *Colourage,* 14: 3–10.

Walmik, A., Saini, R.K. and Banerjee, V.C., 1998. Microbial degradation of Azodyes. *Enz. Microbiol. Technol.,* 22: 185.

Wong, P.K. and Yuen, P.Y., 1996. Decolourization biodegradation of methyl red by *Klebsiella pneumoniae* RS-13. *Water Research,* 30(7): 1736–1744.

Yatome, C., Ogawa and, T. and Hayashi, H., 1991. Microbial reduction of azodyes by several strains. *J. Environ. Sci. Hlth.,* A26: 471–485.

Young, L.Y. and Cerniglia, C.E., 1995. *Microbial Transformation and Degradation of Toxic Organic Chemicals.* Wiley-Liss, New York.

Zimmermann, T., Kulla, H.G. and Leisinger, T., 1982. Properties of purified orange II azoreductase, the enzyme initiating azodye degradation by *Pseudomonas* KF46. *Eur. J. Biochem.,* 129: 197–203.

Index